TECHNIQUE
MATHEMATICAL

TECHNIQUES OF MATHEMATICAL ANALYSIS

C. J. TRANTER
C.B.E., M.A., D.Sc., F.I.M.A.

Professor Emeritus and formerly Bashforth Professor of Mathematical Physics, Royal Military College of Science, Shrivenham

Edward Arnold
A division of Hodder & Stoughton
LONDON BALTIMORE MELBOURNE AUCKLAND

© 1957 C.J. Tranter

First published in Great Britain 1957
Nineteenth Impression 1990

ISBN 0 340 11641 2

All rights reserved. No part of this publication may be reproduced or transmitted in any form or by any means, electronically or mechanically, including photocopying, recording or any information storage or retrieval system, without either prior permission in writing from the publisher or a licence permitting restricted copying. In the United Kingdom such licences are issued by the Copyright Licensing Agency: 33-34 Alfred Place, London WC1E 7DP.

Printed and bound in Hong Kong for Edward Arnold, a division of Hodder and Stoughton Limited, Mill Road, Dunton Green, Sevenoaks, Kent TN13 2YA by Colorcraft Ltd.

GENERAL EDITOR'S FOREWORD
By
SIR GRAHAM SUTTON, C.B.E., D.Sc., F.R.S.

THE present volume is one of a number planned to extend the Physical Science Texts beyond the Advanced or Scholarship levels of the General Certificate of Education. The earlier volumes in this series were prepared as texts for class teaching or self-study in the upper forms at school, or in the first year at the university or technical college. In this next stage, the treatment necessarily assumes a greater degree of maturity in the student than did the earlier volumes, but the emphasis is still on a strongly realistic approach aimed at giving the sincere reader technical proficiency in his chosen subject. The material has been carefully selected on a broad and reasonably comprehensive basis, with the object of ensuring that the student acquires a proper grasp of the essentials before he begins to read more specialized texts. At the same time, due regard has been paid to modern developments, and each volume is designed to give the reader an integrated account of a subject up to the level of an honours degree of any British or Commonwealth university, or the graduate membership of a professional institution.

A course of study in science may take one of two shapes. It may spread horizontally rather than vertically, with greater attention to the security of the foundations than to the level attained, or it may be deliberately designed to reach the heights by the quickest possible route. The tradition of scientific education in this country has been in favour of the former method, and despite the need to produce technologists quickly, I am convinced that the traditional policy is still the sounder. Experience shows that the student who has received a thorough unhurried training in the fundamentals reaches the stage of productive or original work very little, if at all, behind the man who has been persuaded to specialize at a much earlier stage, and in later life there is little doubt who is the better educated man. It is my hope that in these texts we have provided materials for a sound general education in the physical sciences, and that the student who works conscientiously through these books will face more specialized studies with complete confidence.

PREFACE

My previous book in this series was designed to give a course in Pure Mathematics to Advanced Level standard in the General Certificates of Education. This volume is intended to extend the course to the first year at university, and the contents should be useful to undergraduates taking an Honours degree in Mathematics and to those reading Pure Mathematics as a subject in a General degree and also to mathematical specialists in their final years at school.

The contents comprise the appropriate parts of Algebra, Trigonometry and Calculus and the book may therefore be considered as providing a course in the Techniques of Mathematical Analysis. A more sophisticated approach than was employed in my previous book has been adopted when discussing such matters as convergence, continuity, etc., but I have not lost sight of the needs of candidates taking the mathematical papers for scholarships in Natural Science.

Among nearly a thousand examples and exercises, I have included a large number taken from recent papers set to candidates for Mathematical Scholarships and Exhibitions in the Universities of Oxford and Cambridge. My thanks are due to the authorities of the various Colleges for permission to use their questions. I am grateful also to the London Inter-Collegiate Scholarships Board for allowing me to use questions set in examinations for Scholarships, Exhibitions and Bursaries and to the Oxford and Cambridge Schools Examinations Board, the Senate of the University of London, the Joint Matriculation Board of the Universities of Manchester, Liverpool, Leeds, Sheffield and Birmingham and the Welsh Joint Education Committee for questions set at Scholarship Level in the General Certificates of Education administered by these bodies.

Acknowledgement is made of the advice given by many friends and colleagues. I am particularly grateful to Dr. D. R. Dickinson, Senior Mathematics Master, Bristol Grammar School, Mr. H. K. Prout, Head of the Department of Mathematics, Royal Naval College, Dartmouth, and Dr. E. T. Davies, Professor of Mathematics, University of Southampton, all of whom made most useful suggestions when the book was in its first draft.

C. J. TRANTER

Royal Military College of Science,
Shrivenham

CONTENTS

Chap.   *Page*

1 REAL NUMBERS. FUNDAMENTAL INEQUALITIES. FUNCTIONS OF REAL VARIABLES   1

    Rational numbers. Irrational numbers. Real numbers. Inequalities. The continuous real variable. The idea of a function. Functional terminology. The decomposition of a rational function into partial fractions.

2 FINITE SERIES   22

    The summation of finite series. The difference method. The method of induction. Series involving the binomial coefficients. Recurring series and recurrence relations.

3 THE CONVERGENCE OF SEQUENCES AND SERIES   39

    Convergent sequences. Divergent and oscillating sequences. Monotonic sequences. Some useful theorems on limits. Some important limits. Infinite series. Some general theorems on infinite series. Tests for convergence for series of positive terms. Alternating series. Absolute convergence. The multiplication of infinite series. The convergence of the binomial, exponential and logarithmic series. The summation of some infinite series. The generating function of a recurring power series.

4 COMPLEX NUMBERS   75

    Introduction. The geometrical representation of complex numbers. Conjugate complex numbers. Some remarks on the manipulation of complex numbers. The cube roots of unity. Addition of complex numbers in the Argand diagram. Products and quotients of complex numbers in the Argand diagram. Rational functions of the complex variable. Transformations. Infinite series of complex terms.

5 THE THEORY OF EQUATIONS   107

    Introduction. Some remarks on the position of the real roots of an equation. Rolle's theorem. Descartes' rule of signs. The relations between the roots and co-

Chap.		Page
	efficients in an equation. Symmetric functions of the roots of an equation. The sums of powers of the roots of an equation. The transformation of equations. Reciprocal equations. The condition for common roots. Repeated roots. Newton's method of approximation to the roots of equations. The cubic equation. The quartic equation.	
6	ANALYTICAL TRIGONOMETRY	139
	De Moivre's theorem. Fractional powers of complex numbers. Powers of $\cos \theta$ and $\sin \theta$ expressed in multiple angles. Expressions for $\cos n\theta$, $\sin n\theta$, etc., in terms of powers of $\cos \theta$, $\sin \theta$. Factorization of $x^{2n} - 2x^n a^n \cos n\theta + a^{2n}$, etc. De Moivre's property of the circle. Symmetrical functions of $\cos(r\pi/n)$, $\sin(r\pi/n)$, etc. Trigonometrical functions of a complex variable. The logarithmic function of a complex variable. The hyperbolic functions. The real and imaginary parts of $\sin(x + iy)$, etc. The inverse hyperbolic functions. The summation of trigonometrical series.	
7	DETERMINANTS	174
	Introduction. Determinantal notation. Some properties of determinants. Determinantal equations. Differentiation of a determinant. Further examples of the evaluation of determinants. Minors and cofactors. Multiplication of determinants. The solution of simultaneous equations. The consistency of sets of simultaneous equations. Elimination.	
8	FUNCTIONS OF A CONTINUOUS VARIABLE. GENERAL THEOREMS OF THE DIFFERENTIAL CALCULUS	207
	Introduction. Limits of functions of a continuous variable. Continuous functions of a real variable. Types of discontinuity. Continuous functions of more than one variable. Differentiability. Some important general theorems. Mean value theorems. Repeated differentiation. Leibnitz' theorem. The general mean value theorem. Taylor's and Maclaurin's series. Indeterminate forms.	
9	PARTIAL DIFFERENTIATION	236
	Partial derivatives of functions of more than one variable. Higher partial derivatives. The differential	

CONTENTS

Chap.

coefficient of a function of two functions. The mean value theorem for a function of two variables. Differentials. Differentiation of implicit functions. Change of variables. Euler's theorems on homogeneous functions. Exact differentials. Taylor's theorem for a function of two variables. Maxima and minima—two independent variables.

10 MORE ADVANCED METHODS OF INTEGRATION. FURTHER THEOREMS IN THE INTEGRAL CALCULUS — 260

Introduction. Some integrals involving the hyperbolic functions. The integration of $(a + b \cos x)^{-1}$ and similar functions. The integration of $1/(X\sqrt{Y})$ where X and Y are linear or quadratic functions. The integration of $e^{ax} \cos bx$ and $e^{ax} \sin bx$. Reduction formulae. Some general properties of the definite integral. Infinite and improper integrals. Differentiation of definite integrals.

11 SOME FURTHER GEOMETRICAL APPLICATIONS OF THE CALCULUS — 303

Introduction. Tangents and normals. Asymptotes. Curvature. Double points. The nature of the origin. Envelopes. Curve sketching. Some further formulae for plane areas. Further formulae for length of arc. Volumes and surface areas of figures of revolution.

12 ELEMENTARY DIFFERENTIAL EQUATIONS — 349

Introduction. First order differential equations with separable variables. Exact differential equations. The linear first order differential equation. Equations reducible to linear form. Homogeneous first order differential equations. Some artifices for reducing first order equations to standard forms. Second order linear equations with constant coefficients. An elementary method of finding the particular integral. Higher order linear differential equations. The homogeneous linear equation. The solution of linear differential equations by means of the Laplace transform.

ANSWERS TO THE EXERCISES — 379

INDEX — 392

The sources from which some of the examples and exercises have been taken are indicated by the following abbreviations:

O. Examinations for Mathematical Scholarships and Exhibitions set by Colleges in the University of Oxford;

C. Examinations for Scholarships and Exhibitions in Mathematics (and, in a few cases, in Natural Science) set by Colleges in the University of Cambridge;

L.I.C. Examinations for Entrance Scholarships, Exhibitions and Bursaries set by the London Inter-Collegiate Scholarships Board;

O.C. General Certificate Examination at Scholarship Level set by the Oxford and Cambridge Schools Examination Board;

L.U. General Certificate Examination at Scholarship Level set by the University of London;

N.U. General Certificate Examination at Scholarship Level set by the Joint Matriculation Board of the Universities of Manchester, Liverpool, Leeds, Sheffield and Birmingham;

W. General Certificate Examination at Scholarship Level set by the Welsh Joint Education Committee.

CHAPTER 1

REAL NUMBERS. FUNDAMENTAL INEQUALITIES. FUNCTIONS OF REAL VARIABLES

1.1. Rational numbers

At an early stage in the development of mathematics it was found convenient to extend the class of positive integers (the *natural* numbers) 1, 2, 3, ... to include the negative integers -1, -2, -3, ... and the number zero. Without this extension there would be no value of x satisfying an equation such as $x + 6 = 4$ and the operation of subtraction could only be performed subject to inconvenient restrictions.

A second extension, to give a class of numbers in which the operations of subtraction and division can both be performed, is provided by introducing the *rational* numbers p/q where p and q belong to the previous class of numbers but with the proviso that $q \neq 0$. It is, in fact, possible to make further restrictions on p and q. It can be supposed that p and q have no common factor and q may be taken as being positive since

$$p/(-q) = (-p)/q \text{ and } (-p)/(-q) = p/q.$$

Examples of rational numbers are $\dfrac{1}{3}$, $\dfrac{4}{1}$ (or briefly, 4), $\dfrac{-17}{9}$, ... and such numbers enable us to say that there is a value of x satisfying an equation such as $7x = 13$.

It is sometimes convenient to make use of geometrical illustrations and the following representation of a rational number r is useful as an introduction to a second wide class of numbers. Suppose on a straight line (Fig. 1), O is taken as origin, A_1 a fixed point to the

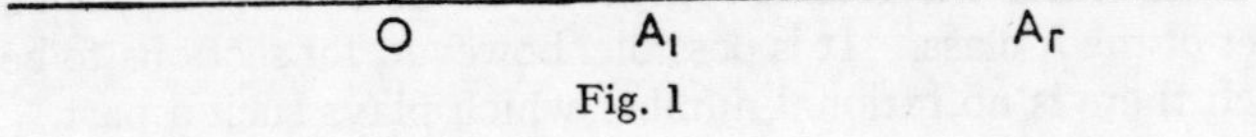

Fig. 1

right of O and we measure from O a length OA_r such that the ratio OA_r/OA_1 is equal to r (A_r being taken on the right or left of O according as r is positive or negative). If the length OA_1 be regarded as the unit of length, the length OA_r can then be regarded as representing the rational number r.

Example 1. *Show that there is no rational number such that its square is 2.*

If possible, let 2 be the square of a rational fraction p/q in its lowest terms. Then $(p/q)^2 = 2$ and $2q^2 - p^2 = 0$.

Hence p^2, and therefore also p, are even and it is possible to write $p = 2r$. It now follows that $(2r/q)^2 = 2$ so that $2r^2 = q^2$ and therefore q is also even. But p and q cannot both be even without contradicting the assumption that p/q is in its lowest terms and hence there can be no rational number whose square is 2.

1.2. Irrational numbers

All the rational numbers can be represented by points on a line (Fig. 1) but the converse is not true. There are, for example, various geometrical constructions which enable a point B to be assigned to the line such that $OB^2 = 2$. But (Example 1) there is no rational number whose square is 2 and thus there are points on the line which do not correspond to rational numbers. Such numbers are said to be *irrational*. The extension of the idea of number to include irrationals also enables us to say that there are values of x satisfying an equation such as $x^2 = 2$ and such an extension is clearly desirable.

A geometrical explanation such as has been given above of the existence of irrational numbers is not sufficient for the logical development of mathematical analysis and the following theory (originally due to Dedekind in 1858) can be established on a purely arithmetical basis.

Dedekind's theory of irrational numbers. The starting point of the theory evolved by Dedekind is a follow-up of the geometrical property that, if all the points on a straight line are separated into two classes such that every point of one class is to the right of every point of the other class, then there is one, and only one, point at which the line is thus divided. Dedekind considered the rules by which a separation (or section) into two classes of all rational numbers could be made. The two classes, the L (or left) class and the R (or right) class, are to possess the following two properties:

(i) there is at least one member of each class,
(ii) every member of the L class is less than every member of the R class.

It is clear that a section of this type is made by any rational number r. r is either the greatest member of the L class or the smallest member of the R class. It is possible, however, for sections to be made in which there is no rational number which plays such a part.

For example, in the case of the number whose square is 2, we may form a section in which the R class is made up of the positive rational numbers whose squares are greater than 2 and in which the L class is made up of all other rational numbers. This section is one in which the R class has no smallest member and the L class no greatest member. For suppose a is any positive rational number whose square is greater than 2 so that a belongs to the R class. Defining β by the relation*

---

* This example is due to E. T. Whittaker and G. N. Watson.

$$\beta = \frac{a(a^2+6)}{3a^2+2},$$

it follows that

$$\beta - a = \frac{2a(2-a^2)}{3a^2+2} \text{ and } \beta^2 - 2 = \frac{(a^2-2)^3}{(3a^2+2)^2}.$$

Since $a^2 > 2$, it can de deduced that

$$2 < \beta^2 < a^2$$

and hence that β is a smaller member of the R class. Similar reasoning will show that there is no greatest member of the L class. Sections such that the R class has no smallest member and the L class no greatest member are said to determine irrational numbers.

1.3. Real numbers

Numbers determined by Dedekind section are called *real* numbers. The aggregate of all real numbers, rational and irrational, is referred to as the *arithmetical continuum*.

The square root of 2 has so far been given as the only example of an irrational number, but the class of irrationals is a very wide one. As well as including quadratic and higher surds, it includes more complicated quantities involving root extraction such as $\sqrt[3]{(2+\sqrt{13})} + \sqrt[3]{(2-\sqrt{13})}$ and the roots of equations of the type

$$a_0 x^n + a_1 x^{n-1} + \ldots + a_{n-1} x + a_n = 0,$$

where $a_0, a_1, \ldots, a_n$ are integers. This latter type of irrational is known as an *algebraical* number.

The algebraical numbers do not by any means exhaust all the kinds of irrational numbers contained in the continuum. It is possible, for example, to draw a circle of unit diameter and suppose that the circumference of this circle (usually denoted by π) to be capable of measurement. It can be shown (but the proof is too intricate to be given here) that this number π is not the root of an algebraical equation with integral coefficients and therefore not an algebraical number. The number π belongs to another sub-class of irrationals. This sub-class is made up of *transcendental* numbers, these being defined as those irrational numbers which are not algebraical.

It should be noted that the fundamental laws of algebra are obeyed by real numbers and that the introduction of the irrationals has made it possible to perform certain operations such as root extraction. All such operations have not, however, been made possible by this extension. It is still not possible, for instance, to extract the square root of -2; in other words, there is no real number whose square is -2. The number system can be extended still further to make such operations possible but this extension is not considered until Chapter 4.

1.4. Inequalities

Certain inequalities are particularly important in mathematical analysis. Many are best proved by methods of the calculus but some important results which can be obtained in an elementary manner are given below.

(a) Elementary considerations

The manipulation of inequalities is performed in a similar way to that of equations. Thus,

if $a > b$, then $a + x > b + x$; (1.1)
if $a > b$, then $ax > bx$ when $x > 0$ but $ax < bx$ when $x < 0$; (1.2)
if $a > b$, and a, b are of the same sign, then $1/a < 1/b$; (1.3)
if $a > b > 0$, then $a^n > b^n$ when $n > 0$ but $a^n < b^n$ when $n < 0$. (1.4)

The reversal of the inequality sign in (1.2) and (1.4) when x and n are respectively negative should be noted. In the inequalities of (1.4) if n is of the form p/q where $q > 0$, it must be understood that $a^{p/q}$, $b^{p/q}$ denote respectively the positive qth root of a^p, b^p.

Example 2. *Prove that, when a, b, c are unequal positive numbers,*

(i) $a^a b^b > a^b b^a$;   (ii) $a^a b^b c^c > a^b b^c c^a$. (O.)

(i) Suppose that $a > b$, then by (1.4), $a^{a-b} > b^{a-b}$.
Multiplying by $a^b b^b$, (1.2) leads to $a^b b^b a^{a-b} > a^b b^b b^{a-b}$, giving $a^a b^b > a^b b^a$.

(ii) Suppose that c is the greatest or least of the numbers. By (1.4) $b^{a-c} > c^{a-c}$, giving

$$\frac{b^a}{b^c} > \frac{c^a}{c^c},$$

so that $b^a c^c > b^c c^a$.
Multiplying the result derived in (i) above by the positive number c^c, $a^a b^b c^c > a^b b^a c^c$. Since it has been shown that $b^a c^c > b^c c^a$, it follows that $a^a b^b c^c > a^b b^c c^a$. The reader should complete the proof to cover all possible positions of the number c.

Example 3. *If a_1, x, a_2 are three positive numbers in order of magnitude, show that*

$$\frac{1}{x} + \frac{1}{a_1 + a_2 - x} < \frac{1}{a_1} + \frac{1}{a_2}.$$ (O.)

Since $a_1 < x < a_2$ it follows that $a_1 - x < 0$ and $a_2 - x > 0$ and therefore
$$(a_1 - x)(a_2 - x) < 0.$$
Hence, $a_1 a_2 - (a_1 + a_2) x + x^2 < 0$, i.e. $a_1 a_2 < x(a_1 + a_2 - x)$.
It follows that,

$$\frac{1}{a_1 a_2} > \frac{1}{x(a_1 + a_2 - x)}$$

and, by multiplying by the positive number $a_1 + a_2$,

$$\frac{a_1 + a_2}{a_1 a_2} > \frac{a_1 + a_2}{x(a_1 + a_2 - x)}.$$

The required result then follows.

1] INEQUALITIES 5

(b) *The arithmetic mean is not less than the geometric mean*

In what follows, $a_1, a_2, \ldots, a_n$ denote n positive real numbers (including, possibly, zero). In elementary work, the arithmetic mean of two numbers a_1, a_2 is $\frac{1}{2}(a_1 + a_2)$ and their geometric mean is $\sqrt{(a_1 a_2)}$. In the case of n numbers, the arithmetic and geometric means are respectively taken as

$$\frac{a_1 + a_2 + \ldots + a_n}{n} \text{ and } (a_1 a_2 \ldots a_n)^{1/n}.$$

Suppose that a_r and a_s are the greatest and least of the n given numbers (if there are more than one greatest or least number any one of them will suffice). Let the geometric mean of the n numbers be denoted by G. G can be taken to be greater than zero for the theorem under discussion is obvious when $G = 0$. Now replace a_r and a_s by

$$b_r = G \text{ and } b_s = a_r a_s/G$$

so that the geometric mean is unaltered in value. It is possible to write

$$b_r + b_s - a_r - a_s = (a_r - G)(a_s - G)/G$$

and, as $a_r \geqslant G$, $a_s \leqslant G$, this is less than or equal to zero. Hence the arithmetic mean is certainly not increased by the substitution of b_r, b_s for a_r, a_s.

This argument may be repeated until each of the n given numbers has been replaced by G, at most $(n-1)$ repetitions being required. The final value of the arithmetic mean is then G and the starting value cannot therefore have been less. Hence

$$\frac{a_1 + a_2 + \ldots + a_n}{n} \geqslant (a_1 a_2 \ldots a_n)^{1/n}, \tag{1.5}$$

the sign of equality holding only when all the a's are equal.

Example 4. *Deduce from* (1.5) *above that, if n is a positive integer and $x > 1$,*
(i) $x^{2n+1} - 1 > (2n + 1)(x - 1)x^n$;
(ii) $\{(1)!(2)! \ldots (n-1)!\}^{1/n} > n/e$. (O.C.)

(i) Since $x^{2n+1} - 1 \equiv (x - 1)(x^{2n} + x^{2n-1} + \ldots + x + 1)$,

$$\frac{x^{2n+1} - 1}{(2n+1)(x-1)} \equiv \frac{1 + x + x^2 + \ldots + x^{2n-1} + x^{2n}}{2n + 1}$$

$$> (1 . x . x^2 \ldots x^{2n-1} . x^{2n})^{\frac{1}{2n+1}},$$

using (1.5) and noting that none of the $(2n + 1)$ numbers 1, x, x^2, $\ldots$, x^{2n-1}, x^{2n} are equal.
Now $1 . x . x^2 . \ldots . x^{2n-1} . x^{2n} = x^{1+2+\ldots+(2n-1)+2n}$ and the index of x is the sum of an arithmetic progression of $2n$ terms whose first term and common ratio are both unity. Thus the index of x in this product is $n(2 + 2n - 1)$, i.e. $n(2n + 1)$ and hence

MATHEMATICAL ANALYSIS

$$\frac{x^{2n+1}-1}{(2n+1)(x-1)} > \{x^{n(2n+1)}\}^{\frac{1}{2n+1}} = x^n.$$

The result required then follows by multiplying by $(2n+1)(x-1)$.

(ii) Since,
$$e = 1 + \frac{1}{(1)!} + \frac{1}{(2)!} + \ldots + \frac{1}{(n-1)!} + \ldots$$
and all the terms in the series are positive, e is certainly greater than the first n terms of this series. It follows that
$$\frac{e}{n} > \frac{1 + \frac{1}{(1)!} + \frac{1}{(2)!} + \ldots + \frac{1}{(n-1)!}}{n}$$
$$> \left(1 \cdot \frac{1}{(1)!} \cdot \frac{1}{(2)!} \cdot \ldots \cdot \frac{1}{(n-1)!}\right)^{1/n}, \text{ by (1.5)}.$$

Inverting this inequality, the required result follows.

Example 5. *If n is a positive integer greater than unity, by using (1.5) and by considering the series Σn and $\Sigma \frac{1}{n(n+1)}$, show that**
$$\left(\frac{n+1}{2}\right)^n > n! > (n+1)^{\frac{n-1}{2}}. \tag{O.}$$

The arithmetic mean of the integers $1, 2, 3, \ldots, n$
$$= \frac{\Sigma n}{n} = \frac{\frac{1}{2}n(n+1)}{n} = \frac{n+1}{2},$$
while their geometric mean $= (1.2.3 \ldots n)^{1/n} = (n!)^{1/n}$. Since these n integers are all different, (1.5) gives
$$\frac{n+1}{2} > (n!)^{1/n},$$
from which it follows that
$$\left(\frac{n+1}{2}\right)^n > n!.$$

The arithmetic mean of the n different quantities $\frac{1}{1.2}, \frac{1}{2.3}, \ldots, \frac{1}{n(n+1)}$
$$= \frac{1}{n}\Sigma\frac{1}{n(n+1)} = \frac{1}{n}\left\{\Sigma\frac{1}{n} - \Sigma\frac{1}{n+1}\right\}$$
$$= \frac{1}{n}\left\{\frac{1}{1} - \frac{1}{2} + \frac{1}{2} - \frac{1}{3} + \ldots + \frac{1}{n} - \frac{1}{n+1}\right\}$$
$$= \frac{1}{n}\left\{1 - \frac{1}{n+1}\right\}, \text{ since alternate terms cancel in pairs,}$$
$$= \frac{1}{n+1}.$$

The geometric mean of these quantities
$$= \left\{\frac{1}{1.2} \cdot \frac{1}{2.3} \cdot \ldots \cdot \frac{1}{n(n+1)}\right\}^{1/n} = \left\{\frac{1}{(n)!(n+1)!}\right\}^{1/n}$$
$$= \frac{1}{(n+1)^{1/n}} \cdot \frac{1}{(n!)^{2/n}}.$$

* See § 2.1 for more detailed remarks on the Σ notation.

Using (1.5),
$$\frac{1}{n+1} > \frac{1}{(n+1)^{1/n}} \cdot \frac{1}{(n!)^{2/n}},$$
so that
$$\frac{1}{(n+1)^{1-1/n}} > \frac{1}{(n!)^{2/n}},$$
leading to $(n!)^{2/n} > (n+1)^{\frac{n-1}{n}}$ or $n! > (n+1)^{\frac{n-1}{2}}$.

(c) *Cauchy's inequality*

The important result
$$(a_1^2 + a_2^2 + \ldots + a_n^2)(b_1^2 + b_2^2 + \ldots + b_n^2)$$
$$\geqslant (a_1 b_1 + a_2 b_2 + \ldots + a_n b_n)^2, \quad (1.6)$$
in which the a's and b's are all real is usually known as *Cauchy's inequality*. The sign of equality occurs when
$$\frac{a_1}{b_1} = \frac{a_2}{b_2} = \ldots = \frac{a_n}{b_n}. \quad (1.7)$$

It can be proved as follows. Consider the expression $\Sigma(a_r x + b_r)^2$ where Σ denotes summation with respect to r from 1 to n.

A value of x such that $a_r x + b_r$ vanishes for all n values of r exists only when the relations (1.7) hold and it is easy to show that (1.6) with the sign of equality is true in this case.

Except therefore in this case, $\Sigma(a_r x + b_r)^2$ is positive for all values of x. Now this expression can be written
$$x^2 \Sigma a_r^2 + 2x \Sigma a_r b_r + \Sigma b_r^2$$
$$= \frac{1}{\Sigma a_r^2}[(x \Sigma a_r^2 + \Sigma a_r b_r)^2 + \Sigma a_r^2 \Sigma b_r^2 - (\Sigma a_r b_r)^2].$$
Since Σa_r^2 is essentially positive, for this expression to be positive for all values of x it is necessary that
$$\Sigma a_r^2 \Sigma b_r^2 > (\Sigma a_r b_r)^2$$
and Cauchy's result has been established.

Example 6. *Real numbers $a_1, a_2, \ldots, a_n$, not all zero, are given, and $x_1, x_2, \ldots, x_n$ are real variables satisfying the equation*
$$a_1 x_1 + a_2 x_2 + \ldots + a_n x_n = 1.$$
Prove that the least value of $x_1^2 + x_2^2 + \ldots + x_n^2$ is
$$(a_1^2 + a_2^2 + \ldots + a_n^2)^{-1}. \quad \text{(O.C.)}$$

By Cauchy's inequality,
$$(x_1^2 + x_2^2 + \ldots + x_n^2)(a_1^2 + a_2^2 + \ldots + a_n^2)$$
$$\geqslant (a_1 x_1 + a_2 x_2 + \ldots + a_n x_n)^2.$$
But the right-hand side of this inequality is unity, so that
$$x_1^2 + x_2^2 + \ldots + x_n^2 \geqslant \frac{1}{a_1^2 + a_2^2 + \ldots + a_n^2},$$
which is the required result.

(d) Tchebychef's inequality

If $a_1, a_2, \ldots, a_n$ and $b_1, b_2, \ldots, b_n$ are two sets of positive numbers, arranged in descending order of magnitude, then

$$(a_1 + a_2 + \ldots + a_n)(b_1 + b_2 + \ldots + b_n) \leqslant n(a_1b_1 + a_2b_2 + \ldots + a_nb_n). \quad (1.8)$$

This result, often known as *Tchebychef's inequality*, can be proved as follows. Since $a_1 \geqslant a_2 \geqslant \ldots \geqslant a_n$, $b_1 \geqslant b_2 \geqslant \ldots \geqslant b_n$, the sum $\sum_{r=1}^{n}(a_r - a_s)(b_r - b_s) \geqslant 0$, the sign of equality occurring only when $a_r = a_s$ or $b_r = b_s$ for all r, s. This can be written as

$$\sum_{r=1}^{n} a_r b_r + n a_s b_s - a_s \sum_{r=1}^{n} b_r - b_s \sum_{r=1}^{n} a_r \geqslant 0.$$

Summing such inequalities for values of s between 1 and n,

$$n \sum_{r=1}^{n} a_r b_r + n \sum_{s=1}^{n} a_s b_s - \sum_{s=1}^{n}\sum_{r=1}^{n} a_s b_r - \sum_{s=1}^{n}\sum_{r=1}^{n} b_s a_r \geqslant 0.$$

Dropping the suffixes and dividing by 2, we have the inequality (1.8) in the equivalent form

$$n \Sigma ab - \Sigma a \, \Sigma b \geqslant 0.$$

A useful extension of this inequality is easily obtained. Excluding the trivial case in which $a_1 = a_2 = \ldots = a_n$ or $b_1 = b_2 = \ldots = b_n$, Tchebychef's inequality can be written

$$\frac{(a_1 + a_2 + \ldots + a_n)}{n} \frac{(b_1 + b_2 + \ldots + b_n)}{n} < \frac{a_1b_1 + a_2b_2 + \ldots + a_nb_n}{n}. \quad (1.9)$$

If $c_1, c_2, \ldots, c_n$ is a set of positive numbers, arranged in descending order and such that $c_r \neq c_s$ for all different r, s, application of (1.9) to the sets $a_1b_1, a_2b_2, \ldots, a_nb_n$, and $c_1, c_2, \ldots, c_n$ gives

$$\frac{(a_1b_1 + a_2b_2 + \ldots + a_nb_n)}{n} \frac{(c_1 + c_2 + \ldots + c_n)}{n} < \frac{a_1b_1c_1 + a_2b_2c_2 + \ldots + a_nb_nc_n}{n}.$$

Combining this with (1.9),

$$\frac{(a_1 + a_2 + \ldots + a_n)}{n} \frac{(b_1 + b_2 + \ldots + b_n)}{n} \frac{(c_1 + c_2 + \ldots + c_n)}{n} < \frac{a_1b_1c_1 + a_2b_2c_2 + \ldots + a_nb_nc_n}{n}.$$

Repeated applications of this type give the result

$$\frac{\Sigma a}{n} \cdot \frac{\Sigma b}{n} \cdot \frac{\Sigma c}{n} \cdot \ldots \cdot \frac{\Sigma l}{n} < \frac{\Sigma abc \ldots l}{n}, \quad (1.10)$$

provided that at least two of the sets do not consist entirely of equal numbers.

Example 7. *If $a_1, a_2, \ldots, a_n$ are positive and not all equal, show that*
$$\frac{\Sigma a^{\alpha+\beta+\gamma}}{n} > \frac{\Sigma a^\alpha}{n} \cdot \frac{\Sigma a^\beta}{n} \cdot \frac{\Sigma a^\gamma}{n},$$
where α, β and γ are positive.

This is a special case of (1.10) for three sets of positive numbers in which a, b, c are replaced respectively by a^α, a^β and a^γ.

EXERCISES 1 (a)

1. If $0 < x < 1$, $0 < y < 1$, prove that $0 < x + y - xy < 1$. (N.U.)
2. Prove that if $a < b < c$, there is no real value of x for which
$$\frac{1}{x-a} - \frac{1}{x-b} + \frac{1}{x-c} = 0. \qquad \text{(N.U.)}$$
3. If a, b are positive and $a + b = 1$, show that $ab \leq \frac{1}{4}$ and deduce that
$$\left(a + \frac{1}{a}\right)^2 + \left(b + \frac{1}{b}\right)^2 \geq \frac{25}{2}.$$
4. If a, b, c are positive and not all equal, prove that
   (i) $\frac{1}{3}(a + b + c) > \sqrt[3]{(abc)}$,
   (ii) $\frac{1}{a} + \frac{1}{b} + \frac{1}{c} > \frac{9}{a + b + c}$.

   Deduce that, if the triangle ABC is not equilateral, then
$$(\tan^2 A + \tan^2 B + \tan^2 C)^3 > 27 (\tan A + \tan B + \tan C)^2.$$
   (L.I.C.)

5. If n is a positive integer, show that
$$\frac{2n-1}{2n} < \sqrt{\left(\frac{2n-1}{2n+1}\right)},$$
and hence, or otherwise, prove that
$$\frac{1}{\sqrt{(2n+1)}} > \frac{1.3.5.\ldots.(2n-1)}{2.4.6.\ldots.2n}.$$
6. If a, b, m, n are all positive and $a \neq b$, show that
$$(ma + nb)^{m+n} > (m + n)^{m+n} a^m b^n.$$
7. If $a, b, c, \ldots, k$ are all positive and all less than unity and if $a + b + c + \ldots + k = s$, show that
   (i) $(1 + a)(1 + b)(1 + c) \ldots (1 + k) > 1 + s$,
   (ii) $(1 - a)(1 - b)(1 - c) \ldots (1 - k) > 1 - s$.

   (These two inequalities are due to *Weierstrass*.)
8. If $a_1, a_2, \ldots, a_n$ are all positive, prove that
$$(a_1 + a_2 + \ldots + a_n)\left(\frac{1}{a_1} + \frac{1}{a_2} + \ldots + \frac{1}{a_n}\right) \geq n^2.$$

1.5. The continuous real variable

There are two ways of regarding the real numbers. We may think of them as an aggregate or individually. When we think of them individually we may think of a particular real number, for example 4, 5/3 or $\sqrt{3}$, or we may think of an unspecified real number x belonging to the aggregate. The number x, used in this sense, is called the *continuous real variable* and individual numbers such as 4, 5/3 or $\sqrt{3}$ are called *values* of the variable.

A variable need not necessarily be continuous. For example, suppose we are considering a sub-class, such as the aggregate of positive integers, of the arithmetical continuum. In this case we might use n to denote the variable, the *positive integral variable*, and individual integers such as 1, 4, 8 would be *values* of this variable.

1.6. The idea of a function

If x and y are two continuous real variables, then y is said to be a *function* of x when to some or all of the values assigned to x there correspond values of y. The relation connecting the two variables may be a simple formula such as $y = 4x^2 + x + 2$ or the relation may be expressed by means of a graph. A common notation expressing such a functional relation between y and x is to write $y = f(x)$, $y = \phi(x)$, etc., and if $f(x) = 4x^2 + x + 2$, $f(3)$ denotes the value of $4x^2 + x + 2$ when $x = 3$, i.e. 41, and similarly for other values of x.

A function such as the example cited in the last paragraph possesses three properties:

(i) y is determined for all values of x; this is sometimes expressed by saying that the function is *defined* for all values of x;

(ii) there is one, and only one, value of y corresponding to each value of x;

(iii) the relationship is expressed by an analytical formula.

Many of the more important functions do in fact possess these properties but they are not in any way essential to a function. For example, if x is positive, the function $y^2 = x$ is not single-valued, for to each value of x there correspond two values $\pm\sqrt{x}$ of y. If $x = 0$ there corresponds only the single value $y = 0$, while if x is negative there is no real value of y satisfying the relation, and this function possesses only property (iii) above and not (i) or (ii).

1.7. Functional terminology

A terminology, similar to that in use for the various sub-classes of the real numbers, is used to describe the chief functions met with in mathematical analysis.

(a) Polynomials

A *polynomial* in x is a function of the form

$$a_0 x^n + a_1 x^{n-1} + \ldots + a_{n-1} x + a_n,$$

where $a_0, a_1, \ldots, a_n$ are constants and $a_0 \neq 0$. Such a polynomial is said to be of *degree* n and n is supposed to be a positive integer.

Example 8. *Find the polynomial $f(x)$ of the fourth degree such that $f(0) = f(1) = 1, f(2) = 13, f(3) = 73$ and $f'(0) = 0$.* (C.)

Since $f(0) = 1$ we can write

$$f(x) = a_0 x^4 + a_1 x^3 + a_2 x^2 + a_3 x + 1.$$

This gives $f'(x) = 4a_0 x^3 + 3a_1 x^2 + 2a_2 x + a_3$, and since $f'(0) = 0$, $a_3 = 0$. Hence $f(x) = a_0 x^4 + a_1 x^3 + a_2 x^2 + 1$.
Inserting $x = 1, 2, 3$ in turn and using $f(1) = 1, f(2) = 13, f(3) = 73$,

$$a_0 + a_1 + a_2 = 0,$$
$$16a_0 + 8a_1 + 4a_2 = 12,$$
$$81a_0 + 27a_1 + 9a_2 = 72.$$

The solution of these simultaneous equations is found to be

$$a_0 = 1, a_1 = 0, a_2 = -1,$$

and hence the required polynomial is $x^4 - x^2 + 1$.

(b) Rational functions

A *rational* function is the quotient of two polynomials. Thus if $P(x)$ and $Q(x)$ are two polynomials in x, the general rational function $R(x)$ is given by

$$R(x) = \frac{P(x)}{Q(x)},$$

and it simplifies matters to include the proviso that $Q(x) \neq 0$ and that $P(x), Q(x)$ do not vanish simultaneously.

When $Q(x)$ is unity (or any other constant), $R(x)$ reduces to a polynomial and the class of rational functions therefore includes the polynomials as a sub-class. The reader should note the similarity of the definition of a rational function to that of a rational number and note also that the class of rational numbers includes that of the integers as a sub-class.

It is usual also to suppose that the polynomials $P(x)$ and $Q(x)$ possess no common factor. It should be noted, however, that the removal of a common factor by division usually changes the character of the function. For example, on removal of the common factor $(x - 1)$ from the rational function $y = (x^2 - 1)/(x - 1)$ the function becomes $y = x + 1$ and, in this form, it is defined for all values of x. In its original form, the function takes the value $0/0$ when $x = 1$ and the original function is therefore undefined for this particular value of x.

One other point is worth noticing here. The definition of a rational function does not imply that the coefficients in the polynomials forming its numerator and denominator are necessarily rational numbers.

Expressions such as $x^4 + \pi$ and $x^2 - 2\sqrt{2}x + 3$ are genuine polynomials and the function
$$\frac{x^4 + \pi}{x^2 - 2\sqrt{2}x + 3}$$
is a rational function. The word rational, in connection with functions, refers only to the way in which the variable (x in this example) enters into the function.

Example 9. *Show that the function*
$$y = \left(\frac{1}{x} + \frac{1}{x-a}\right) \bigg/ \left(\frac{1}{x-b} + \frac{1}{x+b}\right)$$
is a rational function of x.

Summing the algebraical fractions in the numerator and denominator
$$y = \left\{\frac{2x-a}{x(x-a)}\right\} \bigg/ \left\{\frac{2x}{x^2 - b^2}\right\}$$
$$= \frac{(2x-a)(x^2 - b^2)}{2x^2(x-a)}$$
and the function is now in the standard form for a rational function (the quotient of two polynomials). It should be noted that the character of the function has been changed by its reduction to the standard rational form. As given originally, the function is undefined when $x = 0$, a, b and $-b$, while, in the second form it is undefined only when $x = 0$ and a and takes the value 0 when $x = \pm b$.

(c) *Algebraical functions*

The function $y = f(x)$ is an *algebraical* function of x of degree n if y is a root of the equation
$$y^n + R_1 y^{n-1} + \ldots + R_{n-1} y + R_n = 0$$
where $R_1, R_2, \ldots, R_n$ are rational functions of x.

It should be noted that there is no loss of generality in supposing that the coefficient of the first term of the above equation in y is unity, for the remaining coefficients in this equation will remain rational functions of x after division by a rational function. The reader should once again notice the similarity in terminology of functions and numbers by recalling the definition of an algebraical number (§ 1.3). It should be clear that the class of algebraical functions is the widest so far considered and that it contains the rational functions as a sub-class, the rational functions being, in fact, algebraical functions of degree 1.

Example 10. *Show that* $\sqrt{(x + \sqrt{x})}$ *is an algebraical function of degree* 4.

Let $y = \sqrt{(x + \sqrt{x})}$ so that $y^2 = x + \sqrt{x}$ and $(y^2 - x)^2 = x$. Hence y is a root of the quartic equation
$$y^4 - 2xy^2 + x^2 - x = 0$$
and is therefore an algebraical function of degree 4.

(d) Transcendental functions

All functions which are not algebraical are said to be *transcendental*. This is a very wide class of functions and includes the trigonometrical, logarithmic and exponential functions. It includes also many functions (such as those of Bessel and Legendre) which are beyond the scope of this book.

(e) Miscellaneous functional terminology

This section explains a few points in functional terminology which will be useful in what follows.

Functions such as $y = 4x^2 + x + 2$, $y = \sin x$, etc., are termed *explicit* functions. The variable x is called the *independent* and y the *dependent* variable. When the relation between two variables x and y is given in a form such as

$$x^2 + y^2 = 2x \quad \text{or} \quad x + y + \cos y = 4$$

the function is said to be *implicit*. In some cases it is possible to express an implicit function in explicit form. In the first example given above, we can solve for y to obtain $y = \sqrt{(2x - x^2)}$ and, in this form, y is an explicit function of x. Similarly, the algebraical function of Example 10 can be written in explicit form as $y = \sqrt{(x + \sqrt{x})}$, while, as an implicit function, it can be expressed as

$$y^4 - 2xy^2 + x^2 - x = 0.$$

In other cases, such as $x + y + \cos y = 4$, it is not possible to express y as an explicit function of x.

If $f(x) = f(-x)$ for all values of x, $f(x)$ is said to be an *even* function while if $f(x) = -f(-x)$ for all values of x it is said to be an *odd* function. Any function, which is defined for all values of x, is the sum of an even and an odd function of x, for

$$f(x) = \tfrac{1}{2}\{f(x) + f(-x)\} + \tfrac{1}{2}\{f(x) - f(-x)\}$$

and the first function on the right-hand side of this identity is an even function and the second is an odd one.

A function $f(x)$ is said to be *periodic*, with period ω, if $f(x) = f(x + \omega)$ for all values of x for which $f(x)$ is defined. For example the trigonometrical functions $\sin x$ and $\cos x$ are periodic functions with period 2π.

Example 11. *Show that no periodic function can be a rational function unless it is a constant.*

Let $f(x)$ be a periodic function with period ω, so that $f(x) = f(x + \omega)$ for all values of x. Suppose that $f(x)$ is a rational function so that $f(x) = P(x)/Q(x)$ where $P(x)$ and $Q(x)$ are polynomials in x. If $f(0) = c$, then

$$\frac{P(x)}{Q(x)} = c$$

when $x = 0, \omega, 2\omega, 3\omega, \ldots$. Thus, whatever the degree n of the algebraical equation $P(x) - cQ(x) = 0$, it is satisfied by more than n values of x and this is possible only if $P(x) - cQ(x) = 0$ for all values of x, i.e. if $f(x) = c$ (constant).

(*f*) *Functions of several variables*

So far functions containing only one independent variable have been mentioned. Suppose now we have several independent variables $x, y, z, \ldots$, and one dependent variable u. Then u may be given by a relation of the form

$$u = f(x, y, z, \ldots)$$

and u is said to be an explicit function of the several variables $x, y, z, \ldots$. It may happen, however, that the variables $u, x, y, z, \ldots$ are connected by a relation of the form

$$F(u, x, y, z, \ldots) = 0$$

and such a relation defines an implicit function of several variables.

As a simple practical example, take the relation expressing the volume V of a right circular cylinder in terms of its radius r and its height h. Then $V = \pi r^2 h$ and this is an example of an explicit function of the two independent variables r and h. The relation

$$\cos u + ux + 2uy = 0$$

is an implicit function relating the variables u, x, y and it is not possible in this case to express u explicitly in terms of x and y.

EXERCISES 1 (*b*)

1. A polynomial $f(x)$ of the second degree in x is such that $f(1) = 7$, $f(12) = 23$ and $f'(2) = 5$. Find the value of $f(3)$.
2. A sequence of polynomials $f_0(x), f_1(x), f_2(x), \ldots$, is defined by $f_0(x) = 1, f_1(x) = x + 1$, and, when $n \geqslant 1$,

$$f_{n+1}(x) = \{(2n + 1)x + 1\}f_n(x) - n^2x^2f_{n-1}(x).$$

Show, by induction or otherwise, that $f_n(x)$ is a polynomial of degree n in which the coefficient of x^n is $n!$ and the constant term is 1. (O.)
3. If the remainders when a polynomial $f(x)$ is divided by $(x - a)(x - b)$ and by $(x - a)(x - c)$ are equal, prove that

$$(a - b)f(c) + (b - c)f(a) + (c - a)f(b) = 0. \quad (C.)$$

4. If $(ax^2 + bx + c)y + Ax^2 + Bx + C = 0$, show that x is a rational function of y if $(aC - Ac)^2 = (aB - Ab)(bC - Bc)$.
5. Show that no periodic function can be an algebraical function.
6. If $f(m, n) = f(m - 1, n) + f(m, n - 1)$ when $m, n \geqslant 1$ and if $f(m, 0) = 1$ for $m \geqslant 0$ and $f(0, n) = 0$ for $n \geqslant 1$, show that

$$\sum_{m=0}^{\infty} \sum_{n=0}^{\infty} f(m, n)x^m y^n = \frac{1 - y}{1 - x - y},$$

and find a formula for $f(m, n)$.

(x and y may be assumed small enough to ensure convergence.) (O.)

1.8. The decomposition of a rational function into partial fractions

The rules for the decomposition of specific rational functions into partial fractions were given in *Advanced Level Pure Mathematics* (§ 2.8). Here we consider the general rational function $P(x)/Q(x)$ in which $P(x)$ and $Q(x)$ are polynomials in x.

As far as the decomposition into partial fractions is concerned, it will be sufficient to consider only cases in which the degree of $P(x)$ is less than that of $Q(x)$. In other cases, the rational function can be expressed as the sum of a polynomial and a rational function in which the numerator is of lower degree than the denominator by finding the quotient and remainder when $P(x)$ is divided by $Q(x)$.

Suppose that $Q(x)$ contains an unrepeated linear factor $(x - a_1)$ so that

$$\frac{P(x)}{Q(x)} \equiv \frac{P(x)}{(x - a_1)\phi(x)}$$

where $\phi(x)$ is a function of x such that $\phi(a_1) \neq 0$. This can be written in the form

$$\frac{P(x)}{Q(x)} \equiv \frac{A_1}{x - a_1} + \frac{\lambda(x)}{\phi(x)} \equiv \frac{A_1 \phi(x) + (x - a_1)\lambda(x)}{Q(x)}$$

where A_1 is a constant and $\lambda(x)$ some function of x. Comparing the numerators

$$P(x) \equiv A_1 \phi(x) + (x - a_1)\lambda(x),$$

and, writing $x = a_1$,

$$A_1 = P(a_1)/\phi(a_1).$$

Now, if dashes denote differentiation with respect to x,

$$Q'(x) \equiv \frac{d}{dx}\{(x - a_1)\phi(x)\}$$

$$\equiv \phi(x) + (x - a_1)\phi'(x),$$

and $\phi(a_1) = Q'(a_1)$ giving $A_1 = P(a_1)/Q'(a_1)$.

This process can be carried out for each unrepeated linear factor which $Q(x)$ may contain. If $Q(x)$ is made up entirely of n different linear factors $(x - a_r)$, $r = 1, 2, \ldots, n$, we have the important result

$$\frac{P(x)}{Q(x)} \equiv \sum_{r=1}^{n} \frac{P(a_r)}{Q'(a_r)(x - a_r)}. \tag{1.11}$$

When using this formula, it is worth noticing that $Q'(a_r)$ is most easily obtained by writing $x = a_r$ in all the factors of $Q(x)$ with the exception of $(x - a_r)$. For example, suppose that $P(x) = 5$ and $Q(x) = x^2 + x - 6 = (x - 2)(x + 3)$.

Then,
$$Q'(x) = 2x + 1$$

and $Q'(2) = 5$, $Q'(-3) = -5$. These latter values can be obtained very simply by writing $x = 2$ in $(x + 3)$ and $x = -3$ in $(x - 2)$, the expressions for $Q(x)$ less the factors $(x - 2)$ and $(x + 3)$ respectively. Hence

$$\frac{P(2)}{Q'(2)} = \frac{5}{5} = 1 \text{ and } \frac{P(-3)}{Q'(-3)} = \frac{5}{-5} = -1,$$

so that (1.11) gives in this case:

$$\frac{5}{x^2 + x - 6} \equiv \frac{1}{x - 2} - \frac{1}{x + 3}.$$

The result (1.11) applies only to those cases in which the factors of $Q(x)$ are linear and unrepeated. In the cases of quadratic or repeated factors, the corresponding general results are somewhat complicated and will not be given here. One example in which $Q(x)$ contains a repeated factor is given below (Example 13).

Example 12. *If $f(x)$ is a polynomial with n different factors $x - a$, $x - \beta$, $x - \gamma, \ldots$, and if A, B, C, D are constants such that $f(-D/C) \neq 0$, show that*

$$\Sigma \frac{Aa + B}{Ca + D} = n\frac{A}{C} + \frac{(AD - BC)f'(-D/C)}{C^2 f(-D/C)},$$

where the summation extends over the n quantities $a, \beta, \gamma, \ldots$.. (O.)

By division,

$$\Sigma \frac{Aa + B}{Ca + D} = \Sigma \frac{A}{C} + \Sigma \frac{B - AD/C}{Ca + D}$$

$$= n\frac{A}{C} - \frac{AD - BC}{C} \Sigma \frac{1}{Ca + D}. \quad (1.12)$$

Now $f(-D/C) = \left(-\frac{D}{C} - a\right)\left(-\frac{D}{C} - \beta\right) \ldots$ to n factors and logarithmic differentiation with respect to $-D/C$ gives

$$\frac{f'(-D/C)}{f(-D/C)} = \Sigma \frac{1}{(-D/C) - a}.$$

Hence

$$\Sigma \frac{1}{Ca + D} = -\frac{f'(-D/C)}{Cf(-D/C)}$$

and substitution in (1.12) leads to the required result.

Example 13. *$P(x)$ is a polynomial of degree n and*

$$Q(x) \equiv (x - a_1)^2 (x - a_2) \ldots (x - a_n)$$

where $a_1, a_2, \ldots, a_n$ are all different. If

$$\frac{P(x)}{Q(x)} \equiv \frac{A_0}{x - a_1} + \frac{A_1}{(x - a_1)^2} + \frac{A_2}{x - a_2} + \ldots + \frac{A_n}{x - a_n},$$

find $A_0, A_1, A_2, \ldots, A_n$. (C.)

Summing the fractions on the right-hand side of the expression for $P(x)/Q(x)$ and comparing the numerators,

1] DECOMPOSITION INTO PARTIAL FRACTIONS 17

$$P(x) \equiv A_0(x - a_1)(x - a_2) \ldots (x - a_n) + A_1(x - a_2)(x - a_3) \ldots (x - a_n)$$
$$+ \sum_{r=2}^{n} A_r(x - a_1)^2(x - a_2) \ldots (x - a_{r-1})(x - a_{r+1}) \ldots (x - a_n).$$

Hence, $\quad P(a_1) = A_1(a_1 - a_2)(a_1 - a_3) \ldots (a_1 - a_n),$
and
$$P(a_r) = A_r(a_r - a_1)^2(a_r - a_2) \ldots (a_r - a_{r-1})(a_r - a_{r+1}) \ldots (a_r - a_n),$$
$(2 \leqslant r \leqslant n)$, from which the values of $A_1, A_2, \ldots, A_n$ follow immediately.
Differentiating the expression for $P(x)$ with respect to x and setting $x = a_1$,

$$P'(a_1) = (a_1 - a_2) \ldots (a_1 - a_n)\left[A_0 + A_1\left\{\frac{1}{a_1 - a_2} + \ldots + \frac{1}{a_1 - a_n}\right\}\right],$$

and the value of A_0 follows on substitution for A_1.

Example 14. *For all values of x, $F(x)$ satisfies $F(x) = \sum_{r=1}^{n} \frac{a_r}{F(r)(x + r)}$.*

Prove that $\quad F(x)F(-x) = \sum_{r=1}^{n} \frac{2ra_r}{r^2 - x^2}.$

Find $F(x)$ when $n = 2$, $a_1 = a_2 = 1$. (O.)

Let $F(x) = P(x)/Q(x)$ where $P(x)$ is a polynomial and
$$Q(x) = (1 + x)(2 + x) \ldots (n + x).$$
Then, using (1.11),
$$\sum_{r=1}^{n} \frac{a_r}{F(r)(x + r)} = F(x) = \frac{P(x)}{Q(x)} = \sum_{r=1}^{n} \frac{P(-r)}{Q'(-r)(x + r)},$$

and hence $\quad \dfrac{P(-r)}{Q'(-r)} = \dfrac{a_r}{F(r)}.$ \hfill (1.13)

Now $\quad F(x)F(-x) = \dfrac{P(x)P(-x)}{Q(x)Q(-x)}$

$$= \sum_{r=1}^{n} \left\{ \frac{P(-r)P(r)}{[Q'(x)Q(-x) - Q(x)Q'(-x)]_{x=-r}} \cdot \frac{1}{r + x} \right.$$
$$\left. - \frac{P(r)P(-r)}{[Q'(x)Q(-x) - Q(x)Q'(-x)]_{x=r}} \cdot \frac{1}{r - x} \right\}$$

$$= \sum_{r=1}^{n} \left\{ \frac{P(-r)P(r)}{Q'(-r)Q(r)} \cdot \frac{2r}{r^2 - x^2} \right\},$$

since $Q(-r) = 0$ when $r = 1, 2, \ldots, n$.
Using (1.13) and remembering that $F(r) = P(r)/Q(r)$

$$F(x)F(-x) = \sum_{r=1}^{n} \left\{ \frac{a_r}{F(r)} \cdot F(r) \cdot \frac{2r}{r^2 - x^2} \right\} = \sum_{r=1}^{n} \frac{2ra_r}{r^2 - x^2}.$$

When $n = 2$, $a_1 = a_2 = 1$,

$$F(x)F(-x) = \frac{2}{1 - x^2} + \frac{4}{4 - x^2} = \frac{6(2 - x^2)}{(1 - x^2)(4 - x^2)},$$

and since $F(x)$ is the sum of two quantities with denominators $(1 + x)$, $(2 + x)$,

$$F(x) = \frac{\sqrt{6}(\sqrt{2} \pm x)}{(1 + x)(2 + x)}.$$

EXERCISES 1 (c)

1. Express in partial fractions

   (i) $\dfrac{x}{(x-a)(x-b)}$,   (ii) $\dfrac{x^3}{(x-a)(x-b)(x-c)}$.

2. Express in partial fractions
$$\frac{(x-a)(x-b)(x-c)(x-d)}{(x+a)(x+b)(x+c)(x+d)},$$
   (i) when a, b, c, d are all unequal, (ii) when they are all equal. (C.)

3. Use the partial fraction theorem to evaluate
$$\frac{(a-y)(a-z)(a-u)}{(a-b)(a-c)(a-d)(a-x)} + \frac{(b-y)(b-z)(b-u)}{(b-a)(b-c)(b-d)(b-x)}$$
$$+ \frac{(c-y)(c-z)(c-u)}{(c-a)(c-b)(c-d)(c-x)} + \frac{(d-y)(d-z)(d-u)}{(d-a)(d-b)(d-c)(d-x)}.$$
   (O.)

4. Prove that:

   (i) $\dfrac{(n)!}{x(x+1)\ldots(x+n)} = \dfrac{1}{x} - \dfrac{n}{x+1} + \dfrac{\tfrac{1}{2}n(n-1)}{x+2} + \cdots + \dfrac{(-1)^n}{x+n}$,

   (ii) $\dfrac{(2n)!/(n)!}{x(x+1)\ldots(x+2n)} = \dfrac{1}{x(x+1)\ldots(x+n)}$
   $$- \frac{n}{(x+1)(x+2)\ldots(x+n+1)}$$
   $$+ \frac{\tfrac{1}{2}n(n-1)}{(x+2)(x+3)\ldots(x+n+2)} + \cdots$$
   $$+ \frac{(-1)^n}{(x+n)(x+n+1)\ldots(x+2n)}.$$ (O.)

5. Deduce from Exercise 4 (i) that
$$1 + \frac{1}{2} + \frac{1}{3} + \cdots + \frac{1}{n} = n - \frac{n(n-1)}{2(2)!} + \frac{n(n-1)(n-2)}{3(3)!} - \cdots + \frac{(-1)^{n-1}}{n}.$$ (O.)

6. Express in partial fractions the function
$$\frac{x^p}{(x+1)(x+2)\ldots(x+n)}$$
   in the two cases (i) $1 \leq p \leq n-1$, (ii) $p = n$.
   Hence, or otherwise, prove that the expression
$$\frac{1^p}{(1)!(n-1)!} - \frac{2^p}{(2)!(n-2)!} + \cdots + \frac{(-1)^{n-2}(n-1)^p}{(n-1)!(1)!} + \frac{(-1)^{n-1}n^p}{(n)!}$$
   takes the value zero when $p = 1, 2, \ldots, n-1$, and find its value when $p = n$.   (O.)

EXERCISES 1 (d)

1. Show that a rational number, in its lowest terms, can only satisfy the algebraical equation
$$x^n + a_1 x^{n-1} + a_2 x^{n-2} + \ldots + a_{n-1} x + a_n = 0$$
in which $a_1, a_2, \ldots, a_n$ are integers if it is itself an integer. Show also that, in this case, it must be a divisor of a_n.

2. If a, b, c, d are rational numbers, show that $(a\sqrt{2} + b)/(c\sqrt{2} + d)$ is rational only if $ad = bc$.

3. By using the identity
$$(a^3 + b^3) - (a^2 b + ab^2) \equiv (a^2 - b^2)(a - b),$$
or otherwise, show that for any positive numbers a, b, c
$$a^2 b + ab^2 + b^2 c + bc^2 + c^2 a + ca^2 \leqslant 2(a^3 + b^3 + c^3). \quad \text{(L.I.C.)}$$

4. If m and n are odd positive integers and a, b are any unequal real numbers, prove that
$$(a^m + b^m)(a^n + b^n) < 2(a^{m+n} + b^{m+n}).$$
Does the inequality remain valid if m and n are replaced by even positive integers? (L.I.C.)

5. If a, b, c are real positive quantities, prove that
$$\tfrac{1}{2}(a + b + c) - \frac{bc}{b+c} - \frac{ca}{c+a} - \frac{ab}{a+b}$$
can never be negative.

6. Show that the greatest value of
$$\frac{x^n}{1 + x + \ldots + x^{2n}}, \quad (x > 0)$$
is $1/(2n + 1)$. (O.)

7. If n and r are positive integers such that $n > r > 1$, show that
$$\frac{n+1-r}{n-r} > \frac{n+1}{n}.$$
Hence, or otherwise, prove that
$$\left(1 + \frac{1}{n+1}\right)^{n+1} > \left(1 + \frac{1}{n}\right)^n. \quad \text{(L.I.C.)}$$

8. Prove that, if a, b are real, $ab \leqslant \left(\dfrac{a+b}{2}\right)^2$ and deduce that if a, b, c, d are positive, $abcd \leqslant \left(\dfrac{a+b+c+d}{4}\right)^4$, with equality only when the numbers are all equal. By giving d a suitable value in terms of a, b, c, or otherwise, prove that, if a, b, c are positive, $abc \leqslant \left(\dfrac{a+b+c}{3}\right)^3$. (C.)

9. If $a_1, a_2, \ldots, a_n$ are all different positive real numbers, show that
$$a_1^3 + a_2^3 > a_1^2 a_2 + a_2^2 a_1,$$
$$a_1^3 + a_2^3 + a_3^3 > a_1^2 a_2 + a_2^2 a_3 + a_3^2 a_1,$$
and, in general, for $2 \leqslant r \leqslant n$,
$$\sum_{s=1}^{r} a_s^3 > \sum_{s=1}^{r-1} a_s^2 a_{s+1} + a_r^2 a_1. \qquad \text{(O.)}$$

10. Show that if x, y are positive integers and $x - y > 1$, then
$$(x)! + (y)! > (x-1)! + (y+1)!.$$
Hence show that if $x_1, x_2, \ldots, x_n$ are n positive integers such that
$$x_1 + x_2 + \ldots + x_n = np,$$
where p is a positive integer, then
$$(x_1)! + (x_2)! + \ldots + (x_n)! > n(p)!. \qquad \text{(O.)}$$

11. If $a_1, a_2, \ldots, a_n$ are n positive numbers whose sum is unity, prove that
$$\frac{1}{a_1} + \frac{1}{a_2} + \cdots + \frac{1}{a_n} > n^2. \qquad \text{(O.)}$$

12. Show that when the sum of two positive real numbers remains constant, their product increases as their difference decreases. Deduce that if n is a positive integer

   (i) $(n+1)^2 > 1 + 2 \cdot \dfrac{2n}{2n+1} + 3 \cdot \dfrac{2n-1}{2n+1} + \ldots + 1 > 2n + 1,$

   (ii) $(n!)^2 > n^n.$ (O.)

13. Show that if every a, b, c, d is positive,

   (i) $\Sigma a_r^2 \Sigma b_r^2 \geqslant (\Sigma a_r b_r)^2,$

   (ii) $\Sigma a_r^4 \Sigma b_r^4 \Sigma c_r^4 \Sigma d_r^4 \geqslant (\Sigma a_r b_r c_r d_r)^4.$

By giving a suitable value to each d, deduce that
$$\Sigma a_r^3 \Sigma \beta_r^3 \Sigma \gamma_r^3 \geqslant (\Sigma a_r \beta_r \gamma_r)^3. \qquad \text{(O.)}$$

14. If $f(x)$ is a polynomial in x of degree 3, show that the remainder when $f(x)$ is divided by $(x-a)^2$ is $xf'(a) + f(a) - af'(a)$. (L.I.C.)

15. $f(x)$ is a polynomial of the fifth degree, the coefficient of x^5 being 3. $f(x)$ leaves the same remainder when divided by $x^2 + 1$ or $x^2 + 3x + 3$. It leaves the remainder $4x + 5$ when divided by $(x-1)^2(x+1)$. Find $f(x)$. (C.)

16. If x is a real variable and a a real constant, prove that $\tan(x+a)/\tan(x-a)$ assumes all real values, except those lying between two limits, and determine these limits.
Sketch the graph of the function, taking a to be a positive angle less than $\pi/4$. (O.)

17. Prove that the rational function $(x-p)/(x-q)(x-r)$ assumes all real values if p lies between q and r, but that otherwise there are two values between which the expression cannot lie, and that the difference between these values is
$$\frac{4\sqrt{\{(p-q)(p-r)\}}}{(q-r)^2}.$$

Illustrate your answer by sketching roughly the graphs of
$$\frac{x-4}{(x-5)(x-1)} \text{ and } \frac{x-5}{(x-4)(x-1)}.$$ (O.)

18. Polynomials P_n, of degree n, are defined by the relation
$$(n+1)P_{n+1} - (2n+1)xP_n + nP_{n-1} = 0$$
and $P_0 = 1$, $P_1 = x$. Prove by induction that the two terms of P_n of highest degree are
$$\frac{1.3.5.\ldots.(2n-1)}{1.2.3.\ldots.n}\left\{x^n - \frac{n(n-1)}{2(2n-1)}x^{n-2}\right\}.$$ (O.)

19. Prove that, if
$$F(x) \equiv (1-qx)(1-q^2x)\ldots(1-q^{p-1}x) \equiv 1 + A_1x + A_2x^2 + \ldots + A_{p-1}x^{p-1},$$
then $(1-q^px)F(x) = (1-qx)F(qx)$,
and
$$A_r = \frac{(q^p-q)(q^p-q^2)\ldots(q^p-q^r)}{(1-q)(1-q^2)\ldots(1-q^r)} (r > 0).$$ (O.)

20. Resolve
$$\frac{1}{x(x-2)(x-1)^{2n}}$$
into partial fractions when n is a positive integer. (O.C.)

21. The polynomial $f(x)$ is the product of n distinct factors $x-a, \ldots, x-\varkappa$. Prove that
$$\frac{1}{f(x)} = \Sigma \frac{1}{f'(a)} \cdot \frac{1}{x-a},$$
and show that $\Sigma\{a^r/f'(a)\}$ is zero when $r = 0, 1, \ldots, n-2$ and is unity when $r = n-1$. (O.)

22. Prove that
$$\frac{2^n(n)!}{(y+1)(y+3)\ldots(y+2n+1)} = \frac{1}{y+1} - \frac{{}^nC_1}{y+3} + \ldots + \frac{(-1)^n{}^nC_n}{y+2n+1}.$$ (O.)

23. Prove that, for all values of x,
$$\frac{(b-c)(c-a)(a-b)}{(x-a)(x-b)(x-c)}\left\{\frac{1}{x-a} + \frac{1}{x-b} + \frac{1}{x-c}\right\}$$
$$= \frac{c-b}{(x-a)^2} + \frac{a-c}{(x-b)^2} + \frac{b-a}{(x-c)^2}.$$ (O.)

24. Given that $\phi(x) = (x-a_0)(x-a_1)\ldots(x-a_n)$, where $a_0, a_1, \ldots, a_n$ are all distinct, and that $f(x)$ is a polynomial of degree not greater than n, show that
$$\sum_{r=0}^{n} \frac{f(a_r)}{\phi'(a_r)}$$
is the coefficient of x^n in $f(x)$. (O.)

25. Prove that
$$\frac{1}{(x-\alpha)^n(x-\beta)^n} = \frac{(-1)^n}{(\alpha-\beta)^{2n}}\sum_{r=0}^{n-1}C_r\left\{\left(\frac{\beta-\alpha}{x-\alpha}\right)^{n-r} + \left(\frac{\alpha-\beta}{x-\beta}\right)^{n-r}\right\}$$
where C_r is the coefficient of h^r in the expansion of $(1-h)^{-n}$. (O.)

CHAPTER 2

FINITE SERIES

2.1. The summation of finite series

Formulae for the sums of arithmetical and geometrical progressions, the squares of the natural numbers and certain simple series involving the coefficients in the binomial theorem have been discussed in *Advanced Level Pure Mathematics*, §§ 3.2, 3.3, 3.6 and 3.9 (example 20). In the sections which follow some methods of summing other simple finite series are given.

If u_r is the rth term of a finite series consisting of n terms and if
$$s_n = u_1 + u_2 + \ldots + u_r + \ldots + u_n,$$
s_n is called the sum to n terms of the series. It is convenient to use the Σ notation and to write
$$s_n = \sum_{r=1}^{n} u_r \text{ or } s_n = \sum_{1}^{n} u_r,$$
or, when no confusion can arise over the number of terms involved, simply Σu_r. It should be noted that n is necessarily a positive integer and that s_n is a function of the positive integral variable n (see § 1.5).

There is no general method for the summation of series but the methods outlined below are useful in many cases. Often much ingenuity and experience are necessary to determine s_n and in some cases it is not possible to express it as an elementary function of n—for example there is no elementary function of n for the sum of the harmonic series $1 + \dfrac{1}{2} + \dfrac{1}{3} + \ldots + \dfrac{1}{n}$. The summation of finite series of trigonometrical terms is considered later (Chapter 6, § 6.13).

2.2 The difference method

Sometimes it is possible to express the rth term u_r of a finite series as the difference between a function of $(r + 1)$ and the same function of r; in other words, a function $f(r)$ can be found such that
$$u_r \equiv f(r + 1) - f(r).$$
In this case,
$$u_1 = f(2) - f(1),$$
$$u_2 = f(3) - f(2),$$
$$u_3 = f(4) - f(3),$$
$$\text{-----------}$$
$$u_{n-1} = f(n) - f(n - 1),$$
$$u_n = f(n + 1) - f(n).$$

By addition, the sum to n terms s_n is given by
$$s_n = f(n+1) - f(1),$$
for all the other terms on the right-hand side cancel in pairs.

A straightforward instance of this method is given in Example 1, variations of the method being shown in Examples 2 and 3.

Example 1. *Find the sum of the first n terms of the series*
$$\frac{3}{1^2 \cdot 2^2} + \frac{5}{2^2 \cdot 3^2} + \frac{7}{3^2 \cdot 4^2} + \dots \quad \text{(L.I.C.)}$$

The nth term $u_n = \dfrac{2n+1}{n^2(n+1)^2} \equiv \dfrac{1}{n^2} - \dfrac{1}{(n+1)^2}$.

Hence $u_1 = 1 - \dfrac{1}{2^2}$, $u_2 = \dfrac{1}{2^2} - \dfrac{1}{3^2}$, $u_3 = \dfrac{1}{3^2} - \dfrac{1}{4^2}$ and so on. By addition all the terms cancel except the first and last, so that

$$\text{required sum} = 1 - \frac{1}{(n+1)^2} = \frac{n(n+2)}{(n+1)^2}.$$

Example 2. *Express $\dfrac{1}{x(x+2)}$ in partial fractions. Hence find the sum of n terms of the series*
$$\frac{1}{1 \cdot 3} + \frac{1}{2 \cdot 4} + \frac{1}{3 \cdot 5} + \dots \quad \text{(O.C.)}$$

By the ordinary rules for partial fractions,
$$\frac{1}{x(x+2)} \equiv \frac{1}{2x} - \frac{1}{2(x+2)}.$$

The nth term u_n of the given series is given by
$$u_n = \frac{1}{n(n+2)} \equiv \frac{1}{2}\left(\frac{1}{n} - \frac{1}{n+2}\right).$$

Hence $2u_1 = 1 - \dfrac{1}{3}$, $2u_2 = \dfrac{1}{2} - \dfrac{1}{4}$, $2u_3 = \dfrac{1}{3} - \dfrac{1}{5}$, and so on, the last two terms being $2u_{n-1} = \dfrac{1}{n-1} - \dfrac{1}{n+1}$, $2u_n = \dfrac{1}{n} - \dfrac{1}{n+2}$.

On addition it will be seen that all terms cancel in pairs except the first terms of u_1, u_2 and the last terms of u_{n-1}, u_n. Hence, if s_n is the required sum,
$$2s_n = 1 + \frac{1}{2} - \frac{1}{n+1} - \frac{1}{n+2},$$

leading to
$$s_n = \frac{n(3n+5)}{4(n+1)(n+2)}.$$

Example 3. *Sum to n terms the series*
$$\frac{1}{1 \cdot 3 \cdot 5} + \frac{2}{3 \cdot 5 \cdot 7} + \frac{3}{5 \cdot 7 \cdot 9} + \dots \quad \text{(C.)}$$

The rth term u_r is given by
$$u_r = \frac{r}{(2r-1)(2r+1)(2r+3)} \equiv \frac{1}{16}\left[\frac{1}{2r-1} + \frac{2}{2r+1} - \frac{3}{2r+3}\right].$$

by the rule for partial fractions. Hence the sum s_n to n terms is given by

$$16s_n = 16\sum_1^n u_r = \sum_1^n \frac{1}{2r-1} + 2\sum_1^n \frac{1}{2r+1} - 3\sum_1^n \frac{1}{2r+3}$$

$$= \left(\frac{1}{1} + \frac{1}{3} + \frac{1}{5} + \frac{1}{7} + \ldots + \frac{1}{2n-1}\right)$$

$$+ 2\left(\frac{1}{3} + \frac{1}{5} + \frac{1}{7} + \ldots + \frac{1}{2n-1} + \frac{1}{2n+1}\right)$$

$$- 3\left(\frac{1}{5} + \frac{1}{7} + \ldots + \frac{1}{2n-1} + \frac{1}{2n+1} + \frac{1}{2n+3}\right)$$

$$= \frac{1}{1} + \frac{1}{3} + \frac{2}{3} + \frac{2}{2n+1} - \frac{3}{2n+1} - \frac{3}{2n+3},$$

all the other terms cancelling. After reduction, this gives

$$s_n = \frac{n(n+1)}{2(2n+1)(2n+3)}.$$

The difference method can be used to sum the finite series whose rth terms are of either of the following forms:

(a) $u_r = \{a + rd\}\{a + (r+1)d\}\ldots\{a + (r+k-1)d\}$,

(b) $u_r = \dfrac{1}{\{a + rd\}\{a + (r+1)d\}\ldots\{a + (r+k-1)d\}}$, $k > 1$.

Here each term contains k factors which are successive terms of an arithmetical progression, the first factors of the several terms being in the same arithmetical progression.

In case (a), the method consists of forming the difference $v_r - v_{r-1}$ where $v_r = u_r\{a + (r+k)d\}$. Then

$$v_r - v_{r-1} = u_r[\{a + (r+k)d\} - \{a + (r-1)d\}] = u_r(k+1)d.$$

Hence $$u_r = \frac{1}{(k+1)d}(v_r - v_{r-1}),$$

and since, except for the first and last, the terms on the right cancel in pairs on addition, the sum to n terms (s_n) is given by

$$s_n = \sum_1^n u_r = \frac{1}{(k+1)d}(v_n - v_0).$$

This gives the following rule for finding s_n—*write down the nth term of the given series and add the next factor at the end, divide by the number of factors so increased and by the common difference d, then subtract from it the expression obtained by setting n = 0.*

In case (b), the method is similar but now we write $v_r = u_r\{a + rd\}$. In this case

$$v_r - v_{r-1} = u_r[\{a + rd\} - \{a + (r+k-1)d\}] = -u_r(k-1)d.$$

Hence $$u_r = -\frac{1}{(k-1)d}(v_r - v_{r-1})$$

and the sum to n terms is therefore given by
$$s_n = -\frac{1}{(k-1)d}(v_n - v_0).$$
The rule for s_n is in this case—*write down the nth term of the given series and remove the first factor, divide by the reduced number of factors and by the common difference d, change the sign and then subtract from it the expression obtained by setting $n = 0$.*

Example 4. *Find the sum to n terms of the series*
$$\frac{1}{1.5} + \frac{1}{5.9} + \frac{1}{9.13} + \ldots$$

Here $d = 4$, $k = 2$ and the nth term is $\dfrac{1}{(4n-3)(4n+1)}$. The expression given by the above rule is
$$-\frac{1}{(2-1)4} \cdot \frac{1}{(4n+1)} \quad \text{or} \quad \frac{-1}{4(4n+1)}$$
and the value of this when $n = 0$ is $-\frac{1}{4}$. Hence the sum to n terms
$$= \frac{1}{4} - \frac{1}{4(4n+1)} = \frac{n}{4n+1}.$$

Example 5. *Find the rth term of the series*
$$1.2\,n + 2.3.(n-1) + 3.4(n-2) + \ldots + n(n+1).1.$$
Hence show that the sum of the series is
$$\frac{1}{12}n(n+1)(n+2)(n+3). \tag{W.}$$

The rth term $= r(r+1)(n-r+1)$.
This can be written $nr(r+1) - (r-1)r(r+1)$. Hence the sum s_n is given by
$$s_n = n\sum_1^n r(r+1) - \sum_1^n (r-1)r(r+1).$$
For the first series on the right $d = 1$, $k = 2$, the nth term is $n(n+1)$ and the rule for case (a) gives the expression
$$\frac{1}{(2+1)1} \cdot n(n+1)(n+2) \quad \text{or} \quad \frac{1}{3}n(n+1)(n+2)$$
and the value of this when $n = 0$ is zero. Hence
$$\sum_1^n r(r+1) = \frac{1}{3}n(n+1)(n+2).$$
For the second series, $d = 1$, $k = 3$, the nth term is $(n-1)n(n+1)$ and the rule gives the expression
$$\frac{1}{4}(n-1)n(n+1)(n+2),$$
the value of the expression when $n = 0$ being zero. Hence
$$\sum_1^n (r-1)r(r+1) = \frac{1}{4}(n-1)n(n+1)(n+2) \text{ and therefore}$$

$$s_n = \frac{n^2}{3}(n+1)(n+2) - \frac{1}{4}(n-1)n(n+1)(n+2)$$
$$= \frac{1}{12}n(n+1)(n+2)(n+3).$$

2.3. The method of induction

If an expression for the sum of a finite series can be conjectured (or is given), the method of induction can be used to decide the validity (or otherwise) of the conjecture. The general principle of the method has been explained in *Advanced Level Pure Mathematics*, pp. 49, 50. Here we assume the truth of the conjecture for the sum s_n to n terms, add the $(n+1)$th term to give the sum s_{n+1} to $n+1$ terms and show that s_{n+1} is the *same* function of $n+1$ as s_n is of n. If the truth of the conjecture is demonstrated for $n = 1$, etc., it then follows that the result is true generally.

Example 6. *Prove that the sum of n terms of the series*
$$2 + 3 \cdot 2 + 4 \cdot 2^2 + 5 \cdot 2^3 + \ldots \qquad \text{(O.C.)}$$
is $n \cdot 2^n$.

Let the sum to n terms be s_n and assume that $s_n = n \cdot 2^n$.
Then, since the $(n+1)$th term is $(n+2)2^n$,
$$s_{n+1} = n \cdot 2^n + (n+2)2^n$$
$$= 2^n(n+n+2) = (n+1)2^{n+1}.$$

This is the same function of $(n+1)$ as s_n is of n and, since the stated result is true for $n = 1$, it is therefore true generally.

Example 7. *If a_n denotes the nth term of the series which begins*
$$1 + \frac{2^2}{1(1+2\cdot 2^2)} + \frac{3^2}{1(1+2\cdot 2^2)(1+2\cdot 3^2)} + \ldots$$
prove that the sum of the series to n terms is $\frac{1}{2}\left(3 - \frac{a_n}{n^2}\right)$. (O.C.)

Since,
$$a_n = \frac{n^2}{1(1+2\cdot 2^2)\ldots(1+2\cdot n^2)}, \quad a_{n+1} = \frac{(n+1)^2}{1(1+2\cdot 2^2)\ldots\{1+2(n+1)^2\}},$$
$$\frac{a_n}{n^2} = \frac{\{1+2(n+1)^2\}a_{n+1}}{(n+1)^2}. \qquad (2.1)$$

Assuming that the sum to n terms (s_n) is given by $s_n = \frac{1}{2}\left(3 - \frac{a_n}{n^2}\right)$,
$$s_{n+1} = s_n + a_{n+1}$$
$$= \frac{1}{2}\left(3 - \frac{a_n}{n^2}\right) + a_{n+1}$$
$$= \frac{1}{2}\left\{3 - \frac{1+2(n+1)^2}{(n+1)^2}a_{n+1}\right\} + a_{n+1}, \text{ using (2.1)},$$
$$= \frac{1}{2}\left\{3 - \frac{a_{n+1}}{(n+1)^2}\right\}.$$

This is the same function of $n+1$ as s_n is of n and, since the stated result is true for $n = 1$, it is therefore true generally.

2.4. Series involving the binomial coefficients

If c_r denotes the coefficient of x^r in the expansion of $(1 + x)^n$ where n is a positive integer,

$$c_0 + c_1 x + c_2 x^2 + \ldots + c_r x^r + \ldots + c_n x^n \equiv (1 + x)^n, \quad (2.2)$$

and it should be noted that $c_{n-r} = c_r$. If x is given the values $1, -1$ in this identity,

$$c_0 + c_1 + c_2 + \ldots + c_n = 2^n, \quad (2.3)$$
$$c_0 - c_1 + c_2 - \ldots + (-1)^n c_n = 0. \quad (2.4)$$

Differentiation of (2.2) with respect to x gives

$$c_1 + 2c_2 x + \ldots + rc_r x^{r-1} + \ldots + nc_n x^{n-1} \equiv n(1 + x)^{n-1},$$

which, with $x = 1$, gives

$$c_1 + 2c_2 + \ldots + nc_n = n.2^{n-1}, \quad (2.5)$$

and other results can similarly be derived by further differentiation. Integration of (2.2) with respect to x gives

$$c_0 x + c_1 \frac{x^2}{2} + c_2 \frac{x^3}{3} + \ldots + c_r \frac{x^{r+1}}{r+1} + \ldots + c_n \frac{x^{n+1}}{n+1}$$
$$\equiv \frac{(1 + x)^{n+1}}{n+1} + \text{const.},$$

the value of the constant of integration being found to be $-1/(n + 1)$ by setting $x = 0$. Inserting this value for the constant and writing $x = 1$, we have

$$c_0 + \frac{1}{2}c_1 + \frac{1}{3}c_2 + \ldots + \frac{1}{n+1}c_n = \frac{2^{n+1} - 1}{n+1}.$$

Other series can be similarly summed by differentiation or integration of a finite series whose sum is known. For instance, the sum $\sum_{r=0}^{n}(r + 1)c_r$ can be found by differentiating the result

$$c_0 x + c_1 x^2 + c_2 x^3 + \ldots + c_n x^{n+1} \equiv x(1 + x)^n$$

and setting $x = 1$.

Some series involving the binomial coefficients c_r can be summed by constructing a function for which the given series is the coefficient of a certain power of the variable x and then evaluating the coefficient of this power of x independently. This process is shown in Example 8 below.

Example 8. *If c_r is the coefficient of x^r in the expansion of $(1 + x)^n$ where n is a positive integer, and*

$$f(r) = c_0 c_r + c_1 c_{r+1} + \ldots + c_{n-r} c_n$$

prove that   (i) $f(r) = \dfrac{(2n)!}{(n+r)!(n-r)!}$,

(ii) $c_0 f(0) + c_1 f(1) + \ldots + c_n f(n) = \dfrac{(3n)!}{(n)!(2n)!}.$ \hfill (O.)

(i) Since $c_r = c_{n-r}$,
$$(c_0 + c_1 x + \ldots + c_r x^r + \ldots + c_n x^n)(c_n + c_{n-1} x + \ldots + c_{n-r} x^r + \ldots + c_0 x^n) \equiv (1 + x)^n (1 + x)^n \equiv (1 + x)^{2n}.$$

The coefficient of x^{n+r} on the left-hand side is $f(r)$ while the coefficient of x^{n+r} in $(1 + x)^{2n}$ is
$$\frac{(2n)!}{(n + r)!(n - r)!}$$
and (i) has been established.

(ii) It has been established in (i) that
$$(1 + x)^{2n} \equiv \text{terms up to } x^{n-1} + f(0)x^n + f(1)x^{n+1} + \ldots + f(n)x^{2n}.$$

Multiplying by $(1 + x)^n \equiv c_n + c_{n-1} x + \ldots + c_0 x^n$, it is clear that
$$c_0 f(0) + c_1 f(1) + \ldots + c_n f(n)$$
is the coefficient of x^{2n} in the expansion of $(1 + x)^{3n}$. This being
$$\frac{(3n)!}{(2n)!(n)!},$$
the truth of (ii) has been demonstrated.

EXERCISES 2 (a)

1. Show that
$$x(x + 1)(2x + 1) \equiv Ax(x + 1)(x + 2)(x + 3) + B(x - 2)(x - 1)x(x + 1)$$
for certain constant values of A and B and find these values. Hence find the sum of the first n terms of the series
$$1.2.3 + 2.3.5 + 3.4.7 + 4.5.9 + \ldots \qquad \text{(O.C.)}$$

2. Sum to n terms the series
$$\frac{1 + 2 + 3 + 4}{1.2.3.4} + \frac{2 + 3 + 4 + 5}{2.3.4.5} + \frac{3 + 4 + 5 + 6}{3.4.5.6} + \ldots \qquad \text{(O.)}$$

3. Obtain the expression of
$$\frac{1}{(3n - 1)(3n + 2)(3n + 5)}$$
in partial fractions and deduce that the sum of 10 terms of the series
$$\frac{1}{2.5.8} + \frac{1}{5.8.11} + \frac{1}{8.11.14} + \ldots$$
is 37/2240. \qquad (W.)

4. Find the sum of n terms of the series
$$1.2.3 + 2.3.4 + 3.4.5 + \ldots \qquad \text{(C.)}$$

5. Prove that
$$1^2 + 4^2 + 7^2 + \ldots + (3n - 2)^2 = \tfrac{1}{2} n(6n^2 - 3n - 1). \qquad \text{(O.)}$$

6. Write down the nth term of the series
$$\frac{2.1}{2.3} + \frac{2^2.2}{3.4} + \frac{2^3.3}{4.5} + \ldots$$

and prove that the sum of the first n terms of this series is
$$\frac{2^{n+1}}{n+2} - 1. \qquad \text{(O.C.)}$$

7. If $(1 + x)^n \equiv c_0 + c_1 x + c_2 x^2 + \ldots + c_n x^n$, find the value of
$$(c_0 + c_2 + \ldots)^2 + (c_1 + c_3 + \ldots)^2. \qquad \text{(O.)}$$

8. If c_r is the coefficient of x^r in the expansion of $(1 + x)^n$ where n is a positive integer, find the sum of the series
$$\sum_{r=0}^{n} \frac{c_r}{(r+1)(r+2)}. \qquad \text{(C.)}$$

2.5. Recurring series and recurrence relations

The reader will remember that the sum of n terms of a geometrical progression can be found as follows. If
$$s_n = 1 + x + x^2 + \ldots + x^{n-1},$$
$$-x s_n = -x - x^2 - \ldots - x^{n-1} - x^n,$$
giving, on addition, $(1 - x)s_n = 1 - x^n$.

The sum of the slightly more complicated series
$$s_n = a + (a+b)x + (a+2b)x^2 + \ldots + \{a + (n-1)b\}x^{n-1}, \qquad (2.6)$$
can be found by a similar method. Thus,
$$-2x s_n = -2ax - 2(a+b)x^2 - \ldots - 2\{a + (n-2)b\}x^{n-1}$$
$$- 2\{a + (n-1)b\}x^n,$$
$$x^2 s_n = ax^2 + \ldots + \{a + (n-3)b\}x^{n-1} + \{a + (n-2)b\}x^n$$
$$+ \{a + (n-1)b\}x^{n+1},$$
and, by addition,
$$(1 - 2x + x^2)s_n = a + (b-a)x - (a + nb)x^n$$
$$+ \{a + (n-1)b\}x^{n+1},$$
all the other terms cancelling. The success of the method is due to the relation
$$u_{r+2} - 2u_{r+1} + u_r = 0, \; r \geqslant 0 \qquad (2.7)$$
existing between the coefficients
$$u_{r+2} = a + (r+2)b, \; u_{r+1} = a + (r+1)b, \; u_r = a + rb$$
of x^{r+2}, x^{r+1} and x^r in the given series.

The relation
$$u_{r+2} + a_1 u_{r+1} + a_2 u_r = 0, \; r \geqslant 0, \qquad (2.8)$$
in which a_1 and a_2 are constants independent of r, and of which (2.7) is the particular case in which $a_1 = -2$, $a_2 = 1$, is an example of a *recurrence relation*. This particular relation is called a *linear difference equation with constant coefficients, of order* 2. Similarly, the more general recurrence relation
$$u_{r+k} + a_1 u_{r+k-1} + a_2 u_{r+k-2} + \ldots + a_k u_r = 0, \; r \geqslant 0 \qquad (2.9)$$

in which $a_1, a_2, \ldots, a_k$ are constants independent of r, is a *linear difference equation with constant coefficients, of order k*.

The series $\qquad u_0 + u_1 + u_2 + \ldots$

whose terms satisfy a linear difference equation of the type (2.9) is called a *recurring series*, and the difference equation is termed the *scale of relation* of the series. The series

$$u_0 + u_1 x + u_2 x^2 + \ldots$$

whose coefficients form a recurring series is called a *recurring power series* and the scale of relation of the coefficients is known as the scale of relation of the power series. The method used in summing the particular recurring power series (2.6) above can be adapted to the summation of any recurring power series whose scale of relation is a linear difference equation with constant coefficients, but before giving further examples of summation we shall first consider methods of solution of linear difference equations.

First consider the linear difference equation of order 2,

$$u_{r+2} + a_1 u_{r+1} + a_2 u_r = 0.$$

By putting $r = 0$, we have $u_2 = -a_1 u_1 - a_2 u_0$. Since, by writing $r = 1$, $u_3 = -a_1 u_2 - a_2 u_1$, substitution for u_2 gives

$$u_3 = -a_1(-a_1 u_1 - a_2 u_0) - a_2 u_1$$
$$= (a_1^2 - a_2) u_1 + a_1 a_2 u_0.$$

Hence both u_2 and u_3 can be expressed as linear functions of u_1 and u_0, i.e. as sums of multiples of these two quantities. By repeating the process it can be shown that the same is true of u_r and it is possible to write

$$u_r = \lambda_r u_1 + \mu_r u_0$$

where λ_r, μ_r depend on r but not on u_1 and u_0. Since u_1 and u_0 are arbitrary the general solution u_r contains two arbitrary constants which enter the solution in a linear way.

This is a characteristic feature of linear equations of the second order and it is easy to verify by direct substitution that if $u_r = f(r)$ and $u_r = g(r)$ are two particular solutions of the difference equation then

$$u_r = Af(r) + Bg(r),$$

where A, B are two constants independent of r, is also a solution. This solution, containing the same number of arbitrary constants as the order of the difference equation, is called the *general* solution. Similarly the general solution of the third order difference equation

$$u_{r+3} + a_1 u_{r+2} + a_2 u_{r+1} + a_3 u_r = 0$$

will be $\qquad u_r = Af(r) + Bg(r) + Ch(r),$

where A, B and C are constants independent of r and $f(r), g(r), h(r)$ are particular solutions of the equation. In the same way, the general solution of the linear difference equation of order k will contain k arbitrary constants.

Particular solutions $f(r)$, $g(r)$, etc., of this type of difference equation can be found by writing $u_r = x^r$ and solving the resulting "auxiliary" algebraical equation for x. In the case of a difference equation of order 2, the auxiliary equation for x will be a quadratic and if the two roots of this equation are α, β,

$$f(r) = \alpha^r,\ g(r) = \beta^r.$$

The general solution is then

$$u_r = A\alpha^r + B\beta^r$$

where A and B are constants independent of r. If two particular terms of the sequence, say u_0 and u_1, have known values, the appropriate values of A and B can be found by putting $r = 0$ and $r = 1$. For linear difference equations of order 3, the auxiliary equation will be a cubic; if the roots of this cubic are α, β and γ, the general solution of the difference equation is similarly

$$u_r = A\alpha^r + B\beta^r + C\gamma^r$$

with obvious extensions for equations of higher order. When the reader comes to study linear differential equations (§12.8), he will notice many similarities with linear difference equations.*

Example 9. *Find the general solution of the difference equation*

$$u_{r+2} + u_{r+1} - 12u_r = 0,\ r \geqslant 0.$$

Writing $u_r = x^r$, the auxiliary equation is

$$x^r(x^2 + x - 12) = 0,$$

the quadratic equation $x^2 + x - 12 = 0$ having roots $x = 3$ and $x = -4$. Hence the required general solution of the difference equation is

$$u_r = A(3)^r + B(-4)^r,$$

where A and B are constants independent of r.

Example 10. *If $u_r = au_{r-1} + bu_{r-2}$ for every value of $r \geqslant 2$ and if $u_0 = 1$, $u_1 = 3$, $u_2 = 7$, $u_3 = 15$, prove that $u_r = 2^{r+1} - 1$.* (L.I.C.)

The given values of u_0, u_1, u_2 and u_3 enable the values of a and b to be found. Thus, since

$$u_2 = au_1 + bu_0 \text{ and } u_3 = au_2 + bu_1$$

the given values of the first four terms of the sequence give, for the determination of a and b, the simultaneous equations

$$7 = 3a + b \text{ and } 15 = 7a + 3b.$$

* Corresponding to the modern method of solution of linear differential equations by the use of the Laplace transform (see § 12.12), D. F. Lawden (*Mathematical Gazette*, XXXVI, 1952, pp. 193–6) has shown that an entirely parallel method may be employed for the solution of linear difference equations.

These give $a = 3$, $b = -2$ and the recurrence relation is therefore
$$u_r = 3u_{r-1} - 2u_{r-2}.$$
Writing $u_r = x^r$, the auxiliary equation for x is
$$x^{r-2}(x^2 - 3x + 2) = 0$$
leading to $x = 1$ and $x = 2$. Hence the general solution is, since $(1)^r = 1$,
$$u_r = A + B(2)^r,$$
where A and B are independent of r. The values of A and B can now be determined from any pair of u_0, u_1, u_2, u_3. Taking the pair $u_0 = 1$, $u_1 = 3$, and putting in turn $r = 0$ and 1 in the general solution,
$$1 = A + B \text{ and } 3 = A + 2B.$$
These simultaneous equations give $A = -1$, $B = 2$ and hence $u_r = 2^{r+1} - 1$.

Some modification of the above method of solution of linear difference equations is necessary when the auxiliary equation has equal roots. If, for instance, the difference equation under consideration is
$$u_{r+2} - 2au_{r+1} + a^2 u_r = 0,$$
the auxiliary equation, obtained by writing $u_r = x^r$ is $x^2 - 2ax + a^2 = 0$ and both the roots are equal to a. The solution would then appear to be
$$u_r = Aa^r + Ba^r = (A + B)a^r$$
and the two arbitrary constants A and B can be combined into the single constant $A + B$. The general solution containing two independent arbitrary constants is, as can be verified by actual substitution in the difference equation, in this case
$$u_r = (A + Br)a^r.$$
Similarly in the case of a linear difference equation of an order greater than 2, if the auxiliary equation has three roots all equal to a, the terms in the general solution corresponding to this root are $(A + Br + Cr^2)a^r$ with obvious extensions when there are more than three equal roots.

Example 11. *If k is a constant and if, for $r \geqslant 2$, $u_r - 2ku_{r-1} + k^2 u_{r-2} = 0$ and if further, $u_1 = 2k$, $u_2 = 3k^2$, prove that $u_r = (r + 1)k^r$.* (L.I.C.)

The auxiliary equation obtained by writing $u_r = x^r$ in the given difference equation, is $x^2 - 2kx + k^2 = 0$, the roots being both equal to k. The general solution is therefore $u_r = (A + Br)k^r$ where A, B are constants independent of r. Putting in turn $r = 1$ and 2 in the general solution and using the given values $u_1 = 2k$, $u_2 = 3k^2$, A and B are given by the simultaneous equations
$$A + B = 2 \text{ and } A + 2B = 3.$$
These lead to $A = B = 1$ and hence $u_r = (r + 1)k^r$.

So far the difference equations considered have all been linear (in the sense that no powers or products of the u's appear in the given equation) and with constant coefficients. Some non-linear difference equations can be transformed into linear equations which can then be solved by the standard method (see Example 12). Example 13 deals with a linear difference equation in which the right-hand side is a function of r. Here the stated result can be obtained by induction but the general method of solution of such equations will not be considered here.

Example 12. *By means of the substitution* $u_r = \dfrac{v_r}{v_{r-1}} + a$, *show how to reduce the recurrence relation* $u_r u_{r-1} - au_r - bu_{r-1} + c = 0$ *to a form which can be solved by standard methods.*
Given that the sequence $u_1, u_2, u_3, \ldots$ *of finite numbers repeats itself every* k *terms* (i.e. $u_{r+k} = u_r$) *and that no term of the sequence is equal to* a *show that*
$$a\beta(\alpha - \beta)(\alpha^k - \beta^k) = 0$$
where α *and* β *are roots of* $x^2 + (a - b)x + (c - ab) = 0$. (O.C.)

Using the given substitution in the given recurrence relation
$$\left(\frac{v_r}{v_{r-1}} + a\right)\left(\frac{v_{r-1}}{v_{r-2}} + a\right) - a\left(\frac{v_r}{v_{r-1}} + a\right) - b\left(\frac{v_{r-1}}{v_{r-2}} + a\right) + c = 0$$
which reduces to
$$v_r + (a - b)v_{r-1} + (c - ab)v_{r-2} = 0,$$
a linear difference equation which can be solved by standard methods. Writing $v_r = x^r$, the auxiliary equation is
$$x^2 + (a - b)x + (c - ab) = 0$$
and if α, β are the roots of this equation, $v_r = A\alpha^r + B\beta^r$. Since $u_{r+k} = u_r$,
$$\frac{v_{r+k}}{v_{r+k-1}} + a = \frac{v_r}{v_{r-1}} + a$$
and hence $v_{r+k}v_{r-1} = v_{r+k-1}v_r$. Substituting and reducing
$$AB(\alpha^{r+k}\beta^{r-1} + \alpha^{r-1}\beta^{r+k} - \alpha^{r+k-1}\beta^r - \alpha^r\beta^{r+k-1}) = 0,$$
which can be written
$$AB\alpha^{r-2}\beta^{r-2}\{\alpha\beta(\alpha - \beta)(\alpha^k - \beta^k)\} = 0.$$
We assume $k > 1$ (for otherwise all the terms of the u sequence are equal) and $r \geqslant 2$. Since none of the u sequence is equal to a, none of A, B, α, β is zero and the required result follows.

Example 13. *A sequence* $u_1, u_2, \ldots$ *is defined by* $u_1 = 1$ *and* $u_{r+1} = au_r + r + 1 (r \geqslant 1)$ *where* a *is independent of* r. *Find an expression for* u_n *when* $a = 1$.
By induction, show that when $a \neq 1$, u_n *is of the form*
$$u_n = Aa^n + Bn + C$$
where A, B, C *are independent of* n, *and find* A, B *and* C. (O.C.)

When $a = 1$, $u_{r+1} - u_r = r + 1$. Writing in turn $r = n - 1, n - 2, \ldots, 2, 1$ and adding
$$u_n - u_1 = n + (n - 1) + \ldots + 2,$$
giving, since $u_1 = 1$,
$$u_n = n + (n - 1) + \ldots + 2 + 1 = \tfrac{1}{2}n(n + 1).$$

Assuming that $u_n = Aa^n + Bn + C$,
$$u_{n+1} = au_n + n + 1 = a(Aa^n + Bn + C) + n + 1$$
$$= Aa^{n+1} + B(n+1) + C + n(aB - B + 1) + aC - B - C + 1.$$
The assumed result is therefore true also for u_{n+1} if B and C are chosen so that
$$aB - B + 1 = 0 \text{ and } aC - B - C + 1 = 0.$$
This requires $$B = \frac{1}{1-a}, \quad C = \frac{-a}{(1-a)^2}.$$
The assumed result must be true also for u_1. This requires that $Aa + B + C = 1$, leading to
$$A = \frac{1 - B - C}{a} = \frac{a}{(1-a)^2}.$$

Returning now to recurring series and recurring power series, the methods of solution of linear difference equations may be applied to the determination of the general term of a recurring series whose scale of relation is known or can be found and to the summation of such series. The following examples should be sufficient to make the methods clear. Some further points in connection with recurring series will be discussed later (see § 3.14).

Example 14. *Find the nth term and the sum to n terms of the recurring series $2 + 5 + 13 + 35 + \ldots$*

If the linear difference equation giving the scale of relation of a recurring series is of order k, it will contain k constants. These will require k equations for their determination and these can only be obtained if $2k$ terms of the recurring series are given. In the present example in which four terms are given, it is assumed that the scale of relation is of the second order and hence that $u_{r+2} + a_1 u_{r+1} + a_2 u_r = 0$.
Since $u_0 = 2$, $u_1 = 5$, $u_2 = 13$, $u_3 = 35$ this gives when r is taken as 0, 1,
$$13 + 5a_1 + 2a_2 = 0 \text{ and } 35 + 13a_1 + 5a_2 = 0.$$
These simultaneous equations lead to $a_1 = -5$, $a_2 = 6$ and hence the scale of relation is $u_{r+2} - 5u_{r+1} + 6u_r = 0$. Solving this difference equation in the usual way, the auxiliary equation is $x^2 - 5x + 6 = 0$ with roots $x = 2$, $x = 3$. Hence the general solution of the difference equation is $u_r = A(2)^r + B(3)^r$ and since $u_0 = 2$, $u_1 = 5$ it is easy to see that $A = B = 1$. Hence the nth term (since the first term is u_0, the second u_1 and so on) is $2^{n-1} + 3^{n-1}$ and the sum to n terms
$$= \sum_{r=0}^{n-1}(2^r + 3^r) = 2^n - 1 + \tfrac{1}{2}(3^n - 1)$$
on summation of the two geometrical progressions.

Example 15. *Find the sum of n terms of the recurring power series*
$$u_0 + u_1 x + u_2 x^2 + \ldots,$$
the scale of relation of the coefficients being $u_{r+2} + a_1 u_{r+1} + a_2 u_r = 0$.

Let $$s_n = u_0 + u_1 x + u_2 x^2 + \ldots + u_{n-1} x^{n-1},$$
then, $$a_1 x s_n = a_1 u_0 x + a_1 u_1 x^2 + \ldots + a_1 u_{n-2} x^{n-1} + a_1 u_{n-1} x^n,$$

and, $a_2x^2s_n = a_2u_0x^2 + \ldots + a_2u_{n-3}x^{n-1} + a_2u_{n-2}x^n + a_2u_{n-1}x^{n+1}$,
giving, by addition,
$$(1 + a_1x + a_2x^2)s_n = u_0 + (u_1 + a_1u_0)x + (a_1u_{n-1} + a_2u_{n-2})x^n + a_2u_{n-1}x^{n+1},$$
for the coefficient of every other power of x is zero in consequence of the relation
$$u_{r+2} + a_1u_{r+1} + a_2u_r = 0.$$
The sum s_n then follows by division by $(1 + a_1x + a_2x^2)$.

EXERCISES 2 (b)

1. Find the general solution of the difference equation
$u_{r+2} + 3u_{r+1} - 4u_r = 0$, $r \geqslant 0$. Find also the values of the constants in the general solution if $u_0 = 21$ and $u_1 = 1$.

2. If $u_{r+2} - (a+b)u_{r+1} + abu_r = 0$, $r \geqslant 0$ and $u_0 = 1/a$, $u_1 = 1/b$, find u_r.

3. Integers $u_0, u_1, \ldots, u_n$, are defined by
$$u_0 = 0,\ u_1 = 1,\ u_{r+1} = u_r + u_{r-1}\ (r \geqslant 1).$$
Prove that, if r, s are positive integers
$$u_{r+s} = u_r u_{s+1} + u_s u_{r-1},$$
and that, if $s < r$
$$u_{r-s} = (-1)^s(u_r u_{s+1} - u_s u_{r+1}). \tag{C.}$$

4. The sequence of numbers $u_0, u_1, u_2, \ldots$ satisfies the recurrence relation
$$u_{n+2} - 2u_{n+1}\cos\theta + u_n = 0,\ n = 0, 1, 2, \ldots,\ 0 < \theta < \pi;$$
if $v_n = u_n^2$ show that the sequence of numbers $v_0, v_1, v_2, \ldots$ satisfies the recurrence relation
$$v_{n+3} - (2\cos 2\theta + 1)(v_{n+2} - v_{n+1}) - v_n = 0. \tag{C.}$$

5. Find the nth term and the sum of n terms of the recurring series
$$1 + 6 + 24 + 84 + \ldots,$$
assuming that the scale of relation is a linear difference equation with constant coefficients of order 2.

6. Find the sum to n terms of the recurring power series
$$1 + 2x + 3x^2 + 9x^3 + \ldots + u_{n-1}x^{n-1},$$
the scale of relation of the coefficients being
$$u_{n-1} = -u_{n-2} + 10u_{n-3} - 8u_{n-4},\ (n \geqslant 4). \tag{C.}$$

EXERCISES 2 (c)

1. Find the sum of $2n$ terms of the series
$$1^2 - 3^2 + 5^2 - 7^2 + \ldots. \tag{O.C.}$$

2. Sum to n terms the series
$$\frac{2.3}{(5)!} + \frac{3.3^2}{(6)!} + \frac{4.3^3}{(7)!} + \ldots. \tag{O.}$$

3. Find the sum of the first n terms of the series
$$1.4 + 2.9 + 3.16 + 4.25 + \ldots\ldots$$
(O.C.)

4. Write down the nth term of the series
$$1.2 + 2.5 + 3.8 + 4.11 + \ldots$$
and find the sum of the first n terms. (O.C.)

5. By induction, or otherwise, prove that the sum of n terms of the series
$$\frac{2}{3} + \frac{2.4}{3.5} + \frac{2.4.6}{3.5.7} + \ldots$$
is equal to $-2 + \dfrac{2.4\ldots(2n+2)}{1.3\ldots(2n+1)}$. (O.C.)

6. If m is an odd positive integer, prove by induction, or otherwise, that the sum of $\frac{1}{2}(m+1)$ terms of the expression
$$1 + \frac{m-1}{m-2} + \frac{(m-1)(m-3)}{(m-2)(m-4)} + \ldots$$
is m. (O.)

7. If $(1+x)^n \equiv c_0 + c_1 x + \ldots + c_r x^r + \ldots + c_n x^n$, where n is a positive integer, prove that
$$c_0 + 2c_1 + \ldots + (r+1)c_r + \ldots + (n+1)c_n = (n+2)2^{n-1}.$$
(L.I.C.)

8. Write down the formula for the sum of the coefficients in the expansion of $(1+x)^m$, where m is a positive integer. Deduce that
$$\frac{1}{(1)!(2n)!} + \frac{1}{(2)!(2n-1)!} + \frac{1}{(3)!(2n-2)!} + \ldots$$
$$+ \frac{1}{(n)!(n+1)!} = \frac{2^{2n}-1}{(2n+1)!}.$$ (W.)

9. Show that
$$\{1 + (n+1)x\}(1+x)^{n-1} = c_0 + 2c_1 x + 3c_2 x^2 + \ldots + (n+1)c_n x^n,$$
where c_r denotes the coefficient of x^r in the expansion of $(1+x)^n$ by the binomial theorem.
Hence show that $c_0^2 + 2c_1^2 + 3c_2^2 + \ldots + (n+1)c_n^2$ is the coefficient of x^n in the expansion of $\{1+(n+1)x\}(1+x)^{2n-1}$ and that its value is
$$\frac{(n+2)(2n-1)!}{(n)!(n-1)!}.$$ (W.)

10. Prove that if n is an even positive integer and if
$$(1+x)^n \equiv 1 + c_1 x + c_2 x^2 + \ldots + c_n x^n,$$
the sum of $(n-1)$ terms of the series
$$\frac{1}{2}c_1 - \frac{1.3}{2.4}c_2 + \frac{1.3.5}{2.4.6}c_3 - \ldots$$
is unity. (O.)

11. Prove that $\sum_{r=0}^{n} c_r(r - \tfrac{1}{2}n)^2 = 2^{n-2}n$, where n is a positive integer and c_r is the coefficient of x^r in the expansion of $(1 + x)^n$. (C.)

12. If c_r is the coefficient of x^r in the expansion of $(1 + x)^n$, show that
$$\frac{c_1}{c_0} + \frac{2c_2}{c_1} + \frac{3c_3}{c_2} + \ldots + \frac{nc_n}{c_{n-1}} = \frac{n(n+1)}{2}.$$

13. Given that $u_1 = 1$ and $u_2 = 2$ find the value of u_r if
  (i) $4u_{r+2} = u_r$,   (ii) $u_{r+2} + 4u_r = 4u_{r+1}$.

14. By means of the substitution $u_r = v_r + ar + b$ and suitable choice of a, b, reduce the difference equation
$$u_{r+2} - 5u_{r+1} + 6u_r = r$$
to a standard form and hence find its general solution.

15. Given that numbers $u_0, u_1, u_2, \ldots$ satisfy the recurrence formula
$$u_{n+1} = \frac{2}{2 - u_n}$$
for $n = 0, 1, 2, \ldots$, express u_{n+2} and u_{n+3} in terms of u_n and prove that $u_{n+4} = u_n$.
Given that $u_0 = 3$, calculate u_1, u_2, u_3 and sum the following series to $4n$ terms: $u_0 + u_1 x + u_2 x^2 + \ldots + u_r x^r + \ldots$. (O.C.)

16. Given that $a_0 = a_1 \sin^2 \alpha = 2 \cos \alpha$ and that $a_n - 2a_{n+1} + a_{n+2} \sin^2 \alpha = 0$, find an expression for a_r and prove that
$$\sum_{r=1}^{n} a_r = \frac{1}{(1 - \cos \alpha)^n} - \frac{1}{(1 + \cos \alpha)^n}.$$
By putting $t = \tan \tfrac{1}{2}\alpha$, or otherwise, prove that
$$\sum_{r=1}^{n} (1 + t^2)^r \{t^{2n-2r} + t^{2n}\} 2^{n-r} = (1 + t^2)^{n+1} \sum_{r=1}^{n} t^{2r-2}.$$ (O.C.)

17. The numbers x_n satisfy the recurrence formula
$$x_{n+3} = \tfrac{1}{3}(x_n + x_{n+1} + x_{n+2}).$$
Prove that if $y_n = x_n + 2x_{n+1} + 3x_{n+2}$ then $y_{n+1} = y_n = y_1 = 6x$ (say).
Prove also that if
$$z_n = (x_n - x)^2 + 2(x_n - x)(x_{n+1} - x) + 3(x_{n+1} - x)^2,$$
then $z_{n+1} = \tfrac{1}{3} z_n = 3^{-n} z_1$. (O.)

18. Given that $x_1 > 0$ and that
$$x_1 = 1 + \frac{1}{x_2},\ x_2 = 1 + \frac{1}{x_3},\ \ldots,\ x_{n-1} = 1 + \frac{1}{x_n},\ x_n = 1 + \frac{1}{x_1},$$
prove that
  (i) $x_r > 1 (1 \leqslant r \leqslant n)$,   (ii) $x_1 - x_2 = -\dfrac{x_2 - x_3}{x_2 x_3}$,
  (iii) $x_1 = x_2 = \ldots = x_n$,   (iv) $x_1 = \tfrac{1}{2}(1 + \sqrt{5})$. (O.)

19. Verify that the coefficients u_r of x^r in the finite series
$$1^2 + 2^2x + 3^2x^2 + \ldots + (n+1)^2x^n$$
satisfy the difference equation
$$u_{r+3} - 3u_{r+2} + 3u_{r+1} - u_r = 0$$
and hence find an expression for the sum of the series. (O.)

20. By expanding the function $e^{(n-r)x}(e^x - 1)^r$, prove that for positive integral values of s less than r
$$c_0(n-r)^s + c_1(n-r+1)^s + \ldots + c_r n^s = 0,$$
r being a positive integer and $c_0, c_1, c_2, \ldots, c_r$ the coefficients in the expansion of $(1-x)^r$.

Hence, or otherwise, prove that a series whose nth term is a polynomial of degree $(r-1)$ in n is a recurring series whose scale of relation is given by $c_0, c_1, c_2, \ldots, c_r$. (C.)

CHAPTER 3

THE CONVERGENCE OF SEQUENCES AND SERIES

3.1. Convergent sequences

If u_n is a single-valued function of the positive integral variable n, the quantities

$$u_1, u_2, u_3, \ldots, u_n, \ldots$$

are said to form a *sequence* $\{u_n\}$.

When the positive integral variable n assumes successively the values $1, 2, 3, \ldots$, there is no limit to the growth of n. However large a value is given to n (say 1,000 or 1,000,000), the next successive value (1,001 or 1,000,001) will be still larger. It is convenient to have a phrase to express this unending growth of n and the phrase generally used is that 'n *tends to infinity*', or, symbolically, $n \to \infty$, the symbol ∞ being an abbreviation for the word 'infinity'. It should be noted that the phrase 'n tends to infinity' simply means that n assumes a series of values which increase beyond all limit; there is no number infinity and an equation such as $n = \infty$ is meaningless.

Now consider the sequence $\{1/n\}$. By giving n successively the values $1, 2, 3, \ldots$, $1/n$ successively takes the values $1, 1/2, 1/3, \ldots$ and clearly $1/n$ decreases as n increases. It would be more precise to say that $1/n$ tends to zero as n tends to infinity, by which is meant that *if δ is any positive number, however small, then it is possible to find a number n_0, depending on the value of δ, such that $(1/n) < \delta$ for all values of n greater than n_0.* For example, if δ is taken as $0{\cdot}001$, then $n_0 = 1{,}000$ since $(1/n) < 0{\cdot}001$ so long as $n > 1{,}000$. Yet another way of expressing this property of the sequence $\{1/n\}$ is to say that *the limit of the sequence $\{1/n\}$ as n tends to infinity is zero*, or, symbolically

$$\lim_{n \to \infty} \left(\frac{1}{n}\right) = 0,$$

or again,
$$\frac{1}{n} \to 0 \text{ as } n \to \infty.$$

Another phrase is sometimes used with reference to a sequence whose limit as n tends to infinity is zero—the sequence is said to be *convergent* and to *converge to the limit zero*. In a similar way, a sequence $\{u_n\}$ is said to *converge to a limit l* if the sequence

$$u_1 - l, u_2 - l, u_3 - l, \ldots$$

converges to zero. Thus the sequence $\{1 + (1/n)\}$ converges to the limit unity since the sequence

$$\left\{1 + \frac{1}{n} - 1\right\}, \text{ i.e. } \left\{\frac{1}{n}\right\}$$

converges to zero. The formal definition of a convergent sequence $\{u_n\}$ is written most concisely in the form—*if, given any positive number δ, however small, it is possible to find $n_0(\delta)$ so that $|u_n - l| < \delta$ for all values of n greater than $n_0(\delta)$, then $\{u_n\}$ converges to the limit l.* In this definition, the number n_0 is written as $n_0(\delta)$ to show its dependence on δ and the symbol $|u_n - l|$ is used to denote $u_n - l$ or $l - u_n$, whichever is positive.

Example 1. *Find the limits of the sequences $\{u_n\}$ when u_n is given by*

(i) $\dfrac{n^2 + 4}{n^3 - 2}$,   (ii) $\dfrac{3n^2 + 1}{n^2 - 5n}$ $(n > 5)$.

(i) When n is large, the numerator and denominator of u_n may be expected to behave very much like n^2 and n^3 respectively for the constant terms 4 and -2 become less and less important as n increases. u_n may therefore be expected to behave like $1/n$ and this tends to the limit zero as n tends to infinity. Guided by this, a rigorous proof that the limit is in fact zero would run: when $n \geqslant 2$, $n^2 + 4 \leqslant 2n^2$ and $n^3 - 2 > \tfrac{1}{2}n^3$, so that

$$u_n = \frac{n^2 + 4}{n^3 - 2} < \frac{4}{n}.$$

Hence when $n \geqslant 2$, $0 < u_n < (4/n)$ and

$$|u_n| < \delta \text{ when } n \geqslant (4/\delta) \text{ and } n \geqslant 2,$$

showing that the sequence $\{u_n\}$ converges to the limit zero.

(ii) When n is large, 1 and $5n$ are small compared to $3n^2$ and n^2 respectively and it is to be expected that u_n will approach $3n^2/n^2$ or 3. Again being guided by this rough argument, a rigorous proof that in this case the limit is 3 would be:

$$u_n - 3 = \frac{3n^2 + 1}{n^2 - 5n} - 3 = \frac{15n + 1}{n^2 - 5n}.$$

Since $n^2 - 5n > \tfrac{1}{2}n^2$ when $n > 10$,

$$|u_n - 3| < \frac{16n}{(n^2/2)} = \frac{32}{n}$$

for such values of n (the numerator $15n + 1$ being clearly less than $16n$ when $n > 10$). Hence

$$|u_n - 3| < \delta \text{ when } n \geqslant (32/\delta) \text{ and } n > 10$$

showing that the sequence $\{u_n\}$ converges to the limit 3.

3.2. Divergent and oscillating sequences

By no means do all sequences converge to a limit. Consider, for example, the sequence formed by the cubes of the positive integers. The rather crude statement that 'n^3 is large when n is large' is equivalent to the more formal statement that 'it is possible to find a number $n_0(\varDelta)$, depending on $\varDelta$, such that $n^3 > \varDelta$ for all values of n greater than

$n_0(\varDelta)'$. In this case it is natural to say that n^3 tends to infinity as n tends to infinity and to write, symbolically,

$$n^3 \to \infty \text{ as } n \to \infty.$$

Similarly it is natural to say that

$$-n^4 \to -\infty \text{ as } n \to \infty.$$

Another way of expressing the same thing is to say that the sequence $\{n^3\}$ *diverges to positive infinity* and that the sequence $\{-n^4\}$ *diverges to minus infinity*. The formal definition of a sequence $\{u_n\}$ which diverges to positive infinity is—*if, when any number $\varDelta$, however large, is assigned, it is possible to determine $n_0(\varDelta)$ so that $u_n > \varDelta$ for all values of n greater than $n_0(\varDelta)$, then the sequence $\{u_n\}$ diverges to positive infinity.* The reader should have no difficulty in framing the corresponding definition of sequences which diverge to minus infinity.

A sequence which does not converge nor diverge is said to *oscillate*.* A simple example of an oscillating sequence is the sequence $\{(-1)^n\}$; here the terms are equal to $+1$ when n is even and to -1 when n is odd and the values recur cyclically. A continual repetition of a cycle of values is not, however, a necessary property of an oscillating sequence. For example, the sequence whose nth term is

$$\frac{1}{n} + (-1)^n$$

takes values very nearly (but not quite) equal to $+1$ or -1 when n is large. It does not converge to a limit nor does it diverge and the values do not recur cyclically. The numerically greatest term of the sequence is the second and its value is 3/2. Such a sequence is said to *oscillate finitely*. The sequence whose nth term is

$$\frac{n}{2}\left\{1 + (-1)^n\right\}$$

has successive terms 0, 2, 0, 4, 0, 6, ... This sequence oscillates but it is not in this case possible to write down a number greater than any term of the sequence. Such a sequence is said to *oscillate infinitely*. Formal definitions of finite and infinite oscillation are contained in the following—*if a sequence $\{u_n\}$ oscillates, then it is said to oscillate finitely or infinitely according as it is or is not possible to assign a number K such that $|u_n| < K$ for all values of n.*

3.3. Monotonic sequences

When $u_{n+1} \geqslant u_n$ for all values of n, the sequence $\{u_n\}$ is said to be *monotonic increasing* and u_n is said to *increase steadily*. When

---

* The reader should note that some authors include oscillation in their definition of divergence. With such terminology, a sequence converges or diverges; if it diverges, it may diverge to plus or minus infinity, oscillate finitely or oscillate infinitely.

$u_{n+1} > u_n$ for all values of n (i.e. excluding the possibility of $u_{n+1} = u_n$ for some or all values of n) the sequence is said to be monotonic increasing *in the strict sense*. Similarly, if $u_{n+1} \leqslant u_n$ for all values of n, the sequence $\{u_n\}$ is said to be *monotonic decreasing* and u_n is said to *decrease steadily*. Simple examples of sequences which are monotonic increasing and decreasing respectively are 1, 2, 3, ... and 1, 1/2, 1/3, ...

Later work on the convergence of series of positive terms depends on the study of monotonic sequences and the following result (the main point of which is that a monotonic sequence either converges or diverges and cannot merely oscillate) is important:

a sequence which is monotonic increasing either tends to a limit or diverges to positive infinity.

A proof of this result runs thus. Let u_n be the nth term of a monotonic sequence and let U be any rational real number. Then either

(a) $u_n \geqslant U$ for all values of n greater than some value N whose value depends on that of U, or

(b) $u_n < U$ for every value of n.

If (b) above is not true for any value of U, then u_n tends to positive infinity. If, on the other hand, (b) is true, the rational numbers can be divided into two classes, an L-class consisting of those rational numbers for which (a) holds and an R-class made up of those rational numbers for which (b) holds. The section defines a real number l which may be rational or irrational (see §§ 1.2, 1.3). If now δ be some arbitrary positive number, $l - \frac{1}{2}\delta$ belongs to the L-class and it is possible to find a number n_0 such that $u_n \geqslant l - \frac{1}{2}\delta$ whenever $n > n_0$; also $l + \frac{1}{2}\delta$ belongs to the R-class and thus $u_n < l + \frac{1}{2}\delta$. Hence, for $n > n_0$,

$$|u_n - l| < \delta$$

and u_n therefore tends to a limit l.

In the same way it can be shown that a sequence which is monotonic decreasing either tends to a limit or diverges to negative infinity.

Example 2. *If $0 < u_1 < 3$ and $u_{n+1} = 12/(1 + u_n)$ show that the sequences $\{u_{2n+1}\}$ and $\{u_{2n}\}$ are respectively monotonic increasing and decreasing. Show also that the sequence $\{u_n\}$ converges to the limit 3.*

If $u_n < 3$, $u_{n+1} = 12/(1 + u_n) > 3$ and if $u_n > 3$, $u_{n+1} < 3$. But $0 < u_1 < 3$ so that $u_n < 3$ if n is odd and $u_n > 3$ if n is even.

Also,
$$u_{n+2} - u_n = \frac{12}{1 + u_{n+1}} - u_n = \frac{12(1 + u_n)}{(1 + u_n) + 12} - u_n$$
$$= \frac{(4 + u_n)(3 - u_n)}{13 + u_n}.$$

Thus $u_{n+2} > u_n$ if n is odd and $\{u_{2n+1}\}$ is monotonic increasing with all its terms less than 3. Again $u_{n+2} < u_n$ if n is even and $\{u_{2n}\}$ is monotonic decreasing with a lits terms greater than 3.

From this it follows that $\{u_{2n+1}\}$ converges to a limit a and $\{u_{2n}\}$ to a limit b such that $0 < a \leqslant 3 \leqslant b$.

Since $u_{2n+1} \to a$, $u_{2n+3} - u_{2n+1} \to 0$ and, from the expression found for $u_{n+2} - u_n$, it follows that $(4 + a)(3 - a) = 0$, which, since $a > 0$, leads to $a = 3$. Similarly it can be shown that $b = 3$ and the sequence $\{u_n\}$ therefore converges to the limit 3.

Note.—The device of breaking a sequence into two sub-sequences, each of which is monotonic, is often useful in establishing the convergence of the main sequence.

3.4. Some useful theorems on limits

Suppose that the sequence $\{u_n\}$ converges to the limit l and the sequence $\{U_n\}$ converges to the limit L, then it is shown below that:

(i) the sequence $\{u_n \pm U_n\}$ converges to $l \pm L$, (3.1)
(ii) the sequence $\{u_n U_n\}$ converges to lL, (3.2)
(iii) the sequence $\{u_n/U_n\}$ converges to l/L (provided $L \neq 0$). (3.3)

Using the notation $|a|$ to denote the numerical value of a irrespective of sign, it is easy to check that for real numbers a and b

$$|a + b| \leqslant |a| + |b|, \qquad (3.4)$$

the signs of equality and inequality holding respectively when a and b are of the same or different signs. It is also true and easy to check that

$$|ab| = |a| \cdot |b| \quad \text{and} \quad \left|\frac{a}{b}\right| = \frac{|a|}{|b|} \qquad (3.5)$$

(the results (3.4) and (3.5) are true also for a much wider class of numbers than the real ones—see §§ 4.6, 4.7.)

A proof of the result (3.1) with the upper sign can now be concisely given thus: since the sequences $\{u_n\}$, $\{U_n\}$ converge respectively to limits l and L, it is possible to choose n_1 and n_2 so that

$$|u_n - l| < \tfrac{1}{2}\delta \text{ when } n > n_1 \text{ and } |U_n - L| < \tfrac{1}{2}\delta \text{ when } n > n_2.$$

Then, $\quad |u_n + U_n - l - L| \leqslant |u_n - l| + |U_n - L| < \delta$

provided n is greater than n_1 and n_2. This shows that the sequence $\{u_n + U_n\}$ converges to the limit $(l + L)$ and the reader should be able to construct a similar proof for the sequence $\{u_n - U_n\}$.

To prove the second result (3.2), let $u_n = l + u_n'$ and $U_n = L + U_n'$ so that the sequences $\{u_n'\}$ and $\{U_n'\}$ both converge to zero. Then

$$|u_n U_n - lL| = |(l + u_n')(L + U_n') - lL| = |lU_n' + Lu_n' + u_n' U_n'|$$
$$\leqslant |lU_n'| + |Lu_n'| + |u_n' U_n'|.$$

Assuming that neither l nor L is zero, it is possible to choose n_0 so that

$$|u_n'| < \tfrac{1}{3}\delta/|L| \text{ and } |U_n'| < \tfrac{1}{3}\delta/|l| \text{ when } n > n_0.$$

Hence $\quad |u_n U_n - lL| < \tfrac{1}{3}\delta + \tfrac{1}{3}\delta + (\tfrac{1}{9}\delta^2)/|lL|$

and this is certainly less than δ if $\delta < \frac{1}{3}|l||L|$. In other words, we can choose n_0 so that $|u_n U_n - lL| < \delta$ when $n > n_0$ and the result (3.2) follows. It is left to the reader to supply a proof for the case in which at least one of l and L is zero.

For a proof of the third result (3.3), let $U_n = L + U_n'$ so that the sequence $\{U_n'\}$ converges to zero. Then, provided $L \neq 0$,

$$\left|\frac{1}{U_n} - \frac{1}{L}\right| = \left|\frac{1}{L + U_n'} - \frac{1}{L}\right| = \frac{|U_n'|}{|L|.|L + U_n'|}$$

and it is clear, since the sequence $\{U_n'\}$ converges to zero, that n_0 can be chosen so that this is smaller than any assigned quantity δ when $n > n_0$. Hence the sequence $\{1/U_n\}$ converges to $1/L$ and the result (3.3) follows on application of (3.2).

A sequence in which each term is equal to some constant A can be regarded as one which converges to a limit A and immediate corollaries of the foregoing results are that if the sequence $\{u_n\}$ converges to l then the sequence $\{u_n + A\}$ converges to $l + A$, the sequence $\{Au_n\}$ converges to Al and so on. A particularly useful result of this kind is that the sequence whose nth term is

$$A + \frac{B}{n} + \frac{C}{n^2} + \ldots + \frac{K}{n^r},$$

where $A, B, C, \ldots, K$ are constants, converges to the limit A as n tends to infinity.

These theorems on limits provide a simple method for working exercises such as Example 1. This example was previously worked in an elementary, but not very simple, way in § 3.1 and it is reworked, using the foregoing results, below.

Example 3. *Find the limits of the sequences $\{u_n\}$ when u_n is given by*

(i) $\dfrac{n^2 + 4}{n^3 - 2}$,   (ii) $\dfrac{3n^2 + 1}{n^2 - 5n}$.

(i) $\dfrac{n^2 + 4}{n^3 - 2} = \left(\dfrac{1}{n} + \dfrac{4}{n^3}\right) \Big/ \left(1 - \dfrac{2}{n^3}\right)$,

and the numerator tends to zero and the denominator to unity. Hence the complete expression tends to zero.

(ii) $\dfrac{3n^2 + 1}{n^2 - 5n} = \left(3 + \dfrac{1}{n^2}\right) \Big/ \left(1 - \dfrac{5}{n}\right)$,

and the numerator and denominator tend respectively to 3 and 1. The complete expression tends therefore to 3.

3.5. Some important limits

(a) *The behaviour of x^n as n tends to infinity.*

Let $\phi(n) = x^n$. If $x = 1$, $\phi(n) = 1$ and the limit of $\phi(n)$ is clearly unity; if $x = 0$, $\phi(n) = 0$ and the limit of $\phi(n)$ is zero; if $x = -1$, $\phi(n) = (-1)^n$ and $\phi(n)$ oscillates finitely.

Leaving these special cases, suppose first that x is positive. Since
$$\phi(n+1) = x^{n+1} = x \cdot x^n = x\phi(n),$$
$\phi(n)$ increases monotonically with n if $x > 1$ and decreases monotonically if $x < 1$. Taking first the case of $x > 1$, $\phi(n)$ must either converge to a limit or diverge to $+\infty$ (§ 3.3). Suppose it converges to a limit l, then, as an immediate consequence of convergence to a limit,
$$\lim_{n \to \infty} \phi(n+1) = \lim_{n \to \infty} \phi(n) = l.$$
But, by (3.2),
$$\lim_{n \to \infty} \phi(n+1) = \lim_{n \to \infty} \{x\phi(n)\} = x \lim_{n \to \infty} \phi(n) = xl$$
and therefore $l = xl$. As x and l are both greater than unity, this is impossible and x^n in this case diverges to positive infinity. Taking now the case when $x < 1$, $\phi(n)$ decreases monotonically and therefore converges to a limit l or diverges to $-\infty$. Since x^n is positive, the second alternative can be ignored and $\lim_{n \to \infty} x^n = l$. As above $l = xl$ and, as $x \neq 0$, $l = 0$.

To deal with cases in which x is negative, put $x = -y$. Then y is positive and, from what precedes, $\lim_{n \to \infty} y^n = 0$ when $0 < y < 1$ leading to $\lim_{n \to \infty} x^n = 0$ when $-1 < x < 0$. If $x < -1$ and $x = -y$, then $y > 1$ and y^n diverges to $+\infty$. Hence x^n takes values, positive and negative, greater numerically than any assigned number. Hence in this case x^n oscillates infinitely.

To sum up:
$$\left.\begin{array}{l} x^n \to +\infty \ (x > 1), \\ x^n \to 1 (x = 1), \ x^n \to 0 (-1 < x < 1), \\ x^n \text{ oscillates finitely } (x = -1), x^n \text{ oscillates infinitely } (x < -1). \end{array}\right\} (3.6)$$

Example 4. *A sequence is defined by the recurrence formula*
$$u_{n+1} = \frac{6u_n^2 + 6}{u_n^2 + 11}$$
and by the value of the initial term u_0. Show that if u_n converges to a limit l, then l is one of 1, 2, 3.

Show that, if $u_n > 3$, then

(i) $3 < u_{n+1} < u_n$,   (ii) $\dfrac{u_{n+1} - 3}{u_n - 3} < \dfrac{9}{10}$,

and prove that, when $u_0 > 3$, then $u_n \to 3$ (monotonically) as $n \to \infty$.   (O.)

Assuming that $u_n \to l$, then $u_{n+1} \to l$ and the given recurrence relation gives
$$l = \frac{6l^2 + 6}{l^2 + 11}.$$

This reduces to $l^3 - 6l^2 + 11l - 6 = 0$, i.e. $(l-1)(l-2)(l-3) = 0$ showing that l is one of 1, 2, 3.

(i) Using the given recurrence relation

$$u_{n+1} - u_n = \frac{6u_n^2 + 6}{u_n^2 + 11} - u_n = \frac{6u_n^2 + 6 - u_n^3 - 11u_n}{u_n^2 + 11}$$

$$= -\frac{(u_n - 1)(u_n - 2)(u_n - 3)}{u_n^2 + 11} < 0 \text{ as } u_n > 3.$$

Also

$$u_{n+1} - 3 = \frac{6u_n^2 + 6}{u_n^2 + 11} - 3 = \frac{3u_n^2 - 27}{u_n^2 + 11} > 0 \text{ as } u_n > 3.$$

These two inequalities give $3 < u_{n+1} < u_n$.

(ii) From the above expression for $u_{n+1} - 3$,

$$\frac{u_{n+1} - 3}{u_n - 3} = \frac{3(u_n + 3)}{u_n^2 + 11} = \frac{3}{10}\left[3 - \frac{(3u_n - 1)(u_n - 3)}{u_n^2 + 11}\right]$$

$$< \frac{9}{10} \text{ as } u_n > 3.$$

The result (i) shows that u_n decreases monotonically. Multiplying the inequalities (ii) for $n = 0, 1, 2, \ldots, n + 1$,

$$0 < \frac{u_{n+1} - 3}{u_0 - 3} < \left(\frac{9}{10}\right)^{n+1}$$

(the first inequality holding as $u_{n+1} > 3$ and $u_0 > 3$). Since $(9/10) < 1$ this shows that u_n tends to the limit 3.

(b) *The behaviour of $x^n/(n)!$ as n tends to infinity.*

Let $u_n = x^n/(n)!$ and first consider positive values of x. Let p be a fixed positive integer such that $p + 1 > 2x$, then u_n is positive and

$$\frac{u_{p+1}}{u_p} = \frac{x}{p+1} < \frac{1}{2}.$$

Similarly $\quad \dfrac{u_{p+2}}{u_{p+1}} = \dfrac{x}{p+2} < \dfrac{1}{2}, \dfrac{u_{p+3}}{u_{p+2}} = \dfrac{x}{p+3} < \dfrac{1}{2}$

and so on. Hence by multiplication of results of this type, for any positive integer q,

$$\frac{u_{p+q}}{u_p} < \left(\frac{1}{2}\right)^q.$$

u_p being a fixed quantity, $u_{p+q} <$ a constant multiplier $\times (1/2)^q$. Since $(1/2)^q$ converges to zero as q increases, the sequence $\{u_{p+q}\}$ converges to zero and hence

$$\lim_{n \to \infty} (x^n/(n)!) = 0. \tag{3.7}$$

The numerical magnitude of $(-x)^n/(n)!$ is the same as that of $(x)^n/(n)!$ and so the result (3.7) remains true for negative values of x.

(c) *The behaviour of* $\left(1 + \dfrac{1}{n}\right)^n$ *as n tends to infinity.*

From the binomial theorem for a positive integral index,

$$\left(1 + \frac{1}{n}\right)^n = 1 + n \cdot \frac{1}{n} + \frac{n(n-1)}{(2)!} \cdot \frac{1}{n^2} + \cdots$$
$$+ \frac{n(n-1)\cdots(n-n+1)}{(n)!} \cdot \frac{1}{n^n}$$
$$= 1 + 1 + \frac{1}{(2)!}\left(1 - \frac{1}{n}\right) + \frac{1}{(3)!}\left(1 - \frac{1}{n}\right)\left(1 - \frac{2}{n}\right) + \cdots$$
$$+ \frac{1}{(n)!}\left(1 - \frac{1}{n}\right)\left(1 - \frac{2}{n}\right)\cdots\left(1 - \frac{n-1}{n}\right).$$

The $(r+1)$th term in this expression is

$$\frac{1}{(r)!}\left(1 - \frac{1}{n}\right)\left(1 - \frac{2}{n}\right)\cdots\left(1 - \frac{r-1}{n}\right),$$

and each of the terms in brackets is positive and increases as n increases. The number of terms in the expansion also increases with n. Hence $(1 + 1/n)^n$ increases monotonically and so converges to a limit or diverges to positive infinity.

But
$$\left(1 + \frac{1}{n}\right)^n < 1 + 1 + \frac{1}{(2)!} + \frac{1}{(3)!} + \cdots + \frac{1}{(n)!}$$
$$< 1 + 1 + \frac{1}{2} + \frac{1}{2^2} + \cdots + \frac{1}{2^{n-1}}.$$

Since the sum of the geometrical progression

$$1 + \frac{1}{2} + \frac{1}{2^2} + \cdots + \frac{1}{2^{n-1}}$$

is less than 2, $(1 + 1/n)^n < 3$ and it is clearly greater than 2. Hence the expression cannot diverge and

$$\lim_{n\to\infty} \left(1 + \frac{1}{n}\right)^n = e, \tag{3.8}$$

where e is some number such that $2 < e \leqslant 3$.

(d) *If a sequence* $\{u_n\}$ *has a limit* l, *this limit is also that of the sequence* $\{v_n\}$ *where* $v_n = (u_1 + u_2 + \cdots + u_n)/n$.

Write $u_n = l + t_n$; then it has to be shown that if the sequence $\{t_n\}$ converges to zero, so does the sequence whose nth term is $(t_1 + t_2 + \cdots + t_n)/n$.

Suppose first that $t_n \geqslant 0$ for all values of n. Since $\{t_n\}$ converges to zero, it is possible to find m such that $t_n < \tfrac{1}{2}\delta$ whenever $n > m$, δ being any given positive number.

Hence
$$\frac{t_1 + t_2 + \ldots + t_n}{n} = \frac{t_1 + t_2 + \ldots + t_m}{n} + \frac{t_{m+1} + \ldots + t_n}{n}$$
$$< \frac{t_1 + t_2 + \ldots + t_m}{n} + \frac{1}{2}(n-m)\frac{\delta}{n}$$
$$< \frac{t_1 + t_2 + \ldots + t_m}{n} + \frac{1}{2}\delta.$$

Since m is fixed, m' can be chosen, greater than m, so that
$$\frac{t_1 + t_2 + \ldots + t_m}{m'} < \frac{1}{2}\delta,$$
and hence
$$\frac{t_1 + t_2 + \ldots + t_n}{n} < \frac{1}{2}\delta + \frac{1}{2}\delta = \delta \text{ whenever } n > m' > m,$$
showing that the sequence whose nth term is $(t_1 + t_2 + \ldots + t_n)/n$ converges to zero.

If t_n is positive or negative, since the sequence $\{t_n\}$ converges to zero, so does the sequence $\{|t_n|\}$. But
$$|(t_1 + t_2 + \ldots + t_n)/n| \leqslant (|t_1| + |t_2| + \ldots + |t_n|)/n$$
and hence the sequence whose nth term is $|(t_1 + t_2 + \ldots + t_n)|/n$ converges to zero. It follows that the sequence with nth term $(t_1 + t_2 + \ldots + t_n)/n$ also converges to zero.

It should be noted that the converse of this result is false: the convergence of the sequence $\{v_n\}$ does not ensure the convergence of the sequence $\{u_n\}$ (see Example 5 (ii) below).

Example 5. *If the nth term of a sequence $\{v_n\}$ is given by*
$$v_n = (u_1 + u_2 + \ldots + u_n)/n,$$
discuss the behaviour of $\{v_n\}$ as n tends to infinity when

(i) $u_n = 1/n,$ (ii) $u_n = (-1)^n.$

(i) Here $\{u_n\}$ converges to zero and hence by the result (*d*) above so also does $\{v_n\}$.
(ii) In this case $\{u_n\}$ oscillates and we cannot apply result (*d*). If n is even, $v_n = 0$, and if n is odd, $v_n = -1/n$. Hence the sequence $\{v_n\}$ converges to zero.

EXERCISES 3 (*a*)

1. Show that the sequence $\{u_n\}$ where u_n is

(i) $\dfrac{n+3}{n^2+5}$, (ii) $\dfrac{2n+4}{6n-9}$

both converge and find the limits to which they converge.

2. Show that the sequence whose nth term is $\sqrt{(n-5)}$ diverges and that whose nth term is $(-1)^n(3n+4)/n$ oscillates finitely.

3. Show that the sequence given by the recurrence relation
$$u_{n+1} = \frac{4(1+u_n)}{4+u_n}, \ u_1 = 1$$
is monotonic increasing.

4. If $t_n = \sum_{r=1}^{n}\frac{1}{r(r+1)}$ and $s_n = \sum_{r=1}^{n}\frac{1}{r(r+2)}$, prove that $\lim_{n\to\infty}\frac{s_n}{t_n} = \frac{3}{4}$.
(L.I.C.)

5. Given that $a_1 > b_1 > 0$, and that, when n is a positive integer
$$2a_{n+1} = a_n + b_n, \ \frac{2}{b_{n+1}} = \frac{1}{a_n} + \frac{1}{b_n},$$
prove that $b_n < b_{n+1} < a_{n+1} < a_n$.
Prove also that $a_{n+1} - b_{n+1} < \frac{1}{2}(a_n - b_n)$ and deduce that the sequences $\{a_n\}$ and $\{b_n\}$ tend to a common limit l as n tends to infinity, where $l^2 = a_1 b_1$.
(O.C.)

6. The sequence $a_1, a_2, a_3, \ldots$ is defined by means of the relations
$$a_1 = 3, \ a_{p+1} = \frac{a_p^2 + 5}{2a_p} (p > 0).$$
Prove that
$$0 < a_{p+1} - \sqrt{5} < \frac{(3-\sqrt{5})^{2^p}}{(2\sqrt{5})^{2^p-1}} < 6 \times \left(\frac{2}{11}\right)^{2^p}.$$
Hence show that $a_p \to \sqrt{5}$ as $p \to \infty$.
Use this process to calculate $\sqrt{5}$ correct to four places of decimals.
(C.)

3.6. Infinite series

Let s_n be the sum of the first n terms of a series whose rth term is u_r so that
$$s_n = \sum_{r=1}^{n} u_r = u_1 + u_2 + \ldots + u_r + \ldots + u_n. \qquad (3.9)$$
Then s_n is the nth term of a sequence $\{s_n\}$ which may converge to a limit s, diverge to positive or negative infinity, or oscillate (finitely or infinitely). For the case in which $\{s_n\}$ converges to a limit s,
$$s = \lim_{n\to\infty}\sum_{r=1}^{n} u_r,$$
and this equation is generally written in one of the forms
$$s = \sum_{r=1}^{\infty} u_r \quad \text{or} \quad s = u_1 + u_2 + u_3 + \ldots,$$
the dots denoting the indefinite continuance of the u's. In this case the series $u_1 + u_2 + u_3 + \ldots$ is called a *convergent infinite series* and s is called the *sum* of the series. The phrase 'sum to infinity' is

also used to denote the sum of a convergent infinite series, and the phrase 'Σu_r exists' is sometimes used in place of 'Σu_r converges'.

When the sequence $\{s_n\}$, as defined by equation (3.9), diverges or oscillates, the infinite series $u_1 + u_2 + u_3 + \ldots$ is said to *diverge* or *oscillate*.* In such cases the series does not possess a 'sum to infinity', this phrase being restricted to convergent series.

A discussion of the convergence, divergence or oscillation of infinite series for which an analytical expression for the sum (s_n) of the first n terms can be found depends on a discussion of the behaviour of the sequence $\{s_n\}$ as n tends to infinity and this depends on the results of §§ 3.1–3.5. Some illustrative examples follow.

Example 6. *Discuss the behaviour of the infinite series whose nth terms are*

(i) r^{n-1},   (ii) $\dfrac{1}{n(n+1)(n+2)}$,   (iii) n,

and find, in those cases in which the series are convergent, their sums to infinity.

(i) Here s_n is the sum of the first n terms of a geometrical progression with first term unity and common ratio r. Hence
$$s_n = \frac{1-r^n}{1-r} = \frac{1}{1-r} - \frac{r^n}{1-r}.$$
By § 3.5(a), when $-1 < r < 1$, $r^n \to 0$ as $n \to \infty$ and $s_n \to 1/(1-r)$. In this case the series is convergent and its sum is $1/(1-r)$.

If $r = 1$, the series is $1 + 1 + 1 + \ldots$ and $s_n = n$ showing that the series diverges to $+\infty$.

If $r > 1$, s_n is clearly greater than n and again the series diverges to $+\infty$. To discuss the case when $r \leqslant -1$, put $r = -y$ so that $y \geqslant 1$ and
$$s_n = 1 - y + y^2 - \ldots + (-1)^{n-1}y^{n-1}$$
$$= \frac{1 + (-1)^{n+1}y^n}{1 + y}.$$
If n is even and equal to $2m$
$$s_n = s_{2m} = \frac{1 - y^{2m}}{1 + y}$$
and this is zero if $y = 1$ and diverges to $-\infty$ if $y > 1$.

If n is odd and equal to $2m + 1$
$$s_n = s_{2m+1} = \frac{1 + y^{2m+1}}{1 + y}$$
which is unity if $y = 1$ and diverges to $+\infty$ if $y > 1$. Hence the series converges and has the sum $1/(1-r)$ when $-1 < r < 1$, it diverges to $+\infty$ when $r \geqslant 1$, it oscillates finitely when $r = -1$ and it oscillates infinitely when $r < -1$.

(ii) Here the nth term
$$\frac{1}{n(n+1)(n+2)} = \frac{1}{2}\left\{\frac{1}{n(n+1)} - \frac{1}{(n+1)(n+2)}\right\}.$$
By writing n successively equal to 1, 2, 3, ..., n, adding and observing

---

* The remarks in the footnote on p. 41 apply also to infinite series.

that all the terms, except the first and last, cancel in pairs, the sum s_n of the first n terms is given by
$$s_n = \frac{1}{2}\left\{\frac{1}{1.2} - \frac{1}{(n+1)(n+2)}\right\}.$$
The sequence $\{s_n\}$ converges to the limit $1/4$ as n tends to infinity and the given series is therefore convergent with sum $1/4$.

(iii) The sum to n terms, s_n is given by
$$s_n = 1 + 2 + 3 + \ldots + n = \tfrac{1}{2}n(n+1)$$
and clearly the sequence $\{s_n\}$, and therefore also the given series, diverges to positive infinity.

3.7. Some general theorems on infinite series

The foregoing definitions of convergence, divergence and oscillation in relation to infinite series and the general theorems on limits (§ 3.4) lead to the following general theorems on infinite series.

(a) *If $\sum_{r=1}^{\infty} u_r$ is convergent with sum s, then $\sum_{r=m+1}^{\infty} u_r$, where m is any given positive integer, is convergent with sum $s - \sum_{r=1}^{m} u_r$.*

This is true because
$$\lim_{n\to\infty}\left(\sum_{r=m+1}^{m+n} u_r\right) = \lim_{n\to\infty}\left(\sum_{r=1}^{m+n} u_r - \sum_{r=1}^{m} u_r\right)$$
$$= \lim_{n\to\infty}\left(\sum_{r=1}^{m+n} u_r\right) - \sum_{r=1}^{m} u_r$$
$$= s - \sum_{r=1}^{m} u_r.$$

Similarly if $\sum_{r=1}^{\infty} u_r$ diverges or oscillates, so also does $\sum_{r=m+1}^{\infty} u_r$.

It follows that any *finite* number of terms at the beginning of a series may be disregarded when discussing questions of convergence.

(b) *If $\sum_{r=1}^{\infty} u_r$ is convergent with sum s, then $\sum_{r=1}^{\infty} k u_r$, where k is any constant, is convergent with sum ks.*

This follows because
$$\lim_{n\to\infty}\left(\sum_{r=1}^{n} k u_r\right) = k \lim_{n\to\infty}\left(\sum_{r=1}^{n} u_r\right) = ks.$$
Similarly, provided $k \neq 0$, $\sum_{r=1}^{\infty} u_r$ and $\sum_{r=1}^{\infty} k u_r$ diverge or oscillate together.

(c) *If $\sum_{r=1}^{\infty} u_r$ and $\sum_{r=1}^{\infty} v_r$ are both convergent, with sums s and t respectively, then $\sum_{r=1}^{\infty} (u_r \pm v_r)$ is also convergent with sum $s \pm t$.*

A proof of this runs—

$$\lim_{n\to\infty}\left[\sum_{r=1}^{n}(u_r \pm v_r)\right] = \lim_{n\to\infty}\left(\sum_{r=1}^{n}u_r\right) \pm \lim_{n\to\infty}\left(\sum_{r=1}^{n}v_r\right)$$
$$= s \pm t.$$

This result can, of course, be generalised to cases involving any finite number of series.

(d) *If $\sum_{r=1}^{\infty}u_r$ is convergent with sum s, then $\lim_{n\to\infty}u_n = 0$.*

This follows because

$$\lim_{n\to\infty}u_n = \lim_{n\to\infty}\left(\sum_{r=1}^{n}u_r\right) - \lim_{n\to\infty}\left(\sum_{r=1}^{n-1}u_r\right) = s - s = 0.$$

The reader is warned that the converse of this result is not true—if $\lim_{n\to\infty}u_n = 0$ the series $\sum_{r=1}^{\infty}u_r$ is not necessarily convergent as can be seen from the following example.

Consider the harmonic series $1 + \frac{1}{2} + \frac{1}{3} + \frac{1}{4} + \ldots$. The sum of the first four terms is

$$1 + \frac{1}{2} + \frac{1}{3} + \frac{1}{4} > 1 + \frac{1}{2} + \frac{2}{4} = 1 + \frac{1}{2} + \frac{1}{2}.$$

The sum of the next four terms is

$$\tfrac{1}{5} + \tfrac{1}{6} + \tfrac{1}{7} + \tfrac{1}{8} > \tfrac{4}{8} = \tfrac{1}{2};$$

the sum of the next eight terms is greater than 8/16 and so on. Hence the sum of the first $4 + 4 + 8 + 16 + \ldots + 2^n$ (i.e. 2^{n+1}) terms is greater than

$$2 + \tfrac{1}{2} + \tfrac{1}{2} + \tfrac{1}{2} + \ldots + \tfrac{1}{2} = \tfrac{1}{2}(n+3)$$

and this diverges to positive infinity as n tends to infinity. Hence the harmonic series diverges to positive infinity although the limit as n tends to infinity of the nth term $(1/n)$ of the series is zero.

Some points concerning these general theorems are illustrated in the following example.

Example 7. *Show that the series whose rth term is $\{r(r+1)\}^{-1}$ is convergent and that its sum to infinity is unity.*

Let s_n be the sum of the first n terms of the given series, so that

$$s_n = \sum_{r=1}^{n}\frac{1}{r(r+1)} = \sum_{r=1}^{n}\frac{1}{r} - \sum_{r=1}^{n}\frac{1}{r+1}$$
$$= \left(1 + \frac{1}{2} + \frac{1}{3} + \ldots + \frac{1}{n}\right) - \left(\frac{1}{2} + \frac{1}{3} + \ldots + \frac{1}{n} + \frac{1}{n+1}\right)$$
$$= 1 - \frac{1}{n+1}.$$

Hence the sequence $\{s_n\}$ converges to the limit unity and so also does the given series.

[The reader should note that it is quite wrong to work as follows, although this method apparently gives the correct result:

$$\sum_{r=1}^{\infty}\frac{1}{r(r+1)} = \sum_{r=1}^{\infty}\frac{1}{r} - \sum_{r=1}^{\infty}\frac{1}{r+1}$$
$$= (1 + \tfrac{1}{2} + \tfrac{1}{3} + \ldots) - (\tfrac{1}{2} + \tfrac{1}{3} + \ldots) = 1.$$

The fallacy lies in the assumption that the general result (c) above remains valid when the series Σu_r and Σv_r (here the harmonic series) diverge].

EXERCISES 3 (b)

1. Find for what values of x the series

   (i) $\dfrac{1}{1+x} - \dfrac{1-x}{(1+x)^2} + \dfrac{(1-x)^2}{(1+x)^3} - \dfrac{(1-x)^3}{(1+x)^4} + \ldots,$

   (ii) $\dfrac{1}{1+3x} + \dfrac{1+x}{(1+3x)^2} + \dfrac{(1+x)^2}{(1+3x)^3} + \dfrac{(1+x)^3}{(1+3x)^4} + \ldots,$

   converge, and prove that the sum to infinity of the first series is 1/2. (L.U.)

2. Find the sum to n terms and the sum to infinity of the series

   $$\frac{1}{1.2.4} + \frac{1}{2.3.5} + \frac{1}{3.4.6} + \frac{1}{4.5.7} + \ldots.$$ (N.U.)

3. Find an expression for the finite sum

   $$\frac{1+1^2}{1.2.3.4} + \frac{1+2^2}{2.3.4.5} + \frac{1+3^2}{3.4.5.6} + \ldots + \frac{1+n^2}{n(n+1)(n+2)(n+3)},$$

   and discuss the convergence of this series as n tends to infinity. (O.)

4. By induction, or otherwise, prove that the sum of the finite series

   $$\frac{12}{1.3.4} + \frac{18}{2.4.5} + \frac{24}{3.5.6} + \ldots + \frac{6(n+1)}{n(n+2)(n+3)}$$

   is

   $$\frac{17}{6} - \frac{1}{n+1} - \frac{1}{n+2} - \frac{4}{n+3}.$$

   Deduce that as n tends to infinity the series converges and find its sum to infinity.

5. If $-1 < r < 1$, show that $nr^n \to 0$ as $n \to \infty$. Deduce that the infinite series

   $$a + (a+b)r + (a+2b)r^2 + (a+3b)r^3 + \ldots$$

   is convergent and that its sum to infinity is

   $$\frac{a}{1-r} + \frac{br}{(1-r)^2}.$$

6. If the series $\sum_{r=1}^{\infty} u_r$ is convergent, then so also is any series formed by grouping the terms in any way to form new single terms. Show also that the sums of the two series are the same.

3.8. Tests for convergence for series of positive terms

There are comparatively few series for which it is possible to find the sum (s_n) to n terms and hence to discuss their behaviour by considering the behaviour of the sequence $\{s_n\}$. There are, however, large classes of series whose convergence (or otherwise) may be decided without the necessity of actually obtaining analytical expressions for s_n. This is made possible by certain theorems which are usually known as *convergence tests*. Certain of these tests apply only to series in which all the terms are positive and these are discussed first (it is worth noting that, in view of the result (*a*) of § 3.7, such tests can be used also for series which contain only a *finite* number of negative terms).

(*a*) *The infinite series of positive terms* $\sum_{r=1}^{\infty} u_r$ *is convergent if a constant* K *can be found such that* $s_n = \sum_{r=1}^{n} u_r < K$ *for all* n. *In such a case, the sum to infinity of the series is less than or equal to* K. *If no such constant* K *exists, the series is divergent.*

Since all the terms are positive, the sequence $\{s_n\}$ is monotonically increasing. Hence if $s_n < K$ for all n, there exists a limit s of s_n such that
$$\lim_{n\to\infty} s_n = s \leqslant K,$$
and the first part of the above result has been demonstrated.

If, when any constant K has been chosen, there is a value of n for which $s_n > K$, then $s_m > K$ whenever $m > n$ and hence the sequence $\{s_n\}$, and therefore also the given series, diverges to positive infinity.

The real point of this result is that, under such conditions, the series *cannot oscillate*.

(*b*) *The comparison test.* *If* $\sum_{r=1}^{\infty} u_r$ *and* $\sum_{r=1}^{\infty} v_r$ *are two infinite series of positive terms and if* $\sum_{r=1}^{\infty} v_r$ *is convergent with sum to infinity* V *and if* $u_n \leqslant cv_n$ *for all* n, *where* c *is a fixed positive constant, then* $\sum_{r=1}^{\infty} u_r$ *is also convergent and its sum to infinity is less than or equal to* cV.

If $t_n = \sum_{r=1}^{n} v_r$, then the sequence $\{t_n\}$ is monotonic increasing. Since $\sum_{r=1}^{\infty} v_r = V$, then $t_n < V$ for all n. Hence, since $u_n \leqslant cv_n$ for all n,
$$s_n = \sum_{r=1}^{n} u_r \leqslant c \sum_{r=1}^{n} v_r = ct_n < cV.$$
But the sequence $\{s_n\}$ is monotonic increasing so that s_n tends to a

SERIES OF POSITIVE TERMS

limit s less than or equal to cV; in other words, the series $\sum_{r=1}^{\infty} u_r$ is convergent and its sum to infinity is less than or equal to cV.

It is left as an exercise for the reader to show that if $\sum_{r=1}^{\infty} v_r$ is divergent and if $u_n \geqslant cv_n$ for all values of n, then $\sum_{r=1}^{\infty} u_r$ is also divergent.

(c) *The comparison test in a more useful form.*

For working examples the following form of the comparison test is more useful than that given in (b) above—*if $\sum_{r=1}^{\infty} u_r$ and $\sum_{r=1}^{\infty} v_r$ are two infinite series of positive terms and if*

$$\lim_{n \to \infty} \frac{u_n}{v_n} = l > 0,$$

then the two series are either both convergent or both divergent.

In the case in which $\sum_{r=1}^{\infty} v_r$ converges, the proof runs—since $\lim_{n \to \infty} \frac{u_n}{v_n} = l$, a fixed integer n_0 exists such that

$$\frac{u_n}{v_n} < l + 1, \text{ (say), whenever } n > n_0.$$

It then follows from the result (b), with $c = l + 1$, that $\sum_{r=1}^{\infty} u_r$ converges.

Again it is left as an exercise to the reader to show that when the v series diverges, so also does the u series.

A useful series for using as the v series in the comparison tests is the geometrical progression $1 + r + r^2 + \ldots$, which is known to converge when $0 < r < 1$ and to diverge when $r \geqslant 1$. Another series of use in this connection is

$$\sum_{r=1}^{\infty} \frac{1}{r^p} = \frac{1}{1^p} + \frac{1}{2^p} + \frac{1}{3^p} + \ldots \quad (3.10)$$

In the particular case $p = 1$, this becomes the harmonic series which (see § 3.7(d)) diverges. When $p < 1$, $r^p < r$ and $(1/r^p) > (1/r)$ and clearly each term is greater than the corresponding term in the harmonic series. Hence in this case the series again diverges. If $p > 1$, then

$$\frac{1}{2^p} + \frac{1}{3^p} < \frac{2}{2^p} = 2^{1-p},$$

$$\frac{1}{4^p} + \frac{1}{5^p} + \frac{1}{6^p} + \frac{1}{7^p} < \frac{4}{4^p} = 2^{2-2p},$$

$$\frac{1}{8^p} + \frac{1}{9^p} + \ldots + \frac{1}{15^p} < \frac{8}{8^p} = 2^{3-3p},$$

and so on. Hence the sum of the first $1 + 2 + 4 + 8 + \ldots + 2^n$ (i.e. $2^{n+1} - 1$) terms of the series is less than

$$1 + 2^{1-p} + 2^{2-2p} + \ldots + 2^{n-np} = \frac{1 - (\frac{1}{2})^{q(n+1)}}{1 - (\frac{1}{2})^q} < \frac{1}{1 - (\frac{1}{2})^q},$$

where $p = 1 + q$ and hence $q > 0$. Since this last expression is independent of n, there is a number $K \ [= 1/\{1 - (\frac{1}{2})^q\}]$ such that the sum of the first $2^{n+1} - 1$ terms is less than K for all n and the series is therefore convergent. Hence *the series of* (3.10) *is convergent if $p > 1$ and divergent if $p \leqslant 1$.*

Example 8. *Investigate the convergence of the series whose general term is*

$$\frac{n}{n^3 + 1}. \tag{O.}$$

When n is large, the general term is 'about as big' as $1/n^2$ so we compare the series whose nth terms u_n and v_n are given by

$$u_n = \frac{n}{n^3 + 1}, \ v_n = \frac{1}{n^2}.$$

Then, $$\frac{u_n}{v_n} = \frac{n^3}{n^3 + 1} = \frac{1}{1 + (1/n^3)}$$

and the limit as n tends to infinity of u_n/v_n is unity.
Since the series $\Sigma(1/n^2)$ converges, the comparison test shows that the series under discussion also converges.

(*d*) *D'Alembert's ratio test.* The infinite series of positive terms $\sum_{r=1}^{\infty} u_r$ is convergent if

$$\lim_{n \to \infty} \frac{u_n}{u_{n+1}} = l > 1.$$

Since, as $n \to \infty$, u_n/u_{n+1} tends to a limit l greater than unity, then, if c is a constant such that $l > c > 1$,

$$\frac{u_n}{u_{n+1}} > c > 1$$

for all values of n greater than or equal to some fixed integer m. Hence

$$u_{m+1} < \frac{u_m}{c}, \ u_{m+2} < \frac{u_{m+1}}{c} < \frac{u_m}{c^2},$$

and so on. Thus each term of the series $\sum_{r=m+1}^{\infty} u_r$ is less than the corresponding term of the geometrical progression

$$\frac{u_m}{c} + \frac{u_m}{c^2} + \frac{u_m}{c^3} + \ldots$$

Since the common ratio $(1/c)$ is less than unity, this progression is convergent and (remembering that the removal of the first $m - 1$

terms of the given series does not alter its behaviour) the result § 3.7 (b) shows that the series in question converges.

The proof of the similar result that $\sum_{r=1}^{\infty} u_r$ is divergent if

$$\lim_{n \to \infty} \frac{u_n}{u_{n+1}} = l < 1$$

is left as an exercise. The reader should note that these tests are not applicable if

$$\lim_{n \to \infty} \frac{u_n}{u_{n+1}} = 1,$$

or if the limit does not exist.

Example 9. *Discuss the behaviour of the infinite series*
$$1^2 + 2^2 x + 3^2 x^2 + 4^2 x^3 + \ldots$$
for positive values of x.

Here, if u_n denotes the nth term

$$\frac{u_n}{u_{n+1}} = \frac{n^2 x^{n-1}}{(n+1)^2 x^n} = \left(\frac{n}{n+1}\right)^2 \cdot \frac{1}{x} = \left(\frac{1}{1+1/n}\right)^2 \cdot \frac{1}{x}$$

and the limit of this as n tends to infinity is $1/x$. Hence if $x < 1$ the series converges and, if $x > 1$ it diverges. If $x = 1$, the test fails. For this value of x, the series becomes $1^2 + 2^2 + 3^2 + \ldots$ the sum s_n of the first n terms of this series is (see *Advanced Level Pure Mathematics*, p. 43),

$$\frac{n(n+1)(2n+1)}{6}$$

and the sequence $\{s_n\}$, and therefore also the given series, diverges.

(e) *Raabe's test.* *The infinite series of positive terms* $\sum_{r=1}^{\infty} u_r$ *is convergent if*

$$\lim_{n \to \infty} \left\{ n \left(\frac{u_n}{u_{n+1}} - 1 \right) \right\} = l > 1.$$

If the limit stated above exists, then

$$n \left(\frac{u_n}{u_{n+1}} - 1 \right) \geq \tfrac{1}{2}(l+1) > 1$$

for all values of n greater than or equal to some fixed integer m.

Hence
$$\frac{u_n}{u_{n+1}} \geq 1 + \frac{l+1}{2n}$$

for such values of n. Therefore if $r \geq m$,

$$r u_r \geq r u_{r+1} + \tfrac{1}{2}(l+1) u_{r+1},$$

giving $\quad r u_r - (r+1) u_{r+1} \geq \tfrac{1}{2}(l-1) u_{r+1}.$

Putting $r = m, m+1, \ldots, n-1$ in turn and adding,

$$m u_m - n u_n \geq \tfrac{1}{2}(l-1)(u_{m+1} + u_{m+2} + \ldots + u_n)$$

so that
$$u_{m+1} + u_{m+2} + \ldots + u_n < \frac{2(mu_m - nu_n)}{l-1} \leqslant \frac{2mu_m}{l-1}.$$

Hence if $s_n = \sum_{r=1}^{n} u_r$, the sequence $\{s_n\}$ is monotonic increasing and
$$s_n - s_m = u_{m+1} + u_{m+2} + \ldots + u_n \leqslant \frac{2mu_m}{l-1}$$
so that
$$s_n < s_m + \frac{2mu_m}{l-1}.$$

The right-hand side of this inequality being a constant independent of n, the sequence $\{s_n\}$ and therefore the series $\sum_{r=1}^{\infty} u_r$ is convergent.

The corresponding result that $\sum_{r=1}^{\infty} u_r$ is divergent if
$$\lim_{n \to \infty} \left\{ n\left(\frac{u_n}{u_{n+1}} - 1\right) \right\} = l < 1$$
can be proved as follows. If the above limit exists, then
$$n\left(\frac{u_n}{u_{n+1}} - 1\right) < 1$$
for all values of n greater than or equal to some fixed integer m.

Hence
$$\frac{u_n}{u_{n+1}} < 1 + \frac{1}{n}$$
for such values of n. Therefore if $r > m$
$$ru_r < (r+1)u_{r+1}$$
and thus $\quad nu_n > (n-1)u_{n-1} > \ldots > mu_m (n > m)$.

Hence $u_n > (mu_m)/n$ and, since the harmonic series $\Sigma(1/n)$ diverges, the comparison test shows that the u series diverges.

Raabe's test is sometimes useful when D'Alembert's ratio test fails. An instance when this is so is shown in Example 10.

Example 10. *Discuss the convergence (or otherwise) of the series $\Sigma(1/n^2)$.*

If u_n denotes the nth term,
$$\frac{u_n}{u_{n+1}} = \left(\frac{n+1}{n}\right)^2 = \left(1 + \frac{1}{n}\right)^2.$$

As $n \to \infty$, the limit of this ratio is unity and D'Alembert's test fails.

However, $\quad n\left(\dfrac{u_n}{u_{n+1}} - 1\right) = n\left\{\left(1 + \dfrac{1}{n}\right)^2 - 1\right\} = 2 + \dfrac{1}{n}.$

The limit as $n \to \infty$ of this being $2 \, (> 1)$, Raabe's test shows that the series under discussion is convergent.

(f) *Cauchy's test.* If $\lim_{n \to \infty} \sqrt[n]{u_n} = l$, the infinite series of positive terms $\sum_{r=1}^{\infty} u_r$ is convergent or divergent according as l is less or greater than unity.

First suppose that $l < 1$ and let x be any definite number such that $l < x < 1$. Then there exists a fixed integer m such that
$$\sqrt[n]{u_n} < x \text{ or } u_n < x^n$$
for all values of $n \geqslant m$. Since $x < 1$, the series $\sum_{r=1}^{\infty} x^r$ is convergent, and hence, by the comparison test, the series $\sum_{r=1}^{\infty} u_r$ is convergent.

If $l > 1$, there exists a fixed integer m such that
$$\sqrt[n]{u_n} > 1 \text{ or } u_n > 1$$
for all values of $n \geqslant m$. Hence u_n does not tend to zero and the series $\sum_{r=1}^{\infty} u_r$ does not converge. But a series of positive terms must converge or diverge and hence it diverges.

Example 11. *Examine for convergence the series whose nth term is*
$$\frac{1}{(1+1/n)^{n^2}}.$$
If u_n denotes the nth term
$$\sqrt[n]{u_n} = \frac{1}{(1+1/n)^n}$$
and the limit of this as $n \to \infty$ is $1/e$ where $2 < e < 3$ (see § 3.5(c)). Hence the limit is less than unity and the series therefore converges.

The reader should note that Cauchy's test is more general than D'Alembert's ratio test. An example of a series whose behaviour can be decided by Cauchy's test but where the ratio test fails will be found in Exercise 3 (f) 13.

(g) *The condensation test.* If $\phi(n) > 0$ and the sequence $\{\phi(n)\}$ is monotonic decreasing, then the two series
$$\sum_{r=1}^{\infty} \phi(r) \text{ and } \sum_{r=1}^{\infty} 2^r \phi(2^r)$$
converge or diverge together.

This result can be proved by an argument similar to that used in discussing the behaviour of the harmonic series $\Sigma(1/n)$. In the first place since $\phi(n)$ is monotonic decreasing
$$\phi(2) + \phi(3) \leqslant 2\phi(2),$$
$$\phi(4) + \phi(5) + \phi(6) + \phi(7) \leqslant 4\phi(4) = 2^2\phi(2^2),$$

and so on. This set of inequalities shows that if $\Sigma 2^r \phi(2^r)$ converges, then so does $\Sigma \phi(r)$.

On the other hand,
$$\phi(3) + \phi(4) \geqslant 2\phi(4) = \tfrac{1}{2}.2^2\phi(2^2),$$
$$\phi(5) + \phi(6) + \phi(7) + \phi(8) \geqslant 4\phi(8) = \tfrac{1}{2}.2^3\phi(2^3),$$
and so on. These inequalities show that if $\Sigma 2^r \phi(2^r)$ diverges, then so does $\Sigma \phi(r)$.

This test is well suited to series involving logarithms, see, for instance, Example 12.

Example 12. *Show that the series whose rth term is* $\dfrac{1}{r(\log r)^2}$ *is convergent.*

(O.)

Here we take
$$\phi(r) = \frac{1}{r(\log r)^2}$$
and hence
$$\phi(2^r) = \frac{1}{2^r(\log 2^r)^2} = \frac{1}{2^r r^2 (\log 2)^2}.$$
Thus
$$\sum_{r=1}^{\infty} 2^r \phi(2^r) = \frac{1}{(\log 2)^2} \sum_{r=1}^{\infty} \frac{1}{r^2}$$
and, the series $\Sigma \left(\dfrac{1}{r^2}\right)$ being convergent, the condensation test shows that the series under discussion is also convergent.

EXERCISES 3 (c)

Discuss the convergence (or otherwise) of the series whose nth terms are:

1. $(n+1)/n^2$.

2. $\sqrt{(n^2+1)} - n$.

3. $n^2/(n)!$.

4. $\dfrac{1.3.5.\ldots.(2n-3)x^{2n-1}}{2.4.6.\ldots.(2n-2)(2n-1)}(x > 0)$.

5. $\left(\dfrac{nx}{n+1}\right)^n (0 < x < 1)$.

6. $\dfrac{(\log n)^2}{n^2}$.

7. Prove that the series
$$\sum_{n=1}^{\infty} \frac{1.2.3.\ldots.n}{4.7.10.\ldots.(3n+1)} x^n$$
is divergent when $x \geqslant 3$ and convergent when $0 < x < 3$.

8. Show that the series
$$x + \frac{2^2 x^2}{(2)!} + \frac{3^3 x^3}{(3)!} + \frac{4^4 x^4}{(4)!} + \cdots$$
is convergent when $|x| < e^{-1}$.

9. If a and b are positive constants, show that the series
$$\sum_{n=1}^{\infty} \frac{1}{(an+b)^p}$$
diverges for $p \leqslant 1$ and converges for $p > 1$.

10. If, in the series Σu_n of positive terms,
$$\frac{u_n}{u_{n+1}} = \frac{n^s + an^{s-1} + bn^{s-2} + \cdots}{n^s + a'n^{s-1} + b'n^{s-2} + \cdots}$$
where s is a positive integer, show that the series converges when $a - a' - 1 > 0$.

3.9. Alternating series

The tests for convergence so far discussed have all referred to series in which all the terms are positive. A series such as
$$u_1 - u_2 + u_3 - u_4 + \cdots \quad (3.11)$$
where each u is positive and in which the signs preceding the u's are alternatively positive and negative is called an *alternating series* and the following simple convergence test is applicable to such series. *If, in an alternating series, $u_n > u_{n+1}$ for all values of n and if $\lim\limits_{n \to \infty} u_n = 0$, then the series is convergent.*

Let s_{2n} denote the sum of $2n$ terms of the series (3.11). Then
$$s_{2n} = (u_1 - u_2) + (u_3 - u_4) + \cdots + (u_{2n-1} - u_{2n})$$
and since each of the bracketed terms is positive, the sequence $\{s_{2n}\}$ is monotonic increasing and therefore converges to a limit or diverges to positive infinity. But
$$s_{2n} = u_1 - (u_2 - u_3) - (u_4 - u_5) - \cdots - (u_{2n-2} - u_{2n-1}) - u_{2n},$$
showing, since each bracketed term is again positive, that $s_{2n} < u_1$. Hence the sequence $\{s_{2n}\}$ cannot diverge to positive infinity and therefore a limit l exists such that $\lim\limits_{n \to \infty} s_{2n} = l \leqslant u_1$. Finally $s_{2n+1} = s_{2n} + u_{2n+1}$ and thus
$$\lim_{n \to \infty} s_{2n+1} = \lim_{n \to \infty} s_{2n} + \lim_{n \to \infty} u_{2n+1}$$
and, as $\lim\limits_{n \to \infty} u_{2n+1} = 0$, $\lim\limits_{n \to \infty} s_{2n+1} = \lim\limits_{n \to \infty} s_{2n} = l$, showing that the series is convergent.

It is useful to notice the three essentials for the convergence of this type of series. These are

(i) successive terms alternate in sign,
(ii) the terms steadily decrease in numerical value,
(iii) $\lim\limits_{n \to \infty} u_n = 0$.

Thus the series $1 - \frac{1}{2} + \frac{1}{3} - \frac{1}{4} + \cdots$ converges because all three of these essential conditions hold, but the series $2 - \frac{3}{2} + \frac{4}{3} - \frac{5}{4} + \cdots$ is not convergent since condition (iii) is violated.

3.10. Absolute convergence

If, as previously, the notation $|u_r|$ is used to denote the numerical value of u_r, irrespective of its sign, the series Σu_r is said to be *absolutely convergent* if the series $\Sigma |u_r|$ is convergent.

A series which is absolutely convergent is also convergent. To prove this important result, let the series $\Sigma |u_r|$ be convergent and let

$$p_r = u_r \text{ when } u_r \geqslant 0, \ p_r = 0 \text{ when } u_r \leqslant 0,$$
$$q_r = -u_r \text{ when } u_r \leqslant 0, \ q_r = 0 \text{ when } u_r \geqslant 0.$$

Then $p_r \geqslant 0$, $q_r \geqslant 0$ and

$$|u_r| = p_r + q_r, \ u_r = p_r - q_r.$$

From the first of these relations it follows that $p_r \leqslant |u_r|$, $q_r \leqslant |u_r|$ and, since $\Sigma|u_r|$ is convergent, the comparison test (§ 3.8 (*b*)) shows that both Σp_r and Σq_r are convergent. Hence by § 3.7 (*c*), the series whose terms are $p_r - q_r$ (that is, the series Σu_r), is convergent. This result enables the behaviour of some series, previously outside the scope of the tests so far discussed, to be decided. Thus the series

$$1 - \frac{1}{2^3} + \frac{1}{3^3} + \frac{1}{4^3} - \frac{1}{5^3} + \frac{1}{6^3} + \frac{1}{7^3} - \ldots$$

(in which the signs do not alternate strictly) is convergent because it is absolutely convergent in the sense that the series

$$1 + \frac{1}{2^3} + \frac{1}{3^3} + \frac{1}{4^3} + \frac{1}{5^3} + \frac{1}{6^3} + \frac{1}{7^3} + \ldots$$

converges.

In some cases a series of positive and negative terms is convergent but the series formed by changing the sign of every negative term in the original series is divergent. An example is the series

$$1 - \tfrac{1}{2} + \tfrac{1}{3} - \tfrac{1}{4} + \tfrac{1}{5} - \tfrac{1}{6} + \ldots;$$

this series (§ 3.9) converges but the series

$$1 + \tfrac{1}{2} + \tfrac{1}{3} + \tfrac{1}{4} + \tfrac{1}{5} + \tfrac{1}{6} + \ldots,$$

diverges (§ 3.7 (*d*)). In such cases the original series is said to be *semi-convergent* or *conditionally convergent*.

An important property of convergent series of positive terms and of absolutely convergent series is that the sums to infinity of such series are not affected by the order in which the terms are taken. This is usually known as *Dirichlet's theorem* and, in the case of a convergent series of positive terms, can be proved as follows. Suppose Σu_r is a convergent series of positive terms and that another series Σv_r is formed out of the same terms by taking them in a different order. Let s be the sum to infinity of the u series. Then the sum of any number of terms taken from the u series is not greater than s. Since every v is a u, the sum of any number of terms selected from the v series is also not greater than s. Thus the v series is convergent and its sum to infinity (say t) is not greater than s. Similarly it can be shown that s is not greater than t and hence that $s = t$. It is left to the reader to show that the result applies also to absolutely convergent series.

As distinct from this property of convergent series of positive terms and of absolutely convergent series, when the terms of a semi-convergent series are rearranged it may have a different sum to infinity or it may fail to converge at all. As an example, consider the semi-convergent series

$$1 - \tfrac{1}{2} + \tfrac{1}{3} - \tfrac{1}{4} + \tfrac{1}{5} - \tfrac{1}{6} + \cdots$$

and let its sum to infinity be s. Suppose now that the terms are rearranged as

$$1 - \tfrac{1}{2} - \tfrac{1}{4} + \tfrac{1}{3} - \tfrac{1}{6} - \tfrac{1}{8} + \cdots$$

and let the sum of r terms of this series be denoted by t_r. Then

$$t_{3n} = \left(1 - \frac{1}{2}\right) - \frac{1}{4} + \left(\frac{1}{3} - \frac{1}{6}\right) - \frac{1}{8} + \cdots + \left(\frac{1}{2n-1} - \frac{1}{4n-2}\right) - \frac{1}{4n}$$

$$= \frac{1}{2} - \frac{1}{4} + \frac{1}{6} - \frac{1}{8} + \cdots + \frac{1}{4n-2} - \frac{1}{4n}$$

$$= \frac{1}{2}\left(1 - \frac{1}{2} + \frac{1}{3} - \frac{1}{4} + \cdots + \frac{1}{2n-1} - \frac{1}{2n}\right).$$

Hence $\lim\limits_{n \to \infty} t_{3n} = \tfrac{1}{2}s$.

Also, $t_{3n+1} = t_{3n} + \dfrac{1}{2n+1}$, $t_{3n+2} = t_{3n} + \dfrac{1}{2n+1} - \dfrac{1}{4n+2}$

so that $\lim\limits_{n \to \infty} t_{3n+1} = \lim\limits_{n \to \infty} t_{3n+2} = \lim\limits_{n \to \infty} t_{3n} = \tfrac{1}{2}s$

and $\lim\limits_{r \to \infty} t_r = \tfrac{1}{2}s$

for all r.

There is a general theorem (due to Riemann) which states that, by an appropriate rearrangement of terms, a semi-convergent series can be made to converge to any given number, diverge or oscillate. This theorem, although of considerable theoretical interest, is not of great practical importance and a proof will not be given here. It is, however, useful as a warning against attempting to sum alternating series by regrouping.

The tests for series of positive terms and the important result that an absolutely convergent series is also convergent are together of great use in discussing the behaviour of general series. Some worked examples follow.

Example 13. *Discuss, for all values of x, the behaviour of the series*

$$1^2 + 2^2x + 3^2x^2 + 4^2x^3 + \ldots$$

This is the series discussed for positive values only of x in Example 9.
If u_n denotes the nth term

$$\left|\frac{u_n}{u_{n+1}}\right| = \left(\frac{n}{n+1}\right)^2 \cdot \frac{1}{|x|}$$

and the limit of this as n tends to infinity is $1/|x|$. Hence if $|x| < 1$ the series $\Sigma|u_n|$ and hence the series Σu_n converges. Similarly if $|x| > 1$ the series diverges. When $x = 1$, it has been shown in Example 9 that the series diverges and it is clear that when $x = -1$ the series oscillates infinitely.

Example 14. *Prove that when $|x| < 1$ and when $x = -1$, the series*

$$\sum_{n=1}^{\infty} \frac{n+1}{(n+2)(n+3)} x^n$$

converges.

If u_n denotes the nth term

$$\left|\frac{u_n}{u_{n+1}}\right| = \frac{(n+1)(n+4)}{(n+2)^2} \cdot \frac{1}{|x|}$$

and the limit of this as n tends to infinity is $1/|x|$ showing that the series converges when $|x| < 1$.
When $x = -1$, the series is an alternating one in which

$$u_n - u_{n+1} = \frac{n+1}{(n+2)(n+3)} - \frac{n+2}{(n+3)(n+4)}$$

$$= \frac{n}{(n+2)(n+3)(n+4)} > 0$$

and $\lim_{n \to \infty} u_n = 0$. Hence, by § 3.9, the series converges for this value of x.

3.11. The multiplication of infinite series

As a corollary from Dirichlet's theorem, a useful result on the multiplication of series of positive terms or of absolutely convergent series can be obtained. It is sometimes known as *Cauchy's multiplication theorem* and states that—if Σu_r and Σv_r are convergent series of positive terms or absolutely convergent series with sums to infinity s and t respectively, then the series

$$u_0v_0 + (u_1v_0 + u_0v_1) + (u_2v_0 + u_1v_1 + u_0v_2) + \ldots$$

is absolutely convergent and its sum to infinity is st.

To prove this result, take the series whose sum to n terms is

$$(u_0 + u_1 + \ldots + u_{n-1})(v_0 + v_1 + \ldots + v_{n-1}).$$

This can be written

$$u_0v_0 + (u_0v_1 + u_1v_1 + u_1v_0) + \ldots,$$

the rth term containing all the terms involving u_{r-1} or v_{r-1} but no terms with a suffix higher than $(r-1)$. Since Σu_r and Σv_r converge, this series converges and its sum to infinity is st. Since

$$|u_0 v_0| + |u_0 v_1| + |u_1 v_1| + |u_1 v_0| + \ldots \text{ to } m \text{ terms}$$
$$\leqslant \{|u_0| + |u_1| + \ldots + |u_m|\}\{|v_0| + |v_1| + \ldots$$
$$+ |v_m|\} \leqslant s't'$$

where s' and t' are the sums to infinity of $\Sigma|u_r|$ and $\Sigma|v_r|$ respectively, the series

$$u_0 v_0 + u_0 v_1 + u_1 v_1 + u_1 v_0 + \ldots,$$

or any rearrangement of it, is absolutely convergent. One such rearrangement is the series

$$u_0 v_0 + (u_1 v_0 + u_0 v_1) + (u_2 v_0 + u_1 v_1 + u_0 v_2) + \ldots$$

and hence this series is also convergent with sum to infinity st. Further, its convergence is absolute since the series

$$|u_0 v_0| + |u_1 v_0 + u_0 v_1| + |u_2 v_0 + u_1 v_1 + u_0 v_2| + \ldots$$

is a series of positive terms whose sum to n terms is not greater than $s't'$ (see § 3.8 (a)).

EXERCISES 3 (d)

1. Show that the series

    (i) $1 - \dfrac{1}{3} + \dfrac{1}{5} - \dfrac{1}{7} + \ldots$

    (ii) $1 - \dfrac{1}{2} + \dfrac{1.3}{2.4} - \dfrac{1.3.5}{2.4.6} + \ldots$

    are convergent.

2. Show that the series $1 + \tfrac{1}{2} - \tfrac{1}{3} + \tfrac{1}{4} + \tfrac{1}{5} - \tfrac{1}{6} + \ldots$ diverges.

3. Show that the series

    $$1 + \frac{p}{q}x + \frac{2(p+1)}{q^2}x^2 + \frac{3(p+2)}{q^3}x^3 + \ldots$$

    is convergent when $|x| < |q|$.

4. If Σu_r is a convergent series of positive terms, prove that the series $\Sigma u_r x^r$ is absolutely convergent when $|x| \leqslant 1$.

5. If $\sum\limits_{r=1}^{\infty}\left(\dfrac{1}{r^2}\right) = s$, show that

    $$1 + \frac{1}{3^2} + \frac{1}{5^2} + \frac{1}{7^2} + \ldots = \frac{3}{4}s.$$

6. If u_n and v_n are positive for all values of n and Σu_n, Σv_n are both convergent, show that $\Sigma u_n v_n$ is convergent.

3.12. The convergence of the binomial, exponential and logarithmic series

The purpose of this section is to discover under what circumstances the binomial, exponential and logarithmic series converge. These series were discussed briefly in *Advanced Level Pure Mathematics*, pp. 52 and 254, but no questions of the convergence of the series were then considered.

(a) *The binomial series.*

This is the series
$$1 + nx + \frac{n(n-1)}{(2)!}x^2 + \frac{n(n-1)(n-2)}{(3)!}x^3 + \ldots$$
and, if u_r denotes the rth term of the series,
$$\left|\frac{u_r}{u_{r+1}}\right| = \left|\frac{r}{n-r+1} \cdot \frac{1}{x}\right|.$$
The limit of this as r tends to infinity is $1/|x|$. Hence the series is absolutely convergent when $|x| < 1$ and the discussion is not here pursued for other values of x.

(b) *The exponential series.*

Here the series in question is
$$1 + x + \frac{x^2}{(2)!} + \frac{x^3}{(3)!} + \ldots$$
and, if the rth term is denoted by u_r,
$$\frac{u_r}{u_{r+1}} = \frac{r}{x}.$$
Hence $u_r > 2u_{r+1}$ for all values of r greater than $2x$. Thus, by the argument used in the proof of D'Alembert's ratio test (§ 3.8 (d)), the series converges for all positive values of x. When x is negative let $x = -y$, so that the series to be discussed is
$$1 - y + \frac{y^2}{(2)!} - \frac{y^3}{(3)!} + \ldots$$
It follows from the above that this series is absolutely convergent and so is convergent. Hence the exponential series converges for all values of x.

(c) *The logarithmic series.*

The series to be considered is
$$x - \frac{x^2}{2} + \frac{x^3}{3} - \ldots$$

and if u_r is the rth term
$$\left|\frac{u_r}{u_{r+1}}\right| = \left(\frac{r+1}{r}\right) \cdot \frac{1}{|x|}.$$
The limit of this as r tends to infinity being $1/|x|$, the series converges absolutely when $|x| < 1$.

When $x = 1$, the series reduces to
$$1 - \tfrac{1}{2} + \tfrac{1}{3} - \ldots$$
and this (by § 3.9) converges. When $x = -1$, the series becomes
$$-1 - \tfrac{1}{2} - \tfrac{1}{3} - \ldots$$
and this (§ 3.7 (d)) diverges to negative infinity. When $|x| > 1$, the ratio test shows that the series is not convergent and it is easy to see that when $x < -1$ the series diverges to $-\infty$ while it oscillates infinitely if $x > 1$.

3.13. The summation of some infinite series

When series are absolutely convergent, the terms can be rearranged and the terms of two (or more) such series can be added, subtracted or multiplied. This fact and the known* sums of the binomial, exponential and logarithmic series when absolutely convergent, viz.

$$1 + nx + \frac{n(n-1)}{(2)!}x^2 + \frac{n(n-1)(n-2)}{(3)!}x^3 + \ldots$$
$$= (1+x)^n, \qquad (3.12)$$

$$1 + x + \frac{x^2}{(2)!} + \frac{x^3}{(3)!} + \ldots = e^x, \qquad (3.13)$$

$$x - \frac{x^2}{2} + \frac{x^3}{3} - \frac{x^4}{4} + \ldots = \log_e(1+x), \qquad (3.14)$$

enable the sums to infinity of certain other absolutely convergent series to be found. The following worked examples illustrate some of the devices which can be employed.

Example 15. *Find the sum to infinity of the following series*

(i) $\quad 1 + \dfrac{2x}{(1)!} + \dfrac{3x^2}{(2)!} + \dfrac{4x^3}{(3)!} + \ldots,$

(ii) $\quad \dfrac{x}{1.2} + \dfrac{x^2}{2.3} + \dfrac{x^3}{3.4} + \ldots \ (0 < x < 1).$ \hfill (L.U.)

(i) From (3.13), for all values of x,
$$1 + \frac{x}{(1)!} + \frac{x^2}{(2)!} + \frac{x^3}{(3)!} + \ldots + \frac{x^n}{(n)!} + \ldots = e^x.$$
$$x + \frac{x^2}{(1)!} + \frac{x^3}{(2)!} + \ldots + \frac{x^{n+1}}{(n)!} + \ldots = xe^x.$$

* These results have been given in *Advanced Level Pure Mathematics*. They have not yet been proved rigorously.

Since, $\dfrac{1}{(n-1)!} + \dfrac{1}{(n)!} = \dfrac{n+1}{(n)!}$

addition of these results shows that

$$1 + \frac{2x}{(1)!} + \frac{3x^2}{(2)!} + \frac{4x^3}{(3)!} + \ldots + \frac{(n+1)x^n}{(n)!} + \ldots = (1+x)e^x.$$

(ii) Writing $-x$ in place of x in (3.14) and dividing by x to obtain the second result below, if $0 < x < 1$,

$$x + \frac{x^2}{2} + \frac{x^3}{3} + \ldots + \frac{x^n}{n} + \ldots = -\log_e(1-x),$$

$$1 + \frac{x}{2} + \frac{x^2}{3} + \frac{x^3}{4} + \ldots + \frac{x^n}{n+1} + \ldots = -\frac{1}{x}\log_e(1-x).$$

By subtraction

$$-1 + \frac{x}{1.2} + \frac{x^2}{2.3} + \frac{x^3}{3.4} + \ldots + \frac{x^n}{n(n+1)} + \ldots = \left(\frac{1}{x} - 1\right)\log_e(1-x),$$

so that the sum of the series in question is

$$1 + \frac{(1-x)}{x}\log_e x.$$

Example 16. *Find the sum of the infinite series*

$$1 + \frac{5}{3} + \frac{5.7}{3.6} + \frac{5.7.9}{3.6.9} + \ldots \quad \text{(L.U.)}$$

Putting $n = -\tfrac{5}{2}$ and writing $-x$ for x in (3.12), if $|x| < 1$

$$1 + \frac{5}{2}x + \frac{\left(\frac{5}{2}\right)\left(\frac{7}{2}\right)}{(2)!}x^2 + \frac{\left(\frac{5}{2}\right)\left(\frac{7}{2}\right)\left(\frac{9}{2}\right)}{(3)!}x^3 + \ldots = (1-x)^{-5/2},$$

i.e., $\quad 1 + 5\left(\dfrac{x}{2}\right) + \dfrac{5.7}{1.2}\left(\dfrac{x}{2}\right)^2 + \dfrac{5.7.9}{1.2.3}\left(\dfrac{x}{2}\right)^3 + \ldots = (1-x)^{-5/2}.$

The given series results by writing

$$\frac{x}{2} = \frac{1}{3}$$

so that

$$1 + \frac{5}{3} + \frac{5.7}{3.6} + \frac{5.7.9}{3.6.9} + \ldots = \left(1 - \frac{2}{3}\right)^{-5/2} = 3^{5/2} = 9\sqrt{3}.$$

3.14. The generating function of a recurring power series

In § 2.5 a recurring power series was defined as one in which the coefficients u_r multiplying x^r satisfied a linear difference equation. Suppose now that a function $f(x)$ can be expanded, at least for some values of x, in an infinite recurring power series which is convergent. Then, for such values of x,

$$f(x) = u_0 + u_1 x + u_2 x^2 + \ldots + u_r x^r + \ldots$$

and $f(x)$ is called the *generating function* of the series. $f(x)$ is also the sum to infinity of the series for values of x for which it is convergent. Two simple examples are

RECURRING POWER SERIES

$$\frac{1}{1-x} \quad \text{and} \quad \frac{1}{(1+x)^2}$$

which generate respectively the power series

$$1 + x + x^2 + \ldots + x^r + \ldots$$

and $\quad 1 - 2x + 3x^2 - \ldots + (-1)^r(r+1)x^r + \ldots,$

both series converging when $|x| < 1$.

It can be shown that the generating function of a power series is a rational function. Here, for brevity, a proof is given only for the case in which the difference equation satisfied by the coefficients is of the second order but the method applies when the difference equation is of any order. Assume, for some positive quantity x_0, the infinite recurring series $u_0 + u_1 x + u_2 x^2 + \ldots + u_r x^r + \ldots$ is convergent when $|x| < x_0$ and that the scale of relation of its coefficients is $u_{r+2} + a_1 u_{r+1} + a_2 u_r = 0$ $(r \geqslant 0)$. In Example 15 of Chapter 2 it was shown that the sum (s_n) to n terms of this series is given by

$$(1 + a_1 x + a_2 x^2)s_n = u_0 + (u_1 + a_1 u_0)x + (a_1 u_{n-1} + a_2 u_{n-2})x^n + a_2 u_{n-1} x^{n+1}.$$

Since for $|x| < x_0$, $\Sigma u_r x^r$ is convergent

$$\lim_{n \to \infty} (a_1 u_{n-1} + a_2 u_{n-2})x^n = 0 \text{ and } \lim_{n \to \infty} a_2 u_{n-1} x^{n+1} = 0,$$

so that if s is the sum to infinity of the series,

$$(1 + a_1 x + a_2 x^2)s = \nu_1 + (u_1 + a_1 u_0)x$$

and hence

$$u_0 + u_1 x + u_2 x^2 + \ldots = \frac{u_0 + (u_1 + a_1 u_0)x}{1 + a_1 x + a_2 x^2}.$$

The expression on the right being a rational function, it only remains to be shown that the assumption that there are values of x for which the series converges is valid. To do this, suppose that

$$1 + a_1 x + a_2 x^2 \equiv (1 - px)(1 - qx)$$

so that the generating function

$$\frac{u_0 + (u_1 + a_1 u_0)x}{1 + a_1 x + a_2 x^2}$$

can be written, by using partial fractions, as

$$\frac{P}{1 - px} + \frac{Q}{1 - qx}$$

where P and Q are certain constants. This, by the binomial theorem, can be expanded into the convergent power series

$$P(1 + px + p^2 x^2 + \ldots) + Q(1 + qx + q^2 x^2 + \ldots)$$

provided that $|x|$ is the smaller of $1/|p|$ and $1/|q|$.

Example 17. *Find the generating function of the recurring series*
$$1 - 7x - x^2 - 43x^3 - \ldots$$
and the values of x for which the above series is convergent.

Let the scale of relation be $u_{r+2} + a_1 u_{r+1} + a_2 u_r = 0$. Then
$$-43 - a_1 - 7a_2 = 0 \text{ and } -1 - 7a_1 + a_2 = 0$$
giving $a_1 = -1$, $a_2 = -6$.

Let s denote the sum to infinity (the generating function) for values of x for which the series converges. Then
$$s = 1 - 7x - x^2 - 43x^3 - \ldots$$
$$-xs = -x + 7x^2 + x^3 + \ldots$$
$$-6x^2 s = -6x^2 + 42x^3 + \ldots$$
and, by addition, $(1 - x - 6x^2)s = 1 - 8x$.

Hence the generating function is
$$\frac{1 - 8x}{1 - x - 6x^2}$$
and since the factors of $1 - x - 6x^2$ are $1 + 2x$ and $1 - 3x$, the series converges when $|x| < 1/3$.

EXERCISES 3 (e)

1. Prove that
$$1 - \frac{1}{2} \cdot \frac{1}{2} + \frac{1.3}{2.4} \cdot \frac{1}{2^2} - \frac{1.3.5}{2.4.6} \cdot \frac{1}{2^3} + \frac{1.3.5.7}{2.4.6.8} \cdot \frac{1}{2^4} - \ldots = \sqrt{\frac{2}{3}}.$$

2. Show that
$$1 + 3x + \frac{7x^2}{(2)!} + \ldots + \frac{r^2 + r + 1}{(r)!} x^r + \ldots = (x + 1)^2 e^x.$$

3. Show that the coefficient of x^n in the expansion of $e^{(e^x)}$ is
$$\frac{1}{(n)!} \left\{ \frac{1^n}{(1)!} + \frac{2^n}{(2)!} + \frac{3^n}{(3)!} + \ldots + \frac{r^n}{(r)!} + \ldots \right\}.$$
Hence find the sum of the infinite series
$$\frac{1^3}{(1)!} + \frac{2^3}{(2)!} + \frac{3^3}{(3)!} + \ldots \quad \text{(L.U.)}$$

4. If $|x| < 1$, find the sum of the series
$$\tfrac{1}{2} x^2 + \tfrac{2}{3} x^3 + \tfrac{3}{4} x^4 + \tfrac{4}{5} x^5 + \ldots$$

5. Find the generating function of the infinite recurring series
$$1 + 2^2 x + 3^2 x^2 + 4^2 x^3 + \ldots + r^2 x^{r-1} + \ldots$$
given that the scale of relation is a linear difference equation of order 3.

6. By taking the scale of relation as a linear difference equation of as low an order as possible, find the generating function of the recurring power series
$$x + 3x^2 + 7x^3 + 15x^4 + 31x^5 + \ldots + (2^r - 1)x^r + \ldots$$

EXERCISES 3 (f)

1. Examine the behaviour, as n tends to infinity of the sequence $\{u_n\}$ when

   (i) $u_n = \dfrac{a}{n+1}$, (ii) $u_n = \dfrac{a^n}{1 + a^{n+1}}$. (O.)

2. Show that x^n and nx^n both tend to zero as n tends to infinity if $0 < x < 1$. Prove that $(1 + a)^n < (1 - na)^{-1}$ if n is a positive integer, $a > 0$ and $na < 1$, and deduce that, if $0 < x < 1$, $(1 + x^n)^n$ tends to 1 as n tends to infinity. (O.)

3. Show that, if α and β, the roots of the equation $ax^2 + (b + c)x + d = 0$ are unequal, the relation
   $$au_n u_{n+1} + bu_n + cu_{n+1} + d = 0$$
   may be written in the form
   $$\frac{u_{n+1} - \alpha}{u_{n+1} - \beta} = \lambda \left(\frac{u_n - \alpha}{u_n - \beta} \right)$$
   where λ is a constant.
   Show that, if $k > 0$, the sequence $\{u_n\}$ defined by the recurrence formula
   $$u_n u_{n+1} - 2u_{n+1} + 1 - k^2 = 0$$
   tends to the limit $(1 - k)$ for all values of u_1 other than $1 + k$. (O.)

4. Two infinite sequences $a_1, a_2, a_3, \ldots$ and $b_1, b_2, b_3, \ldots$ are connected by the relations
   $$a_{n+1} = (a_n^2 + b_n^2)/(a_n + b_n), \quad b_{n+1} = \tfrac{1}{2}(a_n + b_n).$$
   Given that $a_1 > b_1 > 0$, prove that
   $$a_n > a_{n+1} > b_{n+1} > b_n > 0$$
   for all $n \geqslant 1$ and hence, or otherwise, show that a_n and b_n tend to a common limit as n tends to infinity. (O.)

5. The two infinite sequences of positive terms $a_1, a_2, \ldots, b_1, b_2, \ldots$ are connected by the relations
   $$a_n b^2_{n+1} = b_n a^2_{n+1} = \tfrac{1}{2}(a_n^3 + b_n^3) \text{ and } a_1 < b_1.$$
   Prove
   (i) $(b_n/a_n) \to 1$ as $n \to \infty$,
   (ii) for all sufficiently large n, $a_n < a_{n+1} < b_{n+1} < b_n$,
   (iii) the sequences $\{a_n\}$ and $\{b_n\}$ have a common limit. (O.)

6. Given that u_1 and a are both positive and that
   $$2u_n u_{n+1} - u_n^2 = a^2 (n \geqslant 1),$$
   express $(u_{n+1} - a)/(u_{n+1} + a)$ in terms of u_1 and a. Hence show that u_n tends to a as n tends to infinity.

7. A sequence of real numbers is defined by the relations
   $$u_{n+1} = a + \frac{b}{u_n}.$$

Show that, if $a^2 + 4b > 0$, the sequence tends to a limit which is the numerically greater root of the quadratic equation $\theta^2 - a\theta - b = 0$.

(O.)

8. If $-1 < x < 1$ and $t_n = \sum_{r=1}^{n} \dfrac{1}{r(r+1)}$, find the sum to infinity of the series
$$1 + t_1 x + t_2 x^2 + \ldots + t_n x^n + \ldots.$$
(L.I.C.)

9. Find the sum to n terms and to infinity of the series
$$\frac{5}{1.2}\cdot\frac{1}{3} + \frac{7}{2.3}\cdot\frac{1}{3^2} + \frac{9}{3.4}\cdot\frac{1}{3^3} + \frac{11}{4.5}\cdot\frac{1}{3^4} + \ldots.$$

10. By first finding the sum to n terms show that, if $x > 0$, the series
$$\sum_{n=1}^{\infty} \frac{1}{(1+nx)\{1+(n+1)x\}}$$
is convergent.

11. Prove that:

(i) $\displaystyle\sum_{n=1}^{\infty} \frac{1}{n(n+1)}$ and $\displaystyle\sum_{n=1}^{\infty} \frac{1}{n^2}$ are convergent;

(ii) $\displaystyle\sum_{n=1}^{\infty} \frac{1}{n^2(n+1)^2} = 2\sum_{n=1}^{\infty} \frac{1}{n^2} - 3$;

(iii) $\dfrac{3}{2} < \displaystyle\sum_{n=1}^{\infty} \frac{1}{n^2} < 2$. (O.)

12. Show that, if u_n is the nth term of a convergent series of positive decreasing terms, nu_n tends to zero as n tends to infinity (*Pringsheim's or Abel's theorem*). Hence show that the series $\Sigma\dfrac{1}{n}$ diverges. (O.)

13. Using Cauchy's test (§ 3.8 (*f*)), discuss the behaviour of the series whose nth term is
$$\frac{1}{3^{n+(-1)^n}}.$$
Show also that, in the case of this series, D'Alembert's ratio test fails.

14. If the sequence $\{u_n\}$ is monotonic decreasing with limit zero, show that the series
$$u_1 - \tfrac{1}{2}(u_1 + u_2) + \tfrac{1}{3}(u_1 + u_2 + u_3) - \ldots$$
is convergent.

15. Discuss the convergence of the series whose nth term is
$$\frac{1}{n^\alpha (\log n)^\beta}.$$
(O.)

16. Discuss the convergence of the series whose nth term is
$$\frac{ax^n + b}{\alpha x^n + \beta}$$
for all real positive values of x. Consider particularly the cases in which one, or more, of the constants a, b, α, β vanishes. (O.)

17. If u_n and v_n are positive for all values of n, and the series $\Sigma(u_n^2 + v_n^2)$ is convergent, prove that the series $\Sigma(u_n^3 + v_n^3)$ and $\Sigma u_n v_n$ are both convergent. (O.)

18. Find the sum s_n of the first n terms of the series
$$\frac{1}{2} + \frac{1}{1.3} + \frac{1}{1.2.4} + \frac{1}{1.2.3.5} + \cdots$$
Show that the sum of the infinite series
$$s_1 x + s_2 x^2 + s_3 x^3 + \cdots$$
is
$$\frac{1}{x(1-x)} - \frac{e^x}{x} \text{ if } 0 < x < 1.$$ (L.I.C.)

19. Show that the series
$$1^n + 2^n + \frac{3^n}{(2)!} + \frac{4^n}{(3)!} + \cdots$$
converges for all real values of n. Find the sum of the series when $n = 2$. (O.)

20. Prove that the series
$$^mC_n + \frac{^{m+1}C_n}{(1)!} + \frac{^{m+2}C_n}{(2)!} + \cdots + \frac{^{m+r}C_n}{(r)!} + \cdots \quad (n < m)$$
is convergent, and, by equating powers of x in the identity
$$e^x(1+x)^m \equiv e^{-1} \sum_{r=0}^{\infty} \frac{(1+x)^{m+r}}{(r)!},$$
or otherwise, show that its sum is
$$e\left\{ ^mC_n + {}^mC_{n-1} + \frac{^mC_{n-2}}{(2!)} + \cdots + \frac{^mC_0}{(n)!} \right\}.$$ (O.)

21. Find numerical values of a, b, c, d such that identically
$$n^4 \equiv an + bn(n-1) + cn(n-1)(n-2) + dn(n-1)(n-2)(n-3).$$
Deduce that
$$1 + \frac{2^4}{(2)!} + \frac{3^4}{(3)!} + \frac{4^4}{(4)!} + \frac{5^4}{(5)!} + \cdots = 15e.$$ (W.)

22. Sum to infinity the series
$$\frac{2.3}{(1)!} + \frac{3.4}{(2)!} + \frac{4.5}{(3)!} + \cdots$$ (O.C.)

23. Prove that, for certain values of x,
$$\frac{x^2 - 1}{2} - \frac{1}{1.3} \cdot \frac{(x-1)^3}{x+1} - \frac{1}{3.5} \cdot \frac{(x-1)^5}{(x+1)^3} - \cdots = x \log_e x.$$ (O.)

24. Show that the series whose nth term is $2^n n^2 x^n$ is a recurring series and find its generating function. For what range of values of x does the latter represent the sum to infinity of the series? (C.)

25. Given that the numbers of the sequence $a_0, a_1, a_2, \ldots$ are connected by the recurrence relation
$$a_{n+2} + pa_{n+1} + qa_n = 0 \ (n \geqslant 0)$$
and that $a_0 = 2$, $a_1 = -p$, prove that
$$a_0 + a_1 t + \ldots + a_n t^n + \ldots = \frac{2 + pt}{1 + pt + qt^2}.$$
By considering the value of $a_{n+3}{}^2 + qa_{n+2}{}^2$ find the recurrence relation satisfied by the sequence $a_0{}^2, a_1{}^2, a_2{}^2, \ldots$ and hence find the sum of the series
$$a_0{}^2 + a_1{}^2 t + \ldots + a_n{}^2 t^n + \ldots.$$
(It may be assumed that the value of t is sufficiently small to ensure the convergence of both series.) (O.)

CHAPTER 4

COMPLEX NUMBERS

4.1. Introduction

The introduction of rational and irrational numbers (Chapter 1) makes possible the solution of certain algebraical equations which would otherwise possess no solution. Thus the equation $7x = 13$ has no solution in an algebra which admits only the natural numbers and the equation $x^2 = 2$ has no solution in one admitting only rational numbers. In an algebra admitting only real numbers, equations such as $x^2 + 2 = 0$ and $x^2 - 4x + 13 = 0$ do not possess solutions and, in this chapter, a further generalisation of the notion of number is made in order that such equations are satisfied by numbers of the new type.

There are several methods of introducing numbers of the type mentioned above—that adopted here is the construction of an algebra of *ordered pairs* of real numbers. Such a pair is called a *complex number* and, if a and b are the real numbers involved, the complex number is, for the present, denoted by the symbol $[a, b]$. The idea of an ordered pair of numbers may perhaps be made clearer by the geometrical illustration afforded by the rectangular Cartesian co-ordinates which fix the position of a point in a plane. Thus the point $(3, 7)$ is one with abscissa 3 and ordinate 7 while the point $(7, 3)$ has abscissa 7 and ordinate 3. In the same way, although of course the actual numbers need have no geometrical significance, the complex number $[a, b]$ differs from the complex number $[b, a]$ unless $a = b$.

The generalisation of number by the introduction of complex numbers of the above type is analogous to the introduction of the rational numbers p/q—these too are ordered pairs of numbers. An algebra of rational numbers is subject to the rules that, if p/q and p'/q' are two such numbers, then

(i) $\dfrac{p}{q} = \dfrac{p'}{q'}$ if and only if $pq' = p'q$,

(ii) $\dfrac{p}{q} + \dfrac{p'}{q} = \dfrac{p + p'}{q}$,

(iii) $\left(\dfrac{p}{q}\right) \times \left(\dfrac{p'}{q'}\right) = \dfrac{pp'}{qq'}$.

Similarly in a logical introduction of complex numbers, it is necessary to start by defining certain elementary operations such as equality, addition, multiplication, etc.

In contrast to the rule of equality for rational numbers, two complex numbers are said to be equal if, and only if, they are identical—thus

$$[a, b] = [a', b'] \text{ if and only if } a = a' \text{ and } b = b'. \qquad (4.1)$$

The meanings assigned to addition and subtraction are that

$$[a, b] \pm [a', b'] = [a \pm a', b \pm b'], \qquad (4.2)$$

and, for multiplication by a real number k,

$$k[a, b] = [a, b]k = [ka, kb]. \qquad (4.3)$$

An immediate consequence of these definitions is that any complex number $[a, b]$ may be expressed in the form

$$[a, b] = a[1, 0] + b[0, 1]. \qquad (4.4)$$

Just as the product of two real numbers is found from a multiplication table so it is possible, from the representation of a complex number in the form (4.4), to determine the product of any two complex numbers from a multiplication table for the special complex numbers $[1, 0]$, $[0, 1]$.

The commutative, associative and distributive laws of multiplication are postulated to apply in the algebra of complex numbers and, to link it with the algebra of real numbers, the special complex number $[1, 0]$ is made to behave like the real number unity, so that multiplication by $[1, 0]$ leaves any number unchanged. Thus

$$[1, 0][0, 1] = [0, 1][1, 0] = [0, 1] \text{ and } [1, 0]^2 = [1, 0] = 1. \qquad (4.5)$$

It follows from (4.3) that the complex number $[a, 0]$, with the second term zero, behaves like the real number a. If the other special complex number $[0, 1]$ is denoted by the letter i, any complex number may be expressed, from (4.4),* as $a + bi$.

The special complex number $[1, 0]$ has been chosen, from the second of equations (4.5), so that its square is unity; if now the other special complex number $[0, 1]$ is made to obey

$$i^2 = [0, 1]^2 = [-1, 0] = -1, \qquad (4.6)$$

a genuine extension of the concept of number is obtained. Using (4.6), the product of two complex numbers is then given by

$$\begin{aligned}[a, b] \times [a', b'] &= (a + bi)(a' + b'i) \\ &= aa' + bb'i^2 + (ab' + ba')i \\ &= aa' - bb' + (ab' + ba')i \\ &= [aa' - bb', ab' + ba']. \end{aligned} \qquad (4.7)$$

If x is a real number, $x^2 - 4x + 13 = (x - 2)^2 + 3^2 > 0$ and the equation $x^2 - 4x + 13 = 0$ has no solution in an algebra admitting

---

* This is the usual form when a, b are given numerically, but the same complex number is often written in the equivalent form $a + ib$.

only real numbers. If, however, x is the complex number $a + bi$, since (using the relation $i^2 = -1$)
$$(a + bi)^2 - 4(a + bi) + 13 = a^2 - 4a + 13 - b^2 + (2ab - 4b)i,$$
the equation is satisfied if
$$b^2 = a^2 - 4a + 13 \text{ and } 4b - 2ab = 0.$$
These latter equations are satisfied, since it can be assumed that $b \neq 0$, by $a = 2$, $b = \pm 3$ so that, in an algebra admitting complex numbers, the equation $x^2 - 4x + 13 = 0$ is satisfied by the two complex numbers $2 \pm 3i$. The same result is obtained by applying the usual formula for the roots of a quadratic equation and using the relation $i^2 = -1$.

The reader should be able to satisfy himself that the foregoing definitions of equality, addition, subtraction and multiplication of ordered pairs have been so devised that the ordinary operations of algebra using complex numbers may be performed in the same way as in the algebra of real numbers. It is convenient at this point to drop the notation $[a, b]$ for a complex number and to use the alternative $a + bi$, treating i as an ordinary number but replacing i^2 by -1 whenever it occurs.

4.2. The geometrical representation of complex numbers

It has already been pointed out that an ordered pair of real numbers is used to define the position of a point in a plane and this fact can be used to give a geometrical representation of a complex number. In Fig. 1 (para. 1.1) a rational number r was represented by a point A_r on a straight line at distance r units from some fixed point O on the line. A complex number $x + yi$ may be represented in a similar way—taking Ox, Oy as rectangular axes with origin O, the point P whose co-ordinates referred to these axes are x and y may be regarded as a representation of the complex number in question. The usual conventions as to sign are observed so that if x, y are both positive the representative point P lies in the first of the four quadrants into which the plane has been divided by the axes; if $x < 0$, $y > 0$, P lies in the second quadrant; if $x < 0$, $y < 0$, P lies in the third quadrant and if $x > 0$, $y < 0$, the representative point lies in the fourth quadrant. Examples are given in Fig. 2 in which P_1, P_2 represent respectively the complex numbers $3 + 2i$, $-2 - i$.

In this way, to every point of the plane there corresponds one complex number and, conversely, to every complex number there corresponds one, and only one, point of the plane. The complex number $x + yi$ may be denoted by a single letter z; the point P is called the *representative point* of the complex number z and the number z is often referred to as the *affix* of the point P. The representation of complex numbers afforded by this method was due to J. R. Argand

Fig. 2

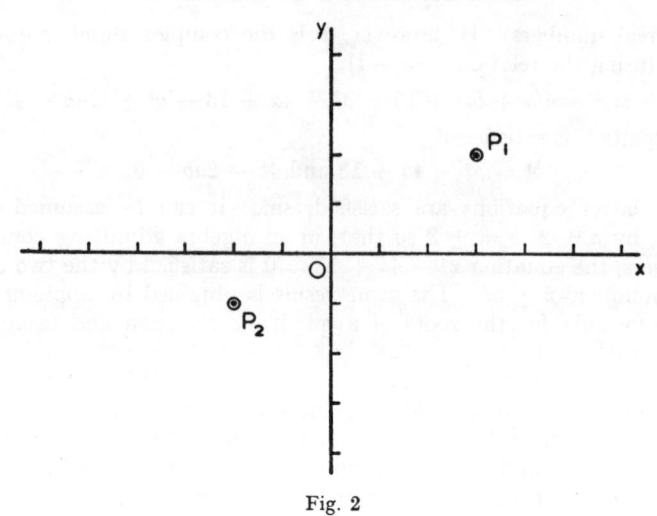

and a figure such as Fig. 2 above is often called *Argand's diagram* (or *the Argand diagram*).

Using the letter z to denote the complex number $x + yi$, so that

$$z = x + yi, \qquad (4.8)$$

it is convenient (although not altogether desirable) to call x and y respectively the *real* and *imaginary* parts of z and the notation

$$x = \mathrm{R}(z),\ y = \mathrm{I}(z) \qquad (4.9)$$

is often useful. Complex numbers for which $y = 0$ are often said to be 'purely real' and those for which $x = 0$ to be 'purely imaginary'. The representative points for such numbers will lie respectively on the axes Ox, Oy of the Argand diagram and these axes are sometimes referred to as the 'real axis' and the 'imaginary axis'. The terminology used above emanates from the time when the theory of complex numbers was not well understood. It is a very convenient terminology and is now sanctioned by usage but it is ill-chosen. The reader should note that there is nothing imaginary about y (or the axis Oy)—it is just as real as x (or the axis Ox).

If in the Argand diagram (Fig. 3), P represents the complex number $z = x + yi$, if the length OP is denoted by r and OP makes an angle θ with the axis Ox, it is clear that

$$\left.\begin{array}{l} x = r\cos\theta,\ y = r\sin\theta, \\ r = +\sqrt{(x^2 + y^2)}, \\ \cos\theta = \dfrac{x}{r},\ \sin\theta = \dfrac{y}{r}. \end{array}\right\} \qquad (4.10)$$

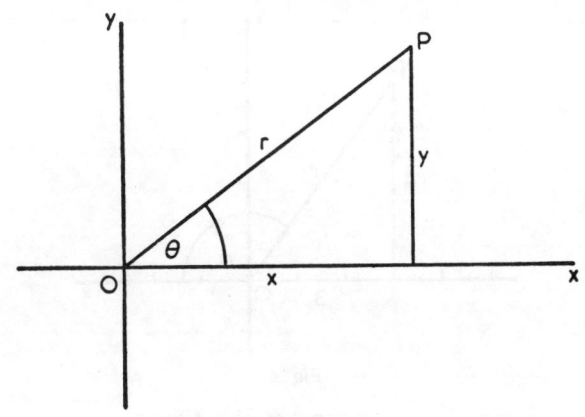

Fig. 3

r is called the *modulus* of the complex number $x + yi$ and is denoted by the symbol $|x + yi|$ or $|z|$, this symbol being due to Weierstrass; its use in connection with real numbers has already been explained in Chapter 3 (§ 3.4). The angle θ is called the *argument* or *amplitude* of the complex number z and is denoted by arg z or am z. The complex number $x + yi$ can be expressed in terms of its modulus r and argument θ by the relation

$$x + yi = r(\cos \theta + i \sin \theta). \tag{4.11}$$

The transformation can be effected by means of equations (4.10) and it should be noted that r is always taken as being positive. The last two of equations (4.10) show that there is one value of θ for which $-\pi < \theta \leqslant \pi$, but any value of θ differing from this by $2n\pi$ (n an integer) would give the same point P in fig. 3. It is usual to take the value of θ given by $-\pi < \theta \leqslant \pi$ as the *principal value* of arg z. The reader should realise that, when determining the principal value, it is essential for θ to satisfy both the equations

$$\cos \theta = \frac{x}{r}, \quad \sin \theta = \frac{y}{r};$$

the single equation $\tan \theta = y/x$, into which the above equations can be combined, is not sufficient for this purpose for it leads to two possible values of θ in the range.

Example 1. *Find the modulus and argument of the complex number* $-3 + 4i$.

In the Argand diagram (Fig. 4), the representative point P has co-ordinates $(-3, 4)$. From the figure, if r and θ are the required modulus and argument

$$r \cos \theta = -3, \, r \sin \theta = 4.$$

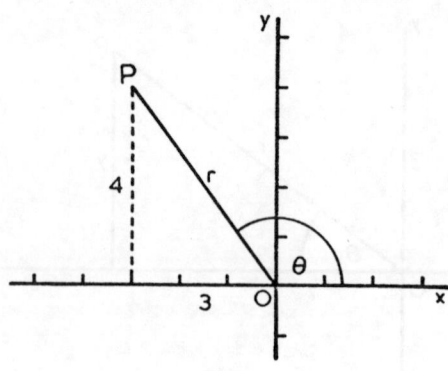

Fig. 4

Hence
$$r = +\sqrt{(3^2 + 4^2)} = 5,$$
$$\cos \theta = -\tfrac{3}{5}, \sin \theta = \tfrac{4}{5}.$$
From the latter, $\theta = 126° 52' = 2·214$ radians.

4.3. Conjugate complex numbers

The two complex numbers $x + yi$, $x - yi$ are said to be *conjugate*. If z denotes the complex number $x + yi$, it is convenient to denote the conjugate number by $\bar{z}$ so that
$$z = x + yi, \; \bar{z} = x - yi. \tag{4.12}$$
Immediate consequences are that
$$z + \bar{z} = x + yi + x - yi = 2x,$$
$$z\bar{z} = (x + yi)(x - yi) = x^2 + y^2, \text{ since } i^2 = -1.$$
Hence both *the sum and product of two conjugate complex numbers are real*. From the definition of the modulus of a complex number, it follows that
$$z\bar{z} = x^2 + y^2 = |z|^2 = |\bar{z}|^2. \tag{4.13}$$
Also if $z = 0$, it follows that
$$x + yi = 0 = 0 + 0i,$$
so that $x = y = 0$ and consequently $\bar{z}$ also vanishes.

The fact that the product of two conjugate complex numbers is real is useful in the reduction of an expression of the form $(x_1 + y_1 i)/(x_2 + y_2 i)$ into the form $u + vi$. If the numerator and denominator are multiplied by $x_2 - y_2 i$ (the conjugate of $x_2 + y_2 i$), the real quantity $x_2^2 + y_2^2$ is obtained as denominator.

Example 2. *Express* $\dfrac{3 + 2i}{1 - 4i}$ *in the form* $a + bi$.

$$\frac{3 + 2i}{1 - 4i} = \frac{(3 + 2i)(1 + 4i)}{(1 - 4i)(1 + 4i)} = \frac{3 + 12i + 2i + 8i^2}{1 - 16i^2}$$
$$= \frac{-5}{17} + \frac{14}{17}i \text{ (since } i^2 = -1\text{)}.$$

Example 3. *Complex numbers z_1 and z_2 are given by the formulae*

$$z_1 = R_1 + i\omega L, \quad z_2 = R_2 - \frac{i}{\omega C}$$

and z is given by

$$\frac{1}{z} = \frac{1}{z_1} + \frac{1}{z_2}.$$

Find the value of ω for which z is a real number. (O.C.)

$$\frac{1}{z_1} = \frac{1}{R_1 + i\omega L} = \frac{R_1 - i\omega L}{R_1^2 + \omega^2 L^2},$$

multiplying numerator and denominator by $R_1 - i\omega L$ (the conjugate of $R_1 + i\omega L$). Similarly

$$\frac{1}{z_2} = \frac{1}{R_2 - \dfrac{i}{\omega C}} = \frac{R_2 + \dfrac{i}{\omega C}}{R_2^2 + \dfrac{1}{\omega^2 C^2}}.$$

Hence,

$$\frac{1}{z} = \frac{1}{z_1} + \frac{1}{z_2} = \frac{R_1 - i\omega L}{R_1^2 + \omega^2 L^2} + \frac{R_2 + \dfrac{i}{\omega C}}{R_2^2 + \dfrac{1}{\omega^2 C^2}}$$

$$= \frac{R_1}{R_1^2 + \omega^2 L^2} + \frac{R_2}{R_2^2 + \dfrac{1}{\omega^2 C^2}} + i\left\{ \frac{\dfrac{1}{\omega C}}{R_2^2 + \dfrac{1}{\omega^2 C^2}} - \frac{\omega L}{R_1^2 + \omega^2 L^2} \right\}.$$

If z is to be a real number, the term in $\{\ \}$ must vanish; this requires that

$$\frac{\dfrac{1}{\omega C}}{R_2^2 + \dfrac{1}{\omega^2 C^2}} = \frac{\omega L}{R_1^2 + \omega^2 L^2},$$

from which it follows that

$$\omega = \left\{ \frac{L - R_1^2 C}{LC(L - R_2^2 C)} \right\}^{1/2}$$

EXERCISES 4 (a)

1. Show in the Argand diagram the representative points of the following complex numbers:

    (a) -1, (b) i, (c) $3 + 4i$, (d) $-i - \sqrt{3}$.

2. Find the modulus and the principal value of the argument of the four complex numbers of Exercise 1 above.

3. If z_1, z_2, z_3 are complex numbers, prove that

    (i) $|z_1 + z_2|^2 + |z_1 - z_2|^2 = 2|z_1|^2 + 2|z_2|^2$,

    (ii) $|2z_1 - z_2 - z_3|^2 + |2z_2 - z_3 - z_1|^2 + |2z_3 - z_1 - z_2|^2$
    $= 3\{|z_2 - z_3|^2 + |z_3 - z_1|^2 + |z_1 - z_2|^2\}$. (O.C.)

4. Prove that, if a quadratic equation with real coefficients has one root which is a complex number, the other is the conjugate complex number. Given that $x = 1 + 3i$ is one solution of the equation $x^4 + 16x^2 + 100 = 0$, find all the solutions. (O.C.)

5. Simplify

(i) $(2 + i)^2 + (2 - i)^2$,   (ii) $\dfrac{a + bi}{a - bi} - \dfrac{a - bi}{a + bi}$.

6. Express the complex number $\dfrac{3 + 5i}{2 - 3i}$ in the form $a + bi$ and show that the modulus of $\dfrac{(2 - 3i)(3 + 4i)}{(6 + 4i)(15 - 8i)}$ is $\dfrac{5}{34}$.

4.4. Some remarks on the manipulation of complex numbers

Complex numbers obey all the ordinary rules of algebra if i is treated as an ordinary number and use is made of the relation $i^2 = -1$. Thus it is possible to write

$$i^3 = i \times i^2 = -i, \; i^4 = i^2 \times i^2 = -1 \times -1 = 1,$$
$$i^5 = i^4 \times i = i, \text{ etc.,}$$

and these relations are often useful in the manipulation of complex numbers.

The equality of two complex numbers $a + bi$, $a' + b'i$ implies (from equation (4.1)) that $a = a'$ and $b = b'$. This implication is often stated in the form—*if two complex numbers are equal, their real and imaginary parts can be equated.*

Some worked examples which show the application of these remarks follow.

Example 4. *Simplify* $(3 + i)^4 - (3 - i)^4$.

The quantity in question

$$= 3^4 + 4 \cdot 3^3 i + 6 \cdot 3^2 i^2 + 4 \cdot 3 i^3 + i^4$$
$$- 3^4 + 4 \cdot 3^3 i - 6 \cdot 3^2 i^2 + 4 \cdot 3 i^3 - i^4$$
$$= 216i + 24i^3 = 216i - 24i = 192i.$$

Example 5. *Express* $\sqrt{(5 + 12i)}$ *in the form* $a + bi$.

If $\sqrt{(5 + 12i)} = a + bi$, then

$$5 + 12i = (a + bi)^2 = a^2 - b^2 + 2abi.$$

Equating the real and imaginary parts

$$a^2 - b^2 = 5, \; ab = 6.$$

The solution of these simultaneous algebraical equations for a and b gives

$$a = \pm 3, b = \pm 2, \text{ or } a = \pm 2i, b = \mp 3i$$

and both solutions lead to $\pm(3 + 2i)$ for the quantity $a + bi$.
Hence $\sqrt{(5 + 12i)} = \pm(3 + 2i)$.

4.5. The cube roots of unity

There is only one real number whose cube is unity but it is possible to find also two complex numbers which possess this property. Suppose that x is a cube root of unity, then $x^3 - 1 = 0$ and this equation can be written in the form
$$(x - 1)(x^2 + x + 1) = 0.$$
Thus either $x = 1$ or x satisfies the quadratic equation $x^2 + x + 1 = 0$. Using the ordinary formula for the roots of a quadratic, possible values of x are given by
$$x = \frac{-1 \pm \sqrt{(1 - 4)}}{2} = \frac{1}{2}(-1 \pm i\sqrt{3}).$$
Hence, in an algebra admitting complex numbers, each of the quantities
$$1,\ \tfrac{1}{2}(-1 + i\sqrt{3}),\ \tfrac{1}{2}(-1 - i\sqrt{3}),$$
is a cube root of unity. If ω is used to denote the complex root $\tfrac{1}{2}(-1 + i\sqrt{3})$, then
$$\omega^2 = \tfrac{1}{4}(-1 + i\sqrt{3})^2 = \tfrac{1}{4}(1 - 2i\sqrt{3} - 3) = \tfrac{1}{2}(-1 - i\sqrt{3}).$$
This is the other complex number satisfying the equation $x^3 - 1 = 0$, and the three cube roots of unity can be written in the form
$$1,\ \omega,\ \omega^2 \text{ where } \omega = \tfrac{1}{2}(-1 + i\sqrt{3}).$$
The reader should note that
$$\omega^3 = 1,\ \omega^4 = \omega \times \omega^3 = \omega,\ \omega^5 = \omega^2 \times \omega^3 = \omega^2,\ \text{etc.},$$
and that
$$1 + \omega + \omega^2 = 1 + \tfrac{1}{2}(-1 + i\sqrt{3}) + \tfrac{1}{2}(-1 - i\sqrt{3}) = 0; \quad (4.14)$$
these relations are often of considerable use when working examples.

Example 6. *It is given that ω is a complex cube root of unity and that a, b and c are real. Show that $(a + \omega b + \omega^2 c)^3$ is only real if a, b and c are not all different.* (C.)

Since $1 + \omega + \omega^2 = 0$, $\omega^2 = -1 - \omega$ and hence
$$(a + \omega b + \omega^2 c)^3 = (a + \omega b - c - \omega c)^3 = \{(a - c) + \omega(b - c)\}^3$$
$$= (a - c)^3 + 3(a - c)^2(b - c)\omega + 3(a - c)(b - c)^2\omega^2 + (b - c)^3,$$
the relation $\omega^3 = 1$ having been used in obtaining the last term on the right. Now ω and ω^2 denote respectively $\tfrac{1}{2}(-1 + i\sqrt{3})$, $\tfrac{1}{2}(-1 - i\sqrt{3})$, and these being conjugate complex numbers, $(a + \omega b + \omega^2 c)^3$ is real only if
$$(a - c)^2(b - c) = (a - c)(b - c)^2.$$
This can be written
$$(a - c)(b - c)\{(a - c) - (b - c)\} = 0,$$
i.e.
$$(a - c)(c - b)(b - a) = 0.$$
Hence $(a + \omega b + \omega^2 c)^3$ is real only if $a = b$, $b = c$, or $c = a$, i.e. if a, b and c are not all different.

EXERCISES 4 (b)

1. If $(x + yi)^4 = a + bi$, show that $a^2 + b^2 = (x^2 + y^2)^4$.
2. Find, in the form $a + bi$,
    (i) $\sqrt{(-5 + 12i)}$, (ii) $\sqrt{i}$.
3. Find two real numbers a and b such that
$$(1 + i)a + 2(1 - 2i)b - 3 = 0.$$
4. If ω is a complex cube root of unity, show that
$$(1 + \omega - \omega^2)^3 - (1 - \omega + \omega^2)^3 = 0.$$
5. If ω is a complex cube root of unity and if $x = a + b$, $y = a\omega + b\omega^2$, $z = a\omega^2 + b\omega$, show that
    (i) $xyz = a^3 + b^3$, (ii) $x^2 + y^2 + z^2 = 6ab$.
6. Show that
    (i) $a^2 + b^2 + c^2 - bc - ca - ab = (a + b\omega + c\omega^2)(a + b\omega^2 + c\omega)$,
    (ii) $a^3 + b^3 + c^3 - 3abc$
    $= (a + b + c)(a + b\omega + c\omega^2)(a + b\omega^2 + c\omega)$,
where ω is a complex cube root of unity.

4.6. Addition of complex numbers in the Argand diagram

Suppose P and Q (Fig. 5) are the representative points in the Argand diagram of the complex numbers $z_1 = x_1 + y_1 i$ and $z_2 = x_2 + y_2 i$. If R is the remaining vertex of the parallelogram of which OP and OQ are adjacent sides, it is a matter of elementary geometry to show that the coordinates of the point R are $(x_1 + x_2, y_1 + y_2)$. Hence R is the representative point of the complex number $x_1 + x_2 + (y_1 + y_2)i$, i.e. of the number $z_1 + z_2$. If QO is produced to a point T such that $TO = OQ$, the coordinates of T will be $(-x_2, -y_2)$ and T will be the representative point of the complex number $-z_2$. If now the parallelogram $OTSP$ be completed, S will be the representative point of the complex number $z_1 + (-z_2)$, i.e. of $z_1 - z_2$.

The above constructions give a useful method of fixing in the Argand diagram the points representing the sum or difference of two given complex numbers and the following important result can be deduced. Since the moduli of z_1, z_2, $z_1 + z_2$ are given respectively by the lengths OP, OQ (which is clearly equal in length to RP), OR and since the length of OR is not greater than the sum of the lengths of OP, RP it follows that

$$|z_1 + z_2| \leqslant |z_1| + |z_2| \qquad (4.15)$$

and hence that *the modulus of the sum of two complex numbers is not greater than the sum of the moduli of the separate numbers.* The sign of equality occurs when Q lies on OP, that is, when the arguments of the

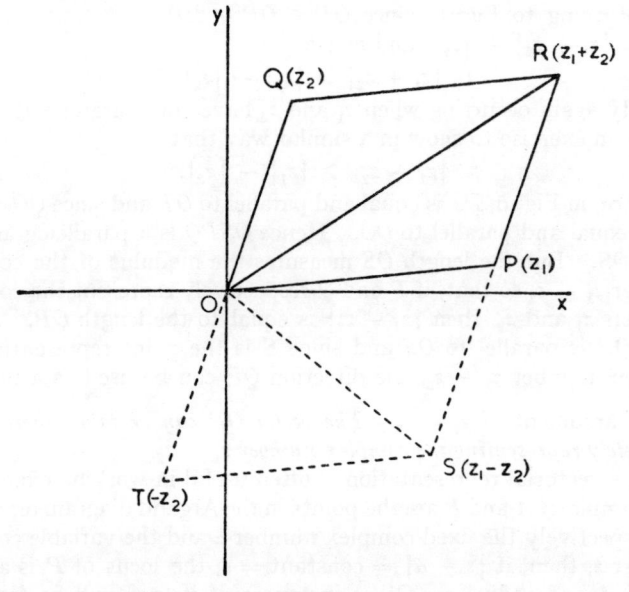

Fig. 5

two numbers z_1, z_2 are equal. The important result (4.15) is the generalisation for a wider class of numbers of the result (3.4) of Chapter 3.

The result (4.15) can be proved without any reference to the Argand diagram and therefore without any appeal to geometry as follows. The sum of the two complex numbers $z_1 = x_1 + y_1 i$, $z_2 = x_2 + y_2 i$ is $z_1 + z_2 = x_1 + x_2 + (y_1 + y_2)i$ and its modulus is $\{(x_1 + x_2)^2 + (y_1 + y_2)^2\}^{\frac{1}{2}}$, or

$$\{(x_1^2 + y_1^2) + (x_2^2 + y_2^2) + 2(x_1 x_2 + y_1 y_2)\}^{1/2}.$$

But

$$\begin{aligned}\{|z_1| + |z_2|\}^2 &= \{(x_1^2 + y_1^2)^{1/2} + (x_2^2 + y_2^2)^{1/2}\}^2 \\ &= (x_1^2 + y_1^2) + (x_2^2 + y_2^2) \\ &\quad + 2(x_1^2 + y_1^2)^{1/2}(x_2^2 + y_2^2)^{1/2} \\ &= (x_1^2 + y_1^2) + (x_2^2 + y_2^2) \\ &\quad + 2\{(x_1 x_2 + y_1 y_2)^2 + (x_1 y_2 - x_2 y_1)^2\}^{1/2}\end{aligned}$$

and this latter expression is greater than (or at least equal to)

$$(x_1^2 + y_1^2) + (x_2^2 + y_2^2) + 2(x_1 x_2 + y_1 y_2).$$

The result (4.15) then follows and it can be further deduced (by induction) that the modulus of *any* number of complex quantities cannot be greater than the sum of their moduli.

Returning to Fig. 5, since $OP \leqslant OR + RP$, $|z_1| \leqslant |z_1 + z_2| + |z_2|$, and hence

$$|z_1 + z_2| \geqslant |z_1| - |z_2|, \qquad (4.16)$$

equality again occurring when z_1 and z_2 have equal arguments. It is left as an exercise to show in a similar way that

$$|z_1 - z_2| \geqslant |z_1| - |z_2|.$$

Since, in Fig. 5, PS is equal and parallel to OT and since $QO = OT$, PS is equal and parallel to QO. Hence $OSPQ$ is a parallelogram and $QP = OS$. But the length OS measures the modulus of the complex number $z_1 - z_2$ so that, if P and Q respectively represent the complex numbers z_1 and z_2, then $|z_1 - z_2|$ is equal to the length QP. Again, since QP is parallel to OS and since S is the point representing the complex number $z_1 - z_2$, the direction QP can be used as a measure of the argument of $z_1 - z_2$. *The vector $\overrightarrow{QP}$ can be taken therefore as completely representing the complex number $z_1 - z_2$.*

This vectorial representation is often useful in working exercises— for example, if A and P are the points in the Argand diagram representing respectively the fixed complex number a and the variable complex number z, then, if $|z - a| = \text{constant} = c$, the locus of P is a circle centre A and radius c. Other instances of its use will be found in Examples 8, 9 and 10 of this chapter.

Example 7. *If a is a real and b is a complex number, prove that the points in the Argand diagram which represent the complex numbers $az_1 + b$, $az_2 + b$, $az_3 + b$ form a triangle similar to that formed by the points representing z_1, z_2, z_3.*

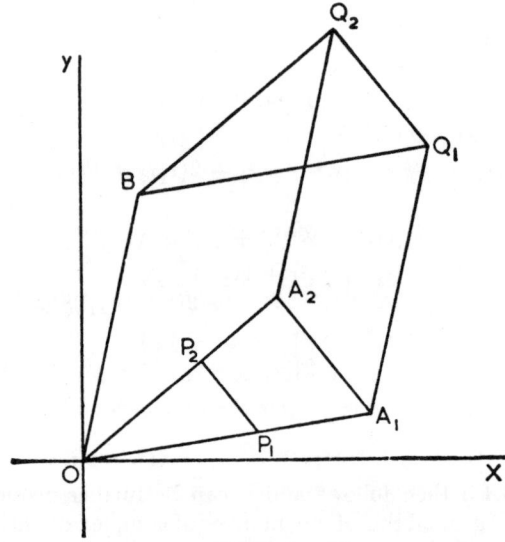

Fig. 6

Show also that the points representing the roots of the cubic $(az + b)^3 = 1$ *form an equilateral triangle and calculate the length of its side.* (O.)

In Fig. 6, P_1, P_2 represent the numbers z_1, z_2 and B represents the number b. The points A_1, A_2 representing the numbers az_1, az_2 will lie on OP_1, OP_2 respectively such that $\dfrac{OP_1}{OA_1} = \dfrac{OP_2}{OA_2} = \dfrac{1}{a}$. Hence A_1A_2 is parallel to P_1P_2 and $A_1A_2 = aP_1P_2$.

If Q_1 represents the number $az_1 + b$, its position will be given by completing the parallelogram OA_1Q_1B and hence A_1Q_1 is equal to and parallel to OB. Similarly, if Q_2 represents the number $az_2 + b$, A_2Q_2 is equal to and parallel to OB. Hence Q_1Q_2 is equal to and parallel to A_1A_2 and therefore parallel to P_1P_2 and equal in length to $a \cdot P_1P_2$. Similarly, if P_3, Q_3 represent respectively the numbers z_3, $az_3 + b$, Q_2Q_3, Q_3Q_1 will be parallel respectively to P_2P_3, P_3P_1 and equal in length to $a \cdot P_2P_3$, $a \cdot P_3P_1$. Hence the triangle $Q_1Q_2Q_3$ will be similar to the triangle $P_1P_2P_3$, the sides of the former triangle being a times those of the latter.

The roots of the equation $(az + b)^3 = 1$ are given by $az + b = 1, \omega, \omega^2$ where ω is a complex cube root of unity. Hence the roots are

$$\frac{1-b}{a}, \frac{\omega-b}{a}, \frac{\omega^2-b}{a}$$

and the points in the Argand diagram representing these roots will form a triangle similar to that formed by the points representing the complex

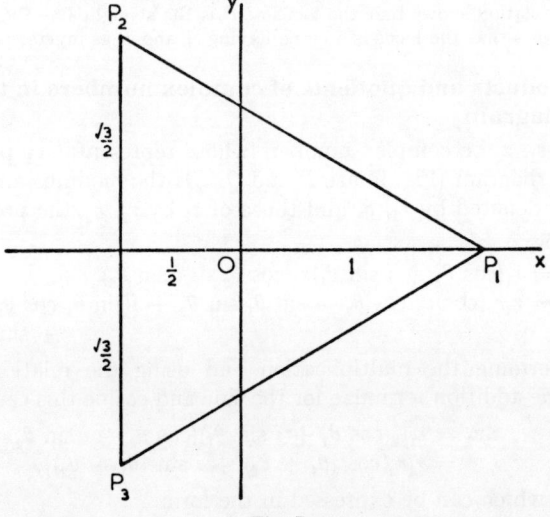

Fig. 7

numbers $1, \omega, \omega^2$, the sides of the former triangle being $1/a$ times those of the latter. The points representing the complex numbers $1, \omega, \omega^2$, i.e. $1 + 0i$, $\tfrac{1}{2}(-1 + i\sqrt{3})$, $\tfrac{1}{2}(-1 - i\sqrt{3})$ are shown as P_1, P_2, P_3 in Fig. 7. It is easy to see that the triangle $P_1P_2P_3$ is equilateral and of side $\sqrt{3}$, so that the triangle formed by the points representing the roots of the given equation is equilateral and of side $\sqrt{3}/a$.

Example 8. *When z is a variable complex number and k is a positive real constant, show that the locus*

$$|z - 1| = k |z + 1|$$

is a circle having $z = 1$ and $z = -1$ as inverse points if $k \neq 1$ and is a straight line if $k = 1$. (O.)

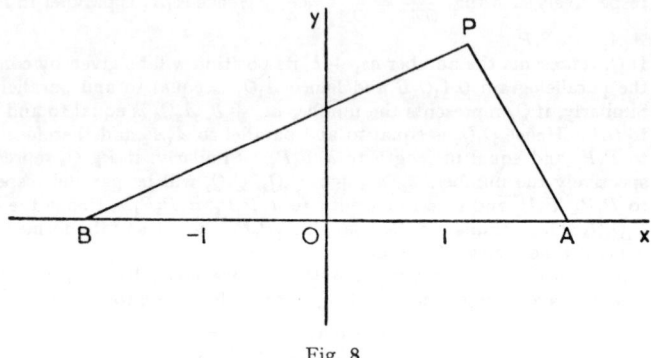

Fig. 8

Let P represent (Fig. 8) the complex number z and let A and B represent the numbers 1 and -1. Then $|z - 1|$ is equal to the length AP and $|z + 1|$ is equal to BP. Hence the point P moves so that $PA = k.PB$. If $k = 1$, this relation shows that the locus of P is the straight line Oy, and if $k \neq 1$ it shows that the locus is a circle having A and B as inverse points.

4.7. Products and quotients of complex numbers in the Argand diagram

Let z_1, z_2 be complex numbers whose representative points in the Argand diagram (Fig. 9) are P and Q. If the modulus and argument of z_1 be denoted by r_1, θ_1 and those of z_2 by r_2, θ_2, the product $z_1 z_2$ is given by

$$z_1 z_2 = r_1(\cos \theta_1 + i \sin \theta_1) r_2(\cos \theta_2 + i \sin \theta_2)$$
$$= r_1 r_2 \{\cos \theta_1 \cos \theta_2 - \sin \theta_1 \sin \theta_2 + i(\sin \theta_1 \cos \theta_2 + \sin \theta_2 \cos \theta_1)\},$$

on performing the multiplication and using the relation $i^2 = -1$. Using the addition formulae for the sine and cosine this can be written

$$z_1 z_2 = r_1 r_2 (\cos \theta_1 + i \sin \theta_1)(\cos \theta_2 + i \sin \theta_2)$$
$$= r_1 r_2 \{\cos (\theta_1 + \theta_2) + i \sin (\theta_1 + \theta_2)\}, \qquad (4.17)$$

a result which can be expressed in the form

$$|z_1 z_2| = |z_1| \cdot |z_2|, \; \arg(z_1 z_2) = \arg(z_1) + \arg(z_2). \qquad (4.18)$$

Thus *the modulus of the product of two complex numbers is equal to the product of their moduli* and *the argument of the product is equal to the sum of the arguments*; the reader should note, however, that the second of these results is not necessarily true of the principal values of the arguments (for example, $\theta_1 + \theta_2$ may exceed π).

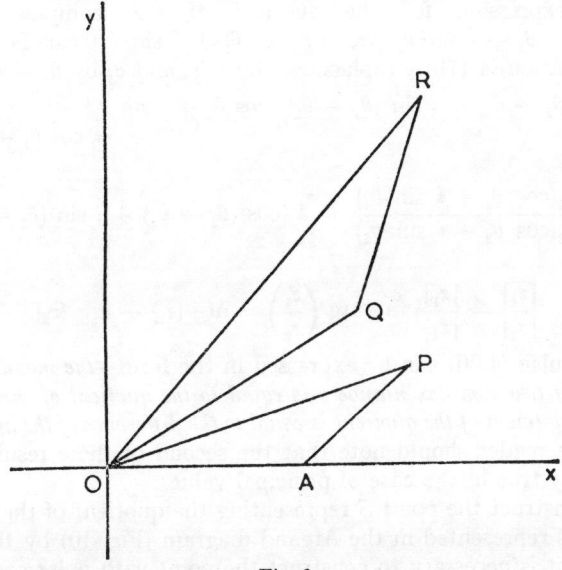

Fig. 9

To construct the point R representing the product of the complex quantities represented in the Argand diagram by the points P and Q it is necessary to construct the point with polar coordinates $(r_1 r_2, \theta_1 + \theta_2)$. This can be done by taking A as the point $(1, 0)$ and drawing the triangle OQR directly similar to the triangle OAP; for with this construction,

$$\frac{OR}{OQ} = \frac{OP}{OA} \text{ giving } \frac{OR}{r_2} = \frac{r_1}{1}, \text{ i.e. } OR = r_1 r_2,$$

and, angle xOR = angle xOQ + angle QOR
= angle xOQ + angle $AOP = \theta_2 + \theta_1$.

Hence R has polar coordinates $(r_1 r_2, \theta_1 + \theta_2)$.

It is worth noticing that repeated applications of formula (4.17) give, for the product of n complex numbers $z_1, z_2, \ldots, z_n$,

$$z_1 z_2 \ldots z_n = r_1 r_2 \ldots r_n \{\cos(\theta_1 + \theta_2 + \ldots + \theta_n) + i \sin(\theta_1 + \theta_2 + \ldots + \theta_n)\},$$

where the modulus and argument of z_s are respectively r_s, θ_s. By putting $r_1 = r_2 = \ldots = r_n = 1$, $\theta_1 = \theta_2 = \ldots = \theta_n = \theta$, so that $z_1 = z_2 = \ldots = z_n = \cos \theta + i \sin \theta$, it follows that

$$(\cos \theta + i \sin \theta)^n = \cos n\theta + i \sin n\theta \tag{4.19}$$

where n is any positive integer. This is a special case of a general theorem which is discussed more fully in Chapter 6.

An expression for the quotient of two complex numbers $z_1 = r_1(\cos\theta_1 + i\sin\theta_1)$, $z_2 = r_2(\cos\theta_2 + i\sin\theta_2)$ can be obtained from equation (4.17) by replacing r_1 by r_1/r_2 and θ_1 by $\theta_1 - \theta_2$ so that
$$\{\cos(\theta_1 - \theta_2) + i\sin(\theta_1 - \theta_2)\}(\cos\theta_2 + i\sin\theta_2)$$
$$= \cos\theta_1 + i\sin\theta_1.$$
Hence
$$\frac{z_1}{z_2} = \frac{r_1(\cos\theta_1 + i\sin\theta_1)}{r_2(\cos\theta_2 + i\sin\theta_2)} = \frac{r_1}{r_2}\{\cos(\theta_1 - \theta_2) + i\sin(\theta_1 - \theta_2)\},$$
and thus
$$\left|\frac{z_1}{z_2}\right| = \frac{|z_1|}{|z_2|} \text{ and } \arg\left(\frac{z_1}{z_2}\right) = \arg(z_1) - \arg(z_2). \quad (4.20)$$

The formulae (4.20) can be expressed in the form—*the modulus of the quotient of two complex numbers is equal to the quotient of their moduli* and *the argument of the quotient is equal to the difference of the arguments*. Again the reader should note that the second of these results is not necessarily true in the case of principal values.

To construct the point S representing the quotient of the complex quantities represented in the Argand diagram (Fig. 10) by the points P and Q it is necessary to construct the point with polar coordinates

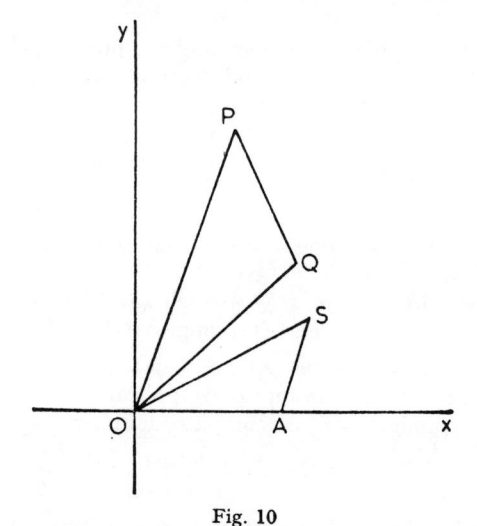

Fig. 10

$(r_1/r_2, \theta_1 - \theta_2)$. This can be done by taking A as the point $(1, 0)$ and drawing the triangle OAS directly similar to the triangle OQP; with this construction

$$\frac{OS}{OA} = \frac{OP}{OQ} \text{ giving } \frac{OS}{1} = \frac{r_1}{r_2}, \text{ i.e. } OS = \frac{r_1}{r_2},$$

ARGAND DIAGRAM

and

angle xOS = angle QOP = angle xOP − angle xOQ = $\theta_1 - \theta_2$,
so that the point S has the required polar coordinates.

Example 9. *In the Argand diagram the points P, Q, R represent respectively the complex numbers z_1, z_2 and $(1-t)z_1 + tz_2$.*

(i) *if t is real, R lies on the line PQ and is such that*

$$\frac{PR}{RQ} = \frac{t}{1-t};$$

(ii) *if t is not real, PQR is a triangle whose sides PR, RQ, QP are in the ratios $|t|:|1-t|:1$.*

Show also that in the case (ii) the triangle PQR is directly similar to the triangle OIT, O being the origin and I, T the representative points of the numbers 1 and t.

(i) If the real and imaginary parts of z_1, z_2 are x_1, y_1 and x_2, y_2 respectively, P and Q have coordinates (x_1, y_1), (x_2, y_2) and R is the point with coordinates $\{(1-t)x_1 + tx_2, (1-t)y_1 + ty_2\}$ so that, when t is real, R lies on PQ and divides it in the ratio $t:1-t$.

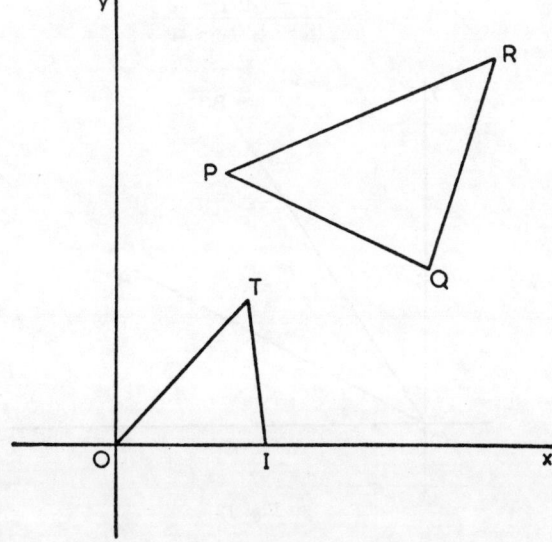

Fig. 11

(ii) Let t be complex. Since R represents the number $(1-t)z_1 + tz_2$ and P the number z_1, the vector $\overrightarrow{PR}$ represents (in the sense that the length PR is a measure of the modulus and that the direction of PR is a measure of the argument) the difference of these numbers. Hence $\overrightarrow{PR}$ represents the number $t(z_2 - z_1)$.

Similarly, the vector $\overrightarrow{PQ}$ represents the complex number $(z_2 - z_1)$.

Since the modulus of the product of two complex numbers is equal to the product of the moduli and the argument of the product is the sum of the arguments, the vector $\overrightarrow{PR}$ is obtained from the vector $\overrightarrow{PQ}$ by turning the latter through an angle arg t and multiplying its length by $|t|$.

Similarly the vector $\overrightarrow{QR}$ represents $(1-t)z_1 + tz_2 - z_2 = (1-t)(z_1 - z_2)$ and the vector $\overrightarrow{QP}$ represents $(z_1 - z_2)$, $\overrightarrow{QR}$ is obtained from $\overrightarrow{QP}$ by turning the latter through an angle arg $(1-t)$ and multiplying its length by $|1-t|$. The ratio of the sides PR, RQ, QP are therefore in the ratios $|t| : |1-t| : 1$. Since I, T represent the complex numbers 1 and t, the vectors $\overrightarrow{OI}, \overrightarrow{IT}, \overrightarrow{TO}$ represent the numbers $1, t-1, -t$ while $\overrightarrow{PQ}, \overrightarrow{QR}, \overrightarrow{RP}$ represent $z_2 - z_1, (1-t)(z_1 - z_2), -t(z_2 - z_1)$. Hence the triangles PQR, OIT are directly similar.

Example 10. *If a, b are the complex numbers represented by points A, B in the Argand diagram, what geometrical quantities correspond to the modulus and argument of b/a?*

Show that, if the four points representing the complex numbers z_1, z_2, z_3, z_4 are concyclic, the fraction

$$\frac{(z_1 - z_2)(z_3 - z_4)}{(z_3 - z_2)(z_1 - z_4)}$$

must be real. (O.)

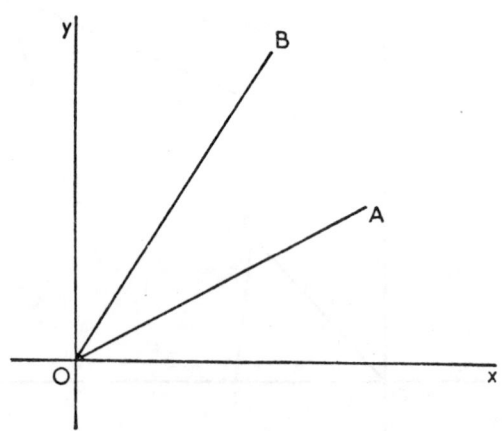

Fig. 12

In Fig. 12 with origin O, the lengths of the vectors $\overrightarrow{AO}, \overrightarrow{BO}$ are equal respectively to $|a|$ and $|b|$ and the angles xOA, xOB are equal to arg a and arg b.

Since
$$\left|\frac{b}{a}\right| = \frac{|b|}{|a|} = \frac{OB}{OA},$$

the modulus of b/a corresponds to the ratio OB/OA. Again, since

$$\arg\left(\frac{b}{a}\right) = \arg(b) - \arg(a) = \text{angle } xOB - \text{angle } xOA$$
$$= \text{angle } AOB,$$

the argument of b/a corresponds to the angle AOB.
Let the representative points of the four complex numbers z_1, z_2, z_3, z_4 be P, Q, R, S. If these four points are concyclic, either Fig. 13 (a) or 13 (b) will apply.

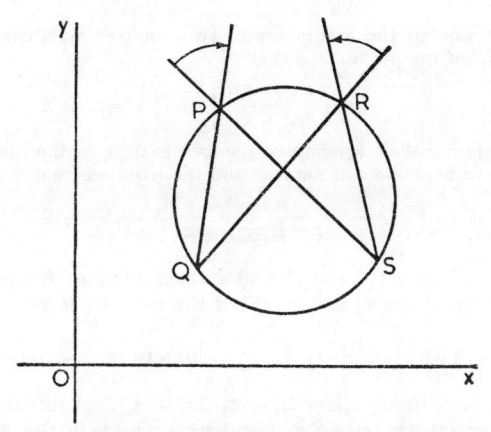

Fig. 13 (a)

In Fig. 13 (a), the angle of turn from the vector $\vec{SP}$ to the vector $\vec{QP}$ is equal in magnitude but opposite in sign to the angle of turn from the vector $\vec{QR}$ to the vector $\vec{SR}$ and the sum of these angles is zero. In Fig. 13 (b),

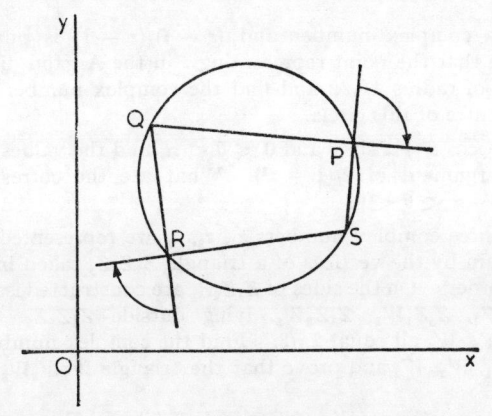

Fig. 13 (b)

the sum of the angles is π. Now the angle of turn from the vector $\vec{SP}$ to the vector $\vec{QP}$ is, since the vectors $\vec{SP}$, $\vec{QP}$ represent respectively the complex numbers $z_4 - z_1$, $z_2 - z_1$,

$$\arg\left(\frac{z_2 - z_1}{z_4 - z_1}\right).$$

Similarly the angle of turn from $\overrightarrow{QR}$ to $\overrightarrow{SR}$ is $\arg\left(\dfrac{z_4 - z_3}{z_2 - z_3}\right)$. Hence

$$\arg\left(\frac{z_2 - z_1}{z_4 - z_1}\right) + \arg\left(\frac{z_4 - z_3}{z_2 - z_3}\right) = 0 \text{ or } \pi.$$

But the sum of the arguments of two complex numbers is equal to the argument of the product, so that

$$\arg\left\{\frac{(z_2 - z_1)(z_4 - z_3)}{(z_4 - z_1)(z_2 - z_3)}\right\} = 0 \text{ or } \pi.$$

If the argument of a complex number is 0 or π, the number is either a positive or negative real number and the required result follows.

EXERCISES 4 (c)

1. Two fixed points A and B, and a variable point P represent the complex numbers z_1, z_2 and z. Find the locus of P if

    (i) $\left|\dfrac{z - z_1}{z - z_2}\right| = 3$, (ii) $\arg(z - z_1) = \arg z_2$.

2. If $(z_2 - z_3)/(z_3 - z_1) = (z_3 - z_1)/(z_1 - z_2)$, show that the points representing the complex numbers z_1, z_2, z_3 in the Argand diagram form an equilateral triangle.

3. The complex numbers a, b, c are of unit modulus, and correspond to the points A, B, C in the Argand diagram. Prove that, if $h = a + b + c$, and h corresponds to the point H, then AH is perpendicular to BC and H is the orthocentre of the triangle ABC.
(O.)

4. z is a complex number and $(z - i)/(z - 1)$ is purely imaginary. Prove that the point representing z in the Argand diagram lies on a circle of radius $1/\sqrt{2}$ and find the complex number represented by the centre of this circle.

5. If $z = \cos\theta + i\sin\theta$ and $0 < \theta < \pi$, find the values of the modulus and argument of $2/(1 - z^2)$. What are the corresponding values when $-\pi < \theta < 0$?

6. The three complex numbers z_1, z_2, z_3 are represented in the Argand diagram by the vertices of a triangle $Z_1Z_2Z_3$ taken in counter clockwise order. On the sides of $Z_1Z_2Z_3$ are constructed isosceles triangles $Z_2Z_3W_1$, $Z_3Z_1W_2$, $Z_1Z_2W_3$, lying outside $Z_1Z_2Z_3$. The angles at W_1, W_2, W_3 all equal $2\pi/3$. Find the complex numbers represented by W_1, W_2, W_3 and prove that the triangle $W_1W_2W_3$ is equilateral.
(C.)

[Hint—use the results of Example 9 (ii) and Exercise 4 (c), 2.]

4.8. Rational functions of the complex variable

A *rational function of the complex variable* z ($= x + yi$) is defined, in the same way as a rational function of the real variable (§ 1.7), as the quotient of two polynomials in z. Suppose that $R(z)$ is such a function and $P(z)$, $Q(z)$ are the two polynomials to whose quotient it is equivalent. The polynomial $P(z) = P(x + yi)$ can be reduced, by the actual expansion of its several terms, to the form $A + Bi$ where A, B are polynomials in x and y with real coefficients; $Q(z)$ can similarly be reduced to the form $C + Di$. Hence

$$R(z) = \frac{P(x+yi)}{Q(x+yi)} = \frac{A+Bi}{C+Di}$$
$$= \frac{(A+Bi)(C-Di)}{(C+Di)(C-Di)} = \frac{AC+BD}{C^2+D^2} + \frac{BC-AD}{C^2+D^2}i.$$

Since A, B, C, D are polynomials in x and y,

$$\frac{AC+BD}{C^2+D^2} \quad \text{and} \quad \frac{BC-AD}{C^2+D^2}$$

are each rational functions of x and y. Thus *any rational function of $z (= x + yi)$ can be reduced to the form $X + Yi$ where X and Y are rational functions of x and y with real coefficients.*

An important theorem in the theory of functions of the complex variable is:

If $R(x + yi) = X + Yi$, where R is a rational function of $x + yi$ with REAL coefficients, then $R(x - yi) = X - Yi$.

The theorem is easily verified when the function R is a simple power $(x + yi)^n$ by actual expansion. It follows by addition that the theorem is true when R is a polynomial with real coefficients. Using the notation of the previous paragraph, the working being as before but with the sign of i changed,

$$R(x-yi) = \frac{A-Bi}{C-Di} = \frac{AC+BD}{C^2+D^2} - \frac{BC-AD}{C^2+D^2}i,$$

and the theorem follows for a rational function with real coefficients.

It follows from this theorem that if $x + yi$ is a root of the equation

$$a_0 z^n + a_1 z^{n-1} + \ldots + a_n = 0$$

in which the coefficients $a_0, a_1, \ldots, a_n$ are all real, then so also is $x - yi$. This important result can be stated thus: *in an equation with real coefficients, complex roots occur in conjugate pairs.*

4.9. Transformations

Suppose that two complex variables $w (= u + vi)$ and $z (= x + yi)$ are connected by a functional relation. For example, suppose that

$$w = z + b. \tag{4.21}$$

Let P be the point representing z in one Argand diagram (called the z diagram or the z-plane) and let P' be the point representing w in a second Argand diagram (called the w diagram or the w-plane). As z varies, w varies in accordance with equation (4.21) and P, P' will vary their positions in their respective diagrams. If P describes a locus in the z-plane, P' will describe a related locus in the w-plane. Since $w = u + vi$, $z = x + yi$, and if b is the complex number $a + \beta i$, equating real and imaginary parts in (4.21) leads to

$$u = x + a, v = y + \beta. \qquad (4.22)$$

To every value of z corresponds one value of w and conversely; there is a *one to one* correspondence between the point P (x, y) of the z-plane and the point P' (u, v) of the w-plane. A passage of this kind from a figure in the z-plane to a related figure in the w-plane is called a *transformation*. In the particular transformation given by the relation (4.21), the relations (4.22) between the coordinates (x, y), (u, v) of the related points P, P' follow and these show that the figure in the w-plane is the same in size, shape and orientation as that in the z-plane,

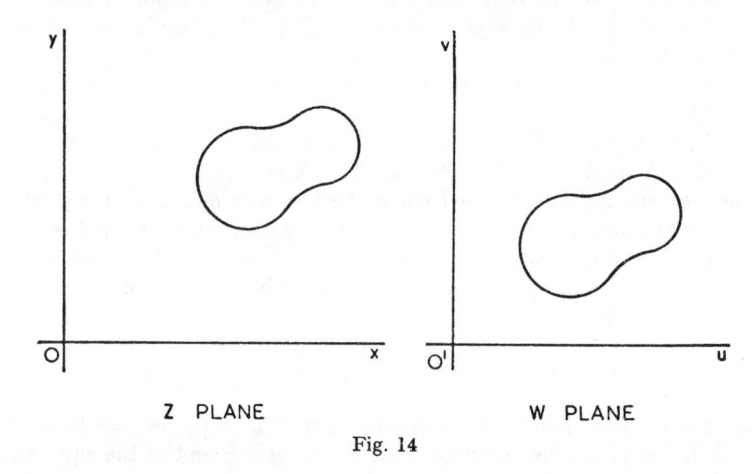

Z PLANE   W PLANE

Fig. 14

but it is moved through a distance a to the left and through a distance β downwards (Fig. 14). The transformation $w = z + b$ is accordingly called a *translation*.

The transformation given by

$$w = rz, \qquad (4.23)$$

where r is a positive real number is called a *magnification*, since in this case, equating real and imaginary parts leads to

$$u = rx, \ v = ry, \qquad (4.24)$$

and any figure in the z-plane described by the point (x, y) will give a figure in the w-plane described by the point (u, v) such that the two figures are similar and similarly situated about their respective origins O, O' but the scale of the figure in the w-plane is r times that of the figure in the z-plane (Fig. 15).

Z PLANE   W PLANE

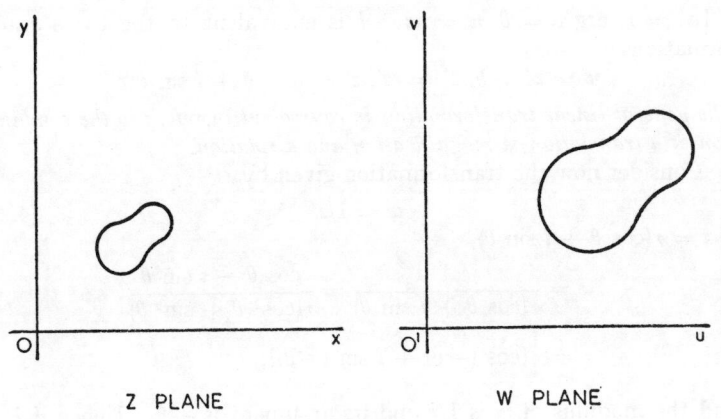

Fig. 15

The transformation
$$w = (\cos\theta + i\sin\theta)z, \qquad (4.25)$$
leads, by equating real and imaginary parts, to
$$u = x\cos\theta - y\sin\theta,\ v = x\sin\theta + y\cos\theta, \qquad (4.26)$$
showing that this transformation is equivalent to *rotating* a figure in the z-plane about the origin through an angle θ in the anti-clockwise direction (Fig. 16).

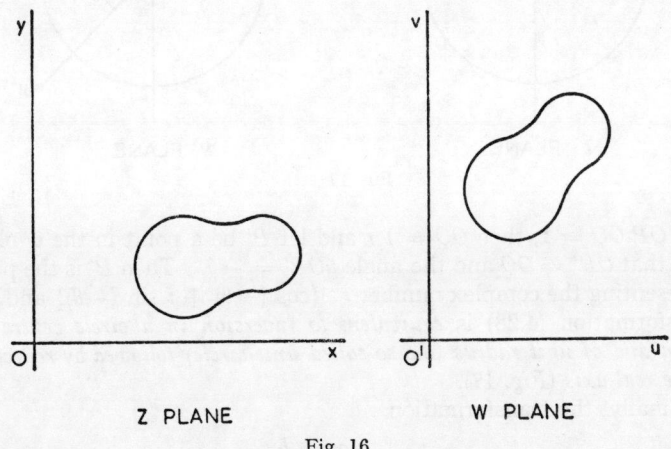

Z PLANE   W PLANE

Fig. 16

The so-called *general linear transformation*
$$w = az + b, \qquad (4.27)$$
is a combination of the three transformations considered above. For

if $|a| = r$, arg $a = \theta$, $w = az + b$ is equivalent to the three transformations
$$w = z'' + b,\ z'' = rz',\ z' = (\cos \theta + i \sin \theta)z.$$
The general linear transformation is equivalent therefore to the combination of a translation, a magnification and a rotation.

Consider now the transformation given by
$$w = 1/z. \qquad (4.28)$$
If $z = r(\cos \theta + i \sin \theta)$,
$$w = \frac{1}{r(\cos \theta + i \sin \theta)} = \frac{\cos \theta - i \sin \theta}{r(\cos^2 \theta + \sin^2 \theta)}$$
$$= \frac{1}{r}\{\cos(-\theta) + i \sin(-\theta)\},$$

and the modulus of w is $1/r$ and its argument is $-\theta$. Hence if P is the point representing the complex number $r(\cos \theta + i \sin \theta)$ in the z-plane, $OP = r$ and the angle $xOP = \theta$. Take a point Q on OP such

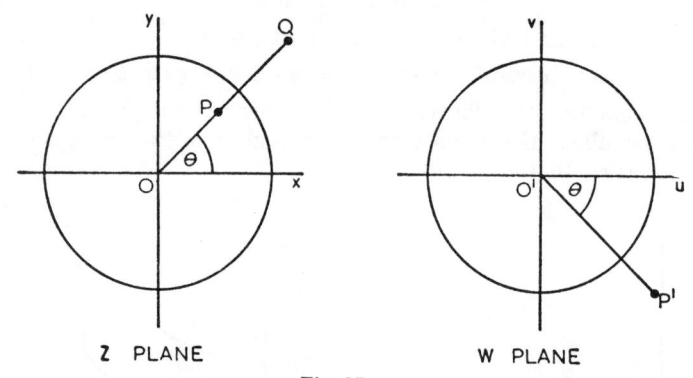

Z PLANE   W PLANE

Fig. 17

that $OP.OQ = 1$, then $OQ = 1/r$ and let P' be a point in the w-plane such that $OP' = OQ$ and the angle $uOP' = -\theta$. Then P' is the point representing the complex number $r^{-1}\{\cos(-\theta) + i \sin(-\theta)\}$ and the transformation (4.28) is *equivalent to inversion in a circle centre the origin and of unit radius (the so-called unit circle) followed by reflection in the real axis* (Fig. 17).

Finally, the transformation
$$w = \frac{az + b}{cz + d}, \qquad (4.29)$$
usually called the *general bilinear transformation*, can be considered as a combination of the separate transformations considered above. It is equivalent to the combination of the three transformations

4] TRANSFORMATIONS 99

$$w = \frac{a}{c} + \left(\frac{bc - ad}{c}\right) z'', \; z'' = \frac{1}{z'}, \; z' = cz + d.$$

Since inversion transforms circles into circles and straight lines into straight lines and since translations, rotations and magnifications leave circles as circles and straight lines as straight lines, the *general bilinear transformation transforms circles into circles and straight lines into straight lines*. This general result does not, however, give precise details of the corresponding regions of the two diagrams—it does not, for instance, give any details such as the centre and radius of the circle of the w-plane into which a given circle of the z-plane transforms. For such details, it is necessary to work on the lines shown in the following examples.

Example 11. *Prove that the bilinear transformation $w = (2z + i)/(2 - iz)$ transforms the interior of the unit circle in the z-plane into the interior of the unit circle in the w-plane.*

For all points on the unit circle in the z-plane, $|z| = 1$ and any such point is given by $z = \cos\theta + i\sin\theta$. The corresponding points in the w-plane are given by

$$|w| = \left|\frac{2z + i}{2 - iz}\right| = \frac{|2z + i|}{|2 - iz|} = \frac{|2\cos\theta + i(2\sin\theta + 1)|}{|2 + \sin\theta - i\cos\theta|}$$

$$= \left\{\frac{4\cos^2\theta + (2\sin\theta + 1)^2}{(2 + \sin\theta)^2 + \cos^2\theta}\right\}^{1/2} = \left\{\frac{4 + 4\sin\theta + 1}{4 + 4\sin\theta + 1}\right\}^{1/2} = 1,$$

and they therefore lie on the unit circle in the w-plane.

An alternative, and possibly more elegant method of showing that $|w| = 1$ when $|z| = 1$ is to use the notation of conjugate complex quantities and the result (4.13). Thus

$$|w|^2 = w\bar{w} = \left(\frac{2z + i}{2 - iz}\right)\left(\frac{2\bar{z} - i}{2 + i\bar{z}}\right) = \frac{4z\bar{z} + 1 + 2i(\bar{z} - z)}{4 + z\bar{z} + 2i(\bar{z} - z)} = 1,$$

as $z\bar{z} = |z|^2 = 1$.

So far it has been shown only that the unit circles transform into one another. To show that an interior point z_0 of the unit circle $|z| = 1$ transforms into an interior point w_0 of the circle $|w| = 1$, it has to be shown that if $|z_0| < 1$ and $w_0 = (2z_0 + i)/(2 - iz_0)$, then $|w_0| < 1$. This can be done as follows: let z_0, z_1 be two interior points of the circle $|z| = 1$, then $|z_0| < 1$ and $|z_1| < 1$. Let the transformed points be w_0, w_1 and let $|w_0| < 1$. Suppose, however, that $|w_1| > 1$ and join the points w_0, w_1 by some curve C'. Since the points w_0, w_1 lie respectively inside and outside the unit circle in the w-plane, C' must cut this circle at some point w_2. If C is the unique curve in the z-plane corresponding to C' in the w-plane, the point z_2 corresponding to w_2 must lie on the unit circle of the z-plane and hence the curve C must cut this circle. This is true for every curve joining w_0, w_1 and it follows that every curve joining z_0, z_1 must cut the circle. This is clearly false as z_0, z_1 are both interior points and hence a contradiction arises. Hence $|w_1| < 1$ and the proof that interior points correspond is complete, provided one point can be found such that $|w_0| < 1$. This can certainly be done by taking $z_0 = 0$, so that $w_0 = (2z_0 + i)/(2 - iz_0)$ $= i/2$, and $|w_0| = \frac{1}{2} < 1$. It should be noted that the centres of the two circles are not corresponding points since the centres of the two circles are the points for which z and w are both zero and it has been shown that corresponding points are given by $z = 0$, $w = i/2$.

Example 12. *If $w = z/(iz + 1)$ and the locus of the point representing w is a straight line, show that the locus of the point representing z is, in general, a circle. In what circumstances is the locus of the point representing z a straight line?*

Let $w = u + vi$, $z = x + yi$, so that

$$u + vi = \frac{x + yi}{1 - y + ix} = \frac{(x + yi)(1 - y - ix)}{(1-y)^2 + x^2}$$

$$= \frac{x + i(y - y^2 - x^2)}{(1-y)^2 + x^2},$$

and hence $\quad u = \dfrac{x}{(1-y)^2 + x^2}, \quad v = \dfrac{y - y^2 - x^2}{(1-y)^2 + x^2}.$

If the locus of the point representing w is a straight line $v = mu + c$, the corresponding locus in the z-plane is given by

$$\frac{y - y^2 - x^2}{(1-y)^2 + x^2} = \frac{mx}{(1-y)^2 + x^2} + c,$$

which reduces to

$$(c + 1)(x^2 + y^2) + mx - (2c + 1)y + c = 0.$$

This is the equation to a circle except when $c = -1$ in which case it is that to a straight line. When $c = -1$, the line in the w-plane becomes $v = mu - 1$; this line passes through the point $(0, -1)$, that is, through the point representing $-i$ in the w-plane.

Only the bilinear transformation has been discussed here. Other relations between two complex variables w and z have important applications, particularly in mathematical physics. Certain transformations involving transcendental functions will be touched upon later (Chapter 6) and below a single example is given of a transformation of the type $w^m = z$. Transformations of this type differ from those so far considered in that a one to one correspondence between points in the w- and z-planes no longer exists and it can be shown, but this is not attempted here, that to each point in the z-plane correspond m points in the w-plane.

Example 13. *The coordinates (x, y) of a point P are expressible in terms of real variables u and v by the formula $x + yi = (u + vi)^2$. Prove that the locus of P is a parabola (i) when u varies and v is constant and also (ii) when v varies and u is constant. Prove also that all the parabolae have a common focus and a common axis.* (O.C.)

$$x + yi = (u + vi)^2 = u^2 - v^2 + 2uvi.$$

Equating real and imaginary parts,

$$x = u^2 - v^2, \ y = 2uv.$$

Eliminating u, $\quad x = \dfrac{y^2}{4v^2} - v^2$, or $y^2 = 4v^2(x + v^2)$, showing that, when v is constant, the locus of P is a parabola with its focus at the origin and axis along $y = 0$.

Similarly, eliminating v,

$$x = u^2 - \frac{y^2}{4u^2}, \text{ or } y^2 = 4u^2(u^2 - x),$$

so that, when u is constant, the locus of P is a parabola with the same focus and axis.

4.10. Infinite series of complex terms

Consider the infinite series Σu_n where $u_n = a_n + ib_n$ and a_n, b_n are real. The consideration of such series does not introduce any really new difficulties. The series is convergent if, and only if, the two series Σa_n, Σb_n are each convergent and the series is said to be *absolutely convergent* if the series Σa_n, Σb_n are each absolutely convergent. If Σu_n is absolutely convergent, then both the series $\Sigma |a_n|$, $\Sigma |b_n|$ are, from the above definition, convergent and hence $\Sigma \{|a_n| + |b_n|\}$ is convergent. But

$$|u_n| = \sqrt{(a_n^2 + b_n^2)} \leqslant |a_n| + |b_n|$$

and hence $\Sigma |u_n|$ is convergent. Also

$$|a_n| \leqslant \sqrt{(a_n^2 + b_n^2)}, \; |b_n| \leqslant \sqrt{(a_n^2 + b_n^2)},$$

and so $\Sigma |a_n|$, $\Sigma |b_n|$ are convergent whenever $\Sigma |u_n|$ converges. Thus *a necessary and sufficient condition for the absolute convergence of* Σu_n *is the convergence of* $\Sigma |u_n|$.

An absolutely convergent series is convergent since the series formed by its real and imaginary parts converge separately. Dirichlet's theorem (§ 3.10) may be extended to absolutely convergent series with complex terms by applying it to the series Σa_n and Σb_n separately.

A most important type of series is the power series $\Sigma a_n z^n$ where a_n is real and z is the complex variable $x + yi$. Suppose such a series converges for a particular value z_1 of z. It follows that the limit as n tends to infinity of $a_n z_1^n$ is zero and a number K can therefore be found such that $|a_n z_1^n| < K$ for all values of n. Hence,

$$|a_n z^n| = |a_n z_1^n| \left|\frac{z}{z_1}\right|^n < K \left|\frac{z}{z_1}\right|^n,$$

and it follows, by comparison with the geometrical progression $\Sigma \left|\dfrac{z}{z_1}\right|^n$, that the series $\Sigma a_n z^n$ converges absolutely whenever $|z| < |z_1|$. In other words, *if a series converges for a value of the complex variable represented in the Argand diagram by a point P, then it converges absolutely for values of the variable whose representative points are nearer the origin than P.*

Important deductions from this result, the details of which are not gone into here, are that the power series $\Sigma a_n z^n$ behaves in one of the following ways:

(i) it may converge for $z = 0$ and for no other values,
(ii) it may converge absolutely for all values of z,
(iii) it may converge absolutely for all values of z within a certain circle of radius R (called the *circle of convergence*) and not converge for any value of z outside this circle.

In case (iii) R is called the *radius of convergence* of the power series. It should be noted that nothing has been said about the behaviour of the series *on* the circle of convergence—in fact, various possibilities can arise for such values of z. An illustrative example follows.

Example 14. *Discuss the behaviour of the series*

$$(a) \sum_{n=0}^{\infty} a^n z^n \;(a\;real), \qquad (b) \sum_{n=1}^{\infty} z^n/(n)!$$

(a) The sum of n terms of the series (c) is

$$\frac{1 - (az)^n}{1 - az},$$

except when $z = 1/a$, in which case the sum is n. Hence the series converges only when $|az| < 1$. This series behaves as in case (iii) above and its radius of convergence is $1/a$. It does not converge anywhere on its circle of convergence, and it can be shown (but to do so goes a little outside the scope of the present book) that it diverges at the point representing $z = 1/a$ and oscillates finitely at all other points of its circle of convergence.

(b) If u_n denotes the nth term of series (b),

$$\frac{|u_{n+1}|}{|u_n|} = \frac{|z|}{n + 1},$$

and this tends to zero as n tends to infinity for all values of z. By d'Alembert's test, $\Sigma|u_n|$ is therefore convergent for all values of z and the series (b) is absolutely convergent for all values of z. This series therefore behaves as in case (ii) above.

EXERCISES 4 (d)

1. If $z = x + yi$, express

   (i) z^2,   (ii) $1/z$,   (iii) $(z^2 + 1)/z$,

   in the form $X + Yi$ where X and Y are real functions of x and y.

2. Show that the transformation

$$\frac{w + i}{w - i} = \frac{z + 1}{z - 1}$$

transforms the real axis of the z-plane into the imaginary axis of the w-plane.

EXERCISES

3. Prove that the transformation
$$w = \frac{z-a}{1-az},$$
where a is real, transforms all points inside the unit circle of the z-plane into all points inside or outside the unit circle of the w-plane according as a is less than or greater than unity.

4. Show that with the bilinear transformation
$$w = \frac{az+b}{cz+d},$$
the unit circle in the w-plane is transformed into a straight line in the z-plane if $|a| = |c|$.

5. The complex variables w and z are related by $w = z^2$. Show that the lines $u =$ constant and $v =$ constant transform into rectangular hyperbolae.

6. Show that the radius of convergence of the series
$$\frac{1}{2}z + \frac{1.3}{2.5}z^2 + \frac{1.3.5}{2.5.8}z^3 + \ldots, \quad (z = x + yi),$$
is 3/2.

EXERCISES 4 (e)

1. If $[a, b] \div [c, d]$ is defined as the complex number $[x, y]$ such that $[x, y] \times [c, d] = [a, b]$, show that
$$[a, b] \div [c, d] = \left[\frac{ac + bd}{c^2 + d^2}, \frac{bc - ad}{c^2 + d^2} \right].$$

2. Show that the representative points in the Argand diagram of the complex numbers $1 + 6i$, $3 + 10i$ and $4 + 12i$ are collinear.

3. If z_1, z_2 are two complex numbers such that $|z_1 - z_2| = |z_1 + z_2|$, prove that the difference of their arguments is $\pi/2$ or $3\pi/2$.

4. Show that the modulus of the sum of any number of complex numbers is not less than the sum of their real (or imaginary) parts.

5. If x is real and less than unity, show that
$$|(1 + ix + i^2x^2 + i^3x^3 + \ldots \text{ to inf.})| = (1 + x^2)^{-1/2}.$$

6. Express
$$\frac{(4 + 3i)\sqrt{(3 + 4i)}}{3 + i}$$
in the form $a + bi$.

7. Find the real part of $(1 + i\sqrt{3})^n$ where n is a positive integer and find the two square roots of $(1 + i\sqrt{3})$. (C.)

8. Find the three roots of the equation $8x^3 = (2 - x)^3$, expressing each in the form $a + bi$. (C.)

9. If ω is a complex cube root of unity, form the quadratic equation whose roots are ω and $1/\omega$.
Express $3/(x^3 - 1)$ as the sum of *three* partial fractions each with a linear denominator.

10. If $(1 + x)^n \equiv c_0 + c_1 x + c_2 x^2 + \ldots + c_n x^n$, n being a positive integer, show that
$$(c_0 + c_3 + \ldots)^2 + (c_1 + c_4 + \ldots)^2 + (c_2 + c_5 + \ldots)^2$$
$$= \tfrac{1}{3}(4^n + 2). \quad \text{(O.)}$$
[Hint. $c_0 + c_3 + \ldots$ can be found by setting $x = 1$, ω, ω^2 in turn and adding. $c_1 + c_4 + \ldots$, $c_2 + c_5 + \ldots$ can be found in a rather similar way.]

11. A network of equilateral triangles in the Argand diagram is formed by three sets of parallel lines, the sets being inclined to one another at angles of 120° and the distance between any two consecutive lines of a set the same for all sets. Two vertices of one triangle represent the complex numbers $a + bi$, $c + di$. Prove that any other vertex in the figure represents a number of the form
$$(a + bi)(1 - m + n\omega) + (c + di)(m - n\omega)$$
where m, n are integers, positive or negative, and ω is a complex cube root of unity. (O.)

12. If ω is one of the complex cube roots of unity, describe the position in the Argand diagram of the point $-\omega^2 z_1 - \omega z_2$.
On the sides of any convex plane hexagon, equilateral triangles are constructed external to it. Their outer vertices are joined to form another hexagon. If $PQRSTU$ are the mid-points of its sides, show that PS, QT and RU are equal and inclined at 60° to one another. (C.)

13. Prove that, if $z = x + yi$, $z' = x' + y'i$ and $zz' = 1$, then
$$x' = \frac{x}{x^2 + y^2}, \quad y' = -\frac{y}{x^2 + y^2},$$
and the circle through the points 1, -1 and z also passes through z'. (O.)

14. The notation $\bar{z}$ denotes $r(\cos\theta - i\sin\theta)$ when z denotes $r(\cos\theta + i\sin\theta)$.
Prove that the locus of z in the Argand diagram is a circle with its centre the origin when $z\bar{z} = k^2$ and k is real.
Prove that Z_1, z_1 are inverse points with respect to the circle when $Z_1 \bar{z}_1 = k^2$. (O.)

15. The roots of a cubic equation $f(z) = 0$ are a, b, c; the roots of $f'(z) = 0$ are p and q. Prove that
$$(a - p)(a - q) = \tfrac{1}{3}(a - b)(a - c),$$
$$\left(\frac{b + c}{2} - p\right)\left(\frac{b + c}{2} - q\right) = -\frac{1}{12}(b - c)^2.$$
If a, b, c are represented in the Argand diagram by the vertices A, B, C of a triangle, and p and q are represented by P and Q, prove that PA and QA are equally inclined to the bisectors of the angle BAC; and also prove that if L is the mid-point of BC, PL and QL are equally inclined to BC. (O.)

16. A_1, A_2, A_3 are three distinct points in the Argand diagram whose affixes are the complex numbers z_1, z_2, z_3. The distances A_2A_3, A_3A_1, A_1A_2 are denoted by a_1, a_2, a_3 respectively. Prove that

$$\frac{a_1^2}{z_2 - z_3} + \frac{a_2^2}{z_3 - z_1} + \frac{a_3^2}{z_1 - z_2} = 0.$$

The positive numbers b_1, b_2, b_3 are such that

$$\frac{b_1^2}{z_2 - z_3} + \frac{b_2^2}{z_3 - z_1} + \frac{b_3^2}{z_1 - z_2} = 0.$$

Prove that, if the points A_1, A_2, A_3 are *not* collinear, then $b_1/a_1 = b_2/a_2 = b_3/a_3$; and that, if A_1, A_2, A_3 *are* collinear then one of the three numbers $b_1^2/a_1, b_2^2/a_2, b_3^2/a_3$ is equal to the sum of the other two. (O.)

17. The complex numbers z_1, z_2, z_3 are such that

$$az_1 + bz_2 + cz_3 = 0,$$
$$a'z_1 + b'z_2 + c'z_3 = 0,$$

where a, b, c, a', b', c' are real. Prove that either (i) z_1, z_2, z_3 have all the same amplitude or (ii) $a : b : c = a' : b' : c'$.

Three non-collinear points A, B, C have their centroid at $z = 0$. The points A', B', C' are the inverses of A, B, C with respect to the circle $|z| = 1$. Prove that, if the centroid of $A'B'C'$ is also at $z = 0$, the points A, B, C lie on a circle $|z| = $ constant. What are the angles of the triangle ABC? (O.)

18. The complex numbers w_1, w_2 are represented in the Argand diagram by the points P_1, P_2. Prove that, if t is real, the number $(1 - t)w_1 + tw_2$ is represented by a point P on the line P_1P_2 such that

$$\frac{P_1P}{PP_2} = \frac{t}{1 - t}.$$

The numbers $z_1, z_2, z_3, (1 + k_1)z_1, (1 + k_2)z_2, (1 + k_3)z_3$ (where k_1, k_2, k_3 are real) are represented by $A_1, A_2, A_3, B_1, B_2, B_3$, and none of the lines A_2A_3, A_3A_1, A_1A_2 passes through the origin. Prove that the point C_1 at which A_2A_3, B_2B_3 intersect represents the number

$$\frac{k_3(1 + k_2)z_2 - k_2(1 + k_3)z_3}{k_3 - k_2}.$$

If A_3A_1, B_3B_1 intersect at C_2 and A_1A_2, B_1B_2 intersect at C_3, show that C_1, C_2, C_3 are collinear and that C_1 is the mid-point of C_2C_3 if and only if

$$\frac{1}{k_2} + \frac{1}{k_3} = \frac{2}{k_1}. \qquad \text{(O.)}$$

19. Prove that $|z_1 + z_2| \leq |z_1| + |z_2|$ where z_1, z_2 are complex numbers. Show that if $|a_n| < 2$ for $1 \leq n \leq N$ then the equation

$$1 + a_1z + \ldots + a_Nz^N = 0$$

has no solution such that $|z| < \tfrac{1}{3}$. (C.)

20. Prove that if the real part of the polynomial

$$a_0 + a_1 z + \ldots + a_n z^n, \ z = x + yi,$$

where $a_0, a_1, \ldots, a_n$ are complex numbers, is never negative for any value (real or complex) of z then $a_1 = a_2 = \ldots = a_n = 0$. Deduce that if the real part of a polynomial is always greater than the imaginary part then the polynomial is a constant. (C.)

21. If $w = 4/z$ and if the point which represents z in the Argand diagram describes a circle of unit radius and whose centre represents the complex number $1 + i$, show that the point representing w describes a circle of radius 4.

22. Show that with the transformation $w = z/(z - 1)$, the straight line $x = \frac{1}{2}$ of the z-plane transforms into the unit circle in the w-plane.

23. Complex variables w and z are related by $z = \frac{1}{2}\{w + w^{-1}\}$. Show that to the circle, in the w-plane, whose centre is the origin and whose radius is R, corresponds, in the z-plane, the ellipse

$$\frac{4x^2}{\left(R + \dfrac{1}{R}\right)^2} + \frac{4y^2}{\left(R - \dfrac{1}{R}\right)^2} = 1.$$

24. Show that the series $a_1 + 2a_2 z + 3a_3 z^2 + 4a_4 z^3 + \ldots$ has the same circle of convergence as the series $a_0 + a_1 z + a_2 z^2 + a_3 z^3 + a_4 z^4 + \ldots$.

25. $\Sigma a_n z^n$, $\Sigma b_n z^n$ are two power series and each converges to the same value for all values of z such that $|z| < a$ where $a > 0$. Show that $a_n = b_n$ for all integral values of n.

CHAPTER 5

THE THEORY OF EQUATIONS

5.1. Introduction

In this chapter a discussion is given of the theory of algebraical equations with real coefficients. The equation studied is $f(x) = 0$ where $f(x)$ is the polynomial in x of degree n given by

$$f(x) \equiv a_0 x^n + a_1 x^{n-1} + a_2 x^{n-2} + \ldots + a_{n-1} x + a_n, \qquad (5.1)$$

the coefficients $a_0, a_1, a_2, \ldots, a_n$ all being supposed to be real. Such an equation is said to be of the *nth degree* and any value of x which makes $f(x)$ vanish is said to be a *root* of the equation. It can be shown* that every such equation has a root, real or complex.

Assuming that the equation $f(x) = 0$ has a root a_1, then $f(x)$ is divisible by $(x - a_1)$ and hence

$$f(x) \equiv (x - a_1)\phi_1(x),$$

where $\phi_1(x)$ is a polynomial in x of degree $(n - 1)$. Again, $\phi_1(x) = 0$ has a root, say a_2, and

$$\phi_1(x) \equiv (x - a_2)\phi_2(x),$$

where $\phi_2(x)$ is a polynomial of degree $(n - 2)$. Proceeding in this way and equating the coefficients of x^n,

$$f(x) \equiv a_0(x - a_1)(x - a_2)\ldots(x - a_n). \qquad (5.2)$$

Hence the equation $f(x) = 0$ has n roots, since $f(x)$ vanishes when x takes any of the values $a_1, a_2, \ldots, a_n$. The equation cannot have more than n roots, for if x assumes a value different from any of the quantities $a_1, a_2, \ldots, a_n$, the factors on the right of (5.2) are all different from zero and therefore $f(x)$ does not vanish for that value of x. In the above, of course, some of the quantities $a_1, a_2, \ldots, a_n$ may be equal; in this case the equation is still said to possess n roots although these are not all different. If r of $a_1, a_2, \ldots, a_n$ are equal the equation is said to possess an *r-multiple root* or *a root of order r* (but such roots are often simply referred to as *repeated* roots). To sum up, allowing an r-multiple root to count as r roots, *an equation of the nth degree possesses n roots, and no more*.

It has been shown in § 4.8 of Chapter 4 that, *in an equation with*

---

* The theorem that every algebraical equation has a root is sometimes called 'the fundamental theorem of algebra'. The proof of the theorem is not easy and requires an appeal to considerations of continuity. Such a proof can be found in the standard books which go a little beyond the scope of the present work; see, for instance, *A Course of Pure Mathematics*, G. H. Hardy, Cambridge University Press, 1946, Appendix 2.

real coefficients, complex roots occur in conjugate pairs. Combining this with the fact that an equation of the nth degree has exactly n roots, it can be deduced that *any equation of odd degree must have at least one real root.* These results also enable certain statements to be made with regard to the roots of an equation of given degree. For example, the three roots of an equation of the third degree may be all real or one may be real and the other two complex; the four roots of an equation of the fourth degree may be all real, two real and two complex or all four may be complex; the roots of an equation of the fifth degree may be all real, three real and two complex or one real and four complex, and so on.

Example 1. *Given that one root of the equation*

$$2x^4 + x^3 + 5x^2 + 4x - 12 = 0$$

is a purely imaginary number, find all the roots. (L.I.C.)

If one root is ia (a real) then a second root is the conjugate quantity $-ia$. Substituting in turn $x = \pm ia$ and using the relation $i^2 = -1$,

$$2a^4 - ia^3 - 5a^2 + 4ia - 12 = 0,$$
$$2a^4 + ia^3 - 5a^2 - 4ia - 12 = 0.$$

By subtraction, it follows that $a^3 - 4a = 0$. Since it is clear that $x = 0$ does not satisfy the given equation, $a \neq 0$ and hence $a^2 - 4 = 0$ giving $a = \pm 2$ and two roots as $\pm 2i$. Factors of the left-hand side of the given equation are therefore $(x - 2i)$, $(x + 2i)$. Dividing $2x^4 + x^3 + 5x^2 + 4x - 12$ by $(x - 2i)(x + 2i)$, that is, by $x^2 + 4$, the other two roots are given by $2x^2 + x - 3 = 0$. The roots of this quadratic equation being 1 and $-3/2$, the four roots of the original equation are $1, -3/2, 2i, -2i$.

5.2. Some remarks on the position of the real roots of an equation

Suppose the equation $f(x) = 0$ is of degree n and that it has $2m$ complex roots and therefore $n - 2m$ real roots. The complex roots occur in conjugate pairs $p \pm iq$ and, to each pair of such roots, will correspond a factor $\{(x - p)^2 + q^2\}$ of $f(x)$. If $a_1, a_2, \ldots, a_{n-2m}$ are the real roots, $f(x)$ can therefore be expressed in the form

$$f(x) \equiv (x - a_1)(x - a_2) \ldots (x - a_{n-2m})g(x),$$

where $g(x)$ is a product of m factors of the type $(x - p)^2 + q^2$. If x is real, $g(x)$ is therefore essentially positive (if there are no complex roots, $g(x)$ is unity). If then a and b are two real values of x such that $f(a), f(b)$ have opposite signs, the linear factors of $f(x)$ cannot all be absent and

$$(a - a_1)(a - a_2) \ldots (a - a_{n-2m}), \quad (b - a_1)(b - a_2) \ldots (b - a_{n-2m})$$

must have opposite signs. But $a - a_s$, $b - a_s$ where $1 \leqslant s \leqslant n - 2m$, have the same sign unless a_s lies between a and b. Hence an odd number of a_s must lie between a and b and it can be concluded that *if*

$f(a)$, $f(b)$ *are not zero and have opposite signs, then an odd number of real roots of the equation* $f(x) = 0$ *lie between a and b.*

Conversely, if an odd number of real roots of the equation $f(x) = 0$ lie between a and b, then $f(a)$, $f(b)$ have opposite signs. Hence it follows that *if* $f(a)$, $f(b)$ *are not zero and have the same signs, then an even number of roots (or no root) of* $f(x) = 0$ *lie between a and b.*

Example 2. $f(x)$ *denotes the polynominal* $\sum_{r=0}^{n} a_r x^{n-r}$ *and* $a_0 > 0$. *Show that*
  (i) *if* n *is odd, the equation* $f(x) = 0$ *has at least one real root whose sign is opposite to that of* a_n;
  (ii) *if* n *is even and* $a_n < 0$, $f(x) = 0$ *has at least two real roots, one positive and one negative.*

When x is large, the sign of $f(x)$ is determined by the sign of its first term $a_0 x^n$. When $x = 0$, $f(x) = a_n$ and, since a_0 is positive, it follows that
(i) n odd:
  when x is: large and negative, 0, large and positive,
  $f(x)$ is: large and negative, a_n, large and positive.
  Hence if $a_n < 0$, there is at least one positive real root and if $a_n > 0$, there is at least one negative real root.
(ii) n even:
  when x is: large and negative, 0, large and positive,
  $f(x)$ is: large and positive, a_n, large and positive.
  If $a_n < 0$, there is thus at least one negative and one positive real root.

5.3. Rolle's theorem

Rolle's theorem is important in analysis and it will be discussed again later (Chapter 8). As an aid to discussing the existence and position of the real roots of algebraical equations, the theorem is required in the form: *if a and b are consecutive real roots of the equation* $f(x) = 0$, *then the equation* $f'(x) = 0$ *(where the prime denotes the derivative with respect to x) has an odd number of real roots between a and b.*

To prove the theorem, let a, b be consecutive r-multiple and s-multiple roots respectively so that

$$f(x) \equiv (x - a)^r (x - b)^s g(x)$$

where $g(x)$ has the same sign throughout the interval a to b. Hence

$$\log f(x) = r \log (x - a) + s \log (x - b) + \log g(x),$$

and, differentiating with respect to x,

$$\frac{f'(x)}{f(x)} = \frac{r}{x-a} + \frac{s}{x-b} + \frac{g'(x)}{g(x)}.$$

Cross multiplying and substituting for $f(x)$,

$$f'(x) \equiv (x - a)^{r-1}(x - b)^{s-1} h(x),$$

where

$$h(x) \equiv \{r(x - b) + s(x - a)\}g(x) + (x - a)(x - b)g'(x).$$

The values taken by $h(x)$ when $x = a$, $x = b$ respectively are
$$h(a) = r(a - b)g(a), \quad h(b) = s(b - a)g(b).$$
Since g is one-signed, these expressions are of opposite signs. It follows that $h(x)$, and therefore also $f'(x)$, vanishes for an odd number of values of x between a and b.

It is a simple matter to deduce that *between two consecutive real roots of $f'(x) = 0$ there is at most one real root of $f(x) = 0$*. For suppose that $f'(a) = 0$, $f'(\beta) = 0$ and that $f(\gamma) = 0$, $f(\delta) = 0$ where $a < \gamma < \delta < \beta$. Then by Rolle's theorem it follows that between γ and δ there must be a real root of $f'(x) = 0$. But this is contrary to hypothesis and hence there cannot be more than one real root. If $f(a)$, $f(\beta)$ are of opposite signs, the result of § 5.2 shows that there is an odd number of real roots of $f(x) = 0$ between $x = a$ and $x = \beta$. Combining these results it can be concluded that *if a, β are consecutive real roots of $f'(x) = 0$ and if $f(x)$ has opposite signs at $x = a$, $x = \beta$, then the equation $f(x) = 0$ has exactly one real root between $x = a$, $x = \beta$*.

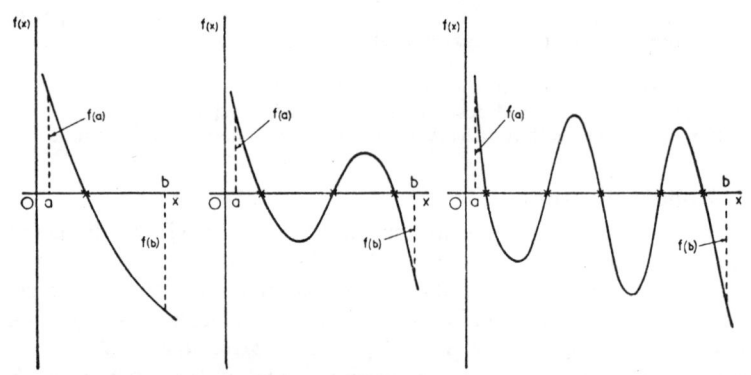

Fig. 18

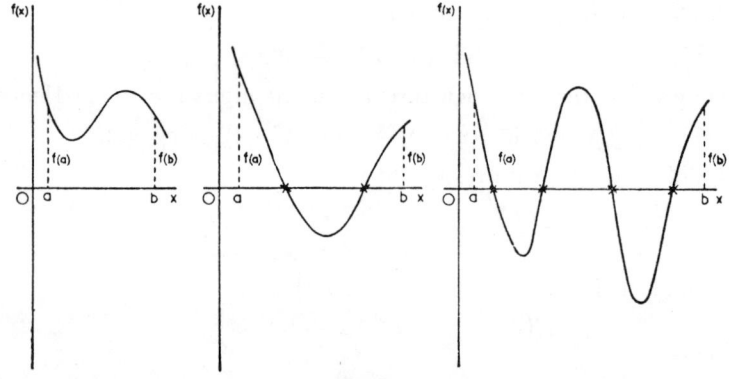

Fig. 19

Example 3. *Assuming that a polynomial is a continuous function,* draw rough diagrams to illustrate the results printed in italics in §§ 5.2 and 5.3.*

The roots of $f(x) = 0$ are the abscissae of the points represented by crosses and the diagrams of Fig. 18 illustrate the result that an odd number of real roots of the equation $f(x) = 0$ lie between $x = a$ and $x = b$ when $f(a)$, $f(b)$ are not zero and have opposite signs.

Fig. 19 illustrates in a similar way that an even number of roots (or no root) of $f(x) = 0$ lie between $x = a$, $x = b$ when $f(a)$, $f(b)$ are not zero and have the same signs.

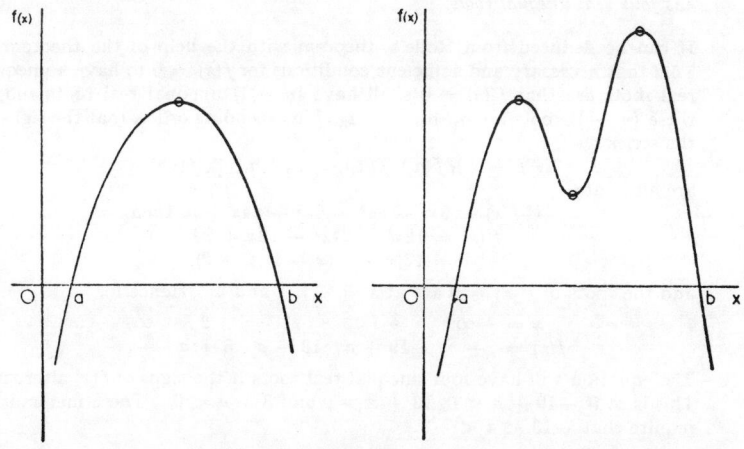

Fig. 20

When $f'(x) = 0$, the tangent to the curve $y = f(x)$ will be parallel to the x-axis and the real roots of the equation $f'(x) = 0$ are the abscissae of the points shown by circles in Fig. 20. The diagrams of this figure illustrate Rolle's theorem—if a and b are consecutive real roots of $f(x) = 0$, then the equation $f'(x) = 0$ has an odd number of real roots between a and b.

Fig. 21 illustrates that between two consecutive real roots of $f'(x) = 0$

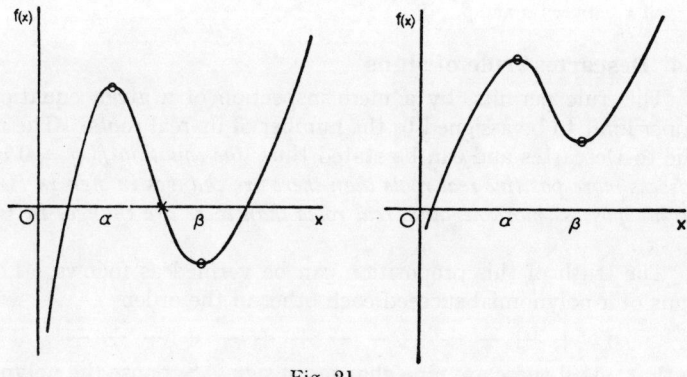

Fig. 21

* See Chapter 8, § 8.3.

there is at most one real root of $f(x) = 0$. The first diagram of this figure shows that if α and β are consecutive real roots of $f'(x) = 0$ and if $f(x)$ has opposite signs at $x = \alpha$, $x = \beta$, then $f(x) = 0$ has exactly one real root between α and β.

Example 4. *What information about the real roots of the equation $f(x) = 0$ is given by a knowledge of the roots of $f'(x) = 0$?*
Determine the range of values of a for which the equation
$$3x^4 - 8x^3 - 6x^2 + 24x + a = 0$$
has four real unequal roots. (O.)

It can be deduced from Rolle's theorem with the help of the theorem of § 5.2 that necessary and sufficient conditions for $f(x) = 0$ to have n unequal real roots are that $f'(x) = 0$ shall have $(n-1)$ unequal real roots and, if these $(n-1)$ roots are $a_1, a_2, \ldots, a_{n-1}$ in ascending order, that the signs of the series
$$f(-\infty), f(a_1), f(a_2), \ldots, f(a_{n-1}), f(\infty)$$
are alternate.

If $f(x) \equiv 3x^4 - 8x^3 - 6x^2 + 24x + a$, then
$$f'(x) = 12x^3 - 24x^2 - 12x + 24$$
$$= 12(x+1)(x-1)(x-2),$$
and the roots of $f'(x) = 0$ are at $x = -1, 1$ and 2. Hence for

$x =$	$-\infty$	-1	1	2	∞,
$f(x) =$	$+$	$-19+a$	$13+a$	$8+a$	$+$.

The equation will have four unequal real roots if the signs of $f(x)$ alternate. This is so if $-19 + a < 0$, $13 + a > 0$ and $8 + a < 0$. These inequalities require that $-13 < a < -8$.

Example 5. *Given that α and β are two real roots of the equation $f(x) = 0$ where $f(x)$ is a polynomial in x, prove that there is at least one real root of the equation $f'(x) + f(x) = 0$ which lies between α and β.* (O.C.)

Let $\phi(x) = e^x f(x)$ so that $\phi'(x) = e^x\{f'(x) + f(x)\}$. Since $f(x)$ vanishes at $x = \alpha$, $x = \beta$ so does $\phi(x)$ and hence, by Rolle's theorem, there is at least one real root of $\phi'(x)$ between α and β.
Since $e^x \neq 0$, it follows that $f'(x) + f(x)$ vanishes for at least one real value of x between α and β.

5.4. Descartes' rule of signs

This rule permits, by a mere inspection of a given equation, an upper limit to be assigned to the number of its real roots. The rule is due to Descartes and can be stated thus: *the equation $f(x) = 0$ cannot possess more positive real roots than there are changes in sign in $f(x)$ and cannot possess more negative real roots than there are changes in sign in $f(-x)$.*

The truth of this proposition can be verified as follows. Let the signs of a polynomial succeed each other in the order

$$+ \; + \; - \; + \; - \; - \; - \; + \; + \; - \; + \; - \; + \; -$$

so that, in all there are nine changes of sign. Suppose the polynomial is multiplied by a binomial expression whose signs (corresponding to a positive root) are $+$ $-$. The signs occurring in the multiplication process will be

$$+\ +\ -\ +\ -\ -\ -\ +\ +\ -\ +\ -\ +\ -$$
$$-\ -\ +\ -\ +\ +\ +\ -\ -\ +\ -\ +\ -\ +$$
$$\overline{+\ \pm\ -\ +\ -\ \mp\ \mp\ +\ \pm\ -\ +\ -\ +\ -\ +}$$

the signs of the third line being those of the polynomial resulting from the multiplication, an ambiguous sign being shown wherever there are two terms of opposite sign to be added. It should be noted that ambiguous signs arise wherever like signs follow one another in the original polynomial. The number of sign variations is never diminished and there is one further variation added at the end. Thus the multiplication of a polynomial by a factor like $(x - a)$ introduces at least one additional change of sign.

Suppose now that the real positive roots of an equation $f(x) = 0$ are $a, \beta, \gamma, \ldots$, and that $\phi(x)$ is the polynomial corresponding to any real negative or imaginary roots which the equation may possess, so that

$$f(x) \equiv \phi(x)(x - a)(x - \beta)(x - \gamma) \ldots$$

The effect of multiplying $\phi(x)$ by $(x - a)$ is to introduce at least one change of sign additional to those already existing in $\phi(x)$. The effect of multiplying $\phi(x)(x - a)$ by $(x - \beta)$ is to introduce at least one more change of sign and so on. Hence the polynomial $f(x)$ has at least as many changes of sign as there are real positive roots $a, \beta, \gamma, \ldots$ of the equation $f(x) = 0$.

To demonstrate the second part of Descartes' rule it is first necessary to show that if $-x$ be substituted for x in the equation $f(x) = 0$, the resulting equation will have the same roots as the original equation except that their signs will be changed. To show this let $a_1, a_2, \ldots, a_n$ be the roots of the original equation so that

$$f(x) \equiv (x - a_1)(x - a_2) \ldots (x - a_n).$$

Hence $\qquad f(-x) \equiv (-1)^n(x + a_1)(x + a_2) \ldots (x + a_n)$

and it is now clear that the roots of the equation $f(-x) = 0$ are $-a_1, -a_2, \ldots, -a_n$. Hence the negative roots of $f(x) = 0$ are the positive roots of $f(-x) = 0$ and the verification of Descartes' rule is now complete.

If $f(x) = 0$ is an equation of the nth degree there will be up to $n + 1$ terms in the polynomial $f(x)$. The equation is said to be *complete* or *incomplete* according as all these $n + 1$ terms are or are not present. In the case of incomplete equations, Descartes' rule is often of use in detecting the existence of imaginary roots. For example, consider the equation

$$x^6 + 4x^3 + x - 2 = 0.$$

Here there is only one variation in sign and the equation cannot therefore possess more than one real positive root. Changing x into $-x$ the resulting equation is

$$x^6 - 4x^3 - x - 2 = 0$$

and again there is only one sign variation showing that the original equation cannot have more than one negative real root. Thus the equation of this example cannot possess more than two real roots and there are, therefore, at least four imaginary roots.

Example 6. *Show that, if n is even, the equation $x^n = 1$ has two and only two real roots, one positive and one negative.*

Let $f(x) \equiv x^n - 1$, so that, since n is even, $f(x)$ is positive for large positive and negative values of x and $f(0) = -1$. It follows that there are at least two real roots, one positive and one negative.
Since $f(x)$ and $f(-x)$ contain only one change in sign, Descartes' rule shows that there are at most one positive and one negative root.

EXERCISES 5 (a)

1. Solve the equation $x^4 + 4x^3 + 6x^2 + 4x + 5 = 0$ given that one root is i.

2. Prove what you can about the number of real roots of the equation
$$(x - a_1)(x - a_2) \ldots (x - a_n) + (x - b_1)(x - b_2) \ldots (x - b_n) = 0,$$
where $a_1 > b_1 > a_2 > b_2 \ldots > a_n > b_n$. (C.)

3. By considering the turning values of the function $y \equiv x^3 - 3bx + c$, show that the equation $x^3 - 3bx + c = 0$ has three real roots, or only one, according as $4b^3 - c^2$ is positive or negative. (W.)

4. Deduce from Example 5 above that if α, β, γ are real roots of the equation $f(x) = 0$ and are such that $\alpha < \beta < \gamma$, then there is a real root of the equation
$$f''(x) + 2f'(x) + f(x) = 0$$
which lies between α and γ. (O.C.)

5. Show that the equation $f(x) + kf'(x) = 0$ has at least as many real roots as the equation $f(x) = 0$.
Show by induction that, if the roots of the two equations $f(x) = 0$, $x^r + p_1 x^{r-1} + \ldots + p_r = 0$ are all real, so also are all the roots of the equation
$$f(x) + p_1 f'(x) + p_2 f''(x) + \ldots + p_r f^{(r)}(x) = 0.$$ (O.)

6. Show that the equation
$$1 - x + \frac{x^3}{2} - \frac{x^3}{3} + \ldots + (-1)^n \frac{x^n}{n} = 0$$
has one and only one real root if n is odd and no real root if n is even. (C.)

5.5. The relations between the roots and coefficients in an equation

Suppose the equation is $f(x) = 0$, where

$$f(x) \equiv a_0 x^n + a_1 x^{n-1} + a_2 x^{n-2} + \ldots + a_{n-1} x + a_n, \quad (5.3)$$

and let its n roots be $a_1, a_2, \ldots, a_n$. Then, as in equation (5.2),

$$f(x) \equiv a_0(x - a_1)(x - a_2) \ldots (x - a_n). \quad (5.4)$$

The terms involving x^{n-1} in this product are formed by taking x from any $(n-1)$ of the factors and one of $-a_1, -a_2, \ldots, -a_n$ from the remaining factor. Thus the coefficient of x^{n-1} is minus the sum of the roots. This is conveniently denoted by Σ_1. The terms involving x^{n-2} are formed by taking x from any $(n-2)$ of the factors and two of $-a_1, -a_2, \ldots, -a_n$ from the remaining two factors. The coefficient of x^{n-2} is therefore the sum of the products of the roots taken two at a time and this can be similarly denoted by Σ_2. Proceeding in this way, (5.4) can be written

$$f(x) \equiv a_0 \{x^n - \Sigma_1 x^{n-1} + \Sigma_2 x^{n-2} - \ldots + (-1)^n \Sigma_n\},$$

where Σ_r denotes the sum of the n roots taken r at a time, Σ_n being simply the product of the n roots. Comparing this expression with that given in (5.3) and equating coefficients,

$$\left. \begin{array}{l} \Sigma_1 = \text{sum of roots taken one at a time} = -a_1/a_0, \\ \Sigma_2 = \text{sum of roots taken two at a time} = a_2/a_0, \\ \Sigma_3 = \text{sum of roots taken three at a time} = -a_3/a_0, \\ \ldots\ldots\ldots\ldots\ldots\ldots\ldots\ldots\ldots\ldots\ldots\ldots\ldots, \\ \Sigma_n = \text{product of roots} = (-1)^n a_n/a_0. \end{array} \right\} \quad (5.5)$$

This important result is sometimes also written in the alternative notation, where now $a, \beta, \gamma, \ldots, \varkappa$ denote the roots of the equation,

$$\left. \begin{array}{l} \Sigma a = -a_1/a_0, \ \Sigma a\beta = a_2/a_0, \ \Sigma a\beta\gamma = -a_3/a_0, \ldots, \\ a\beta\gamma \ldots \varkappa = (-1)^n a_n/a_0. \end{array} \right\} \quad (5.6)$$

It must not be assumed that these n relations between the n roots of an equation provide an advantageous method of solving the equation. For suppose that the root a is required—this can only be obtained by eliminating $\beta, \gamma, \ldots, \varkappa$ from the n relations of (5.6), and the result of this elimination will simply be the original equation with a substituted for x. The relations are, however, often of great use in facilitating the solution of equations when particular relations between the roots are known to exist. They are also of use in establishing the relations which must exist between the coefficients in an equation corresponding to known relations among its roots. Some illustrative examples follow.

Example 7. *If one root of the equation $x^3 + ax + b = 0$ is twice the difference of the other two, prove that one root is $13b/3a$.* (C.)

Let the roots be a, β and $2(a - \beta)$. Since the coefficient of x^2 in the given equation is zero, the sum of the roots taken one at a time vanishes and hence $a + \beta + 2(a - \beta) = 0$, so that $\beta = 3a$ and the three roots can be taken as a, $3a$ and $-4a$. The sum of the roots taken two at a time and the product of the three roots are respectively a and $-b$, so that

$$3a^2 - 12a^2 - 4a^2 = a, \quad -12a^3 = -b.$$

Thus, $-13a^2 = a$, $12a^3 = b$ and, by division, $a = -13b/12a$. But $-4a$ is a root and hence one root of the equation is $13b/3a$.

Example 8. *If the equation $x^3 + ax^2 + bx + c = 0$ has three distinct real roots, find the value of c so that the roots may be in geometrical progression. If c has this value, prove that the arithmetic mean between the first and last terms of the progression is $(b - a^2)/2a$.* (L.I.C.)

Let the roots be a/r, a and ar. Then

$$\frac{a}{r} + a + ar = -a,$$

$$\frac{a^2}{r} + a^2 r + a^2 = b,$$

$$\frac{a}{r} \cdot a \cdot ar = -c.$$

Division of the first two of these relations gives $a = -b/a$ and substitution of this value of a in the third relation yields $c = b^3/a^3$.

The arithmetic mean of the first and last terms of the progression is $\frac{1}{2}\left(\frac{a}{r} + ar\right)$, which, from the first of the above relations reduces to $\frac{1}{2}(-a - a)$. Putting $a = -b/a$, the arithmetic mean is equal to

$$\frac{1}{2}\left(-a + \frac{b}{a}\right), \text{ i.e. } \frac{1}{2a}(b - a^2).$$

5.6. Symmetric functions of the roots of an equation

Symmetric functions of the roots are those functions in which all the roots are involved in the same way and in which the functions are unaltered when any two of the roots are interchanged. For example, all the functions of equations (5.5) are of this nature and these functions are the simplest symmetric functions of the roots, each root appearing in the first degree only in any term of any one of them.

It is possible to obtain, in terms of the coefficients of an equation, many other symmetric functions of its roots by means of the relations (5.5) or (5.6) without a knowledge of the separate values of the several roots. It can be shown, in fact, that *any* symmetric function of the roots can be so found but here only a few representative examples will be given. In this connection, some relations connecting certain symmetric functions with the simpler ones appearing in the relations (5.5), (5.6) are first given and these (and similar ones which can be similarly derived) are often of great use in working exercises. The notation adopted is that used in the relations (5.6) so that Σa^2 denotes the sum of the squares of the roots taken one at a time, $\Sigma a^2\beta^2$ denotes the sum of the squares of the roots taken two at a time and so on. With this notation, it is easily verified that

$$\left.\begin{array}{l}\Sigma a^2 = (\Sigma a)^2 - 2\Sigma a\beta, \\ \Sigma a^2\beta = (\Sigma a)(\Sigma a\beta) - 3\Sigma a\beta\gamma, \\ \Sigma a^3 = (\Sigma a)^3 - 3\Sigma a^2\beta - 6\Sigma a\beta\gamma.\end{array}\right\} \quad (5.7)$$

Other symmetric functions of the roots can be similarly expressed in terms of the simpler functions of (5.6).

It should be noted that some care is necessary in using the above notation. Thus, for example, in the case of three roots a, β, γ,

$$\Sigma a^2\beta = a^2\beta + a\beta^2 + \beta^2\gamma + \beta\gamma^2 + \gamma^2 a + \gamma a^2,$$

and the notation is to be interpreted as the sum of all the six terms which can be formed as the product of the square of one root by another; since $a\beta = \beta a$, etc., the sum $\Sigma a\beta$ will contain, in this case, only the three terms $a\beta + \beta\gamma + \gamma a$.

Example 9. *If a, β, γ are the roots of the equation $x^3 + ax^2 + bx + c = 0$, find the values in terms of a, b, c, of* (i) $\Sigma a^2\beta^2$, (ii) $\Sigma a(\beta - \gamma)^2$.

(i) $\Sigma a^2\beta^2 = (\Sigma a\beta)^2 - 2a\beta\gamma \Sigma a$
$= (b)^2 - 2(-c)(-a) = b^2 - 2ac$.

(ii) $\Sigma a(\beta - \gamma)^2 = \Sigma a\beta^2 - 6a\beta\gamma$
$= (\Sigma a)(\Sigma a\beta) - 3a\beta\gamma - 6a\beta\gamma$
$= (-a)(b) - 9(-c) = 9c - ab$.

Example 10. *The roots of the equation $ax^3 + bx^2 + cx + d = 0$ ($a \neq 0$, $d \neq 0$) are a, β, γ. Form the equation whose roots are $\beta\gamma/a$, $\gamma a/\beta$ and $a\beta/\gamma$. Deduce a condition that the product of two of the roots of the original equation may be equal to the third root.* (C.)

The sum of the roots of the required equation taken one at a time

$$= \Sigma\frac{\beta\gamma}{a} = \frac{1}{a\beta\gamma}\Sigma\beta^2\gamma^2 = \frac{1}{a\beta\gamma}\{(\Sigma\beta\gamma)^2 - 2a\beta\gamma\Sigma a\}$$

$$= -\frac{a}{d}\left\{\left(\frac{c}{a}\right)^2 - 2\left(-\frac{d}{a}\right)\left(-\frac{b}{a}\right)\right\} = \frac{2bd - c^2}{ad}.$$

The sum of the roots taken two at a time

$$= \Sigma\left(\frac{\beta\gamma}{a} \cdot \frac{\gamma a}{\beta}\right) = \Sigma\gamma^2 = (\Sigma\gamma)^2 - 2\Sigma a\beta = \left(-\frac{b}{a}\right)^2 - 2\left(\frac{c}{a}\right) = \frac{b^2 - 2ac}{a^2}.$$

The product of the roots

$$= \frac{\beta\gamma}{a} \cdot \frac{\gamma a}{\beta} \cdot \frac{a\beta}{\gamma} = a\beta\gamma = -\frac{d}{a}.$$

Hence the equation with roots $\beta\gamma/a$, $\gamma a/\beta$, $a\beta/\gamma$ is

$$x^3 - \left(\frac{2bd - c^2}{ad}\right)x^2 + \left(\frac{b^2 - 2ac}{a^2}\right)x + \frac{d}{a} = 0,$$

i.e. $\qquad a^2dx^3 + a(c^2 - 2bd)x^2 + d(b^2 - 2ac)x + ad^2 = 0.$

T.M.A.—5

If the product of two of α, β, γ is equal to the third quantity, one of the roots of the equation just derived will be unity. The required condition can thus be obtained by writing $x = 1$ and is

$$a^2d + a(c^2 - 2bd) + d(b^2 - 2ac) + ad^2 = 0.$$

EXERCISES 5 (b)

1. Given that the sum of the reciprocals of two of the roots of the equation $6x^3 - 11x^2 - 22x + 12 = 0$ is equal to $5/2$, solve the equation. (W.)

2. Solve the equation $24x^3 - 14x^2 - 63x + 45 = 0$ given that one root is double another.

3. Solve the equation $81x^4 + 54x^3 - 189x^2 - 66x + 40 = 0$ given that the roots are in arithmetic progression. (C.)

4. If α, β, γ are the roots of the equation $x^3 - ax^2 + bx - c = 0$, find the value of

   (i) $\dfrac{1}{\alpha^2} + \dfrac{1}{\beta^2} + \dfrac{1}{\gamma^2}$,   (ii) $\dfrac{1}{\beta^2\gamma^2} + \dfrac{1}{\gamma^2\alpha^2} + \dfrac{1}{\alpha^2\beta^2}$.

5. If α, β, γ, δ are the roots of the equation $x^4 + ax^3 + bx^2 + cx + d = 0$ find the value of

   $(\beta + \gamma + \delta - \alpha)(\gamma + \delta + \alpha - \beta)(\delta + \alpha + \beta - \gamma)(\alpha + \beta + \gamma - \delta).$

6. The equation $x^3 + px + q = 0$ has roots α, β, γ. Find the equation whose roots are $(\beta - \gamma)^2$, $(\gamma - \alpha)^2$, $(\alpha - \beta)^2$.
   Hence, or otherwise, deduce the condition for the equation $a_0x^3 + 3a_1x + a_2 = 0$ to have a pair of equal roots in the form

   $$a_0 a_2^2 + 4a_1^3 = 0.$$ (C.)

5.7. The sums of powers of the roots of an equation

Let $a_1, a_2, \ldots, a_n$ be the n roots of the equation $f(x) = 0$ where

$$f(x) \equiv a_0 x^n + a_1 x^{n-1} + a_2 x^{n-2} + \ldots + a_{n-1} x + a_n, \quad (5.8)$$

and let the sum of the rth power of the roots be denoted by S_r so that

$$S_r = a_1^r + a_2^r + \ldots + a_n^r. \quad (5.9)$$

In terms of $a_1, a_2, \ldots, a_n$, $f(x)$ can be written

$$f(x) \equiv a_0(x - a_1)(x - a_2) \ldots (x - a_n)$$

and logarithmic differentiation with respect to x gives

$$\frac{f'(x)}{f(x)} = \frac{1}{x - a_1} + \frac{1}{x - a_2} + \ldots + \frac{1}{x - a_n}.$$

Since $f'(x) = na_0 x^{n-1} + (n-1)a_1 x^{n-2} + (n-2)a_2 x^{n-3} + \ldots + a_{n-1}$, this gives, when x is replaced by $1/y$,

$$\frac{na_0 + (n-1)a_1 y + (n-2)a_2 y^2 + \ldots + a_{n-1} y^{n-1}}{a_0 + a_1 y + a_2 y^2 + \ldots + a_{n-1} y^{n-1} + a_n y^n} = \sum_{r=1}^{n} \frac{1}{1 - a_r y}.$$

$$(5.10)$$

SUMS OF POWERS OF ROOTS

But by division, for any positive integer p,
$$\frac{1}{1-a_r y} = 1 + a_r y + a_r^2 y^2 + \ldots + a_r^{p-1} y^{p-1} + \frac{a_r^p y^p}{1-a_r y},$$
and hence,
$$\sum_{r=1}^{n} \frac{1}{1-a_r y} = \sum_{r=1}^{n} 1 + y\sum_{r=1}^{n} a_r + y^2 \sum_{r=1}^{n} a_r^2 + \ldots + y^{p-1} \sum_{r=1}^{n} a_r^{p-1}$$
$$+ y^p \sum_{r=1}^{n} \frac{a_r^p}{1-a_r y}$$
$$= n + S_1 y + S_2 y^2 + \ldots + S_{p-1} y^{p-1} + y^p \sum_{r=1}^{n} \frac{a_r^p}{1-a_r y}.$$

Substituting in (5.10)
$$na_0 + (n-1)a_1 y + (n-2)a_2 y^2 + \ldots + a_{n-1} y^{n-1}$$
$$= (a_0 + a_1 y + a_2 y^2 + \ldots + a_{n-1} y^{n-1} + a_n y^n) \times$$
$$\left(n + S_1 y + S_2 y^2 + \ldots + S_{p-1} y^{p-1} + y^p \sum_{r=1}^{n} \frac{a_r^p}{1-a_r y}\right) \quad (5.11)$$

and, by equating the coefficients of like powers of y,
$$a_0 S_1 + na_1 = (n-1)a_1,$$
$$a_0 S_2 + a_1 S_1 + na_2 = (n-2)a_2,$$
$$\ldots\ldots\ldots\ldots\ldots\ldots\ldots\ldots,$$
$$a_0 S_r + a_1 S_{r-1} + a_2 S_{r-2} + \ldots + a_{r-1} S_1 + na_r = (n-r)a_r,$$
$$\ldots\ldots\ldots\ldots\ldots\ldots\ldots\ldots.$$

On simplification these become
$$\left.\begin{aligned} a_0 S_1 + a_1 &= 0, \\ a_0 S_2 + a_1 S_1 + 2a_2 &= 0, \\ \ldots\ldots\ldots\ldots\ldots\ldots\ldots, \\ a_0 S_r + a_1 S_{r-1} + a_2 S_{r-2} + \ldots + a_{r-1} S_1 + ra_r &= 0, \end{aligned}\right\} \quad (5.12)$$

and these relations are called *Newton's formulae for the sums of powers of the roots*.

The relation (5.11) is true for any positive integer p but the form of the relations (5.12) changes when $r > n$. These relations remain true, however, for any r if it be assumed that $a_{n+1} = a_{n+2} = \ldots = 0$, so that, for example,
$$a_0 S_{n+1} + a_1 S_n + \ldots + a_n S_1 = 0,$$
$$a_0 S_{n+3} + a_1 S_{n+2} + \ldots + a_n S_3 = 0.$$

Newton's formulae can be solved successively to give $S_1, S_2, \ldots$ but considerable labour is, of course, involved if the value of S_r is separately required when r is large.

Example 11. *The roots of the equation $x^3 = ax + b$ are α, β, γ and $S_r = \alpha^r + \beta^r + \gamma^r$. Prove that $S_2 = 2a$ and $S_3 = 3b$. Show also that $6S_5 = 5S_2 S_3$.* (C.)

The equation can be written $x^3 - ax - b = 0$ so that in the notation of (5.12), $a_0 = 1, a_1 = 0, a_2 = -a, a_3 = -b, a_4 = a_5 = \ldots = 0$.

Hence (5.12) gives

$S_1 = 0$,
$S_2 - 2a = 0$,
$S_3 - aS_1 - 3b = 0$,
$S_4 - aS_2 - bS_1 = 0$,
$S_5 - aS_3 - bS_2 = 0$.

From the first three of these relations $S_2 = 2a$, $S_3 = 3b$. Substitution in the last yields $S_5 - 3ab - 2ab = 0$ so that $S_5 = 5ab$. Hence $S_5 = 5ab = \tfrac{5}{6}(2a.3b) = \tfrac{5}{6}S_2S_3$ and the required result follows immediately.

Example 12. *If x, y, z are real and if*
$$x + y + z = 0,\ x^4 + y^4 + z^4 = 50,\ x^5 + y^5 + z^5 = -50,$$
show that $x^6 + y^6 + z^6 = 262$ and find the value of $x^2 + y^2 + z^2$. (O.)

There is no need to find x, y and z separately by solving the given equations. Since $x + y + z = 0$, x, y and z can be considered as the roots of the equation $t^3 + pt + q = 0$. If $S_r = x^r + y^r + z^r$, then $S_1 = 0$, $S_4 = 50$, $S_5 = -50$ and Newton's formulae give, with $a_0 = 1$, $a_1 = 0$, $a_2 = p$, $a_3 = q$, $a_4 = a_5 = \ldots = 0$,

$S_2 + 2p = 0$, $\qquad -50 + pS_3 + qS_2 = 0$,
$S_3 + 3q = 0$, $\qquad S_6 + 50p + qS_3 = 0$.
$50 + pS_2 = 0$,

Eliminating p between the first and third of these relations, $S_2{}^2 = 100$; since x, y, z are real and S_2 is the sum of their squares, it follows that S_2 is positive and hence
$$x^2 + y^2 + z^2 = S_2 = 10.$$
The first of the above relations now gives $p = -\tfrac{1}{2}S_2 = -5$ and the second relation gives $S_3 = -3q$. Substitution in the fourth relation yields
$$-50 + 15q + 10q = 0$$
so that $q = 2$. The last relation then gives
$$x^6 + y^6 + z^6 = S_6 = -50p - qS_3 = -50p + 3q^2$$
$$= -50(-5) + 3(2)^2 = 262.$$

EXERCISES 5 (c)

1. The roots of the equation $x^4 + px^2 + qx + r = 0$ are α, β, γ, δ and $S_r = \Sigma \alpha^r$. Prove the formulae
$$S_3 = -3q,\ S_4 = 2p^2 - 4r$$
and determine the value of S_5. (O.)

2. If $\alpha + \beta + \gamma = a$, $\alpha^2 + \beta^2 + \gamma^2 = b$, $\alpha^3 + \beta^3 + \gamma^3 = c$, find $\alpha\beta\gamma$ and $\alpha^4 + \beta^4 + \gamma^4$ in terms of a, b and c. Verify that when $a = 0$, they are respectively $\tfrac{1}{3}c$ and $\tfrac{1}{2}b^2$. (C.)

3. If rs_r denotes the sum of the rth powers of the roots of the equation
$$x^5 - ax^2 - bx - c = 0,$$
prove that $s_{11} = 2s_5s_6 + s_4s_7$. (O.)

4. The roots of the equation $x^n + p_1x^{n-1} + \ldots + p_n = 0$ are $\alpha_1, \alpha_2, \ldots, \alpha_n$, and S_r denotes $\alpha_1{}^r + \alpha_2{}^r + \ldots + \alpha_n{}^r$. Express $\Sigma \alpha_1{}^2\alpha_2\alpha_3$ in terms of (i) S_r, (ii) p_r. (O.)

5] TRANSFORMATION OF EQUATIONS 121

5. S_r denotes the sum of the rth powers of the roots of an algebraical equation. If the equation is of the fourth degree and has four real roots and if $S_1 = S_5 = 0$, prove that the equation is of the form
$$x^4 + ax^2 + b = 0.$$
What other form is possible if the roots are not necessarily real? (O.)

6. Writing S_r for the sum of the rth powers of the roots of the equation
$$x^n + a_1 x^{n-1} + \ldots + a_n = 0,$$
show that $1 + a_1 t + \ldots + a_n t^n = \exp(-S_1 t - \tfrac{1}{2} S_2 t^2 - \tfrac{1}{3} S_3 t^3 - \ldots)$, when t is sufficiently small.
Find the equation of the third degree in which $S_1 = -1$, $S_2 = -2$, $S_3 = -3$ and show that $S_4 = 8\tfrac{1}{6}$. (O.)

5.8. The transformation of equations

The discussion of an algebraical equation is often aided by transforming it into another whose roots bear an assigned relation to those of the original equation. Some of the more important cases are discussed below.

(i) *Roots with signs changed.* To transform the equation $f(x) = 0$ where
$$f(x) \equiv a_0 x^n + a_1 x^{n-1} + \ldots + a_{n-1} x + a_n,$$
into another whose roots are the same as those of $f(x) = 0$ but with opposite signs it is only necessary to substitute $-y$ for x. The required equation is therefore $f(-y) = 0$, i.e.
$$a_0 y^n - a_1 y^{n-1} + \ldots + (-1)^{n-1} a_{n-1} y + (-1)^n a_n = 0.$$

(ii) *Roots multiplied by a given quantity k.* Let the given equation be $f(x) = 0$ and put $y = kx$. Then $x = y/k$ and the transformed equation will be $f(y/k) = 0$.

Example 13. *Transform the equation*
$$x^3 - \tfrac{1}{2} x^2 + \tfrac{2}{3} x - 1 = 0$$
into one whose roots are six times those of the given equation.

Let $y = 6x$ so that $x = y/6$ and the required equation is
$$\left(\frac{y}{6}\right)^3 - \frac{1}{2}\left(\frac{y}{6}\right)^2 + \frac{2}{3}\left(\frac{y}{6}\right) - 1 = 0;$$
this reduces to $y^3 - 3y^2 + 24y - 216 = 0$.

(iii) *Roots increased by a given quantity h.* To effect this transformation let $y = x + h$, so that $x = y - h$ and the equation $f(x) = 0$ becomes $f(y - h) = 0$. The chief use of this transformation is to remove an assigned term from an equation, for such a procedure often facilitates its solution. If the original equation is
$$f(x) \equiv a_0 x^n + a_1 x^{n-1} + \ldots + a_{n-1} x + a_n = 0,$$

the transformed equation (with roots increased by h) is
$$f(y - h) \equiv a_0(y - h)^n + a_1(y - h)^{n-1} + \ldots + a_{n-1}(y - h) + a_n = 0.$$
This equation reduces to
$$a_0 y^n + (a_1 - na_0 h)y^{n-1} + \{a_2 - (n - 1)a_1 h + \frac{n(n - 1)}{2}a_0 h^2\}y^{n-2} + \ldots = 0,$$
and the second term can be removed by choosing h so that
$$a_1 - na_0 h = 0. \tag{5.13}$$
If h be either of the values satisfying the quadratic equation
$$\tfrac{1}{2}n(n - 1)a_0 h^2 - (n - 1)a_1 h + a_2 = 0, \tag{5.14}$$
the third term will be absent from the transformed equation. The removal of the fourth term will require the solution of a third degree equation to give a suitable value to h and so on.

Example 14. *Solve the equation $x^4 - 12x^3 + 48x^2 - 72x + 35 = 0$ by first transforming it into an equation lacking a second term.* (C.)

The equation whose roots are those of the given equation increased by h is
$$(y - h)^4 - 12(y - h)^3 + 48(y - h)^2 - 72(y - h) + 35 = 0$$
and the term in y^3 is missing if $-12 - 4h = 0$, i.e. if $h = -3$.
With this value of h, the transformed equation is
$$(y + 3)^4 - 12(y + 3)^3 + 48(y + 3)^2 - 72(y + 3) + 35 = 0$$
and this reduces to $\quad y^4 - 6y^2 + 8 = 0.$
This can be written $\quad (y^2 - 2)(y^2 - 4) = 0$
so that $y = \pm\sqrt{2}, \pm 2$. But the roots of the original equation are related to these by $y = x + (-3)$ so that $x = y + 3$ and the required roots are $3 \pm \sqrt{2}$, 1 and 5.

(iv) *Reciprocal roots.* To transform an equation $f(x) = 0$ into an equation whose roots are the reciprocals of those of $f(x) = 0$, let $y = 1/x$ so that $x = 1/y$ and the required equation is $f(1/y) = 0$. One of the main uses of this transformation is to evaluate expressions involving symmetric functions of negative powers of the roots.

Example 15. *If S_r denotes the sum of the rth powers of the roots of the equation $x^4 - 4x^3 - 2x^2 + 1 = 0$, find the value of S_{-4}.*

The equation whose roots are the reciprocals of those of the given equation is
$$\frac{1}{y^4} - \frac{4}{y^3} - \frac{2}{y^2} + 1 = 0,$$
or, $y^4 - 2y^2 - 4y + 1 = 0$. The sum of the rth powers of the roots of this equation is therefore S_{-r} and Newton's formulae (5.12) give

$S_{-1} = 0,$ $\quad\quad S_{-3} - 2S_{-1} - 12 = 0,$
$S_{-2} - 4 = 0,$ $\quad\quad S_{-4} - 2S_{-2} - 4S_{-1} + 4 = 0.$

The first two and last of these give $S_{-4} = 4$.

RECIPROCAL EQUATIONS

Further examples involving the transformation of algebraical equations will be found below. The reader should note the various devices employed and work similar exercises for himself.

Example 16. *Find the equation whose roots are the squares of the roots of* $x^4 + x^3 + 2x^2 + x + 1 = 0$.

The required equation is obtained by writing $y = x^2$, or $x = \sqrt{y}$, and is
$$y^2 + y\sqrt{y} + 2y + \sqrt{y} + 1 = 0.$$
This can be written $(y + 1)^2 = -(y + 1)\sqrt{y}$,
which, after squaring, reduces to $y^4 + 3y^3 + 4y^2 + 3y + 1 = 0$.

Example 17. *If α, β, γ are the roots of the equation $x^3 + ax^2 + bx + c = 0$, find the equation whose roots are $\beta + \gamma - 2\alpha$, $\gamma + \alpha - 2\beta$, $\alpha + \beta - 2\gamma$.*

Since $\alpha + \beta + \gamma = -a$, the roots of the required equation can be written $-a - 3\alpha$, $-a - 3\beta$, $-a - 3\gamma$. Hence the equation required is obtained by substituting for x in the given equation through the relation $y = -a - 3x$. Hence
$$\left(\frac{-a-y}{3}\right)^3 + a\left(\frac{-a-y}{3}\right)^2 + b\left(\frac{-a-y}{3}\right) + c = 0,$$
which reduces to $y^3 + (9b - 3a^2)y - 2a^3 + 9ab - 27c = 0$.

5.9. Reciprocal equations

If an equation remains unaltered when x is changed into $1/x$, it is called a *reciprocal equation*. If the original equation is
$$a_0 x^n + a_1 x^{n-1} + a_2 x^{n-2} + \ldots + a_{n-1} x + a_n = 0,$$
the equation obtained by writing $1/x$ for x and clearing of fractions is
$$a_n x^n + a_{n-1} x^{n-1} + a_{n-2} x^{n-2} + \ldots + a_1 x + a_0 = 0.$$
If these equations are the same,
$$\frac{a_0}{a_n} = \frac{a_1}{a_{n-1}} = \frac{a_2}{a_{n-2}} = \ldots = \frac{a_{n-1}}{a_1} = \frac{a_n}{a_0},$$
and hence, using the first and last relation, $a_0/a_n = \pm 1$. There are, therefore, two types of reciprocal equations:

(a) if $a_0/a_n = 1$, then
$$a_1 = a_{n-1},\ a_2 = a_{n-2},\ \ldots,$$
and the coefficients of the terms equidistant from the beginning and end of the left-hand side of the given equation are equal;

(b) if $a_0/a_n = -1$, then
$$a_1 = -a_{n-1},\ a_2 = -a_{n-2},\ \ldots,$$
and the coefficients of the terms equidistant from the beginning and end are equal in magnitude but opposite in sign; in addition, for an equation of this type of even degree $n = 2m$, $a_m = -a_m$ so that $a_m = 0$ and the middle term is absent.

Every reciprocal equation can be reduced to a reciprocal equation of type (a) of even degree and such an equation can be considered as the *standard form* of this type of equation. For if the given equation is of type (a) and its degree $n = 2m$, it is already in the required form. If it is of type (a) and $n = 2m + 1$, it can be written

$$a_0 x^{2m+1} + a_1 x^{2m} + a_2 x^{2m-1} + \ldots + a_2 x^2 + a_1 x + a_0 = 0,$$

or, $a_0(x^{2m+1} + 1) + a_1 x(x^{2m-1} + 1) + a_2 x^2(x^{2m-3} + 1) + \ldots = 0$,

and, on division by $(x + 1)$, a reciprocal equation of the standard form is obtained. If the given equation is of type (b), it can be written

$$a_0 x^n + a_1 x^{n-1} + a_2 x^{n-2} + \ldots - a_2 x^2 - a_1 x - a_0 = 0,$$

or, $a_0(x^n - 1) + a_1 x(x^{n-2} - 1) + a_2 x^2(x^{n-4} - 1) + \ldots = 0$.

Factors of the left-hand side are $(x^2 - 1)$ and $(x - 1)$ in the cases in which n is even and odd respectively and division by these factors gives a reciprocal equation of the required form.

Example 18. *Reduce $6x^6 - 25x^5 + 31x^4 - 31x^2 + 25x - 6 = 0$ to a reciprocal equation of the standard form.*

The given equation is of type (b) and of even degree. The left-hand side therefore contains a factor $(x^2 - 1)$ and, after division by this, the required equation is $6x^4 - 25x^3 + 37x^2 - 25x + 6 = 0$, which, being of type (a) and of even degree, is of the standard form.

A reciprocal equation of type (a) and of degree $2m$ (i.e. of standard form) can be reduced to an equation of degree m, and this fact is often of use in effecting the solution of such equations. To prove this, let the equation be

$$a_0 x^{2m} + a_1 x^{2m-1} + a_2 x^{2m-2} + \ldots + a_2 x^2 + a_1 x + a_0 = 0.$$

Then, on division by x^m and rearrangement,

$$a_0\left(x^m + \frac{1}{x^m}\right) + a_1\left(x^{m-1} + \frac{1}{x^{m-1}}\right) + a_2\left(x^{m-2} + \frac{1}{x^{m-2}}\right) + \ldots = 0.$$

Now, $x^{p+1} + \dfrac{1}{x^{p+1}} = \left(x^p + \dfrac{1}{x^p}\right)\left(x + \dfrac{1}{x}\right) - \left(x^{p-1} + \dfrac{1}{x^{p-1}}\right)$,

so that, if $y = x + 1/x$, writing $p = 1, 2, 3, \ldots$ in turn, gives

$$x^2 + \frac{1}{x^2} = y^2 - 2,$$

$$x^3 + \frac{1}{x^3} = y(y^2 - 2) - y = y^3 - 3y,$$

$$x^4 + \frac{1}{x^4} = y(y^3 - 3y) - (y^2 - 2) = y^4 - 4y^2 + 2,$$

. . . .,

and so on. In general, $x^m + (1/x^m)$ is of degree m and the equation in y is therefore of degree m.

EXERCISES

Example 19. *Solve the equation* $6x^6 - 25x^5 + 31x^4 - 31x^2 + 25x - 6 = 0$.

It has already been shown in Example 18 above that two roots of this equation are ± 1 and that the remaining roots are those of the reciprocal equation $6x^4 - 25x^3 + 37x^2 - 25x + 6 = 0$. This can be written, after division by x^2,

$$6\left(x^2 + \frac{1}{x^2}\right) - 25\left(x + \frac{1}{x}\right) + 37 = 0.$$

Writing $y = x + (1/x)$, $y^2 - 2 = x^2 + (1/x^2)$, this becomes

$$6(y^2 - 2) - 25y + 37 = 0.$$

This can be written $6y^2 - 25y + 25 = 0$, or, $(3y - 5)(2y - 5) = 0$ so that $y = 5/2, 5/3$. Since $y = x + 1/x$, the remaining roots are given by

$$x + \frac{1}{x} = \frac{5}{2} \text{ and } x + \frac{1}{x} = \frac{5}{3}.$$

These reduce to the quadratic equations $2x^2 - 5x + 2 = 0$, $3x^2 - 5x + 3 = 0$ with roots $2, 1/2$ and $(5 \pm i\sqrt{11})/6$ respectively.

EXERCISES 5 (d)

1. Prove that the equation $a_0 x^3 + 3a_1 x^2 + 3a_2 x + a_3 = 0$ can be reduced to the form $z^3 + 3Hz + G = 0$ by diminishing the roots by a suitable (positive or negative) constant. Show also that H and G are related to the coefficients of the original equation by

   $$a_0^2 H = a_0 a_2 - a_1^2, \quad a_0^3 G = a_0^2 a_3 - 3a_0 a_1 a_2 + 2a_1^3,$$

   and apply the transformation to the equation $x^3 - 6x^2 + 10x - 3 = 0$.

2. In the equation $8x^3 + 36x^2 + 40x + 12 = 0$ substitute $x = az + \beta$ and choose a, β so that the resulting equation in z shall have (i) the coefficient of z^3 equal to unity and (ii) no term in z^2. Show that $z = 1$ satisfies this equation and hence deduce all the roots of the original equation.

3. The roots of the equation $x^3 - ax^2 + bx - c = 0$ are a, β, γ. Form the equation whose roots are $\beta + \gamma, \gamma + a, a + \beta$.
   Hence, or otherwise, show that $(\beta + \gamma)(\gamma + a)(a + \beta) = ab - c$ and express

   $$\frac{1}{\beta + \gamma} + \frac{1}{\gamma + a} + \frac{1}{a + \beta}$$

   in terms of a, b and c.

4. Solve the equation $x^4 - 10x^3 + 26x^2 - 10x + 1 = 0$.

5. By means of the substitution $x - 1/x = y$, or otherwise, solve the equation
   $$5x^4 - 19x^3 - 34x^2 + 19x + 5 = 0. \quad \text{(W.)}$$

6. Find whether any of the roots of the equation
   $$x^5 + 8x^4 + 6x^3 - 42x^2 - 19x - 2 = 0$$
   are integers, and solve it completely. (C.)

T.M.A.—5*

5.10. The condition for common roots

If two equations $f_1(x) = 0, f_2(x) = 0$ have a common root, they are satisfied by one and the same value of x. If, as is theoretically possible, one equation is solved for x and this value is substituted in the other, the result (called the *eliminant* of $f_1(x) = 0, f_2(x) = 0$) is the required relation satisfied by the coefficients of the two equations when they possess a common root. Some examples are given which show methods of determining the eliminant in special cases—it is not always necessary, nor indeed is it always desirable or easy, actually to solve one of the equations to effect the elimination of x.

(a) If the two equations are $ax^2 + bx + c = 0$ and $Ax + B = 0$, the second equation can be solved to give $x = -B/A$. When this is substituted in the first equation and after some reduction has been made there results

$$aB^2 - bAB + cA^2 = 0$$

as the required condition for a common root.

(b) If the two equations are $ax^2 + bx + c = 0, Ax^2 + Bx + C = 0$ and the common root is x, then the two equations hold simultaneously. They can be solved as simultaneous equations in x^2 and x to give

$$x^2 = \frac{bC - cB}{aB - bA}, \quad x = \frac{aC - cA}{bA - aB}.$$

Squaring the second of these and substituting in the first, the required condition for a common root is

$$(aC - cA)^2 = (aB - bA)(bC - cB).$$

(c) If the two equations are $ax^2 + bx + c = 0$ and $Ax^3 + Bx^2 + Cx + D = 0$, by multiplying respectively by Ax, a and subtracting, the equations $ax^2 + bx + c = 0$ and $(aB - bA)x^2 + (aC - cA)x + aD = 0$ have a common root. x can now be eliminated between these two quadratic equations as in (b) above.

Example 20. *Find the condition that the equations $x^3 + px + q = 0$ and $x^3 + rx + s = 0$ shall have a common root.*

Subtracting the given equations, $(p - r)x + q - s = 0$, so that

$$x = \frac{s - q}{p - r}.$$

Substituting this value of x in the first equation and clearing of fractions, the required condition is

$$(s - q)^3 + p(s - q)(p - r)^2 + q(p - r)^3 = 0.$$

5.11. Repeated roots

If $x = a$ is a root of order r of the equation $f(x) = 0$, then

$$f(x) \equiv (x - a)^r \phi(x),$$

REPEATED ROOTS

where $\phi(x)$ is a function of x such that $\phi(a) \neq 0$. Differentiating with respect to x,

$$f'(x) = (x-a)^r \phi'(x) + r(x-a)^{r-1} \phi(x)$$
$$= (x-a)^{r-1} \{(x-a)\phi'(x) + r\phi(x)\}.$$

The expression in { } does not vanish when $x = a$ for $\phi(a) \neq 0$ so that *if $x = a$ is a root of order r of $f(x) = 0$, then $x = a$ is a root of order $r - 1$ of $f'(x) = 0$.* Continuing this method of argument it follows that if $f^{(s)}(x)$ denotes the sth derivative of $f(x)$ and $s < r$, then a root of order r of $f(x) = 0$ is a root of order $r - s$ of $f^{(s)}(x) = 0$.

As a corollary it follows that the condition for the equation $f(x) = 0$ to have a double root is the same as the condition for the equations $f(x) = 0, f'(x) = 0$ to have a common root. This condition can be found by the method of the preceding section.

Example 21. *Find the values of a for which the equation $x^3 + x^2 - 8x + a = 0$ has a repeated root. Solve the equation for these values of a.* (N.U.)

Let $f(x) \equiv x^3 + x^2 - 8x + a$ so that $f'(x) = 3x^2 + 2x - 8$. The given equation has a repeated root if it has a root common with one of those of the equation $3x^2 + 2x - 8 = 0$. This quadratic equation can be written $(x + 2)(3x - 4) = 0$ and its roots are therefore -2 and $4/3$.
If -2 is also a root of the original equation

$$(-2)^3 + (-2)^2 - 8(-2) + a = 0,$$

leading to $a = -12$; if $4/3$ is a root of the original equation

$$\left(\frac{4}{3}\right)^3 + \left(\frac{4}{3}\right)^2 - 8\left(\frac{4}{3}\right) + a = 0,$$

giving $a = 176/27$, and these two values of a are those required.
When $a = -12$, -2 is a repeated root of the equation and thus $(x + 2)^2$ is a factor. By division, the other root is in this case given by $x - 3 = 0$ and the three roots are therefore $-2, -2, 3$. Similarly when $a = 176/27$ it will be found that the roots are $4/3, 4/3, -11/3$.

Example 22. *If the equation $x^4 - (a+b)x^3 + (a-b)x - 1 = 0$ has a double root, prove that $a^{2/3} - b^{2/3} = 2^{2/3}$.* (C.)

If the equation has a double root, the equations

$$f(x) \equiv x^4 - (a+b)x^3 + (a-b)x - 1 = 0,$$
$$f'(x) \equiv 4x^3 - 3(a+b)x^2 + a - b = 0,$$

will have a common root and the required result will be obtained by eliminating x between these two equations. This can be done as follows: multiply the second equation by x and subtract the first giving

$$a + b = \frac{3x^4 + 1}{2x^3};$$

multiply the second equation by x and subtract three times the first equation giving

$$a - b = \frac{x^4 + 3}{2x}.$$

Adding and subtracting the relations giving $a + b$ and $a - b$,

$$2a = \frac{3x^4 + 1}{2x^3} + \frac{x^4 + 3}{2x} = \frac{3x^4 + 1 + x^6 + 3x^2}{2x^3} = \frac{(x^2 + 1)^3}{2x^3} = \frac{1}{2}\left(x + \frac{1}{x}\right)^3,$$

$$2b = \frac{3x^4 + 1}{2x^3} - \frac{x^4 + 3}{2x} = \frac{3x^4 + 1 - x^6 - 3x^2}{2x^3} = -\frac{(x^2 - 1)^3}{2x^3}$$

$$= -\frac{1}{2}\left(x - \frac{1}{x}\right)^3.$$

Hence $\quad x + \dfrac{1}{x} = 2^{2/3}a^{2/3},\ x - \dfrac{1}{x} = -2^{2/3}b^{2/3}$

and, by squaring and subtracting, the required result follows.

5.12. Newton's method of approximation to the roots of equations

Suppose that a is an approximation to a root of the equation $f(x) = 0$ and that the actual value of this root is $a + h$ where h is small. If then

$$f(x) \equiv a_0 x^n + a_1 x^{n-1} + a_2 x^{n-2} + \ldots + a_{n-1} x + a_n,$$
$$0 = f(a + h) = a_0(a + h)^n + a_1(a + h)^{n-1} + a_2(a + h)^{n-2} + \ldots + a_{n-1}(a + h) + a_n$$
$$= a_0 a^n + a_1 a^{n-1} + a_2 a^{n-2} + \ldots + a_{n-1} a + a_n$$
$$+ h\{na_0 a^{n-1} + (n-1)a_1 a^{n-2} + (n-2)a_2 a^{n-3} + \ldots + a_{n-1}\}$$
$$+ \tfrac{1}{2}h^2\{n(n-1)a_0 a^{n-2} + (n-1)(n-2)a_1 a^{n-3}$$
$$+ (n-2)(n-3)a_2 a^{n-4} + \ldots + a_{n-2}\}$$
$$+ \text{terms involving higher powers of } h.$$

This can be written

$$f(a) + hf'(a) + \tfrac{1}{2}h^2 f''(a) + \ldots = 0.$$

Hence $\quad h = -\dfrac{f(a)}{f'(a)} - \dfrac{1}{2}h^2 \dfrac{f''(a)}{f'(a)} - \ldots$

and it follows that a better approximation to the root $x = a$ is

$$x = a - \frac{f(a)}{f'(a)}.$$

If this approximation be denoted by β, the argument may be repeated and a still better approximation will be given by $\beta - f(\beta)/f'(\beta)$ and the argument may be carried still further until the root is determined as accurately as is required.

From the graphical point of view, suppose (Fig. 22) that PQ represents the function $y = f(x)$ in the neighbourhood of a real root of $f(x) = 0$. Let R be the point on the x-axis immediately below Q and let $OR = a$. Suppose PQ cuts the x-axis at S, so that OS represents the true value of the root and OR is an approximation to it. Let the tangent to the curve at Q meet Ox in T. Then

$$QR = f(a) \text{ and } \tan(\text{angle } QTR) = f'(a).$$

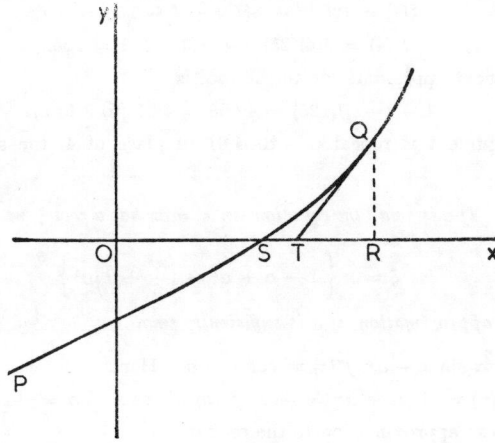

Fig. 22

From the right-angled triangle QTR it follows that
$$TR = QR \cot (\text{angle } QTR) = f(a)/f'(a)$$
and hence $\quad OT = OR - TR = a - f(a)/f'(a).$

It is clear from the diagram that T is nearer to S than is R and hence the value of x corresponding to T gives a better approximation to the root than the value at R.

In applying this method it is necessary to use a fairly close approximation to the root if a good result is to be obtained by one application of the method. The reader should note that the method applies to all types of equation although the foregoing analysis has been given only when $f(x)$ is a polynomial.

In the practical application of Newton's method, it is often more convenient to express the rule in the following equivalent form: if the required root is known to lie between $x = a$ and $x = a + k$, where k is small, a good approximation to the true value of the root is
$$a - \frac{kf(a)}{f(a+k) - f(a)}.$$

The value of this form lies in the fact that a root can often be located approximately by finding two values of x for which $f(x)$ has opposite signs and these two values of x can then be used as a and $a + k$.

Example 23. *Given that a root of the equation*
$$(x^2 + 9)^{3/2} + 8x^2 + x - 258 = 0$$
is close to 4, find the value of this root correct to three significant figures. (O.C.)

Here $f(x) \equiv (x^2 + 9)^{3/2} + 8x^2 + x - 258$ and
$$f'(x) \equiv 3x(x^2 + 9)^{1/2} + 16x + 1.$$

Thus $f(4) = (25)^{3/2} + 8(16) + 4 - 258 = -1,$
and $f'(4) = 3(4)(25)^{1/2} + 16(4) + 1 = 125.$
Thus the next approximation to the root is
$$4 - (-1)/(125) = 4{\cdot}008 = 4{\cdot}01 \text{ (to 3 figs.)}.$$
If the argument is repeated with 4·01 in place of 4, the same result is obtained.

Example 24. *If a is small the equation $\sin x = ax$ has a root ξ, nearly equal to π. Show that*
$$\xi = \pi\left\{1 - a + a^2 - \left(\frac{\pi^2}{6} + 1\right)a^3\right\}$$
is a better approximation, if a is sufficiently small. (C.)

Here $f(x) \equiv \sin x - ax$, $f'(x) \equiv \cos x - a$. Hence
$$f(\pi) = \sin \pi - a\pi = -a\pi, \; f'(\pi) = \cos \pi - a = -1 - a,$$
and the next approximation to the root is
$$\pi - \frac{f(\pi)}{f'(\pi)} = \pi - \frac{a\pi}{1+a}.$$
Now, $f\left(\pi - \frac{a\pi}{1+a}\right) = \sin\left(\pi - \frac{a\pi}{1+a}\right) - a\left(\pi - \frac{a\pi}{1+a}\right)$
$$= \sin\left(\frac{a\pi}{1+a}\right) - \frac{a\pi}{1+a},$$
$$f'\left(\pi - \frac{a\pi}{1+a}\right) = \cos\left(\pi - \frac{a\pi}{1+a}\right) - a = -\cos\left(\frac{a\pi}{1+a}\right) - a.$$
A better approximation to the root is therefore
$$\pi - \frac{a\pi}{1+a} - \frac{f\left(\pi - \frac{a\pi}{1+a}\right)}{f'\left(\pi - \frac{a\pi}{1+a}\right)}$$
$$= \frac{\pi}{1+a} + \frac{\sin\left(\frac{a\pi}{1+a}\right) - \frac{a\pi}{1+a}}{\cos\left(\frac{a\pi}{1+a}\right) + a}.$$

Now (see *Advanced Level Pure Mathematics*, Exercises 13 (d), 5, 6 or this volume, Chapter 8, § 8.12),
$$\sin x = x - \frac{x^3}{6} + \ldots, \; \cos x = 1 - \frac{x^2}{2} + \ldots$$
and if these series, limited to the third order terms, are substituted, the above approximation to the root becomes
$$\frac{\pi}{1+a} - \frac{\dfrac{\pi^3 a^3}{6(1+a)^3}}{1 + a - \dfrac{\pi^2 a^2}{2(1+a)^2}};$$
expansion by the binomial theorem, neglect of terms in a^4 and above, and some reduction then leads to the required result.

EXERCISES 5 (e)

1. The roots of the equation $x^4 + 4bx + c = 0$ are known to be a, a, β, γ. Express a in terms of b and c, assuming that $b \neq 0$. (L.I.C.)

2. Prove that the equation $x^3 + px + q = 0$ has a repeated root if $4p^3 + 27q^2 = 0$. Reduce the equation $x^3 - x^2 - 8x + 12 = 0$ to the form $z^3 + pz + q = 0$ and hence, or otherwise, solve it.

3. Find what values a and b must have in order that the equation $x^5 + ax^3 + bx - 6 = 0$ may have a real repeated root of order 3. (C.)

4. Find, correct to three decimal places, the root of $x^4 - 8x = 60$ which is nearly equal to 3. (C.)

5. If n is a large positive integer, there is a root of the equation $x \sin x = 1$ nearly equal to $2n\pi$. Show that a better approximation is $2n\pi + \dfrac{1}{2n\pi}$. (O.C.)

6. Use Newton's method of approximating to a root of an equation to find the positive square root of 2 correct to three decimal places.

5.13. The cubic equation

So far only equations of a special type (for example, reciprocal equations) or those in which there exist special relations between the roots or coefficients have been considered. In this section, a brief discussion is given of the general solution of the equation of the third degree (the *cubic* equation).

The general form of the cubic equation is
$$a_0 x^3 + 3a_1 x^2 + 3a_2 x + a_3 = 0, \qquad (5.15)$$
but, by writing $x = z - (a_1/a_0)$, this can be transformed into the equation
$$z^3 + 3Hz + G = 0, \qquad (5.16)$$
where
$$a_0^2 H = a_0 a_2 - a_1^2, \quad a_0^3 G = a_0^2 a_3 - 3a_0 a_1 a_2 + 2a_1^3. \qquad (5.17)$$
The solution of the general cubic (5.15) is known therefore whenever that of the simpler equation (5.16) is available and the equation $z^3 + 3Hz + G = 0$ can therefore be taken as the standard form of the cubic equation.

To solve the equation $z^3 + 3Hz + G = 0$, let $z = u + v$ so that
$$z^3 = u^3 + v^3 + 3uv(u+v) = u^3 + v^3 + 3uvz.$$
Hence the standard form of the cubic can be written
$$u^3 + v^3 + 3(uv + H)z + G = 0. \qquad (5.18)$$
At present, u and v are any two quantities subject only to the condition that their sum is equal to one of the roots of the standard cubic. If

they are made to satisfy the further condition that $uv + H = 0$, it follows from (5.18) that

$$u^3 + v^3 = -G, \quad u^3v^3 = -H^3.$$

Hence u^3, v^3 can be regarded as the roots of the quadratic equation

$$t^2 + Gt - H^3 = 0, \qquad (5.19)$$

so that

$$u^3 = -\frac{G}{2} + \sqrt{\left(\frac{G^2}{4} + H^3\right)},$$

$$v^3 = -\frac{G}{2} - \sqrt{\left(\frac{G^2}{4} + H^3\right)},$$

and a root of the standard cubic is given, through the relation $z = u+v$, by

$$z = \left\{-\frac{G}{2} + \sqrt{\left(\frac{G^2}{4} + H^3\right)}\right\}^{1/3} + \left\{-\frac{G}{2} - \sqrt{\left(\frac{G^2}{4} + H^3\right)}\right\}^{1/3}. \qquad (5.20)$$

This solution was obtained by Tartaglia some four hundred years ago; it was first published by Cardan and is usually known as *Cardan's solution* although there seems little doubt that he was not its real inventor.

Considering the solution (5.20) in more detail:

(*a*) If $G^2 + 4H^3 > 0$, u^3 and v^3 are both real and, if $\bar{u}$, $\bar{v}$ are their arithmetical cube roots, possible values for u, v are $\bar{u}$, $\omega\bar{u}$, $\omega^2\bar{u}$ and $\bar{v}$, $\omega\bar{v}$, $\omega^2\bar{v}$, where ω is a complex cube root of unity. Since, in Cardan's solution, u, v have been chosen so that $uv + H = 0$, the product uv must equal the real quantity $-H$ and the only combinations consistent with this give, as the three roots of the cubic,

$$\bar{u} + \bar{v}, \quad \omega\bar{u} + \omega^2\bar{v}, \quad \omega^2\bar{u} + \omega\bar{v}.$$

Hence in this case *the cubic has one real and two complex roots*.

(*b*) If $G^2 + 4H^3 = 0$, u^3 and v^3 are real and both equal to $-G/2$ and, if $\bar{u}$ is the arithmetical cube root of this quantity, the three roots of the cubic are $\bar{u} + \bar{u}$, $\omega\bar{u} + \omega^2\bar{u}$, $\omega^2\bar{u} + \omega\bar{u}$. Since $1 + \omega + \omega^2 = 0$, the three roots can be written

$$2\bar{u}, \quad -\bar{u}, \quad -\bar{u},$$

and, in this case, *the cubic has three real roots, two of which are equal*.

(*c*) If $G^2 + 4H^3 < 0$, u^3 and v^3 are complex and their cube roots will be conjugate complex quantities of the form $p \pm iq$. The roots of the cubic will be

$$p + iq + p - iq, \quad \omega(p + iq) + \omega^2(p - iq) \text{ and } \omega^2(p + iq) + \omega(p - iq).$$

Since $1 + \omega + \omega^2 = 0$ and $\omega - \omega^2 = i\sqrt{3}$, these reduce to

$$2p, \quad -p - q\sqrt{3} \text{ and } -p + q\sqrt{3}$$

so that *all three roots are real*. When Cardan's method of solution was first obtained, complex numbers had not been introduced and this case was accordingly often referred to as *irreducible*. Although calculations to extract the cube root of a complex quantity can now be made, they are usually troublesome and it is better in this case to employ the following trigonometrical method (see also Chapter 6, § 6.2). Since G is real and $G^2 + 4H^3 < 0$, H must be negative. Let $z = 2\sqrt{(-H)} \cos \theta$ so that substitution in the equation $z^3 + 3Hz + G = 0$ gives

$$8(-H)^{3/2} \cos^3 \theta - 6(-H)^{3/2} \cos \theta = -G,$$

which, since $\cos 3\theta = 4 \cos^3 \theta - 3 \cos \theta$, becomes

$$2(-H)^{3/2} \cos 3\theta = -G.$$

In a numerical case, three values of θ can then be found to satisfy this equation.

Example 25. *Solve the cubic* $x^3 - 9x + 28 = 0$.

Here $G = 28$, $H = -3$, so that the quadratic (5.19) giving u^3, v^3 is $t^2 + 28t + 27 = 0$.
The roots of this are -1, -27 and the arithmetical cube roots $\bar{u}$, $\bar{v}$ of these quantities are $\bar{u} = -1$, $\bar{v} = -3$. The required roots of the cubic are $\bar{u} + \bar{v}$, $\omega \bar{u} + \omega^2 \bar{v}$, $\omega^2 \bar{u} + \omega \bar{v}$, or $-1 - 3$, $-\omega - 3\omega^2$, $-\omega^2 - 3\omega$. These reduce to -4, $2 \pm i\sqrt{3}$.

Example 26. *Solve the equation* $x^3 - 3x - 1 = 0$.

Here $G = -1$, $H = -1$ so that $G^2 + 4H^3 = -3$. This is the irreducible case and the required trigonometrical substitution is $x = 2 \cos \theta$ so that $8 \cos^3 \theta - 6 \cos \theta - 1 = 0$. Hence $2 \cos 3\theta = 1$ and $\cos 3\theta = 1/2 = \cos (\pi/3)$. The three values of θ satisfying this equation are $\pi/9$, $7\pi/9$, and $13\pi/9$ so that the required roots are $2 \cos (\pi/9)$, $2 \cos (7\pi/9)$, $2 \cos (13\pi/9)$. To three places of decimals, these are 1.879, -1.532 and -0.347.

5.14. The quartic equation

A general solution of an algebraical equation of the fourth degree (often called a *quartic* or *biquadratic* equation) was first given by Ferrari, one of Cardan's pupils. The quartic is written for convenience in the form

$$a_0 x^4 + 4a_1 x^3 + 6a_2 x^2 + 4a_3 x + a_4 = 0, \qquad (5.21)$$

and this can be written

$$(a_0 x^2 + 2a_1 x + a_2 + 2\lambda)^2 = (2\alpha x + \beta)^2 \qquad (5.22)$$

if λ, α and β are chosen so that

$$4a_1^2 + 2a_0(a_2 + 2\lambda) - 4\alpha^2 = 6a_0 a_2,$$
$$4a_1(a_2 + 2\lambda) - 4\alpha\beta = 4a_0 a_3,$$
$$(a_2 + 2\lambda)^2 - \beta^2 = a_0 a_4.$$

These relations can be rearranged to give $\alpha^2 = a_0 \lambda + a_1^2 - a_0 a_2$, $\alpha\beta = 2a_1 \lambda + a_1 a_2 - a_0 a_3$ and $\beta^2 = (2\lambda + a_2)^2 - a_0 a_4$. Eliminating α, β between these three relations,

$(2a_1\lambda + a_1a_2 - a_0a_3)^2 = (a_0\lambda + a_1^2 - a_0a_2)\{(2\lambda + a_2)^2 - a_0a_4\}$

which reduces to

$4a_0\lambda^3 - a_0(a_0a_4 - 4a_1a_3 + 3a_2^2)\lambda + (a_2^2 - a_0a_4)(a_1^2 - a_0a_2)$
$\qquad - (a_1a_2 - a_0a_3)^2 = 0.$ (5.23)

From this cubic in λ (called the *reducing cubic*), one real root can always be found and values of α and β can be deduced from the relations expressing α^2 and β^2 in terms of λ. Values of λ, α, β being now supposed known, the quartic can be reduced from (5.22) into the two quadratic equations

$$a_0x^2 + 2a_1x + a_2 + 2\lambda = \pm(2\alpha x + \beta). \qquad (5.24)$$

A slightly different solution was given by Descartes. Here the equation is first supposed to be reduced (by removing the term in x^3) to the form

$$x^4 + 6Hx^2 + 4Gx + K = 0, \qquad (5.25)$$

and it is assumed that constants k, l, m can be found so that

$$x^4 + 6Hx^2 + 4Gx + K \equiv (x^2 - 2kx + l)(x^2 + 2kx + m).$$

This requires that

$$l + m - 4k^2 = 6H,\ k(l - m) = 2G,\ lm = K.$$

The first two of these relations give

$$l = 2k^2 + 3H + \frac{G}{k},\ m = 2k^2 + 3H - \frac{G}{k}, \qquad (5.26)$$

so that, substitution in the third relation gives

$$(2k^3 + 3Hk + G)(2k^3 + 3Hk - G) = Kk^2,$$

which reduces to

$$4k^6 + 12Hk^4 + (9H^2 - K)k^2 - G^2 = 0.$$

This is a cubic in k^2 which always (since the last term is essentially negative) has one real positive root. Thus a real value for k can be found; the corresponding values of l and m follow from equations (5.26) and the solution of the original quartic is obtained by solving the two quadratic equations

$$x^2 - 2kx + l = 0,\ x^2 + 2kx + m = 0.$$

Of these two methods of solution, that of Ferrari is probably the better for general use. It does not require the preliminary removal of the term in x^3 and the reducing cubic does not contain a second term.

Example 27. *Solve the quartic equation $x^4 + 32x = 60$.*

Here $a_0 = 1$, $a_1 = a_2 = 0$, $a_3 = 8$, $a_4 = -60$ and the reducing cubic (5.23) becomes, after a little reduction, $\lambda^3 + 15\lambda - 16 = 0$. One root of this cubic is clearly $\lambda = 1$. With the given values of the coefficients the relations giving α, β are

$$\alpha^2 = \lambda,\ \beta^2 = 4\lambda^2 + 60,\ \alpha\beta = -8.$$

With $\lambda = 1$, $a = \pm 1$, $\beta = \pm 8$ and, since $a\beta = -8$, possible values of a, β are $a = 1$, $\beta = -8$ or $a = -1$, $\beta = 8$. With these values of λ, a and β, equation (5.24) shows that the quartic can be written
$$x^2 + 2 = \pm(2x - 8),$$
so that the quartic reduces to the two quadratic equations $x^2 + 2x - 6 = 0$ and $x^2 - 2x + 10 = 0$. Solving these quadratic equations, the required roots of the quartic are $-1 \pm \sqrt{7}$ and $1 \pm 3i$. The reader should note that the work becomes very heavy if it is not easy to spot a root of the reducing cubic.

The foregoing discussion of cubic and quartic equations has deliberately been kept very brief. It is possible to carry the discussion much further and many interesting results will be found in works which devote more space to this topic. The general solution of algebraical equations of a degree higher than the fourth has not been obtained and it is known that such a solution is not possible. In cases where the coefficients of the equation are given *numerically*, numerical methods exist which give the value of any real root to any required degree of accuracy. A full account of these methods will be found, for example, in the *Theory of Equations* by Burnside and Panton.

EXERCISES 5 (f)

1. Solve the cubic equation $x^3 - 6x = 9$.
2. Solve the equation $x^3 - 12x - 16 = 0$.
3. Find the three real roots of the equation $4x^3 - 43x + 21 = 0$.
4. Remove the second term from the equation $x^3 - 8x^2 + 20x - 16 = 0$ and hence solve it.
5. Using Ferrari's method of solution, show that one root of the reducing cubic for the equation $x^4 - 3x^2 - 6x - 2 = 0$ is $1/2$ and hence find all the roots of the given quartic.
6. Use Descartes' method to solve the quartic equation of Ex. 5 above.

EXERCISES 5 (g)

1. A cubic equation $f(x) = 0$ has one real root a and complex roots $\beta \pm i\gamma$. Points A, B, C representing the roots are plotted in the Argand diagram, A representing the real root. Show that the roots of the derived equation $f'(x) = 0$ are complex if A falls inside one of two equilateral triangles described on BC as base. (O.)
2. Prove that the necessary condition that the points representing in the Argand diagram the roots of the equation
$$x^4 + 4ax^3 + 6bx^2 + 4cx + d = 0$$
shall form a square is that $b = a^2$, $c = a^3$. (C.)
3. If $f(x) = 0$ is a cubic equation with real roots a, β, γ in order of magnitude, show that one root of $f'(x) = 0$ lies between $\frac{1}{2}(a + \beta)$ and $\frac{1}{3}(2a + \beta)$ and the other between $\frac{1}{2}(\beta + \gamma)$ and $\frac{1}{3}(\beta + 2\gamma)$. (O.)

136   MATHEMATICAL ANALYSIS   [5

4. By inspection, or otherwise, find all the real roots of each of the equations

(i) $(x - 1)^3 + (x - 2)^3 = 0$,
(ii) $(x - 1)^4 + (x - 2)^4 = 1$.   (C.)

5. Prove that if $a_1, a_2, \ldots, a_n$ are the roots of the equation $x^n + nax = b$, then
$$(a_1 - a_2)(a_1 - a_3) \ldots (a_1 - a_n) = n(a_1^{n-1} + a).$$
Show that the equation whose roots are $a_1^{n-1}, a_2^{n-1}, \ldots, a_n^{n-1}$ is
$$z(z + na)^{n-1} = b^{n-1}.$$
Deduce that the product of the squared differences of the roots of the original question is
$$(-1)^{\frac{1}{2}(n-1)(n+2)} n^n \{(n-1)^{n-1} a^n + b^{n-1}\}.$$ (O.C.)

6. Show that $a + b$ is a root of the equation $x^3 - 3abx - (a^3 + b^3) = 0$. Hence find a root of the equation $4x^3 - 6x - 3 = 0$.   (W.)

7. The roots of the quadratic equation $ax^2 + bx + c = 0$ are α, β; those of the quadratic equation $a'x^2 + b'x + c' = 0$ are γ, δ. Form the equation whose roots are
$$\frac{\alpha}{\gamma} + \frac{\beta}{\delta} \text{ and } \frac{\alpha}{\delta} + \frac{\beta}{\gamma}.$$ (L.U.)

8. The roots of the equation in x, $x^3 + px + q = 0$ are α, β, γ. Express p and q in terms of β and γ only.
If one root is k times another, prove that
$$p^3(k^2 + k)^2 + q^2(k^2 + k + 1)^3 = 0$$
and show that, if $q \neq 0$, one root is then expressible in the form
$$\frac{-q(k^2 + k + 1)}{p(k^2 + k)},$$
and find the other two roots.   (L.I.C.)

9. Solve the equation $x^4 - 10x^3 + 24x^2 + 4x - 4 = 0$ whose roots $\alpha, \beta, \gamma, \delta$ satisfy $\alpha\beta + \gamma\delta = 0$.
Form also the quadratic equation whose roots are $\alpha\gamma + \beta\delta$ and $\alpha\delta + \beta\gamma$.   (O.)

10. The roots of the quartic equation
$$(x^2 + 1)^2 = ax(1 - x^2) + b(1 - x^4),$$
satisfy the equation $x^3 + px^2 + qx + r = 0$.
Prove that $p^2 - q^2 - r^2 + 1 = 0$.   (C.)

11. Given that α, β, γ are the roots of $t^3 + at^2 + bt + c = 0$, find the equation whose roots are
$$\xi = \alpha^2 - \beta\gamma, \; \eta = \beta^2 - \gamma\alpha, \; \zeta = \gamma^2 - \alpha\beta,$$
and prove that the equation whose roots are $\xi^2 - \eta\zeta$, $\eta^2 - \zeta\xi$, $\zeta^2 - \xi\eta$, is
$$t^3 + a\varrho t^2 + b\varrho^2 t + c\varrho^3 = 0$$
where $\varrho = a(3b - a^2)$.

Hence, or otherwise, solve the equations
$$\alpha^2 - \beta\gamma = \xi,\ \beta^2 - \gamma\alpha = \eta,\ \gamma^2 - \alpha\beta = \zeta,$$
for α, β, γ in terms of ξ, η, ζ. (O.)

12. If $\alpha, \beta, \gamma, \delta$ are the roots of the quartic equation
$$ax^4 + 4bx^3 + 6cx^2 + 4dx + e = 0,$$
show that $\alpha\beta + \gamma\delta,\ \alpha\gamma + \beta\delta,\ \alpha\delta + \beta\gamma$ are the roots of $a^3y^3 - 6a^2cy^2 + (16abd - 4a^2e)y - 16ad^2 + 24ace - 16b^2e = 0$.
Find the cubic whose roots are $(\alpha + \beta)(\gamma + \delta)$, $(\alpha + \gamma)(\beta + \delta)$ and $(\alpha + \delta)(\beta + \gamma)$. (O.)

13. The roots of the equation $x^2 + px + q = 0$ are denoted by α and β, while S_n denotes $\alpha^n + \beta^n$. By expressing $S_n{}^2 - S_{2n}$ and $S_nS_{n+1} - S_{2n+1}$ in terms of α and β verify that
$$S_{2n} = S_n{}^2 - 2q^n,\ S_{2n+1} = S_nS_{n+1} + pq^n.$$
Find an expression for S_7 in terms of p and q. (O.C.)

14. Prove that, if $bc + ca + ab = 0$, then
$$3S_4 + S_1{}^4 = 4S_1S_3,$$
where S_r denotes $a^r + b^r + c^r$. (O.)

15. If $S_r = a_1{}^r + a_2{}^r + \ldots + a_n{}^r$ and $a_1, a_2, \ldots, a_n$ are the roots of the equation $x^n + ax + b = 0$, show that $S_r = 0$ for $1 \leqslant r \leqslant n - 2$. Find S_{n-1}, S_n and S_{n^2}. (C.)

16. When $r \geqslant 1$, S_r is defined in terms of ω, x, y, z by the equation $rS_r = \omega^r + x^r + y^r + z^r - 1$.
Prove that, if $S_1 = S_2 = 0$, then
$$S_4 = S_5 \text{ and } S_7 = (S_3 + 1)S_4.$$ (O.)

17. If α, β, γ are the roots of the cubic equation $x^3 - px - q = 0$, show by expanding $\log(1 - px^2 - qx^3)$, or otherwise, that
$$\alpha^m + \beta^m + \gamma^m = m\Sigma\frac{(\lambda + \mu - 1)!}{(\lambda)!(\mu)!}p^\lambda q^\mu,$$
where the summation extends over all values of λ and μ such that $2\lambda + 3\mu = m$. (C.)

18. Form the equation whose roots are $\omega^{-1}p + \omega q,\ p + q,\ \omega p + \omega^{-1}q$ where $\omega^3 = 1 (\omega \neq 1)$.
Solve the equation $x^3 + 6x^2 - 12x + 32 = 0$ by reducing it to the form of equation so obtained. (C.)

19. The equation $x^6 - 6x^5 + 5x^4 + 20x^3 - 14x^2 - 28x - 8 = 0$ is unaltered by replacing x by $2 - x$. Use this information to solve it. (O.)

20. (i) If $\alpha, \beta, \gamma, \delta$ are the roots of the equation
$$x^4 - px^3 + qx^2 - px + r = 0,$$
show that $\dfrac{(\alpha + \beta)(\alpha + \gamma)(\alpha + \delta)}{1 + \alpha^2}$
has the same value whichever root α denotes.

(ii) Solve the equation when $p = 2, q = -1, r = 1$. (O.)

21. If the polynomial $ax^3 + x^2 - 3bx + 3b^2$ has two coincident zeros, show that, in general, it is a perfect cube.
Hence, or otherwise, show that if $x^4 + 4ax^3 + 2x^2 - 4bx + 3b^2$ has three equal zeros then the fourth zero is identical with them, and find for what values of a and b this is the case. (C.)

22. Given that the equation $f(x) = 0$ has two pairs of equal roots, where $f(x)$ is defined by
$$f(x) \equiv x^4 - px^3 + qx^2 - rx + s,$$
with $p > 0$, prove that $r^2 = sp^2$ and $4pq - p^3 = 8r$.
Prove also that $f(x)$ has a stationary value when $x = p/4$ and that, if $p^3 > 16r$, the roots in both pairs are real. (O.C.)

23. Show that between any two consecutive even integers there is one, and only one, real root of the equation
$$\frac{1}{2x} = \tan \frac{\pi x}{2}.$$
Prove that for a large value of n, the root between $2n$ and $2(n+1)$ is approximately $2n + 1/(2\pi n)$.
Prove similar results for the equation $4x = \tan(\pi x/2)$, with the result $2n + 1 - 1/\{2\pi(2n+1)\}$. (C.)

24. If the cubic equation $a_0 x^3 + 3a_1 x^2 + 3a_2 x + a_3 = 0$ is written in the form
$$A(x + p)^3 + B(x + q)^3 = 0,$$
show that p, q are roots of the equation
$$(a_0 a_2 - a_1^2)t^2 - (a_0 a_3 - a_1 a_2)t + (a_1 a_3 - a_2^2) = 0.$$
Hence find the condition that the roots of the cubic may be real. (O.)

25. Find for what values of the constant a the equation $x^3 - 3x + a = 0$ has three distinct real roots.
Show that if $h > 0$, the equation $x^3 - 3x - 2 - 27h = 0$ has just one real root and that, if this is denoted by $2 + 3\xi$, then $0 < \xi < h$; and with the aid of this result obtain the narrower limits
$$\frac{h}{(1 + h)^2} < \xi < h. \qquad \text{(C.)}$$

CHAPTER 6

ANALYTICAL TRIGONOMETRY

6.1. De Moivre's theorem

It has been shown in Chapter 4, equation (4.19), that when n is a positive integer,
$$(\cos\theta + i\sin\theta)^n = \cos n\theta + i\sin n\theta. \tag{6.1}$$
This is a particular case of the more general result that *when n is any rational number, then $\cos n\theta + i\sin n\theta$ is a value of $(\cos\theta + i\sin\theta)^n$*. This general result is of great importance in analytical trigonometry and is usually known as *De Moivre's theorem*. To complete a proof of the theorem it is sufficient to consider here only the cases in which n is a negative integer or a fraction.

Taking first the case when n is a negative integer, put $n = -m$ so that m is a positive integer. Then
$$(\cos\theta + i\sin\theta)^n = (\cos\theta + i\sin\theta)^{-m}$$
$$= \frac{1}{(\cos\theta + i\sin\theta)^m} = \frac{1}{\cos m\theta + i\sin m\theta}, \tag{6.2}$$
using the result (4.19) of Chapter 4 as m is a positive integer. Now
$$\frac{1}{\cos m\theta + i\sin m\theta} = \frac{\cos m\theta - i\sin m\theta}{\cos^2 m\theta - i^2\sin^2 m\theta}$$
$$= \cos m\theta - i\sin m\theta$$
$$= \cos(-m\theta) + i\sin(-m\theta)$$
$$= \cos n\theta + i\sin n\theta.$$
Substitution in (6.2) then completes the proof when n is a negative integer.

Next, suppose that n is a fraction and put $n = p/q$ where p, q are integers and no loss of generality results if q is taken to be positive. By (6.1), since q is a positive integer,
$$\left(\cos\frac{p\theta}{q} + i\sin\frac{p\theta}{q}\right)^q = \cos p\theta + i\sin p\theta.$$
Since p is a positive or negative integer
$$\cos p\theta + i\sin p\theta = (\cos\theta + i\sin\theta)^p,$$
and it follows therefore that
$$\left(\cos\frac{p\theta}{q} + i\sin\frac{p\theta}{q}\right)^q = (\cos\theta + i\sin\theta)^p.$$

Hence* $\cos\dfrac{p\theta}{q} + i\sin\dfrac{p\theta}{q}$ is a qth root of $(\cos\theta + i\sin\theta)^p$
and it follows that

$\cos\dfrac{p\theta}{q} + i\sin\dfrac{p\theta}{q}$ is a value of $(\cos\theta + i\sin\theta)^{p/q}$.

The theorem has therefore been proved for all rational values of n.

It has been shown above that, when n is a fraction p/q, $\cos(p\theta/q) + i\sin(p\theta/q)$ is one of the values of $(\cos\theta + i\sin\theta)^{p/q}$ but it is useful to find also the other values of this quantity. To do this suppose that $\varrho(\cos a + i\sin a)$ represents any value of $(\cos\theta + i\sin\theta)^{p/q}$, then

$$\varrho^q(\cos a + i\sin a)^q = (\cos\theta + i\sin\theta)^p,$$

and, since p and q are integers,

$$\varrho^q(\cos qa + i\sin qa) = \cos p\theta + i\sin p\theta.$$

By equating the real and imaginary parts

$$\varrho^q \cos qa = \cos p\theta,\ \varrho^q \sin qa = \sin p\theta.$$

Squaring and adding, $\varrho^q = 1$ and, since ϱ is the modulus of a complex quantity, it follows that $\varrho = 1$. The above relations between a and θ then reduce to

$$\cos qa = \cos p\theta,\ \sin qa = \sin p\theta$$

so that $\qquad qa = 2r\pi + p\theta,$

where r is an integer or zero. Taking, in succession, $r = 0, 1, 2, \ldots, (q-1)$,

$$\cos\dfrac{p\theta}{q} + i\sin\dfrac{p\theta}{q},\ \cos\left(\dfrac{2\pi}{q} + \dfrac{p\theta}{q}\right) + i\sin\left(\dfrac{2\pi}{q} + \dfrac{p\theta}{q}\right),$$

$$\cos\left(\dfrac{4\pi}{q} + \dfrac{p\theta}{q}\right) + i\sin\left(\dfrac{4\pi}{q} + \dfrac{p\theta}{q}\right), \ldots,$$

$$\cos\left\{\dfrac{2(q-1)\pi}{q} + \dfrac{p\theta}{q}\right\} + i\sin\left\{\dfrac{2(q-1)\pi}{q} + \dfrac{p\theta}{q}\right\},$$

are all values of $(\cos\theta + i\sin\theta)^{p/q}$. These q values are all distinct and there are no further values given by other values of r, since any other integral value of r will differ from any one of $0, 1, 2, \ldots, (q-1)$ by a multiple of q.

A convenient notation† for $\cos\theta + i\sin\theta$ is cis θ, so that the q

* So far no meaning has been assigned to symbols such as $\sqrt[q]{z}$ or $z^{p/q}$ where z is a complex number and p, q are integers. It is natural to adopt the definitions used in elementary algebra for real values of z and $\sqrt[q]{z}$ or $z^{1/q}$ is defined as a number Z which satisfies the equation $Z^q = z$.

† The reader should note that, since the argument of the product of two or more complex numbers is equal to the sum of the arguments,

$$\text{cis } \theta_1 \text{ cis } \theta_2 \ldots \text{ cis } \theta_n = \text{cis } (\theta_1 + \theta_2 + \ldots + \theta_n).$$

values of $(\cos\theta + i\sin\theta)^{p/q}$ can be written

$$\text{cis}\left(\frac{2r\pi}{q} + \frac{p\theta}{q}\right), r = 0, 1, 2, \ldots, (q-1).$$

Since there are q different values of $(\cos\theta + i\sin\theta)^{p/q}$, it is convenient to call one of these the *principal value*. The convention usually adopted is to take cis $(p\theta/q)$ to be the principal value only when $-\pi < \theta \leqslant \pi$. If θ lies outside these limits, and k is a positive or negative integer such that $-\pi < \theta + 2k\pi \leqslant \pi$, the principal value is taken as

$$\text{cis}\left(\frac{2pk\pi}{q} + \frac{p\theta}{q}\right).$$

With this convention, it should be noted that the principal value of $(\cos\theta + i\sin\theta)^{p/q}$ where $-\pi < \theta \leqslant \pi$ is not the same as the principal value of $(\cos p\theta + i\sin p\theta)^{1/q}$ unless $-\pi < p\theta \leqslant \pi$. The principal value of the qth root of $(\cos\theta + i\sin\theta)^p$, sometimes called the *principal qth root* of $(\cos\theta + i\sin\theta)^p$ is taken to mean the principal value of $(\cos\theta + i\sin\theta)^{p/q}$ as defined above.

6.2. Fractional powers of complex numbers

Suppose r is any real positive number, p, q are integers and q is positive. The notation $\sqrt[q]{(r^p)}$ denotes the unique positive qth root of r^p. Since any complex number z can be written in the form $r(\cos\theta + i\sin\theta)$ where $r = |z|$ and $-\pi < \theta \leqslant \pi$, the values of $z^{p/q}$ can be written

$$\sqrt[q]{(r^p)} \text{ cis}\left(\frac{2s\pi}{q} + \frac{p\theta}{q}\right),$$

where $s = 0, 1, 2, \ldots, (q-1)$. The principal value of $z^{p/q}$ is, since $-\pi < \theta \leqslant \pi$, $\sqrt[q]{(r^p)}$ cis $(p\theta/q)$.

As a simple example, consider the two values of $a^{1/2}$, where a is a positive real quantity. Since $|a| = a$, arg $a = 0$, the required values are $\sqrt{(a)}$ cis $(s\pi)$ where $s = 0$ and 1. Since $\cos 0° + i\sin 0° = 1$, $\cos \pi + i\sin \pi = -1$, these reduce to $\pm\sqrt{(a)}$.

Since $|1| = 1$, arg $1 = 0$, the nth roots of unity are given by cis $(2s\pi/n)$, $s = 0, 1, 2, \ldots, n-1$. Written out in full these are

$$1, \cos\frac{2\pi}{n} + i\sin\frac{2\pi}{n}, \cos\frac{4\pi}{n} + i\sin\frac{4\pi}{n}, \ldots,$$

$$\cos\frac{2(n-1)\pi}{n} + i\sin\frac{2(n-1)\pi}{n}.$$

If ω is used to denote $\cos(2\pi/n) + i\sin(2\pi/n)$, De Moivre's theorem shows that the complex nth roots of unity can be written $\omega, \omega^2, \ldots, \omega^{n-1}$. In particular, the three cube roots of unity are $1, \omega, \omega^2$ where now

$$\omega = \cos\frac{2\pi}{3} + i\sin\frac{2\pi}{3} = -\frac{1}{2} + \frac{i\sqrt{3}}{2};$$

these were previously obtained by another method in § 4.5.

Example 1. *Obtain, in the form of numerical surds, the four fourth roots of* $(3 + 4i)/5$.

Let $(3 + 4i)/5 = r(\cos\theta + i\sin\theta)$ so that $r\cos\theta = 3/5$ and $r\sin\theta = 4/5$. By squaring and adding, it follows that $r = 1$, $\cos\theta = 3/5$ and $\sin\theta = 4/5$.

Hence $\left(\dfrac{3+4i}{5}\right)^{1/4} = \text{cis}\left(\dfrac{2s\pi}{4} + \dfrac{\theta}{4}\right)$, where $s = 0, 1, 2, 3$.

These four values reduce respectively to

$$\cos\frac{\theta}{4} + i\sin\frac{\theta}{4}, \ -\sin\frac{\theta}{4} + i\cos\frac{\theta}{4}, \ -\cos\frac{\theta}{4} - i\sin\frac{\theta}{4}, \ \sin\frac{\theta}{4} - i\cos\frac{\theta}{4}.$$

Since $\cos\theta = 3/5$, $\sin\theta = 4/5$, $0 < \theta < 90°$ and

$$\cos\tfrac{1}{2}\theta = \left\{\tfrac{1}{2}(1 + \cos\theta)\right\}^{1/2} = \left\{\tfrac{1}{2}\left(1 + \tfrac{3}{5}\right)\right\}^{1/2} = \left(\tfrac{4}{5}\right)^{1/2} = \frac{2\sqrt{5}}{5},$$

since $\tfrac{1}{2}\theta$ is necessarily less than 90° and hence its cosine is positive. Again,

$$\cos\frac{\theta}{4} = \left\{\tfrac{1}{2}\left(1 + \cos\tfrac{\theta}{2}\right)\right\}^{1/2} = \left\{\tfrac{1}{2}\left(1 + \frac{2\sqrt{5}}{5}\right)\right\}^{1/2} = \sqrt{\left(\frac{5 + 2\sqrt{5}}{10}\right)},$$

$$\sin\frac{\theta}{4} = \left\{\tfrac{1}{2}\left(1 - \cos\tfrac{\theta}{2}\right)\right\}^{1/2} = \left\{\tfrac{1}{2}\left(1 - \frac{2\sqrt{5}}{5}\right)\right\}^{1/2} = \sqrt{\left(\frac{5 - 2\sqrt{5}}{10}\right)},$$

the positive square roots again being taken as $(\theta/4) < 90°$. The required values of the four fourth roots of $(3 + 4i)/5$ are therefore

$$\pm\left\{\sqrt{\left(\frac{5 + 2\sqrt{5}}{10}\right)} + i\sqrt{\left(\frac{5 - 2\sqrt{5}}{10}\right)}\right\},$$

$$\pm\left\{\sqrt{\left(\frac{5 - 2\sqrt{5}}{10}\right)} - i\sqrt{\left(\frac{5 + 2\sqrt{5}}{10}\right)}\right\}.$$

Powers and roots of complex quantities are easily represented geometrically in the Argand diagram. Suppose (Fig. 23) that P_1 represents a complex number of unit modulus and argument θ, O being

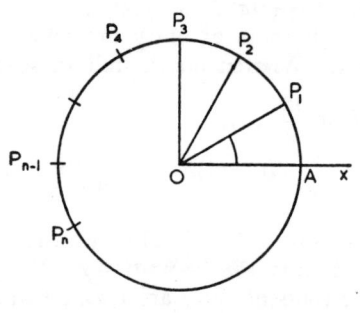

Fig. 23

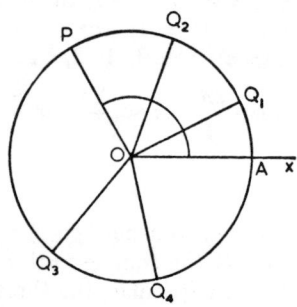

Fig. 24

the origin and Ox the real axis. Then P_1 lies on a circle, centre O and of unit radius and, if Ox meets this circle at A, the angle $P_1OA = \theta$. P_1 is the representative point of the complex number $\cos\theta + i\sin\theta$ and, following the geometrical construction for the product of two complex numbers given in § 4.7, if points $P_2, P_3, \ldots, P_{n-1}, P_n$ are taken on the circle such that each of the triangles $P_2OP_1, P_3OP_2, \ldots, P_{n-1}OP_n$ is similar to the triangle P_1OA, then $P_2, P_3, \ldots, P_n$ will represent respectively the complex numbers $(\cos\theta + i\sin\theta)^2$, $(\cos\theta + i\sin\theta)^3, \ldots, (\cos\theta + i\sin\theta)^n$. If, in a similar diagram (Fig. 24), P represents the complex number $\cos\phi + i\sin\phi$, an nth root of this number will be represented by a point Q on the circle such that the arc $AP = n$. arc AQ. Since the arc AP can be regarded as subtending at the origin O angles $\phi, \phi + 2\pi, \phi + 4\pi, \ldots$, the arc AQ subtends at O angles

$$\frac{\phi + 2s\pi}{n},$$

where s is any integer. This yields n different points $Q_1, Q_2, \ldots, Q_n$, each of which represents one of the nth roots of $\cos\phi + i\sin\phi$ and it is clear that $Q_1Q_2 \ldots Q_n$ is a regular polygon inscribed in the circle. Similarly the nth roots of the complex number $r(\cos\phi + i\sin\phi)$ of modulus r can be represented as the vertices of a regular polygon inscribed in a circle whose centre is the origin and whose radius is $\sqrt[n]{r}$.

To obtain a purely geometrical method of obtaining the nth roots of a complex number it is necessary therefore (i) to divide an angle into n equal parts, (ii) to inscribe a regular polygon of n sides in a circle and (iii) to construct a straight line whose length is the nth root of that of a given line. To obtain the nth roots of unity, only the second of these geometrical problems need be solved since, in this case, the angle to be divided into n equal parts is zero and the modulus of the number is unity. This geometrical problem can be solved by 'ruler and compasses only' when n is a power of 2, when n is a prime number of the form $2^m + 1$ (m a positive integer) and when n is the product of different prime numbers of the form $2^m + 1$ and of any power of 2 but the details are outside the scope of the present book.

EXERCISES 6 (a)

1. Use De Moivre's theorem to show that
$$\frac{(\cos 2\theta - i\sin 2\theta)^7(\cos 3\theta + i\sin 3\theta)^{-5}}{(\cos 4\theta + i\sin 4\theta)^{12}(\cos 5\theta - i\sin 5\theta)^{-6}} = \cos 107\theta - i\sin 107\theta.$$

2. Find the two square roots of $-7 + 24i$.

3. Obtain in the form of numerical surds the four imaginary fifth roots of unity. (O.)

4. Find, in the form $a + bi$, all the roots of the equation $x^4 + 1 = 0$.

5. If ω is an imaginary nth root of unity, prove that
$$1 + \omega + \omega^2 + \ldots + \omega^{n-1} = 0.$$
If $n = 5$, evaluate
$$(1 + \omega)(1 + \omega^2)^2(1 + \omega^3)^3(1 + \omega^4)^4. \qquad \text{(O.)}$$

6. Establish the identity
$$\frac{a^2(x-b)(x-c)}{(a-b)(a-c)} + \frac{b^2(x-c)(x-a)}{(b-c)(b-a)} + \frac{c^2(x-a)(x-b)}{(c-a)(c-b)} = x^2.$$
By writing $x = \text{cis } 2\theta$, $a = \text{cis } 2\alpha$, $b = \text{cis } 2\beta$ and $c = \text{cis } 2\gamma$, deduce that
$$\Sigma \sin 2(\theta + \alpha) \frac{\sin(\theta - \beta)\sin(\theta - \gamma)}{\sin(\alpha - \beta)\sin(\alpha - \gamma)} = \sin 4\theta,$$
where the summation sign refers to cyclic interchanges of α, β, γ. (O.)

6.3. Powers of cos θ and sin θ expressed in multiple angles

A convenient way of evaluating integrals such as $\int \cos^2\theta \, d\theta$ or $\int \sin^3\theta \, d\theta$ is to write them in the forms $\tfrac{1}{2}\int(1 + \cos 2\theta)d\theta$, $\tfrac{1}{4}\int(3\sin\theta - \sin 3\theta)d\theta$ respectively and it is useful to have available a quick method for expressing powers of $\cos\theta$ and $\sin\theta$ in terms of cosines and sines of multiple angles.

Let $\cos\theta + i\sin\theta = z$ so that
$$\cos\theta - i\sin\theta = \cos(-\theta) + i\sin(-\theta) = (\cos\theta + i\sin\theta)^{-1} = \frac{1}{z},$$
and hence, by addition and subtraction,
$$2\cos\theta = z + \frac{1}{z}, \quad 2i\sin\theta = z - \frac{1}{z}. \qquad (6.3)$$
Also, $\qquad \cos n\theta + i\sin n\theta = (\cos\theta + i\sin\theta)^n = z^n$
and
$$\cos n\theta - i\sin n\theta = \cos(-n\theta) + i\sin(-n\theta) = (\cos\theta + i\sin\theta)^{-n} = \frac{1}{z^n},$$
so that $\qquad 2\cos n\theta = z^n + \frac{1}{z^n}, \quad 2i\sin n\theta = z^n - \frac{1}{z^n}. \qquad (6.4)$

Powers of $\cos\theta$ and $\sin\theta$ can be expressed, with the aid of (6.3) and the binomial theorem, as powers of z and $1/z$, and these expressions can then be replaced as cosines and sines of multiple angles by means of (6.4). Details of the method should be clear from the following examples.

Example 2. *Express $\cos^4\theta$ in terms of multiple angles.*

By (6.3), if $z = \cos\theta + i\sin\theta$,
$$(2\cos\theta)^4 = \left(z + \frac{1}{z}\right)^4 = z^4 + 4z^2 + 6 + \frac{4}{z^2} + \frac{1}{z^4}$$
$$= \left(z^4 + \frac{1}{z^4}\right) + 4\left(z^2 + \frac{1}{z^2}\right) + 6$$
$$= 2\cos 4\theta + 8\cos 2\theta + 6, \text{ using (6.4)}.$$
Hence $\qquad \cos^4\theta = \tfrac{1}{8}\{\cos 4\theta + 4\cos 2\theta + 3\}.$

Example 3. *Find an expression for $\cos^5 \theta \sin^7 \theta$ as a sum of sines of multiples of θ. Hence evaluate $\int_0^{\pi/4} \cos^5 \theta \sin^7 \theta \, d\theta$.*

If $z = \cos \theta + i \sin \theta$

$$(2 \cos \theta)^5 (2i \sin \theta)^7 = \left(z + \frac{1}{z}\right)^5 \left(z - \frac{1}{z}\right)^7 = \left(z^2 - \frac{1}{z^2}\right)^5 \left(z - \frac{1}{z}\right)^2$$

$$= \left(z^{10} - 5z^6 + 10z^2 - \frac{10}{z^2} + \frac{5}{z^6} - \frac{1}{z^{10}}\right)\left(z^2 - 2 + \frac{1}{z^2}\right)$$

$$= \left(z^{12} - \frac{1}{z^{12}}\right) - 2\left(z^{10} - \frac{1}{z^{10}}\right) - 4\left(z^8 - \frac{1}{z^8}\right) + 10\left(z^6 - \frac{1}{z^6}\right)$$

$$+ 5\left(z^4 - \frac{1}{z^4}\right) - 20\left(z^2 - \frac{1}{z^2}\right).$$

Using (6.4) this reduces to, since $(i/i^7) = -1$,

$$\cos^5 \theta \sin^7 \theta = -\frac{1}{2^{11}} (\sin 12\theta - 2 \sin 10\theta - 4 \sin 8\theta + 10 \sin 6\theta$$
$$+ 5 \sin 4\theta - 20 \sin 2\theta).$$

The given integral now becomes

$$-\frac{1}{2^{11}} \left[\frac{-\cos 12\theta}{12} + \frac{2 \cos 10\theta}{10} + \frac{4 \cos 8\theta}{8} - \frac{10 \cos 6\theta}{6} - \frac{5 \cos 4\theta}{4} + \frac{20 \cos 2\theta}{2} \right]_0^{\pi/4} = \frac{11}{3840}, \text{ after a little reduction.}$$

6.4. Expressions for $\cos n\theta$, $\sin n\theta$, etc., in terms of powers of $\cos \theta$, $\sin \theta$

If n is a positive integer and if, for brevity, $\cos \theta$ is denoted by c and $\sin \theta$ by s, use of De Moivre's and the binomial theorems gives

$$\cos n\theta + i \sin n\theta = (\cos \theta + i \sin \theta)^n = (c + is)^n$$

$$= c^n + nc^{n-1} is + \frac{n(n-1)}{(2)!} c^{n-2}(is)^2 + \ldots$$

$$= \left\{ c^n - \frac{n(n-1)}{(2)!} c^{n-2} s^2 + \ldots \right\}$$

$$+ i \left\{ nc^{n-1} s - \frac{n(n-1)(n-2)}{(3)!} c^{n-3} s^2 + \ldots \right\}.$$

Hence, equating real and imaginary parts,

$$\cos n\theta = c^n - \frac{n(n-1)}{(2)!} c^{n-2} s^2 + \frac{n(n-1)(n-2)(n-3)}{(4)!} c^{n-4} s^4 - \ldots, \quad (6.5)$$

$$\sin n\theta = nc^{n-1} s - \frac{n(n-1)(n-2)}{(3)!} c^{n-3} s^3 + \ldots. \quad (6.6)$$

By writing $s^2 = 1 - c^2$, $s^4 = (1 - c^2)^2$, etc., it is clear that both $\cos n\theta$ and $\sin n\theta / \sin \theta$ can be expressed as polynomials in $\cos \theta$. Thus

$$\cos n\theta = a_n \cos^n \theta + a_{n-2} \cos^{n-2} \theta + \ldots, \quad (6.7)$$

where the a's are constants. The last term is $a_1 \cos \theta$ if n is odd and a_0 if n is even. Similarly

$$\frac{\sin n\theta}{\sin \theta} = b_{n-1} \cos^{n-1} \theta + b_{n-3} \cos^{n-3} \theta + \ldots \qquad (6.8)$$

where the b's are constants, the last term being b_0 if n is odd and $b_1 \cos \theta$ if n is even.

Example 4. *If a_r is the coefficient of $\cos^r \theta$ when $\cos n\theta$ is expressed as a polynomial in $\cos \theta$, show that*

(i) $(r^2 - n^2)a_r = (r + 2)(r + 1)a_{r+2}$; (ii) $a_n = 2^{n-1}$.

(i) Since $\cos n\theta = \Sigma a_r \cos^r \theta$, differentiation with respect to θ gives,
$$-n \sin n\theta = -\Sigma r a_r \cos^{r-1} \theta \sin \theta.$$
A second differentiation gives
$$-n^2 \cos n\theta = \Sigma\{r(r-1)a_r \cos^{r-2} \theta \sin^2 \theta - r a_r \cos^r \theta\}.$$
Writing $\cos n\theta = \Sigma a_r \cos^r \theta$ and $\sin^2 \theta = 1 - \cos^2 \theta$,
$$-n^2 \Sigma a_r \cos^r \theta = \Sigma\{r(r-1)a_r \cos^{r-2} \theta - r(r-1)a_r \cos^r \theta - r a_r \cos^r \theta\}.$$
Equating coefficients of $\cos^r \theta$,
$$-n^2 a_r = (r+2)(r+1)a_{r+2} - r(r-1)a_r - r a_r,$$
giving, $(r^2 - n^2)a_r = (r+2)(r+1)a_{r+2}.$

(ii) Putting $s^2 = 1 - c^2$ in (6.5),
$$\cos n\theta = c^n - {}^nC_2 c^{n-2}(1-c^2) + {}^nC_4 c^{n-4}(1-c^2)^2 - \ldots,$$
so that $a_n = $ coefficient of $c^n = 1 + {}^nC_2 + {}^nC_4 + \ldots = 2^{n-1}$,
using the results (2.3), (2.4) of § 2.4.

The reader should note that the two results obtained in this example enable all the coefficients in (6.7) to be calculated. The coefficients of (6.8) can then be found by differentiating (6.7) with respect to θ.

An expression giving $\tan n\theta$ in powers of $\tan \theta$ can be obtained by dividing (6.6) by (6.5). Thus

$$\tan n\theta = \frac{\sin n\theta}{\cos n\theta}$$

$$= \frac{nc^{n-1}s - \dfrac{n(n-1)(n-2)}{(3)!}c^{n-3}s^3 + \ldots}{c^n - \dfrac{n(n-1)}{(2)!}c^{n-2}s^2 + \dfrac{n(n-1)(n-2)(n-3)}{(4)!}c^{n-4}s^4 - \ldots},$$

or, dividing numerator and denominator by $\cos^n \theta$ and writing $t = \tan \theta$,

$$\tan n\theta = \frac{nt - \dfrac{n(n-1)(n-2)}{(3)!}t^3 + \ldots}{1 - \dfrac{n(n-1)}{(2)!}t^2 + \dfrac{n(n-1)(n-2)(n-3)}{(4)!}t^4 - \ldots}. \qquad (6.9)$$

FACTORISATION

The result (6.9) is a particular case of a more general result. Let $\tan \theta_1$, $\tan \theta_2$, ..., $\tan \theta_n$ be denoted by $t_1, t_2, \ldots, t_n$ and let Σ_r denote the sum of the products of $t_1, t_2, \ldots, t_n$ taken r at a time so that, if $n = 4$ for example, $\Sigma_1 = t_1 + t_2 + t_3 + t_4$, $\Sigma_2 = t_1 t_2 + t_1 t_3 + t_1 t_4 + t_2 t_3 + t_2 t_4 + t_3 t_4$, $\Sigma_3 = t_1 t_2 t_3 + t_1 t_2 t_4 + t_1 t_3 t_4 + t_2 t_3 t_4$ and $\Sigma_4 = t_1 t_2 t_3 t_4$. Then,

$$\cos(\theta_1 + \theta_2 + \ldots + \theta_n) + i \sin(\theta_1 + \theta_2 + \ldots + \theta_n)$$
$$= (\cos \theta_1 + i \sin \theta_1)(\cos \theta_2 + i \sin \theta_2) \ldots (\cos \theta_n + i \sin \theta_n)$$
$$= \cos \theta_1 \cos \theta_2 \ldots \cos \theta_n (1 + it_1)(1 + it_2) \ldots (1 + it_n)$$
$$= \cos \theta_1 \cos \theta_2 \ldots \cos \theta_n (1 + i\Sigma_1 + i^2 \Sigma_2 + i^3 \Sigma_3 + \ldots).$$

Equating real and imaginary parts,
$$\cos(\theta_1 + \theta_2 + \ldots + \theta_n) = \cos \theta_1 \cos \theta_2 \ldots \cos \theta_n (1 - \Sigma_2 + \Sigma_4 - \ldots)$$
$$\sin(\theta_1 + \theta_2 + \ldots + \theta_n) = \cos \theta_1 \cos \theta_2 \ldots \cos \theta_n (\Sigma_1 - \Sigma_3 + \ldots),$$
and by division,
$$\tan(\theta_1 + \theta_2 + \ldots + \theta_n) = \frac{\Sigma_1 - \Sigma_3 + \ldots}{1 - \Sigma_2 + \Sigma_4 - \ldots}. \tag{6.10}$$

The result (6.9) then follows by writing $\theta_1 = \theta_2 = \ldots = \theta_n = \theta$.

EXERCISES 6 (b)

1. Use the result that $2 \cos n\theta = x^n + x^{-n}$ when $2 \cos \theta = x + x^{-1}$ to solve the equation $5x^4 - 11x^3 + 16x^2 - 11x + 5 = 0$. (O.C.)

2. Prove that, if m and n are positive integers, then $\sin^m \theta \cos^n \theta$ can be expressed either as a sum of sines or else as a sum of cosines of multiples of θ.
   Express $\sin^3 \theta \cos^5 \theta$ as a sum of sines of multiples of θ. (O.C.)

3. Express $\cos^6 \theta$ and $\sin^6 \theta$ as sums of cosines of multiples of θ and hence evaluate $\int_0^{\pi/2} \cos^6 \theta \, d\theta$ and $\int_0^{\pi/2} \sin^6 \theta \, d\theta$.

4. Show that $\cos 4\theta = \cos^4 \theta - 6 \cos^2 \theta \sin^2 \theta + \sin^4 \theta$ and that $\sin 6\theta = 6 \cos^5 \theta \sin \theta - 20 \cos^3 \theta \sin^3 \theta + 6 \cos \theta \sin^5 \theta$.

5. Prove that $\sin 5\alpha / \sin \alpha = 16 \cos^4 \alpha - 12 \cos^2 \alpha + 1 = 5 - 20 \sin^2 \alpha + 16 \sin^4 \alpha$, and that
$$\sin 5\alpha = 16 \sin \alpha (\sin^2 36° - \sin^2 \alpha)(\sin^2 72° - \sin^2 \alpha)$$
$$= 4 \sin \alpha (\cos 2\alpha - \cos 72°)(\cos 2\alpha + \cos 36°). \quad \text{(O.)}$$

6. If α, β and γ are the roots of the equation $x^3 + ax^2 + bx + a = 0$, show that $\tan^{-1} \alpha + \tan^{-1} \beta + \tan^{-1} \gamma = n\pi$ except when $b = 1$.

6.5. Factorisation of $x^{2n} - 2x^n a^n \cos n\theta + a^{2n}$, etc.

If the equation
$$x^{2n} - 2x^n a^n \cos n\theta + a^{2n} = 0 \tag{6.11}$$
is regarded as a quadratic in x^n, its two roots are given by
$$a^n \cos n\theta \pm \sqrt{(a^{2n} \cos^2 n\theta - a^{2n})},$$

i.e. by $a^n(\cos n\theta \pm i \sin n\theta)$. Thus the $2n$ values of x satisfying equation (6.11) are the n values of $a(\cos n\theta \pm i \sin n\theta)^{1/n}$. These can be written as

$$a\left\{\cos\left(\theta + \frac{2r\pi}{n}\right) \pm i \sin\left(\theta + \frac{2r\pi}{n}\right)\right\} \text{ where } r = 0, 1, 2, \ldots, n-1.$$

Hence $x^{2n} - 2x^n a^n \cos n\theta + a^{2n}$ can be resolved into $2n$ factors of which a typical pair is

$$\left[x - a\left\{\cos\left(\theta + \frac{2r\pi}{n}\right) + i \sin\left(\theta + \frac{2r\pi}{n}\right)\right\}\right] \times$$
$$\left[x - a\left\{\cos\left(\theta + \frac{2r\pi}{n}\right) - i \sin\left(\theta + \frac{2r\pi}{n}\right)\right\}\right].$$

Since,

$$\left\{\cos\left(\theta + \frac{2r\pi}{n}\right) + i \sin\left(\theta + \frac{2r\pi}{n}\right)\right\} \times$$
$$\left\{\cos\left(\theta + \frac{2r\pi}{n}\right) - i \sin\left(\theta + \frac{2r\pi}{n}\right)\right\}$$
$$= \cos^2\left(\theta + \frac{2r\pi}{n}\right) + \sin^2\left(\theta + \frac{2r\pi}{n}\right) = 1,$$

this pair of factors can be written

$$x^2 - 2xa \cos\left(\theta + \frac{2r\pi}{n}\right) + a^2,$$

and there results

$$x^{2n} - 2x^n a^n \cos n\theta + a^{2n} = \prod_{r=0}^{n-1}\left\{x^2 - 2xa \cos\left(\theta + \frac{2r\pi}{n}\right) + a^2\right\}, \quad (6.12)$$

the symbol $\prod_{r=0}^{n-1}$ denoting the product for all integral values of r from $r = 0$ to $r = n - 1$ of the expression following it.

Many important results can be deduced as special cases of (6.12). Firstly, putting $a = 1$, $\theta = 0$,

$$(x^n - 1)^2 = \prod_{r=0}^{n-1}\left\{x^2 - 2x \cos \frac{2r\pi}{n} + 1\right\}$$

and, since the factor corresponding to $r = 0$ is $x^2 - 2x + 1$, this can be written

$$(x^n - 1)^2 = (x - 1)^2 \prod_{r=1}^{n-1}\left\{x^2 - 2x \cos \frac{2r\pi}{n} + 1\right\}.$$

If n is odd the factors in the product occur in pairs so that, allowing for this and taking the square root,

$$x^n - 1 = (x - 1) \prod_{r=1}^{\frac{1}{2}(n-1)}\left\{x^2 - 2x \cos \frac{2r\pi}{n} + 1\right\}, \quad (n \text{ odd}). \quad (6.13)$$

FACTORISATION

If n is even, the factors in the product again occur in pairs except for the single factor when $r = \tfrac{1}{2}n$. This factor is $x^2 - 2x \cos \pi + 1$, or, $x^2 + 2x + 1$, and, in this case, extracting the square root gives

$$x^n - 1 = (x-1)(x+1) \prod_{r=1}^{\frac{1}{2}n-1} \left\{ x^2 - 2x \cos \frac{2r\pi}{n} + 1 \right\}, \ (n \text{ even}). \quad (6.14)$$

By taking $a = 1$, $\theta = \pi/n$ and working similarly, there result

$$x^n + 1 = (x+1) \prod_{r=0}^{\frac{1}{2}(n-3)} \left\{ x^2 - 2x \cos \frac{(2r+1)\pi}{n} + 1 \right\}, \ (n \text{ odd}), \quad (6.15)$$

$$x^n + 1 = \prod_{r=0}^{\frac{1}{2}(n-2)} \left\{ x^2 - 2x \cos \frac{(2r+1)\pi}{n} + 1 \right\}, \ (n \text{ even}). \quad (6.16)$$

If in (6.12), $a = 1$, the left-hand side is divided by x^n and each of the n factors on the right is divided by x,

$$x^n + \frac{1}{x^n} - 2 \cos n\theta = \prod_{r=0}^{n-1} \left\{ x + \frac{1}{x} - 2 \cos \left(\theta + \frac{2r\pi}{n} \right) \right\}.$$

Using the result of § 6.3 that if $x + x^{-1} = 2 \cos \alpha$ then $x^n + x^{-n} = 2 \cos n\alpha$, this can be written after division by 2,

$$\cos n\alpha - \cos n\theta = 2^{n-1} \prod_{r=0}^{n-1} \left\{ \cos \alpha - \cos \left(\theta + \frac{2r\pi}{n} \right) \right\}. \quad (6.17)$$

Example 5. *Prove that, for all values of β,*

$$\sin n\beta = 2^{n-1} \prod_{r=0}^{n-1} \sin \left(\beta + \frac{r\pi}{n} \right),$$

without ambiguity of sign. (O.)

This result can be deduced as another special case of (6.12). Putting $x = a = 1$, $\theta = 2\beta$,

$$2(1 - \cos 2n\beta) = \prod_{r=0}^{n-1} \left\{ 2 - 2 \cos 2 \left(\beta + \frac{r\pi}{n} \right) \right\}$$

or, $$4 \sin^2 n\beta = \prod_{r=0}^{n-1} \left\{ 4 \sin^2 \left(\beta + \frac{r\pi}{n} \right) \right\},$$

giving, on taking the square root,

$$\sin n\beta = \pm 2^{n-1} \prod_{r=0}^{n-1} \sin \left(\beta + \frac{r\pi}{n} \right).$$

So long as β lies between 0 and π/n, $\sin n\beta$ and each factor on the right is positive. As β increases or decreases beyond these limits, $\sin n\beta$ changes sign only when β passes through $r\pi/n$ and then one factor on the right also changes sign. Hence the ambiguous sign is always positive and the required result follows.

Example 6. *Show that* $n \cot n\beta = \sum_{r=0}^{n-1} \cot \left(\beta + \frac{r\pi}{n} \right).$ (O.)

This result can be deduced from Example 5 by logarithmic differentiation. Thus, the result of Example 5 can be written

$$\log \sin n\beta = (n-1) \log 2 + \sum_{r=0}^{n-1} \log \sin \left(\beta + \frac{r\pi}{n} \right),$$

which gives, on differentiation with respect to β,
$$n \cot n\beta = \sum_{r=0}^{n-1} \cot \left(\beta + \frac{r\pi}{n}\right).$$

Example 7. *Show that the expression for $1/(z^{2n} - 1)$ in partial fractions with real quadratic denominators can be written*
$$\frac{n}{z^{2n} - 1} = \frac{1}{z^2 - 1} + \sum_{r=1}^{n-1} \frac{z \cos r\alpha - 1}{z^2 - 2z \cos r\alpha + 1}, \text{ where } \alpha = \frac{\pi}{n}.$$
By writing $z = \cos\theta + i \sin\theta$, prove that
$$n \cot n\theta = \cot\theta - \sum_{r=1}^{n-1} \frac{\sin\theta}{\cos\theta - \cos r\alpha}. \quad \text{(O.C.)}$$

By writing $2n$ in place of n and taking $\alpha = \pi/n$, the result (6.14) gives
$$x^{2n} - 1 = (x^2 - 1) \prod_{r=1}^{n-1} \{x^2 - 2x \cos r\alpha + 1\}.$$
Taking logarithms and differentiating with respect to x,
$$\frac{2nx^{2n-1}}{x^{2n} - 1} = \frac{2x}{x^2 - 1} + \sum_{r=1}^{n-1} \frac{2x - 2\cos r\alpha}{x^2 - 2x \cos r\alpha + 1}.$$
Writing $x = 1/z$, dividing by 2 and making a few reductions leads to
$$\frac{n}{z^{2n} - 1} = \frac{1}{z^2 - 1} + \sum_{r=1}^{n-1} \frac{z \cos r\alpha - 1}{z^2 - 2z \cos r\alpha + 1}.$$
This can be written
$$\frac{nz^{-n}}{z^n - z^{-n}} = \frac{z^{-1}}{z - z^{-1}} + \sum_{r=1}^{n-1} \frac{\cos r\alpha - z^{-1}}{z + z^{-1} - 2\cos r\alpha},$$
and, if $z = \cos\theta + i\sin\theta$, $z^{-n} = \cos n\theta - i \sin n\theta$, $z^{-1} = \cos\theta - i \sin\theta$, $z^n - z^{-n} = 2i \sin n\theta$, $z - z^{-1} = 2i \sin\theta$, $z + z^{-1} = 2\cos\theta$, so that
$$\frac{n(\cos n\theta - i \sin n\theta)}{2i \sin n\theta} = \frac{\cos\theta - i \sin\theta}{2i \sin\theta} + \sum_{r=1}^{n-1} \frac{\cos r\alpha - \cos\theta + i \sin\theta}{2\cos\theta - 2\cos r\alpha}.$$
Equating the imaginary parts and multiplying by 2,
$$n \cot n\theta = \cot\theta - \sum_{r=1}^{n-1} \frac{\sin\theta}{\cos\theta - \cos r\alpha}.$$

6.6. De Moivre's property of the circle

Let $A_0, A_1, A_2, \ldots, A_{n-1}$ be the vertices of a regular polygon of n sides which is inscribed in a circle centre O, radius a. Let P be a point such that $OP = x$ and the angle $POA_0 = \theta$ (Fig. 25). Then each of the angles $A_0OA_1, A_1OA_2, \ldots, A_{n-2}OA_{n-1}, A_{n-1}OA_0$ is $2\pi/n$ and the angle POA_r is $\theta + (2r\pi/n)$ for $r = 0, 1, 2, \ldots, n - 1$. Since $OP = x$, $OA_r = a$,
$$PA_r^2 = OP^2 + OA_r^2 - 2OP \cdot OA_r \cos POA_r$$
$$= x^2 + a^2 - 2xa \cos\left(\theta + \frac{2r\pi}{n}\right).$$

DE MOIVRE'S PROPERTY

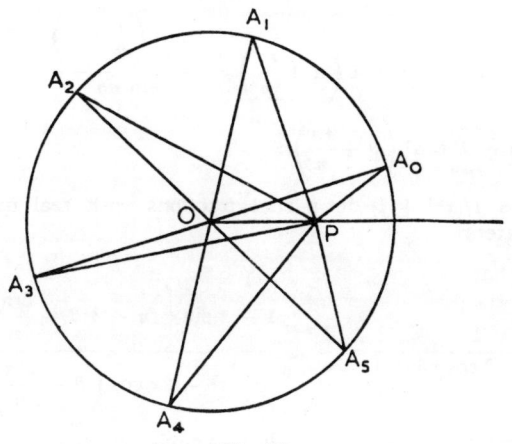

Fig. 25

Hence, from (6.12),

$$x^{2n} - 2x^n a^n \cos n\theta + a^{2n} = \prod_{r=0}^{n-1} \left\{ x^2 - 2xa \cos\left(\theta + \frac{2r\pi}{n}\right) + a^2 \right\}$$
$$= \prod_{r=0}^{n-1} PA_r^2,$$

so that

$$PA_0 . PA_1 . PA_2 . \ldots . PA_{n-1} = (x^{2n} - 2x^n a^n \cos n\theta + a^{2n})^{1/2},$$

a result known as *De Moivre's property* of the circle.

Example 8. $A_0, A_1, A_2, \ldots, A_{n-1}$ *are the vertices of a regular polygon inscribed in a circle of radius a, centre O. P is a point such that $OP = x$. Show that*
(i) *when P lies on OA_0, $PA_0 . PA_1 \ldots PA_{n-1} = |x^n - a^n|$;*
(ii) *when OP bisects the angle $A_{n-1}OA_0$, $PA_0 . PA_1 \ldots PA_{n-1} = x^n + a^n$.*

(i) When P lies on OA_0, the angle θ of § 6.6 is zero and De Moivre's property gives
$$PA_0 . PA_1 \ldots PA_{n-1} = (x^{2n} - 2x^n a^n + a^{2n})^{1/2}$$
$$= x^n - a^n, x > a,$$
$$= a^n - x^n, x < a.$$

(ii) When OP bisects the angle $A_{n-1}OA_0$, $\theta = \pi/n$, $\cos n\theta = \cos \pi = -1$ and De Moivre's property gives
$$PA_0 . PA_1 \ldots PA_{n-1} = (x^{2n} + 2x^n a^n + a^{2n})^{1/2} = x^n + a^n.$$

These two special cases of De Moivre's result are known as *Cote's properties* of the circle.

EXERCISES 6 (c)

1. Show that $\cos n\theta = 2^{n-1} \prod_{r=0}^{n-1} \sin\left(\theta + \frac{2r+1}{2n}\pi\right)$ (C.)

2. Prove that
$$\prod_{r=0}^{n-1} \cot\left(a + \frac{r\pi}{n}\right) = \frac{\sin n\left(\frac{\pi}{2} + a\right)}{\sin na}.$$ (O.)

3. Evaluate $\sum_{r=0}^{n-1} \tan^2\left(\theta + \frac{r\pi}{n}\right).$ (O.)

4. Express $1/(x^{2n} + 1)$ in partial fractions with real quadratic denominators. (C.)

5. Show that
$$\frac{1}{x^{2n} - 2x^n \cos n\theta + 1} = \frac{1}{n}\sum_{r=0}^{n-1} \frac{1 - x \sin\left\{(n-1)\theta - \frac{2r\pi}{n}\right\} \csc n\theta}{x^2 - 2x \cos\left(\theta + \frac{2r\pi}{n}\right) + 1}.$$

6. The n points $A_0, A_1, \ldots, A_{n-1}$ lie on a circle with centre O and radius a, and are the vertices of a regular polygon. C is a point on the circle such that the angle $A_0OC = \theta$ (where $0 < n\theta < \pi$) and P lies on OC between O and C.
Prove that
$$PA_0 . PA_1 \ldots PA_{n-1} \geqslant a^n \text{ if } \tfrac{1}{2}\pi < n\theta < \pi,$$
$$\geqslant a^n \sin n\theta \text{ if } 0 < n\theta < \tfrac{1}{2}\pi.$$ (O.)

6.7. Symmetrical functions of cos $(r\pi/n)$, sin $(r/\pi n)$, etc.

Many results involving cos $(r\pi/n)$, sin $(r\pi/n)$, etc., can be derived from the properties of symmetrical functions of the roots of equations. For example, the result (6.9) can be written

$$\tan n\theta \left\{1 - \frac{n(n-1)}{(2)!}t^2 + \frac{n(n-1)(n-2)(n-3)}{(4)!}t^4 - \ldots\right\}$$
$$= nt - \frac{n(n-1)(n-2)t^3}{(3)!} + \ldots,$$

where $t = \tan \theta$. This equation can be regarded as an equation in $\tan \theta$ of which the roots are

$$\tan \theta, \tan\left(\theta + \frac{\pi}{n}\right), \tan\left(\theta + \frac{2\pi}{n}\right), \ldots, \tan\left\{\theta + \frac{(n-1)\pi}{n}\right\},$$

and may therefore be used for calculating symmetrical functions of these expressions. Similarly the expression for cos $n\theta$ as a polynomial in cos θ (the result of writing $s^2 = 1 - c^2$ in (6.5)) can be used to calculate symmetrical functions of the n cosines

$$\cos\left(\theta + \frac{2r\pi}{n}\right), r = 0, 1, 2, \ldots, n-1,$$

and some illustrations follow as worked examples.

Example 9. *Express $\cos 7\theta$ as a polynomial in $\cos \theta$ and deduce that*

$$\sec^2 \frac{\pi}{14} + \sec^2 \frac{3\pi}{14} + \sec^2 \frac{5\pi}{14} = 8.$$

If $c = \cos \theta$, $s = \sin \theta$, equation (6.5) gives, with $n = 7$,
$$\cos 7\theta = c^7 - 21c^5s^2 + 35c^3s^4 - 7cs^6;$$
writing $s^2 = 1 - c^2$ and making some reductions,
$$\cos 7\theta = 64c^7 - 112c^5 + 56c^3 - 7c.$$
If $\cos 7\theta = 0$, $7\theta = \left(2r + \frac{1}{2}\right)\pi$ and $\theta = (4r + 1)\frac{\pi}{14}$, where r is an integer.
Hence the roots of the equation
$$64c^7 - 112c^5 + 56c^3 - 7c = 0$$
are $\cos (4r + 1)\frac{\pi}{14}$, $r = 0, 1, 2, \ldots, 6$, and the root $c = 0$ corresponds to the value $r = 5$.
Removing this root, the roots of
$$64c^6 - 112c^4 + 56c^2 - 7 = 0$$
are $\cos (\pi/14)$, $\cos (5\pi/14)$, $\cos (9\pi/14)$, $\cos (13\pi/14)$, $\cos (17\pi/14)$ and $\cos (25\pi/14)$. Since $\cos (9\pi/14) = -\cos (5\pi/14)$, $\cos (13\pi/14) = -\cos (\pi/14)$, $\cos (17\pi/14) = -\cos (3\pi/14)$ and $\cos (25\pi/14) = \cos (3\pi/14)$, the roots of the equation
$$64c^6 - 112c^4 + 56c^2 - 7 = 0$$
can be taken as $\pm \cos (\pi/14)$, $\pm \cos (3\pi/14)$, $\pm \cos (5\pi/14)$. Hence the roots of the cubic (obtained by writing $c^2 = x$)
$$64x^3 - 112x^2 + 56x - 7 = 0$$
are $\cos^2 (\pi/14)$, $\cos^2 (3\pi/14)$, $\cos^2 (5\pi/14)$. The relations between the roots and the coefficients then give

$\cos^2 (\pi/14) \cos^2 (3\pi/14) + \cos^2 (\pi/14) \cos^2 (5\pi/14)$
$$+ \cos^2 (3\pi/14) \cos^2 (5\pi/14) = 56/64,$$
$$\cos^2 (\pi/14) \cos^2 (3\pi/14) \cos^2 (5\pi/14) = 7/64.$$
By division,
$$\sec^2 (\pi/14) + \sec^2 (3\pi/14) + \sec^2 (5\pi/14) = 8.$$

Example 10. *Show that $\tan \alpha + \tan \left(\alpha + \frac{\pi}{3}\right) + \tan \left(\alpha + \frac{2\pi}{3}\right) = 3 \tan 3\alpha$.*

Formula (6.9) gives
$$\tan 3\theta = \frac{3t - t^3}{1 - 3t^2}, \text{ where } t = \tan \theta.$$
Since $\tan 3\theta = \tan 3\alpha$ is satisfied by $\theta = \alpha + (r\pi/3)$, where r is an integer, the roots of the equation
$$\frac{3t - t^3}{1 - 3t^2} = \tan 3\alpha,$$
i.e. $t^3 - (3 \tan 3\alpha)t^2 - 3t + \tan 3\alpha = 0$ are $\tan \alpha$, $\tan \left(\alpha + \frac{\pi}{3}\right)$ and $\tan \left(\alpha + \frac{2\pi}{3}\right)$. Since the sum of the roots of the cubic in t is $3 \tan 3\alpha$, the required result follows.

Example 11. *Find the equation whose roots are* $\tan^2(\pi/7)$, $\tan^2(2\pi/7)$, $\tan^2(4\pi/7)$. *Show that the equation whose roots are* $\tan(\pi/7)$, $\tan(2\pi/7)$, $\tan(4\pi/7)$ *is*
$$t^3 + \sqrt{7}t^2 - 7t + \sqrt{7} = 0. \tag{O.}$$

Putting $n = 7$ in (6.9), $\tan 7\theta = 0$ when
$$7t - 35t^3 + 21t^5 - t^7 = 0,$$
t being $\tan \theta$. The values of θ satisfying $\tan 7\theta = 0$ being $r\pi/7$ where r is an integer, if the root $t = 0$ is removed, the roots of
$$t^6 - 21t^4 + 35t^2 - 7 = 0$$
are $\tan(r\pi/7)$ where $r = 1, 2, \ldots, 6$. Since $\tan(\pi - \theta) = -\tan\theta$, the roots of this equation can be taken as $\pm\tan(\pi/7)$, $\pm\tan(2\pi/7)$, $\pm\tan(4\pi/7)$. Writing $t^2 = x$, the equation whose roots are $\tan^2(\pi/7)$, $\tan^2(2\pi/7)$, $\tan^2(4\pi/7)$ is therefore
$$x^3 - 21x^2 + 35x - 7 = 0.$$
Since,
$$t^6 - 21t^4 + 35t^2 - 7 \equiv -(t^3 + \sqrt{7}t^2 - 7t + \sqrt{7})(-t^3 + \sqrt{7}t^2 + 7t + \sqrt{7}),$$
the roots of the equation $t^3 + \sqrt{7}t^2 - 7t + \sqrt{7} = 0$ are three of $\pm\tan(\pi/7)$, $\pm\tan(2\pi/7)$, $\pm\tan(4\pi/7)$. The sum of the products two at a time for this equation is -7 so the roots cannot all be positive or all negative. Possibilities are therefore two positive and one negative root or two negative and one positive root. The latter possibility is excluded as the product of all three roots is $-\sqrt{7}$. Since $\tan(\pi/7)$, $\tan(2\pi/7)$ are positive and $\tan(4\pi/7)$ is negative, the required conditions are fulfilled by this selection and the required result follows.

EXERCISES 6 (d)

1. Express $\tan 7\theta = 1$ as an equation in $\tan \theta$ and show that
$$\tan\frac{\pi}{28} - \tan\frac{3\pi}{28} + \tan\frac{5\pi}{28} + \tan\frac{9\pi}{28} - \tan\frac{11\pi}{28} + \tan\frac{13\pi}{28} = 8,$$
$$\sec\frac{\pi}{7} - \sec\frac{2\pi}{7} + \sec\frac{3\pi}{7} = 4.$$

2. Prove that $\cot\frac{\pi}{7} + \cot\frac{2\pi}{7} - \cot\frac{3\pi}{7} = \sqrt{7}$
and find the value of $\cot\frac{\pi}{7}\cot\frac{2\pi}{7}\cot\frac{3\pi}{7}$. (C.)

3. Prove that the roots of $x^4 - 28x^3 + 70x^2 - 28x + 1 = 0$ are
$\tan^2(\pi/16)$, $\tan^2(3\pi/16)$, $\tan^2(5\pi/16)$ and $\tan^2(7\pi/16)$. (O.)

4. Construct a cubic equation whose roots are $\tan^2(\pi/18)$, $\tan^2(5\pi/18)$, $\tan^2(7\pi/18)$ and prove that
$$\sec\frac{\pi}{18}\sec\frac{5\pi}{18}\sec\frac{7\pi}{18} = \frac{8}{\sqrt{3}}. \tag{O.}$$

5. If $x = (2/\sqrt{7})\sin\theta$, express $(\sin 7\theta)/(\sin\theta)$ as a polynomial in x. Show that the roots of the equation $x^3 - x^2 + (1/7) = 0$ are
$(2/\sqrt{7})\sin(2\pi/7)$, $(2/\sqrt{7})\sin(4\pi/7)$, $(2/\sqrt{7})\sin(8\pi/7)$. (C.)

6. Solve completely the equation $\sin 5x = \cos 4x$.
Deduce that one root of the equation $16s^4 + 8s^3 - 12s^2 - 4s + 1 = 0$ is given by $s = \sin(\pi/18)$. Express the other three roots as sines of certain angles. (C.)

6.8. Trigonometrical functions of a complex variable

Let $z = x + iy$ and let the sum of the series (which by Example 14 of Chapter 4 is absolutely convergent)*

$$1 + \frac{z}{(1)!} + \frac{z^2}{(2)!} + \frac{z^3}{(3)!} + \cdots \qquad (6.18)$$

be denoted by exp (z). When z is real it can be seen from equation (3.13) of Chapter 3 that exp (z) reduces to e^z. If $z = iy$ where y is real and hence z is purely imaginary,

$$\exp(iy) = 1 + \frac{iy}{(1)!} + \frac{(iy)^2}{(2)!} + \frac{(iy)^3}{(3)!} + \frac{(iy)^4}{(4)!} + \frac{(iy)^5}{(5)!} + \cdots$$

$$= \left\{1 - \frac{y^2}{(2)!} + \frac{y^4}{(4)!} - \cdots\right\} + i\left\{y - \frac{y^3}{(3)!} + \frac{y^5}{(5)!} - \cdots\right\}$$

$$= \cos y + i \sin y, \qquad (6.19)$$

if the results given in Exercises 13 (d), 5 and 6 of *Advanced Level Pure Mathematics* are assumed (for a later discussion of these series for the cosine and sine, see § 8.12). Changing y into $-y$, it can be shown similarly that

$$\exp(-iy) = \cos y - i \sin y, \qquad (6.20)$$

and hence, by addition and subtraction,

$$\cos y = \frac{1}{2}\{\exp(iy) + \exp(-iy)\}, \sin y = \frac{1}{2i}\{\exp(iy) - \exp(-iy)\}. \qquad (6.21)$$

From Cauchy's theorem on the multiplication of two absolutely convergent series (§ 3.11),

$$\exp(z_1) \cdot \exp(z_2) = \left(1 + \frac{z_1}{(1)!} + \frac{z_1^2}{(2)!} + \cdots\right)\left(1 + \frac{z_2}{(1)!} + \frac{z_2^2}{(2)!} + \cdots\right)$$

$$= 1 + \frac{z_1 + z_2}{(1)!} + \frac{z_1^2 + 2z_1 z_2 + z_2^2}{(2)!} + \cdots$$

$$= \exp(z_1 + z_2). \qquad (6.22)$$

Hence, using (6.19), (6.22) and exp $(x) = e^x$ when x is real,

$$\exp(x + iy) = \exp(x) \exp(iy)$$

$$= e^x(\cos y + i \sin y). \qquad (6.23)$$

So far only the trigonometrical functions of a real variable have been considered; these have been defined from a right-angled triangle, and various properties, such as the addition theorems, have been developed from these definitions. Equation (6.20) shows the con-

---

* The function exp (z) is sometimes denoted by e^z with the advantage that the law of combination given in (6.22) is more easily remembered. With this notation it must be remembered that e^z is merely a symbol denoting the sum of the series (6.18) and that no attempt has been made to show that the sum is $(2 \cdot 718 \ldots)^z$ except when z is real.

nection between the sine and cosine and the newly introduced 'exp' function. The *generalised* trigonometrical functions are defined by relations which are an extension of (6.21). For any complex quantity z, the sine and cosine are *defined by*

$$\cos z = \frac{1}{2}\{\exp(iz) + \exp(-iz)\}, \sin z = \frac{1}{2i}\{\exp(iz) - \exp(-iz)\}, \quad (6.24)$$

and these, which are often called *Euler's* relations, are taken as the fundamental functions. The remaining trigonometrical functions are then defined in terms of these by the familiar relations

$$\tan z = \frac{\sin z}{\cos z}, \cot z = \frac{\cos z}{\sin z}, \sec z = \frac{1}{\cos z}, \csc z = \frac{1}{\sin z}. \quad (6.25)$$

Immediate consequences of the definitions (6.24) are

$$\cos z + i \sin z = \exp(iz), \cos z - i \sin z = \exp(-iz), \quad (6.26)$$

$$\cos z = 1 - \frac{z^2}{(2)!} + \frac{z^4}{(4)!} - \ldots, \sin z = z - \frac{z^3}{(3)!} + \frac{z^5}{(5)!} - \ldots \quad (6.27)$$

All the well-known properties such as $\cos 0 = 1$, $\sin(-z) = -\sin z$, $\cos^2 z + \sin^2 z = 1$, $\sec^2 z = 1 + \tan^2 z$, $\sin(z_1 + z_2) = \sin z_1 \cos z_2 + \cos z_1 \sin z_2$, etc., can be derived from (6.24), (6.18) and (6.22). Thus, for example, (6.18) shows that $\exp(0) = 1$ and (6.24) then gives

$$\cos 0 = \tfrac{1}{2}\{\exp(0) + \exp(0)\} = \tfrac{1}{2}(1 + 1) = 1.$$

Again, writing $-z$ for z in the second of (6.24),

$$\sin(-z) = \frac{1}{2i}\{\exp(-iz) - \exp(iz)\}$$

$$= -\frac{1}{2i}\{\exp(iz) - \exp(-iz)\} = -\sin z.$$

From (6.26)

$$\cos^2 z + \sin^2 z = (\cos z + i \sin z)(\cos z - i \sin z)$$
$$= \exp(iz)\exp(-iz) = \exp(0) = 1,$$

and, as a final example,

$$\sin z_1 \cos z_2 + \cos z_1 \sin z_2$$

$$= \frac{1}{4i}[\{\exp(iz_1) - \exp(-iz_1)\}\{\exp(iz_2) + \exp(-iz_2)\}$$
$$+ \{\exp(iz_1) + \exp(-iz_1)\}\{\exp(iz_2) - \exp(-iz_2)\}]$$

$$= \frac{1}{4i}[2\exp(iz_1)\exp(iz_2) - 2\exp(-iz_1)\exp(-iz_2)]$$

$$= \frac{1}{2i}[\exp\{i(z_1 + z_2)\} - \exp\{-i(z_1 + z_2)\}]$$

$$= \sin(z_1 + z_2).$$

Other familiar results of the above type should be worked by the reader who should convince himself that Euler's relations lead to no results inconsistent with those satisfied by the trigonometrical functions of a purely real variable—some examples will be found in the next set of exercises. It is possible also to show that the generalised trigonometrical functions defined in this way are periodic functions with period 2π. This is, of course, a most important property but the details are omitted here due to lack of space. Another approach will be found in *A Course of Modern Analysis*, E. T. Whittaker and G. N. Watson, 1950, Appendix.

6.9. The logarithmic function of a complex variable

The complex quantity ζ ($= \xi + i\eta$) is said to be a *logarithm* of another complex quantity $z(= x + iy)$ if $z = \exp(\zeta)$, a convenient way of expressing the inverse relationship being $\zeta = \text{Log } z$. The reader will recall the similarity of these relations to those already in use for real variables, viz. $s = \log_e t$ if $t = e^s$, s and t being real.

To express ζ in terms of z, use of the relation (6.23) gives

$$x + iy = z = \exp(\zeta) = \exp(\xi + i\eta)$$
$$= e^\xi (\cos \eta + i \sin \eta),$$

showing that the modulus of z is e^ξ and that a value of the argument of z is η. Thus $|z| = e^\xi$ giving $\xi = \log_e |z|$ and $\eta = \arg z$, so that

$$\text{Log } z = \zeta = \xi + i\eta$$
$$= \log_e |z| + i \arg z. \quad (6.28)$$

Since $\arg z$ is many-valued, the above result implies that $\text{Log } z$ is many-valued. If θ is the principal value of $\arg z$, so that $-\pi < \theta \leqslant \pi$, then the corresponding value of $\text{Log } z$ is written $\log z$ and termed the *principal value* of $\text{Log } z$. If $z = x + iy$, $|z| = +(x^2 + y^2)^{1/2}$ and $\arg z = 2n\pi + \tan^{-1}(y/x)$, n being an integer. The principal value of $\arg z$ is $\tan^{-1}(y/x)$ when $x > 0$, $\pi + \tan^{-1}(y/x)$ when $x < 0 < y$ and $-\pi + \tan^{-1}(y/x)$ when $x < 0$, $y < 0$ so that

$$\left.\begin{aligned}\log(x+iy) &= \frac{1}{2}\log_e(x^2+y^2) + i\tan^{-1}\left(\frac{y}{x}\right), x > 0, \\ &= \frac{1}{2}\log_e(x^2+y^2) + i\left\{\pi + \tan^{-1}\left(\frac{y}{x}\right)\right\}, x < 0 < y, \\ &= \frac{1}{2}\log_e(x^2+y^2) - i\left\{\pi - \tan^{-1}\left(\frac{y}{x}\right)\right\}, x < 0, y < 0.\end{aligned}\right\} \quad (6.29)$$

If $y = 0$, the principal value of $\arg z$ is 0 or π according as x is positive or negative, so that for real x,

$$\log x = \log_e x, \, x > 0, \, \log x = \log_e |x| + i\pi, \, x < 0. \quad (6.30)$$

158   MATHEMATICAL ANALYSIS   [6

If $x = 0$, the principal value of arg z is $\pm \pi/2$ according as y is positive or negative and, for real y,

$$\log(iy) = \log_e y + \frac{i\pi}{2}, y > 0, \log(iy) = \log_e |y| - \frac{i\pi}{2}, y < 0. \quad (6.31)$$

Example 12. *Write down the values of* (i) *log* 3, (ii) *log* (-3), (iii) *log i*, (iv) *log* $(-2i)$, (v) *log* $(1+i)$.

The first four results follow directly from (6.30) and (6.31). They are

$$\log 3 = \log_e 3, \log(-3) = \log_e 3 + i\pi,$$
$$\log i = i\pi/2, \text{ since } \log_e 1 = 0, \log(-2i) = \log_e 2 - i\pi/2.$$

The fifth result follows from the first of (6.29) and is

$$\log(1+i) = \frac{1}{2}\log_e(1^2 + 1^2) + i \tan^{-1}\left(\frac{1}{1}\right) = \frac{1}{2}\log_e 2 + \frac{i\pi}{4}.$$

It will be observed that when z is real and positive, the principal value of its logarithm as now defined reduces to its logarithm to base e. Certain properties of logarithms to base e remain true for the logarithms at present under consideration. Thus, since the modulus and argument of the product of two complex numbers are equal respectively to the product of their moduli and the sum of their arguments, (6.28) gives

$$\begin{aligned}
\text{Log}(z_1 z_2) &= \log_e |z_1 z_2| + i \arg(z_1 z_2) \\
&= \log_e \{|z_1| \cdot |z_2|\} + i (\arg z_1 + \arg z_2) \\
&= \log_e |z_1| + i \arg z_1 + \log_e |z_2| + i \arg z_2 \\
&= \text{Log } z_1 + \text{Log } z_2.
\end{aligned} \quad (6.32)$$

It should be noted that (6.32) is true in the sense that every value of Log z_1 + Log z_2 is equal to some value of Log $(z_1 z_2)$ and conversely; the relation is not necessarily true of principal values (see § 4.7, Chapter 4). In a similar way it can be shown that, in the same sense,

$$\text{Log}(z_1/z_2) = \text{Log } z_1 - \text{Log } z_2, \quad (6.33)$$

with the particular case Log $(1/z) = -$ Log z.

Example 13. *Show that* $i \log\left(\frac{1-i}{1+i}\right) = \frac{\pi}{2}$.

Since arg $(1-i) = 2m\pi - \pi/4$, arg $(1+i) = 2n\pi + \pi/4$, m and n being integers, and the moduli of these quantities are both $\sqrt{2}$,

$$\begin{aligned}
\text{Log}\left(\frac{1-i}{1+i}\right) &= \text{Log}(1-i) - \text{Log}(1+i) \\
&= \frac{1}{2}\log_e 2 + i\left(2m\pi - \frac{\pi}{4}\right) - \frac{1}{2}\log_e 2 - i\left(2n\pi - \frac{\pi}{4}\right) \\
&= i\left(2m - 2n - \frac{1}{2}\right)\pi.
\end{aligned}$$

The principal value is that for which $-\pi < (2m - 2n - \tfrac{1}{2})\pi \leq \pi$ and this requires that $m = n = 0$. Hence

$$\log\left(\frac{1-i}{1+i}\right) = -\frac{i\pi}{2},$$

so that

$$i\log\left(\frac{1-i}{1+i}\right) = \frac{\pi}{2}.$$

When z is a real number x such that $|x| < 1$, the series $x - \frac{x^2}{2} + \frac{x^3}{3} - \ldots$ is absolutely convergent and its sum is $\log_e(1+x)$, (see (3.14) of Chapter 3). When z is complex it can be shown that when $|z| < 1$,

$$\log(1+z) = z - \frac{z^2}{2} + \frac{z^3}{3} - \ldots, \tag{6.34}$$

the series then being absolutely convergent, but it is not proposed here to do more than state this result.

Example 14. *If x is real show that*

$$\log(1+ix) - \log(1-ix) = 2i\tan^{-1}x$$

and hence express $\tan^{-1}x$ as a series of ascending powers of x when x is numerically less than unity.

From the first of (6.29),

$$\log(1+ix) = \tfrac{1}{2}\log_e(1+x^2) + i\tan^{-1}x,$$
$$\log(1-ix) = \tfrac{1}{2}\log_e(1+x^2) + i\tan^{-1}(-x),$$

so that by subtraction,

$$\log(1+ix) - \log(1-ix) = i\tan^{-1}x - i\tan^{-1}(-x) = 2i\tan^{-1}x.$$

Using (6.34) this gives, as $|x| < 1$,

$$\tan^{-1}x = \frac{1}{2i}\left[ix - \frac{(ix)^2}{2} + \frac{(ix)^3}{3} - \frac{(ix)^4}{4} + \ldots + ix + \frac{(ix)^2}{2} + \frac{(ix)^3}{3} + \frac{(ix)^4}{4} + \ldots\right]$$
$$= x - \frac{x^3}{3} + \frac{x^5}{5} - \ldots.$$

EXERCISES 6 (*e*)

Starting from Euler's relations for the trigonometrical functions, show that:

1. $\sin 0 = 0$, $\cos(-z) = \cos z$, $\sec^2 z = 1 + \tan^2 z$.
2. (i) $\cos(z_1 + z_2) = \cos z_1 \cos z_2 - \sin z_1 \sin z_2$,
   (ii) $\cos 2z = \cos^2 z - \sin^2 z = 2\cos^2 z - 1 = 1 - 2\sin^2 z$.
3. $\sin z_1 + \sin z_2 = 2\sin\tfrac{1}{2}(z_1+z_2)\cos\tfrac{1}{2}(z_1-z_2)$.
4. Write down the values of:
   (i) $\log(-1)$,   (ii) $\log(-i)$,   (iii) $\text{Log}(1+i)$.
5. If $-\pi < 2\theta \leq \pi$, show that
   $$\log(1+\cos 2\theta + i\sin 2\theta) = \log_e(2\cos\theta) + i\theta.$$

6. By expanding $\dfrac{1}{2i} \log \left(\dfrac{1 + ix}{1 - ix} \right)$ in real form where x is real, deduce the expansion of
$$\tan^{-1} \left(\frac{2x \sin \alpha}{1 - x^2} \right)$$
in ascending powers of x when x is numerically less than unity. (O.)

6.10. The hyperbolic functions

If, in Euler's relations (6.24), i is omitted, the two functions involving exp (z) so obtained are called the hyperbolic cosine and sine of z and written respectively cosh z and sinh z. Thus

$$\cosh z = \tfrac{1}{2}\{\exp(z) + \exp(-z)\}, \sinh z = \tfrac{1}{2}\{\exp(z) - \exp(-z)\}. \tag{6.35}$$

Other hyperbolic functions, by analogy with the other trigonometrical functions, are defined by the relations

$$\left. \begin{aligned} \tanh z &= \frac{\sinh z}{\cosh z}, \; \coth z = \frac{\cosh z}{\sinh z} = \frac{1}{\tanh z}, \\ \operatorname{sech} z &= \frac{1}{\cosh z} \text{ and cosech } z = \frac{1}{\sinh z}. \end{aligned} \right\} \tag{6.36}$$

These functions of the complex variable z have important applications in analysis and this section is devoted to the derivation of their chief properties.

The relations (6.35), (6.24) show that

$$\cosh z = \cos iz, \sinh z = -i \sin iz, \tag{6.37}$$

and these, with (6.36) and (6.25), give

$$\left. \begin{aligned} \tanh z &= -i \tan iz, \coth z = i \cot iz, \\ \operatorname{sech} z &= \sec iz, \operatorname{cosech} z = i \operatorname{cosec} iz. \end{aligned} \right\} \tag{6.38}$$

Hence,

$$\left. \begin{aligned} \cosh^2 z - \sinh^2 z &= \cos^2 iz + \sin^2 iz = 1, \\ \operatorname{sech}^2 z &= \sec^2 iz = 1 + \tan^2 iz = 1 - \tanh^2 z, \\ \operatorname{cosech}^2 z &= -\operatorname{cosec}^2 iz = -1 - \cot^2 iz = -1 + \coth^2 z, \end{aligned} \right\} \tag{6.39}$$

and the reader should notice the differences in signs in these and the corresponding relations involving the trigonometrical functions. Since $\cos^2 t + \sin^2 t = 1$, the point whose coordinates are given by $x = \cos t$, $y = \sin t$ lies on the circle $x^2 + y^2 = 1$ and the trigonometrical functions are, for this reason, often called circular functions. The choice of the term 'hyperbolic' for the functions under discussion in this section arises in a similar way from the first of the relations

(6.39)—since $\cosh^2 t - \sinh^2 t = 1$, the point with coordinates $x = \cosh t$, $y = \sinh t$ lies on the rectangular hyperbola $x^2 - y^2 = 1$.

Other properties of the hyperbolic functions may be derived from (6.37), (6.38) and the known properties of the trigonometrical functions. Thus,

$$\left.\begin{array}{l}\sinh(-z) = -i\sin(-iz) = i\sin iz = -\sinh z,\\ \cosh(-z) = \cos(-iz) = \cos iz = \cosh z,\end{array}\right\} \quad (6.40)$$

and $\quad \cosh(0) = \cos(0) = 1,\ \sinh(0) = -i\sin(0) = 0.\quad (6.41)$

Addition formulae are given by

$$\sinh(z_1 \pm z_2) = -i\sin i(z_1 \pm z_2) = -i\sin iz_1 \cos iz_2 \mp i\sin iz_2 \cos iz_1$$
$$= \sinh z_1 \cosh z_2 \pm \cosh z_1 \sinh z_2, \quad (6.42)$$

and, similarly,

$$\cosh(z_1 \pm z_2) = \cosh z_1 \cosh z_2 \pm \sinh z_1 \sinh z_2. \quad (6.43)$$

Division of the last two results leads to

$$\tanh(z_1 \pm z_2) = \frac{\tanh z_1 \pm \tanh z_2}{1 \pm \tanh z_1 \tanh z_2}. \quad (6.44)$$

In a similar way, the following double-angle formulae are easily derived,

$$\left.\begin{array}{l}\sinh 2z = 2\sinh z \cosh z,\\ \cosh 2z = \cosh^2 z + \sinh^2 z = 2\cosh^2 z - 1 = 2\sinh^2 z + 1.\end{array}\right\} \quad (6.45)$$

Again the reader should note certain differences in signs in these and the corresponding relations involving the trigonometrical functions.

Two useful relations which follow from (6.35) by addition and subtraction are

$$\cosh z + \sinh z = \exp(z),\ \cosh z - \sinh z = \exp(-z), \quad (6.46)$$

and these, when z is the real quantity x, reduce to

$$\cosh x + \sinh x = e^x,\ \cosh x - \sinh x = e^{-x}. \quad (6.47)$$

From tables of e^x and e^{-x} it is a simple matter to plot the graphs of the hyperbolic functions of a real variable—those of $\cosh x$, $\sinh x$ and $\tanh x$ are shown in Figs. 26, 27 on page 162.

Finally, the definitions (6.35) and (6.18) give, for the series representation of the hyperbolic functions,

$$\left.\begin{array}{l}\cosh z = 1 + \dfrac{z^2}{(2)!} + \dfrac{z^4}{(4)!} + \cdots,\\[1em] \sinh z = z + \dfrac{z^3}{(3)!} + \dfrac{z^5}{(5)!} + \cdots\end{array}\right\} \quad (6.48)$$

Fig. 26

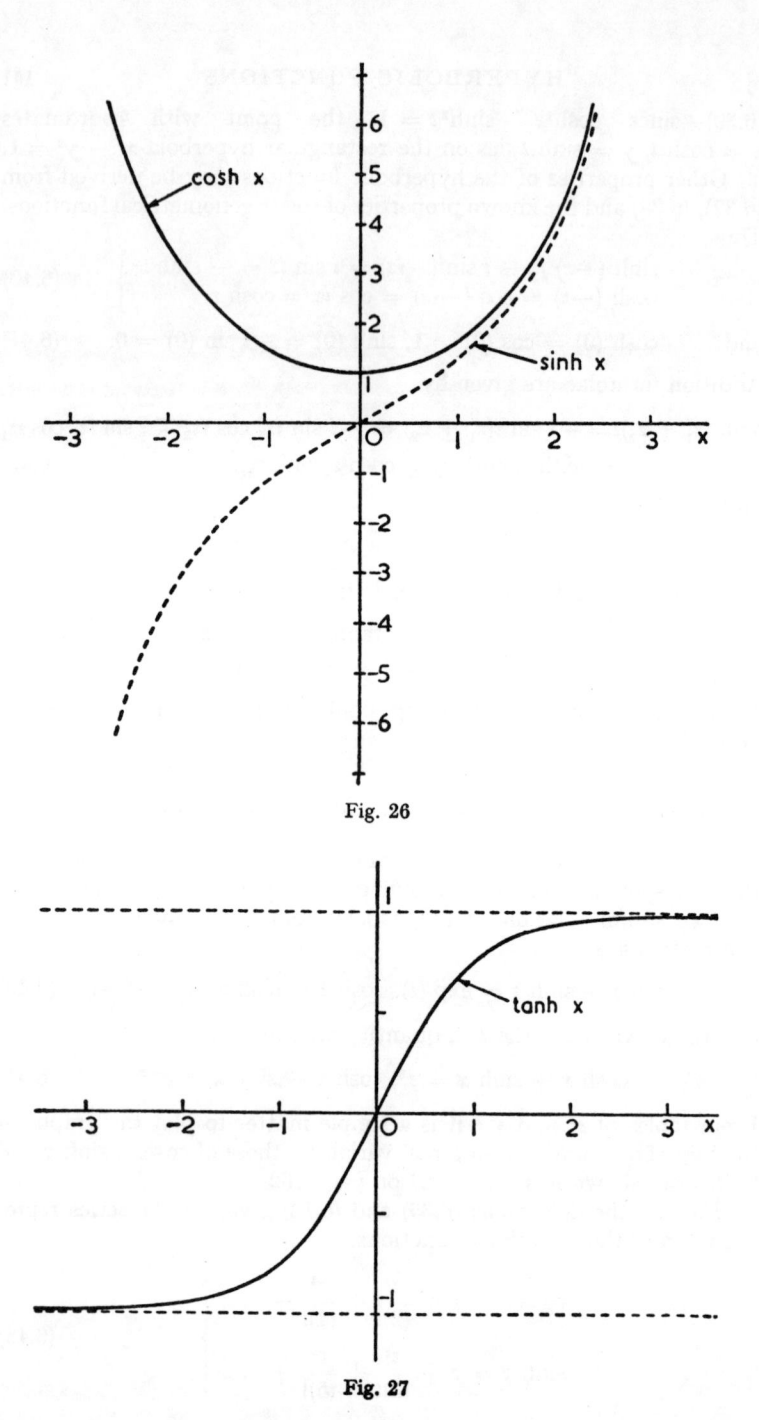

Fig. 27

6.11. The real and imaginary parts of sin $(x + iy)$, etc.

The trigonometrical and hyperbolic functions of a complex variable are easily expressed in the form $A + iB$ by means of the addition formulae and the relations of § 6.10. Thus, if x and y are real,

$$\sin (x + iy) = \sin x \cos iy + \cos x \sin iy$$
$$= \sin x \cosh y + i \cos x \sinh y, \quad (6.49)$$

so that the real and imaginary parts of sin $(x + iy)$ are respectively $\sin x \cosh y$ and $\cos x \sinh y$. Similarly

$$\cos (x + iy) = \cos x \cosh y - i \sin x \sinh y. \quad (6.50)$$

From these two relations and (6.37), it follows that

$$\sinh (x + iy) = -i \sin i(x + iy) = -i \sin (-y + ix)$$
$$= -i\{\sin (-y) \cos x + i \cos (-y) \sinh x\}$$
$$= \sinh x \cos y + i \cosh x \sin y. \quad (6.51)$$

It can be shown in a similar way that

$$\cosh (x + iy) = \cosh x \cos y + i \sinh x \sin y. \quad (6.52)$$

Example 15. *Prove that, if x and y are real and $\sin (x + iy) = x + iy$,*

(i) $\cos x > 0$, (ii) $y = \pm \cos x \left(\dfrac{x^2}{\sin^2 x} - 1\right)^{1/2}$. (O.C.)

Using (6.49),

$$x + iy = \sin (x + iy) = \sin x \cosh y + i \cos x \sinh y,$$

so that, by equating real and imaginary parts,

$$x = \sin x \cosh y, \ y = \cos x \sinh y. \quad (6.53)$$

Hence, $\cos x = y/\sinh y$ and, since y and $\sinh y$ are either both positive or both negative, it follows that $\cos x > 0$.
Equations (6.53) can be written

$$\cosh y = \frac{x}{\sin x}, \ \sinh y = \frac{y}{\cos x}. \quad (6.54)$$

The first of (6.39) can be written $\sinh^2 y = \cosh^2 y - 1$, so that

$$\frac{y^2}{\cos^2 x} = \frac{x^2}{\sin^2 x} - 1,$$

leading to $y = \pm \cos x \left(\dfrac{x^2}{\sin^2 x} - 1\right)^{1/2}$.

Example 16. *If $x + iy = c \cosh (u + iv)$, where x, y, u, v, c are real, prove that the curves $u = $ constant are confocal ellipses and that the curves $v =$ constant are the corresponding confocal hyperbolas.*
If w and z are complex variables connected by the relation $w = z^2 - 2$ and the point that represents z in the Argand diagram describes an ellipse with foci at $z = 2, z = -2$, prove that the point that represents w describes a confocal ellipse. (O.)

Using (6.52),

$$x + iy = c \cosh (u + iv) = c (\cosh u \cos v + i \sinh u \sin v),$$

so that $\quad x = c \cosh u \cos v, \ y = c \sinh u \sin v. \quad (6.55)$

Hence
$$\frac{x}{c\cosh u} = \cos v, \quad \frac{y}{c\sinh u} = \sin v,$$
and, on squaring and adding, since $\cos^2 v + \sin^2 v = 1$,
$$\frac{x^2}{c^2\cosh^2 u} + \frac{y^2}{c^2\sinh^2 u} = 1.$$
When u is constant, the point (x, y) therefore lies on an ellipse with semi-axes $a = c \cosh u$, $b = c \sinh u$. Its eccentricity ε is given by $c^2 \sinh^2 u = c^2 \cosh^2 u\,(1 - \varepsilon^2)$,

so that
$$\varepsilon^2 = \frac{c^2(\cosh^2 u - \sinh^2 u)}{c^2 \cosh^2 u} = \operatorname{sech}^2 u.$$

Thus $\varepsilon = \operatorname{sech} u$ and the coordinates $(\pm a\varepsilon, 0)$ of its foci are $(\pm c, 0)$. Hence the curves $u = $ constant are confocal ellipses.
Equations (6.55) can also be written in the form
$$\frac{x}{c\cos v} = \cosh u, \quad \frac{y}{c\sin v} = \sinh u,$$
so that, squaring and subtracting, since $\cosh^2 u - \sinh^2 u = 1$,
$$\frac{x^2}{c^2\cos^2 v} - \frac{y^2}{c^2\sin^2 v} = 1.$$
When v is constant, the point (x, y) therefore lies on a hyperbola and it can be shown as above that the foci of this hyperbola are also at the points $(\pm c, 0)$. Hence the curves $v = $ constant are the corresponding confocal hyperbolas.

If $z = x + iy = 2\cosh(u + iv)$, when u is constant, the point z lies on an ellipse whose foci are at $z = \pm 2$. Now
$$w = z^2 - 2 = 4\cosh^2(u + iv) - 2 = 2\cosh 2(u + iv),$$
showing that, when u is constant, the point w lies on an ellipse with foci at $w = \pm 2$, i.e. on a confocal ellipse.

6.12. The inverse hyperbolic functions

Inverse hyperbolic functions can be defined in a similar way to inverse trigonometrical functions. Thus if z and w are complex variables and $z = \tanh w$, w is said to be the inverse hyperbolic tangent of z and the inverse relation is written $w = \operatorname{Tanh}^{-1} z$, with a similar meaning and notation for the other hyperbolic functions. These inverse functions can all be expressed in terms of logarithms. Thus
$$z = \tanh w = \frac{\sinh w}{\cosh w} = \frac{\exp(w) - \exp(-w)}{\exp(w) + \exp(-w)} = \frac{\exp(2w) - 1}{\exp(2w) + 1},$$
multiplying numerator and denominator by $\exp(w)$ and using the relations $\exp(w)\exp(w) = \exp(2w)$, $\exp(w)\exp(-w) = \exp(0) = 1$. Solving for $\exp(2w)$,
$$\exp(2w) = \frac{1+z}{1-z},$$
giving
$$2w = \operatorname{Log}\left(\frac{1+z}{1-z}\right),$$

6] INVERSE HYPERBOLIC FUNCTIONS

hence, $\quad \text{Tanh}^{-1} z = w = \frac{1}{2} \text{Log}\left(\frac{1+z}{1-z}\right)$,

so that the inverse hyperbolic tangent is a many-valued function of z. This is true also of the other functions of this type. It is usual to denote the principal value of such a function by the same notation but using a small initial letter (as was done for the generalised logarithmic function of § 6.9) and hence to write

$$\tanh^{-1} z = \frac{1}{2} \log\left(\frac{1+z}{1-z}\right). \qquad (6.56)$$

These functions provide a convenient way of expressing the values of certain integrals (see Chapter 10, § 10.2) and, for the purposes of this book, the functions are required only when the variable is real. No further work involving the complex variable will therefore be carried out here and in what follows all variables are supposed to be real. A glance at Fig. 27 shows that, if x is real, the function $\tanh^{-1} x$ is defined only if $-1 < x < 1$ and (6.56) shows, since the logarithmic function reduces to the ordinary logarithm to base e when the variable is real, that

$$\tanh^{-1} x = \frac{1}{2} \log_e\left(\frac{1+x}{1-x}\right), \; (-1 < x < 1). \qquad (6.57)$$

To derive an expression for $y = \cosh^{-1} x$, let $x = \cosh y$ so that $x = \frac{1}{2}(e^y + e^{-y})$ leading to the quadratic equation

$$e^{2y} - 2xe^y + 1 = 0.$$

This gives $\quad e^y = x \pm (x^2 - 1)^{1/2}$,

and, since $\{x + (x^2 - 1)^{1/2}\}\{x - (x^2 - 1)^{1/2}\} = 1$, it follows that

$$e^y = x + (x^2 - 1)^{1/2} \text{ or } e^y = \frac{1}{x + (x^2 - 1)^{1/2}},$$

and hence $y = \pm \log_e\{x + (x^2 - 1)^{1/2}\}$. Referring to Fig. 26, it will be seen that $\cosh^{-1} x$ is defined only when $x \geq 1$ and that it is two-valued. It is conventional to adopt the positive sign and to write

$$\cosh^{-1} x = \log_e\{x + (x^2 - 1)^{1/2}\}, \; (x \geq 1). \qquad (6.58)$$

It is left as an exercise for the reader to show in a similar way that

$$\sinh^{-1} x = \log_e\{x + (x^2 + 1)^{1/2}\}, \; (x \geq 0 \text{ and } x \leq 0), \qquad (6.59)$$

$$\coth^{-1} x = \frac{1}{2} \log_e\left(\frac{x+1}{x-1}\right), \; (|x| > 1), \qquad (6.60)$$

$$\text{sech}^{-1} x = \log_e\left\{\frac{1 + (1 - x^2)^{1/2}}{x}\right\}, \; (0 < x \leq 1), \qquad (6.61)$$

$$\text{cosech}^{-1} x = \log_e\left\{\frac{1 \pm (1 + x^2)^{1/2}}{x}\right\}, \qquad (6.62)$$

the alternative signs in the last formula being taken according as $x > 0$ or $x < 0$.

Example 17. *Find real values of x, in logarithmic form, which satisfy the equation $a \cosh^2 x + \sinh^2 x = 1$ when $-1 < a \leqslant 1$.*
Prove that, if $a > 1$, the solutions of the equation are

$$x = n\pi i \pm i \cos^{-1}\left(\frac{2}{1+a}\right)^{1/2},$$

where n is any integer. (O.C.)

Since $\cosh^2 x - \sinh^2 x = 1$, $\sinh^2 x = \cosh^2 x - 1$ and the given equation can be written

$$a \cosh^2 x + \cosh^2 x - 1 = 1$$

giving $(a+1)\cosh^2 x = 2$. Hence

$$\cosh x = \pm\left(\frac{2}{a+1}\right)^{1/2}.$$

The right-hand side being purely real, x must be purely real or, since $\cos ix = \cosh x$, purely imaginary. For real values of x, $\cosh x > 1$ and possible real values are given by

$$\cosh x = \left(\frac{2}{a+1}\right)^{1/2}, \; (-1 < a \leqslant 1).$$

Hence
$$x = \pm \cosh^{-1}\left(\frac{2}{a+1}\right)^{1/2},$$

the alternative signs being permissible since, if x is a root of the given equation, so also is $-x$. Using (6.58), the required real values of x are given by

$$x = \pm \log_e \left\{\left(\frac{2}{a+1}\right)^{1/2} + \left(\frac{2}{a+1} - 1\right)^{1/2}\right\}.$$

If $a > 1$, the roots of the equation are imaginary and are given by

$$\cos ix = \pm\left(\frac{2}{a+1}\right)^{1/2},$$

both signs being now permissible. Hence,

$$ix = 2r\pi \pm \cos^{-1}\left\{\pm\left(\frac{2}{a+1}\right)^{1/2}\right\},$$

where r is any integer or zero. Since

$$\cos^{-1}\left\{-\left(\frac{2}{a+1}\right)^{1/2}\right\} = \pi - \cos^{-1}\left(\frac{2}{a+1}\right)^{1/2},$$

this can be written $\quad x = n\pi i \pm i \cos^{-1}\left(\frac{2}{a+1}\right)^{1/2},$

where n is any integer or zero.

EXERCISES 6 (*f*)

1. Prove that

    (i) $\sinh 3z = 4 \sinh^3 z + 3 \sinh z$,

    (ii) $\tanh nz = \dfrac{{}^nC_1 T + {}^nC_3 T^3 + {}^nC_5 T^5 + \cdots}{1 + {}^nC_2 T^2 + {}^nC_4 T^4 + \cdots}$,

    where $T = \tanh z$, n is a positive integer and ${}^nC_r$ is the coefficient of x^r in the expansion of $(1+x)^n$. (O.C.)

2 Prove that
$$\tan^{-1}\left(\frac{\tan 2\alpha + \tanh 2\beta}{\tan 2\alpha - \tanh 2\beta}\right) + \tan^{-1}\left(\frac{\tan \alpha - \tanh \beta}{\tan \alpha + \tanh \beta}\right)$$
$$= \tan^{-1}(\cot \alpha \coth \beta). \quad \text{(C.)}$$

3. If x, y, u, v, c are all real and $x + iy = c \cosh(u + iv)$, prove that with a certain convention about the sign of the square roots
$$\{(x + c)^2 + y^2\}^{1/2} + \{(x - c)^2 + y^2\}^{1/2} = 2c \cosh u. \quad \text{(O.)}$$

4. Show that if $x + iy = \cot(u + iv)$ then
$$x^2 + y^2 - 2x \cot 2u - 1 = 0$$
and
$$x^2 + y^2 + 2y \coth 2v + 1 = 0.$$

Show also that for any constant values of u, v, these two circles cut one another orthogonally; and that, in the first circle, $2u$ is the angle between the lines joining the point (x, y) to the points $(0, \pm 1)$. (C.)

5. Find the real root of the equation $5 \sinh x - 3 \cosh x = 3$. (O.C.)

6. Prove that the solutions of $\tan(\tanh u) = \cot(\coth u)$ are given by
$$\tanh u = \tfrac{1}{4}[(2n+1)\pi \pm \{(2n+1)^2\pi^2 - 16\}^{1/2}]. \quad \text{(O.)}$$

6.13. The summation of trigonometrical series

A useful application of some of the results derived in this chapter can be found in the following method which is often successful in effecting the summation of trigonometrical series. Suppose the sum of the finite series $\sum_{r=1}^{n} a_r \cos r\theta$, where the coefficient a_r is a given function of r, is required. Let

$$\left.\begin{array}{l} C = a_1 \cos \theta + a_2 \cos 2\theta + a_3 \cos 3\theta + \ldots + a_n \cos n\theta, \\ S = a_1 \sin \theta + a_2 \sin 2\theta + a_3 \sin 3\theta + \ldots + a_n \sin n\theta, \end{array}\right\} \quad (6.63)$$

so that, multiplying S by i and adding,
$$C + iS = a_1(\cos \theta + i \sin \theta) + a_2(\cos 2\theta + i \sin 2\theta) +$$
$$+ a_3(\cos 3\theta + i \sin 3\theta) + \ldots + a_n(\cos n\theta + i \sin n\theta).$$

Let $z = \cos \theta + i \sin \theta$ so that, by De Moivre's theorem, $z^r = \cos r\theta + i \sin r\theta$, and $C + iS$ can be written as the sum of a power series, viz.,
$$C + iS = a_1 z + a_2 z^2 + a_3 z^3 + \ldots + a_n z^n. \quad (6.64)$$

The success of the method depends upon the possibility of summing this power series. If this summation can be effected and if the sum is $u + iv$, equating real and imaginary parts in (6.64) gives $C = u$, $S = v$. Hence the sum (C) of the first of the series (6.63) will have been found and, as a by-product, the sum (S) of the second series will also be available.

A similar method applies also in the case of infinite trigonometrical series of this type. In such cases it is essential that the infinite series corresponding to that in (6.64), i.e. $a_1 z + a_2 z^2 + a_3 z^3 + \ldots$ should be convergent: such cases occur when the infinite series $a_1 + a_2 + a_3 + \ldots$ is absolutely convergent. Details of the work in particular cases are shown in the examples which follow.

Example 18. *Show that*
$$\cos x \cos x + \cos^2 x \cos 2x + \ldots + \cos^n x \cos nx = \sin nx \cos^n x \cot x.$$
(C.)

Let
$$C = \cos x \cos x + \cos^2 x \cos 2x + \ldots + \cos^n x \cos nx,$$
$$S = \cos x \sin x + \cos^2 x \sin 2x + \ldots + \cos^n x \sin nx,$$
so that, if $z = \cos x + i \sin x$,
$$C + iS = z \cos x + z^2 \cos^2 x + \ldots + z^n \cos^n x$$
$$= \frac{z \cos x (1 - z^n \cos^n x)}{1 - z \cos x},$$
on summing the n terms of the geometrical progression with common ratio $z \cos x$. Since $z = \cos x + i \sin x$, $z^{n+1} = \cos(n+1)x + i \sin(n+1)x$, this can be written
$$C + iS = \frac{\cos x [\cos x + i \sin x - \cos^n x \{\cos(n+1)x + i \sin(n+1)x\}]}{1 - \cos x (\cos x + i \sin x)}.$$
The denominator $= 1 - \cos^2 x - i \cos x \sin x = \sin x (\sin x - i \cos x)$ so that multiplying numerator and denominator by $\sin x + i \cos x$,
$$C + iS = \frac{\cos x [\cos x + i \sin x - \cos^n x \{\cos(n+1)x + i \sin(n+1)x\}](\sin x + i \cos x)}{\sin x (\sin x - i \cos x)(\sin x + i \cos x)}$$
and, equating the real parts,
$$C = \cot x [\cos x \sin x - \cos^n x \sin x \cos(n+1)x - \cos x \sin x$$
$$+ \cos^{n+1} x \sin(n+1)x]$$
$$= \cot x \cos^n x [\sin(n+1)x \cos x - \cos(n+1)x \sin x]$$
$$= \cot x \cos^n x \sin nx,$$
which is the required result.

Example 19. *Prove that, for $|r| < 1$,*
$$\sum_{n=1}^{\infty} \frac{r^n}{n} \sin n\theta = \tan^{-1} \left\{ \frac{r \sin \theta}{1 - r \cos \theta} \right\},$$
$$\sum_{n=1}^{\infty} \frac{r^n}{n} \cos n\theta = -\frac{1}{2} \log_e (1 - 2r \cos \theta + r^2).$$
(O.C.)

Let
$$C = r \cos \theta + \frac{r^2}{2} \cos 2\theta + \frac{r^3}{3} \cos 3\theta + \ldots,$$
$$S = r \sin \theta + \frac{r^2}{2} \sin 2\theta + \frac{r^3}{3} \sin 3\theta + \ldots,$$
so that, if $z = \cos \theta + i \sin \theta$,
$$C + iS = rz + \tfrac{1}{2} r^2 z^2 + \tfrac{1}{3} r^3 z^3 + \ldots$$
$$= -\log(1 - rz), \text{ since } |r| < 1, \text{ (using (6.34))}.$$
Since $z = \cos \theta + i \sin \theta$, this can be written
$$C + iS = -\log(1 - r \cos \theta - ir \sin \theta)$$
$$= -\tfrac{1}{2} \log_e \{(1 - r \cos \theta)^2 + r^2 \sin^2 \theta\} + i \tan^{-1} \left\{ \frac{r \sin \theta}{1 - r \cos \theta} \right\},$$

using the first of (6.29) and remembering that $\tan^{-1}(-x) = -\tan^{-1} x$. Since $(1 - r\cos\theta)^2 + r^2 \sin^2\theta = 1 - 2r\cos\theta + r^2(\cos^2\theta + \sin^2\theta) = 1 - 2r\cos\theta + r^2$, the required results follow on equating real and imaginary parts.

Example 20. *Sum to n terms the series*
$$\sin^3\theta + \tfrac{1}{3}\sin^3 3\theta + \tfrac{1}{9}\sin^3 9\theta + \tfrac{1}{27}\sin^3 27\theta + \ldots \quad \text{(C.)}$$

This series can be summed without recourse to the method discussed in this section. The example is worked here to show another artifice which is sometimes of use. Since $\sin^3 x = \tfrac{1}{4}(3\sin x - \sin 3x)$, the series can be written

$$\frac{3}{4}\left(\sin\theta + \frac{1}{3}\sin 3\theta + \frac{1}{9}\sin 9\theta + \ldots + \frac{1}{3^{n-1}}\sin 3^{n-1}\theta\right)$$
$$-\frac{1}{4}\left(\sin 3\theta + \frac{1}{3}\sin 9\theta + \ldots + \frac{1}{3^{n-2}}\sin 3^{n-1}\theta + \frac{1}{3^{n-1}}\sin 3^n\theta\right).$$

It will be observed that the terms cancel in pairs except for the first and last so that the required sum

$$= \frac{1}{4}\left(3\sin\theta - \frac{1}{3^{n-1}}\sin 3^n\theta\right).$$

EXERCISES 6 (g)

1. Sum the series, n being a positive integer,
$$1 + x\cos\theta + x^2\cos 2\theta + \ldots + x^n\cos n\theta. \quad \text{(C.)}$$

2. Sum the infinite series
$$1 + x^2\cos 2x + x^4\cos 4x + \ldots, \ (x^2 < 1).$$

3. Find the sum of the first n terms of the series
$$\cos^3\theta - \tfrac{1}{3}\cos^3 3\theta + \tfrac{1}{9}\cos^3 9\theta - \tfrac{1}{27}\cos^3 27\theta + \ldots \quad \text{(C.)}$$

4. Prove that the sum of n terms of the series
$$\sin\theta + \sin(\theta + 2\phi) + \sin(\theta + 4\phi) + \sin(\theta + 6\phi) + \ldots$$
is $\sin\{\theta + (n-1)\phi\}\sin n\phi \operatorname{cosec}\phi$. (C.)

5. Find the sum of the infinite series
$$\cos\theta - \frac{1}{2}\cos 3\theta + \frac{1.3}{2.4}\cos 5\theta - \frac{1.3.5}{2.4.6}\cos 7\theta + \ldots \quad \text{(C.)}$$

6. Sum, for any positive integer n,
   (i) $\sin\theta + \sin 2\theta + \sin 3\theta + \ldots + \sin n\theta + \tfrac{1}{2}\sin(n+1)\theta$,
   (ii) $\sin\dfrac{\theta}{2} + \sin\dfrac{3\theta}{2} + \sin\dfrac{5\theta}{2} + \ldots + \sin\dfrac{2n+1}{2}\theta$,

   and show that the ratio of the first sum to the second is $\cos\dfrac{\theta}{2}$. (C.)

EXERCISES 6 (h)

1. Find the value of the real variable t for which
$$|(a + ib)\exp(ipt) - c|$$
has its least value, if a, b, c and p are all real. (C.)

2. Prove that when a, b, c are distinct and k has one of the values, 0, 1, 2 that
$$\Sigma \frac{a^k(x-b)(x-c)}{(a-b)(a-c)} \equiv x^k,$$
the summation being taken over the three expressions obtained by cyclic changes of the variables a, b, c.

Deduce from one or more of these identities (the number depends on the method used) that, when $\exp(i\phi)$ denotes $\cos\phi + i\sin\phi$,
$$\Sigma \exp(2i a) \frac{\sin(\theta-\beta)\sin(\theta-\gamma)}{\sin(\alpha-\beta)\sin(\alpha-\gamma)} = \exp(2i\theta).$$

Prove a corresponding formula for $\exp(3i\theta)$ and four angles α, β, γ, δ. (O.)

3. The sequence $A_0, A_1, \ldots, A_n$ is defined by $A_0 = 0$ and
$$A_{n+1}\cos n\theta - A_n \cos(n+1)\theta = 1.$$
If $\cos n\theta \neq 0$ for any integral value of n, prove that
$$A_{n+2} - 2A_{n+1}\cos\theta + A_n = 0$$
and hence find A_n.

Hence, or otherwise, sum the series $\sum\limits_{r=1}^{n} \sec r\theta \sec(r+1)\theta$. (C.)

4. Show that
$$\frac{1+\cos^5\theta}{1+\cos\theta} = \frac{1}{8}(\cos 4\theta - 2\cos 3\theta + 8\cos 2\theta - 14\cos\theta + 15).$$ (C.)

5. Prove by the use of complex numbers, or otherwise, that, if n is a positive integer, $\cos n\theta$ can be expressed as a polynomial in $\cos\theta$ of the form
$$\cos n\theta = p_0 \cos^n\theta - p_1 \cos^{n-2}\theta + p_2 \cos^{n-4}\theta - \ldots,$$
where $p_0, p_1, p_2, \ldots$ are positive integers. (It is to be understood that the summation continues so long as the indices remain non-negative.)

Show that $p_0 + p_1 + p_2 + \ldots = \frac{1}{2}\{(1+\sqrt{2})^n + (1-\sqrt{2})^n\}$. (C.)

6. Show that $\cos 2r\theta$, $\sin(2r-1)\theta$ can be expressed as polynomials in $\sin\theta$.

Show also that $\{1 - \sin(4r+1)\theta\}/(1-\sin\theta)$ is the square of a polynomial in $\sin\theta$.

7. Resolve $x^{2n+1} + a^{2n+1}$ into a product of real factors of the first and second degree in x and a.

Show that
$$\prod_{r=1}^{n}\left\{\cos\theta - \cos\frac{(2r-1)\pi}{2n+1}\right\} = \frac{\cos\dfrac{(2n+1)\theta}{2}}{2^n \cos\dfrac{\theta}{2}}.$$ (O.)

8. Prove that
$$\cosh nu - \cos na = 2^{n-1}\prod_{r=0}^{n-1}\left\{\cosh u - \cos\left(a + \frac{2r\pi}{n}\right)\right\}.$$

9. Prove that if $\theta_p = a + \dfrac{2p\pi}{n}$, where $n > 2$, then
$$\cos\theta_1 + \cos\theta_2 + \ldots + \cos\theta_n = 0,$$
$$\cos^2\theta_1 + \cos^2\theta_2 + \ldots + \cos^2\theta_n = \dfrac{n}{2}.$$

A regular polygon of n sides is inscribed in a circle of radius a, and a point P lies in the plane of this circle at a great distance r from the centre. Show that, if r^{-2} is negligible, the sum of the distances of P from the vertices of the polygon is approximately $n\left\{r + \dfrac{a^2}{4r}\right\}$. (O.)

10. ABC is an equilateral triangle inscribed in a circle of radius a; P is any point on a concentric circle of radius r. Show that $PA^2 + PB^2 + PC^2$ is constant and that $PA \cdot PB \cdot PC$ lies between $r^3 \sim a^3$ and $r^3 + a^3$. (C.)

11. Prove that if n is a positive integer
$$\left(\dfrac{1 + i\tan\theta}{1 - i\tan\theta}\right)^n = \dfrac{1 + i\tan n\theta}{1 - i\tan n\theta}.$$

Hence, or otherwise, express $\tan n\theta$ in the form $\dfrac{f(\tan\theta)}{g(\tan\theta)}$, where $f(x)$ and $g(x)$ are polynomials in x.

Evaluate

(i) $\displaystyle\sum_{k=0}^{n-1} \cot\dfrac{(4k+1)\pi}{4n}$, (ii) $\displaystyle\sum_{k=0}^{n-1} \cot^2\dfrac{(4k+1)\pi}{4n}$. (C.)

12. If a denotes $\cos\dfrac{2\pi}{7} + i\sin\dfrac{2\pi}{7}$, obtain the quadratic equation whose roots are $a + a^2 + a^4$ and $a^3 + a^5 + a^6$.
Prove that $\sin^3\dfrac{2\pi}{7} + \sin^3\dfrac{4\pi}{7} + \sin^3\dfrac{8\pi}{7} = \dfrac{\sqrt{7}}{2}$. (O.)

13. Express $\tan 5\theta$ in powers of $\tan\theta$.
Show that $\tan 36°$ is a root of the equation $x^4 - 10x^2 + 5 = 0$ and express the other roots as tangents of angles. (C.)

14. If $\theta = 2\pi/7$, prove that
$$\sin\theta + \sin 2\theta + \sin 4\theta = \sqrt{7}/2,$$
$$\tan^2\theta + \tan^2 2\theta + \tan^2 4\theta = 21.$$ (C.)

15. (i) If $z = x + iy$ where x and y are real and $2z = (1 - z)\exp(i\theta)$, prove that $3(x^2 + y^2) = 1 - 2x$.
(ii) If $z = x + iy$ and a is real, express $\log(z^2 - a^2)$ in the form $A + iB$ where A and B are real.

16. Show that $\sinh x - \sinh y = 2\sinh\tfrac{1}{2}(x - y)\cosh\tfrac{1}{2}(x + y)$.
Hence, or otherwise, show that
$$\cosh a + \cosh(a + \beta) + \cosh(a + 2\beta) + \ldots + \cosh\{a + (n-1)\beta\}$$
$$= \dfrac{\cosh\{a + \tfrac{1}{2}(n-1)\beta\}\sinh\tfrac{1}{2}n\beta}{\sinh\tfrac{1}{2}\beta}.$$ (O.C.)

17. Prove that:
   (i) $\tanh x < x$ given that x is positive,
   (ii) $\sinh (2n+1)x - c_1 \sinh (2n-1)x + c_2 \sinh (2n-3)x - \ldots + (-1)^n c_n \sinh x = 2^{2n} \sinh^{2n+1} x$,
   where c_r denotes the binomial coefficient $^{2n+1}C_r$.
   Show that, when $\cosh 2nx$ is expanded in powers of $\cosh x$, the coefficient of $\cosh^4 x$ is $(-1)^n \tfrac{2}{3} n^2(n^2-1)$. (O.C.)

18. If x, y, c, u, v are all real and $x + iy = c \cosh (u + iv)$, what curves in the plane of x, y are given by the equations $u =$ constant, $v =$ constant?
   Two ellipses and two hyperbolas of a confocal system intersect in four points inside one of the right angles formed by the axes, their arcs making a curvilinear quadrilateral. Prove that the straight lines joining opposite vertices of this figure are equal. (O.)

19. Two complex variables $z = x + iy$, $Z = X + iY$ are connected by the relation $Z = \sin(\pi z/2)$.
   Show that to every point in the complex Z-plane there corresponds a point in the strip $|x| \leqslant 1$ of the complex z-plane. Show also that the lines $x =$ constant, $y =$ constant map into certain mutually orthogonal systems of ellipses and hyperbolas in the Z-plane. (C.)

20. Prove that $\displaystyle\sum_{r=0}^{n} 3^r \sin^3\left(\frac{\pi}{3^{r+1}}\right) = \frac{3^{n+1}}{4} \sin\left(\frac{\pi}{3^{n+1}}\right)$. (O.C.)

21. If ω is one of the imaginary $2n$th roots of unity, prove that
$$\sin\theta + \omega \sin 2\theta + \omega^2 \sin 3\theta + \ldots + \omega^{2n-1} \sin 2n\theta = \frac{\sin\theta + \omega \sin 2n\theta - \sin(2n+1)\theta}{1 - 2\omega \cos\theta + \omega^2}.$$
Deduce the sums of the series
$$\sin\theta + \omega^2 \sin 3\theta + \omega^4 \sin 5\theta + \ldots + \omega^{2n-2} \sin(2n-1)\theta$$
and $\sin 2\theta + \omega^2 \sin 4\theta + \omega^4 \sin 6\theta + \ldots + \omega^{2n-2} \sin 2n\theta$. (C.)

22. Prove the identities
   (i) $1 + 2\cos 2\theta + 2\cos 4\theta + \ldots + 2\cos 2n\theta = \operatorname{cosec}\theta \sin(2n+1)\theta$.
   (ii) $\displaystyle\sum_{r=0}^{2n} \cos(\theta + rs a) = 0$, where $a = 2\pi/(2n+1)$ and s is any integer not a multiple of $2n+1$.
   Hence, or otherwise, prove that
$$\sum_{s=0}^{2n} \operatorname{cosec}(\theta + sa) = (2n+1)\operatorname{cosec}(2n+1)\theta.$$ (O.)

23. Prove that
$$\frac{n+1}{2} + n\cos\theta + (n-1)\cos 2\theta + \ldots + \cos n\theta = \frac{1}{2}\left[\frac{\sin\dfrac{(n+1)\theta}{2}}{\sin\dfrac{\theta}{2}}\right]^2$$
(O.)

24. Sum
   (i) $\cos \theta \cos \theta + \cos 2\theta \cos^2 \theta + \cos 3\theta \cos^3 \theta + \ldots,$
   (ii) $\sin \theta \cos \theta + \sin 2\theta \cos^2 \theta + \sin 3\theta \cos^3 \theta + \ldots$
   to n terms and, when $\sin \theta \neq 0$, to infinity.

   Show that, when $\cos \theta \neq 0$,
   $$0 = \sin \theta \sin \theta - \cos 2\theta \sin^2 \theta - \sin 3\theta \sin^3 \theta + \cos 4\theta \sin^4 \theta \\ + \sin 5\theta \sin^5 \theta - \cos 6\theta \sin^6 \theta - \ldots \text{ to inf.} \quad \text{(O.)}$$

25. If $a_n + a_{n-1} + a_{n-2} = 0$ for $n > 2$, show that
   $$a_1 \cos \theta + a_2 \cos 2\theta + \ldots + a_n \cos n\theta \\ = \frac{a_1 + (a_1 + a_2) \cos \theta - a_{n-1} \cos n\theta + a_n \cos (n+1)\theta}{1 + 2\cos\theta}. \quad \text{(C.)}$$

CHAPTER 7

DETERMINANTS

7.1. Introduction

The elimination of n variables from $n + 1$ linear algebraical equations in these variables gives the relation which exists between the coefficients of the various equations if the $n + 1$ equations are to be *consistent* (in the sense that the values of the variables found from any n of the $n + 1$ equations satisfy the remaining equation). Thus for the two homogeneous equations

$$\left.\begin{array}{l} a_1 x + b_1 y = 0, \\ a_2 x + b_2 y = 0, \end{array}\right\} \tag{7.1}$$

in one effective variable (the ratio x/y), the eliminant is

$$a_1 b_2 - a_2 b_1 = 0, \tag{7.2}$$

as is easily found by equating the values of x/y found from each equation. The condition that the three equations of this type (with effectively two variables x/z and y/z),

$$\left.\begin{array}{l} a_1 x + b_1 y + c_1 z = 0, \\ a_2 x + b_2 y + c_2 z = 0, \\ a_3 x + b_3 y + c_3 z = 0, \end{array}\right\} \tag{7.3}$$

be consistent is easily found, by solving the first two equations for x/z, y/z and substituting in the third equation, to be

$$a_1 b_2 c_3 - a_1 b_3 c_2 - a_2 b_1 c_3 + a_3 b_1 c_2 + a_2 b_3 c_1 - a_3 b_2 c_1 = 0. \tag{7.4}$$

Relations similar to (7.2) and (7.4) for four or more equations can be found in a similar way but, of course, their complexity increases as the number of equations increases.

The left-hand sides of relations such as (7.2), (7.4), and the similar ones obtained when the number of algebraical equations involved is greater than three, are conveniently expressed in terms of *determinants*. It is the object of this chapter to give a brief discussion of the notation and some of the more important properties of determinants and to indicate some advantages of their use.

7.2. Determinantal notation

The square array

$$\begin{vmatrix} a_1 & b_1 \\ a_2 & b_2 \end{vmatrix}$$

of two rows and columns and 2^2 *elements* (or *constituents*) a_1, b_1, a_2 and b_2 is called a *determinant of the second order*. It is a notation for the expression $a_1b_2 - a_2b_1$ formed by the difference of the products of the two letters a, b in alphabetical order when suffixes are attached to these letters corresponding to the permutations of the numbers 1 and 2. Thus

$$\begin{vmatrix} a_1 & b_1 \\ a_2 & b_2 \end{vmatrix} = a_1b_2 - a_2b_1. \tag{7.5}$$

In a similar way the array

$$\begin{vmatrix} a_1 & b_1 & c_1 \\ a_2 & b_2 & c_2 \\ a_3 & b_3 & c_3 \end{vmatrix}$$

of three rows and columns and 3^2 elements is a *determinant of the third order*. Here the determinant is a notation for the sum of all the 'signed' products of the three letters a, b, c in alphabetical order when suffixes are attached corresponding to the permutations of the numbers 1, 2, 3. The interpretation of 'signed' product is as follows: the 'natural' order of the suffixes is 1, 2, 3 and the product with suffixes in this order is taken as positive: the signs of the other products are fixed by considering the number of interchanges of the order of the suffixes necessary to obtain the 'natural' order; thus the order 3, 1, 2 requires two interchanges (firstly to 1, 3, 2 and secondly to 1, 2, 3) so that the 'signed' product of $a_3b_1c_2$ is taken to mean $+a_3b_1c_2$; similarly the order 3, 2, 1 requires but one interchange (3 with 1) and the 'signed' product of $a_3b_2c_1$ is $-a_3b_2c_1$. Hence,

$$\begin{vmatrix} a_1 & b_1 & c_1 \\ a_2 & b_2 & c_2 \\ a_3 & b_3 & c_3 \end{vmatrix} = a_1b_2c_3 - a_1b_3c_2 - a_2b_1c_3 + a_3b_1c_2 + a_2b_3c_1 - a_3b_2c_1. \tag{7.6}$$

The expressions on the right of (7.5) and (7.6) are termed the *expansions* of the determinants on the left-hand sides of these relations. Each product such as a_1b_2 in (7.5) or $a_3b_1c_2$ in (7.6) is called a *term* of the expansion. It should be noted that the expansions of the two determinants (of the second and third orders) contain respectively 2! and 3! terms. The elements a_1, b_2 of the determinant of (7.5) and the elements a_1, b_2, c_3 of that of (7.6) are, for a reason which should be self-explanatory, called the elements of the *leading* (or *principal*) *diagonal*.

By writing the right-hand side of (7.6) in the form

$$a_1(b_2c_3 - b_3c_2) - b_1(a_2c_3 - a_3c_2) + c_1(a_2b_3 - a_3b_2),$$

and using relations derived from (7.5) by suitable changes of letters and suffixes, it follows that

$$\begin{vmatrix} a_1 & b_1 & c_1 \\ a_2 & b_2 & c_2 \\ a_3 & b_3 & c_3 \end{vmatrix} = a_1 \begin{vmatrix} b_2 & c_2 \\ b_3 & c_3 \end{vmatrix} - b_1 \begin{vmatrix} a_2 & c_2 \\ a_3 & c_3 \end{vmatrix} + c_1 \begin{vmatrix} a_2 & b_2 \\ a_3 & b_3 \end{vmatrix}. \tag{7.7}$$

It should be noted that the three second order determinants on the right of this relation are given by striking out from the determinant on the left the row and column in which a_1, b_1, c_1 respectively appear. The expression on the right of (7.7) above is called the *expansion of the determinant by its first row*. By writing the right-hand side of (7.6) in the alternative form

$$a_1(b_2c_3 - b_3c_2) - a_2(b_1c_3 - b_3c_1) + a_3(b_1c_2 - b_2c_1),$$

the expansion of the determinant by its first column can be written

$$\begin{vmatrix} a_1 & b_1 & c_1 \\ a_2 & b_2 & c_2 \\ a_3 & b_3 & c_3 \end{vmatrix} = a_1 \begin{vmatrix} b_2 & c_2 \\ b_3 & c_3 \end{vmatrix} - a_2 \begin{vmatrix} b_1 & c_1 \\ b_3 & c_3 \end{vmatrix} + a_3 \begin{vmatrix} b_1 & c_1 \\ b_2 & c_2 \end{vmatrix} \quad (7.8)$$

and the second order determinants on the right are now given by striking out the row and column in which a_1, a_2, a_3 respectively appear.

All that has been said so far can be taken over to apply to determinants of the nth order, with n rows, n columns and n^2 elements. Such a determinant could be written

$$\begin{vmatrix} a_1 & b_1 & c_1 & \ldots & \ldots & k_1 \\ a_2 & b_2 & c_2 & \ldots & \ldots & k_2 \\ a_3 & b_3 & c_3 & \ldots & \ldots & k_3 \\ \ldots & \ldots & \ldots & \ldots & \ldots & \ldots \\ \ldots & \ldots & \ldots & \ldots & \ldots & \ldots \\ a_n & b_n & c_n & \ldots & \ldots & k_n \end{vmatrix} \quad (7.9)$$

but, to save space, the alternative notation $(a_1b_2c_3 \ldots k_n)$ is often used. The value of this determinant is defined to be the sum of all the $n!$ 'signed' products which can be formed by writing the n letters $a, b, c, \ldots, k$ in alphabetical order and attaching suffixes corresponding to the permutations of the n numbers $1, 2, 3, \ldots, n$, the meaning of 'signed' product being as explained in the case of the third order determinant.

The following important property follows immediately from this definition: *if two rows or two columns of a determinant are interchanged, the determinant is unchanged in absolute value but its sign is changed.* Consider first the interchange of two rows; in this case, two suffixes are interchanged throughout the expansion and this alters only the sign of each term of the expansion. If two columns are interchanged, two letters are interchanged throughout the expansion. This is equivalent to keeping the letters in the original order and interchanging the two suffixes so that again the only alteration is in the sign of each term of the expansion.

Just as the third order determinant can be expanded by its first row (or column) in terms of determinants of the second order so can the determinant of the nth order be expanded in terms of determinants of order $n - 1$. Since, from the definition, every term in the expansion of a determinant of any order contains one, and only one, element from each row and from each column, all the terms in the expansion of the determinant (7.9) which contain a_1 as a factor will be obtained by taking the sum of the 'signed' products of the $n - 1$ letters $b, c, \ldots, n$ and the $n - 1$ suffixes $2, 3, \ldots, n$. Thus all the terms with a_1 as a factor will be given by

$$a_1 \begin{vmatrix} b_2 & c_2 & \ldots & \ldots & k_2 \\ b_3 & c_3 & \ldots & \ldots & k_3 \\ \ldots & \ldots & \ldots & \ldots & \ldots \\ \ldots & \ldots & \ldots & \ldots & \ldots \\ b_n & c_n & \ldots & \ldots & k_n \end{vmatrix}.$$

To find the terms containing b_1, interchange the first and second columns: this alters the sign of the determinant. Then arguing as before, the terms with b_1 as a factor will be given by

$$-b_1 \begin{vmatrix} a_2 & c_2 & \ldots & \ldots & k_2 \\ a_3 & c_3 & \ldots & \ldots & k_3 \\ \ldots & \ldots & \ldots & \ldots & \ldots \\ \ldots & \ldots & \ldots & \ldots & \ldots \\ a_n & c_n & \ldots & \ldots & k_n \end{vmatrix}.$$

The terms in c_1 will, by interchanging the first and third columns, be similarly given by

$$-c_1 \begin{vmatrix} b_2 & a_2 & \ldots & \ldots & k_2 \\ b_3 & a_3 & \ldots & \ldots & k_3 \\ \ldots & \ldots & \ldots & \ldots & \ldots \\ \ldots & \ldots & \ldots & \ldots & \ldots \\ b_n & a_n & \ldots & \ldots & k_n \end{vmatrix},$$

and this, on interchanging the first and second columns, becomes

$$c_1 \begin{vmatrix} a_2 & b_2 & \ldots & \ldots & k_2 \\ a_3 & b_3 & \ldots & \ldots & k_3 \\ \ldots & \ldots & \ldots & \ldots & \ldots \\ \ldots & \ldots & \ldots & \ldots & \ldots \\ a_n & b_n & \ldots & \ldots & k_n \end{vmatrix}.$$

Continuing in this way, the determinant can be *expanded by its first row* to give

$$(a_1 b_2 c_3 \ldots k_n) = a_1 \alpha_1 - b_1 \beta_1 + c_1 \gamma_1 - \ldots + (-1)^{n-1} k_1 \varkappa_1, \quad (7.10)$$

where $\alpha_1, \beta_1, \gamma_1, \ldots, \varkappa_1$ are determinants of order $n - 1$ called respectively the *first minors* of $a_1, b_1, c_1, \ldots, k_1$. The minors are given by

178 MATHEMATICAL ANALYSIS [7

$$\alpha_1 = \begin{vmatrix} b_2 & c_2 & .. & .. & k_2 \\ b_3 & c_3 & .. & .. & k_3 \\ .. & .. & .. & .. & .. \\ .. & .. & .. & .. & .. \\ b_n & c_n & .. & .. & k_n \end{vmatrix}, \beta_1 = \begin{vmatrix} a_2 & c_2 & .. & .. & k_2 \\ a_3 & c_3 & .. & .. & k_3 \\ .. & .. & .. & .. & .. \\ .. & .. & .. & .. & .. \\ a_n & c_n & .. & .. & k_n \end{vmatrix},$$

$$\ldots, \varkappa_1 = \begin{vmatrix} a_2 & b_2 & .. & .. & j_2 \\ a_3 & b_3 & .. & .. & j_3 \\ .. & .. & .. & .. & .. \\ .. & .. & .. & .. & .. \\ a_n & b_n & .. & .. & j_n \end{vmatrix}, \quad (7.11)$$

and can be written down from the original determinant (7.9) by omitting the rows and columns in which $a_1, b_1, \ldots, k_1$ respectively appear. In a similar way, the determinant may be expanded *by its first column* in the form

$$(a_1 b_2 c_3 \ldots k_n) = a_1 \alpha_1 - a_2 \alpha_2 + a_3 \alpha_3 - \ldots + (-1)^{n-1} a_n \alpha_n, \quad (7.12)$$

where now the minor α_s is the determinant of order $n-1$ obtained from the original determinant (7.9) by omitting the row and column containing a_s ($s = 1, 2, 3, \ldots, n$).

A determinant of order n can therefore be expressed, by expanding it by its first row or column, in terms of n minor determinants each of order $n-1$. These first minors can each in turn be expressed in terms $n-1$ *second minor* determinants each of order $n-2$, and so on until the original determinant is completely expanded. This method will theoretically give the complete expansion of a determinant of any order, but the process is generally laborious when n is much greater than three. The work entailed can often, in the case of particular determinants, be reduced considerably by making use of some of the properties of determinants given in the next section.

7.3. Some properties of determinants

(a) *The value of a determinant is unaltered by changing rows into columns and columns into rows.*

Let

$$\Delta_1 = \begin{vmatrix} a_1 & b_1 & c_1 & .. & .. & k_1 \\ a_2 & b_2 & c_2 & .. & .. & k_2 \\ a_3 & b_3 & c_3 & .. & .. & k_3 \\ .. & .. & .. & .. & .. & .. \\ .. & .. & .. & .. & .. & .. \\ a_n & b_n & c_n & .. & .. & k_n \end{vmatrix} \text{ and } \Delta_2 = \begin{vmatrix} a_1 & a_2 & a_3 & .. & .. & a_n \\ b_1 & b_2 & b_3 & .. & .. & b_n \\ c_1 & c_2 & c_3 & .. & .. & c_n \\ .. & .. & .. & .. & .. & .. \\ .. & .. & .. & .. & .. & .. \\ k_1 & k_2 & k_3 & .. & .. & k_n \end{vmatrix}$$

where each determinant is of the nth order and Δ_2 is obtained from Δ_1 by changing rows into columns. The first term $a_1 b_2 c_3 \ldots k_n$ is the same in the expansion of each determinant. The other terms in the expansion of Δ_1 are obtained from the term $a_1 b_2 c_3 \ldots k_n$ by keeping fixed the order of the letters and permuting the suffixes, while the other terms in the expansion of Δ_2 are obtained from the same first term by keeping fixed the order of the suffixes and permuting the letters. Each process gives rise to the same $n!$ terms, the signs of corresponding terms being the same since, in the expansion of Δ_1, an interchange of two suffixes gives rise to a change of sign while, in the expansion of Δ_2, a change of sign occurs with each interchange of two letters. Hence $\Delta_1 \equiv \Delta_2$.

This property having been established, it is only necessary to prove the properties which follow in the case of 'rows'. The reader is reminded that the properties remain true for 'columns' by inserting, in the enunciation of the properties, the words 'or columns' after 'rows' whenever the latter word occurs.

(b) If two rows (or columns) of a determinant are interchanged, the determinant is unchanged in absolute value but its sign is changed.

This property has already been established in § 7.2: it is restated here for completeness.

(c) If the elements of one row (or column) of a determinant are identical with the corresponding elements of another row (or column), the value of the determinant is zero.

If Δ is the value of the determinant, since the determinant is unaltered when the two identical rows are interchanged and since, by (b) above, such an interchange alters the sign of the determinant, then it follows that $\Delta = -\Delta$ and hence $\Delta = 0$.

Example 1. *Show that*

$$\begin{vmatrix} 1 & a & a^3 \\ 1 & b & b^3 \\ 1 & c & c^3 \end{vmatrix} = (a-b)(b-c)(c-a)(a+b+c).$$

If $a = b$, the first and second rows of the determinant become identical and hence the determinant vanishes. Hence $(a-b)$ is a factor of the determinant. Similarly $(b-c)$, $(c-a)$ are also factors. Since each term of the expansion of the determinant is of the fourth degree in a, b, c, the remaining factor must be linear in these letters. Further, the determinant changes sign if two letters are interchanged for this is equivalent to the interchange of two rows. Since the product $(a-b)(b-c)(c-a)$ changes sign if two letters are interchanged, the remaining factor must therefore be linear and symmetrical in a, b and c. Hence the value of the determinant can be written

$$\lambda(a-b)(b-c)(c-a)(a+b+c)$$

where λ is a numerical constant. The leading term of the expansion is bc^3 and the term bc^3 in the above expression has coefficient λ. Hence $\lambda = 1$ and the required result follows.

(*d*) *If every element of a given row (or column) is multiplied by the same factor, the value of the determinant is multiplied by that factor.*

This property follows at once from the definition of the expansion of the determinant as the sum of 'signed' products: every term of the expansion contains one, and only one, element from any row so that each term of the expansion contains the multiplying factor once and once only.

It follows as a corollary from properties (*c*) and (*d*) that *if a determinant has two rows (or columns) whose corresponding elements differ only by a constant factor, then the value of the determinant is zero.*

Example 2. *Express the determinant*

$$\begin{vmatrix} b_1c_1 & c_1a_1 & a_1b_1 \\ b_2c_2 & c_2a_2 & a_2b_2 \\ b_3c_3 & c_3a_3 & a_3b_3 \end{vmatrix}$$

in terms of a third order determinant with elements $1/a_1$, $1/b_1$, $1/c_1$, *etc.*

The given determinant can be written

$$\begin{vmatrix} \dfrac{a_1b_1c_1}{a_1} & \dfrac{a_1b_1c_1}{b_1} & \dfrac{a_1b_1c_1}{c_1} \\ \dfrac{a_2b_2c_2}{a_2} & \dfrac{a_2b_2c_2}{b_2} & \dfrac{a_2b_2c_2}{c_2} \\ \dfrac{a_3b_3c_3}{a_3} & \dfrac{a_3b_3c_3}{b_3} & \dfrac{a_3b_3c_3}{c_3} \end{vmatrix},$$

which (removing the common factors $a_1b_1c_1$, $a_2b_2c_2$, $a_3b_3c_3$ from the first, second and third rows respectively) can be expressed as

$$a_1b_1c_1\, a_2b_2c_2\, a_3b_3c_3 \begin{vmatrix} \dfrac{1}{a_1} & \dfrac{1}{b_1} & \dfrac{1}{c_1} \\ \dfrac{1}{a_2} & \dfrac{1}{b_2} & \dfrac{1}{c_2} \\ \dfrac{1}{a_3} & \dfrac{1}{b_3} & \dfrac{1}{c_3} \end{vmatrix}.$$

(*e*) *If equimultiples of the elements of any row (or column) of a determinant are added to the corresponding elements of any other row (or column), the value of the determinant is unaltered.*

In view of property (*b*), there is no loss of generality in taking the first two rows as those involved. Let Δ denote the *n*th order determinant $(a_1b_2c_3 \ldots k_n)$ and let Δ' be the determinant obtained when m times the elements of the second row are added to the corresponding elements of the first row, so that

PROPERTIES OF DETERMINANTS

$$\Delta' = \begin{vmatrix} a_1 + ma_2 & b_1 + mb_2 & c_1 + mc_2 & \ldots & \ldots & k_1 + mk_2 \\ a_2 & b_2 & c_2 & \ldots & \ldots & k_2 \\ a_3 & b_3 & c_3 & \ldots & \ldots & k_3 \\ \ldots & \ldots & \ldots & \ldots & \ldots & \ldots \\ \ldots & \ldots & \ldots & \ldots & \ldots & \ldots \\ a_n & b_n & c_n & \ldots & \ldots & k_n \end{vmatrix}.$$

If $\alpha_1, \beta_1, \gamma_1, \ldots, \varkappa_1$ are the minors of $a_1, b_1, c_1, \ldots, k_1$ respectively in the original determinant, the determinant Δ' can be expanded by its first row to give

$$\Delta' = (a_1 + ma_2)\alpha_1 - (b_1 + mb_2)\beta_1 + (c_1 + mc_2)\gamma_1 - $$
$$\ldots + (-1)^{n-1}(k_1 + mk_2)\varkappa_1$$
$$= a_1\alpha_1 - b_1\beta_1 + c_1\gamma_1 - \ldots + (-1)^{n-1}k_1\varkappa_1 +$$
$$+ m[a_2\alpha_1 - b_2\beta_1 + c_2\gamma_1 - \ldots + (-1)^{n-1}k_2\varkappa_1].$$

The first n terms on the right are the expansion of the original determinant Δ and the n terms in square brackets are the expansion of the determinant obtained from Δ by writing $a_2, b_2, c_2, \ldots, k_2$ in place of $a_1, b_1, c_1, \ldots, k_1$ in its first row. By property (c) such a determinant vanishes and hence $\Delta' = \Delta$.

By repeated application of this result, it follows that a determinant is unaltered in value when to each element of any row (or column) are added equimultiples of any other rows (or columns). Thus, for example,

$$\begin{vmatrix} a_1 & pa_1 + b_1 & ma_1 + nb_1 + c_1 \\ a_2 & pa_2 + b_2 & ma_2 + nb_2 + c_2 \\ a_3 & pa_3 + b_3 & ma_3 + nb_3 + c_3 \end{vmatrix} = \begin{vmatrix} a_1 & b_1 & c_1 \\ a_2 & b_2 & c_2 \\ a_3 & b_3 & c_3 \end{vmatrix}.$$

The reader should observe that when more than one row (or column) is altered by a series of such changes, at least one row (or column) remains unaltered.

Particular cases of the property occur when the factors involved are ± 1 so that *the elements of a given row (or column) of a determinant can be added to, or subtracted from, the corresponding elements of any other row (or column) without altering the value of the determinant.*

Some worked examples showing how the use of these properties (in particular, property (e)) can shorten the work involved in evaluating particular determinants are given below. To indicate the steps taken in passing from one determinant to another in these and subsequent worked examples the notation $r_1, r_2, \ldots, r_n, c_1, c_2, \ldots, c_n$ is used to denote respectively the first, second, ..., nth rows and columns. Dashes are used to denote rows and columns of the transformed determinant. Thus, for example, $(r_3' = r_3 - 4r_2)$ denotes that the elements of the third row of the transformed determinant are obtained by subtracting four times the elements of the second row from the elements of the third row of the original determinant.

Example 3. *Evaluate the determinant*

$$\begin{vmatrix} 1 & 1 & 1 \\ 10 & 12 & 16 \\ 14 & 17 & 21 \end{vmatrix}.$$

If Δ represents the value of the determinant, direct expansion by its first row gives

$$\Delta = 1 \begin{vmatrix} 12 & 16 \\ 17 & 21 \end{vmatrix} - 1 \begin{vmatrix} 10 & 16 \\ 14 & 21 \end{vmatrix} + 1 \begin{vmatrix} 10 & 12 \\ 14 & 17 \end{vmatrix}$$

$= (12 \times 21 - 16 \times 17) - (10 \times 21 - 16 \times 14) + (10 \times 17 - 12 \times 14)$
$= 252 - 272 - 210 + 224 + 170 - 168 = -4.$

The arithmetical work is simplified by using property (*e*) to obtain two zeros in the first row. Thus

$$\Delta = \begin{vmatrix} 1 & 0 & 0 \\ 10 & 2 & 6 \\ 14 & 3 & 7 \end{vmatrix} \quad (c_2' = c_2 - c_1,\; c_3' = c_3 - c_1)$$

$= 2 \times 7 - 6 \times 3 = -4.$

Example 4. *Evaluate the determinant*

$$\begin{vmatrix} 12 & 15 & 18 \\ 11 & 14 & 17 \\ 10 & 13 & 16 \end{vmatrix}$$

Here, $\begin{vmatrix} 12 & 15 & 18 \\ 11 & 14 & 17 \\ 10 & 13 & 16 \end{vmatrix} = \begin{vmatrix} 2 & 2 & 2 \\ 1 & 1 & 1 \\ 10 & 13 & 16 \end{vmatrix} \quad (r_1' = r_1 - r_3,\; r_2' = r_2 - r_3.)$

$= 2 \begin{vmatrix} 1 & 1 & 1 \\ 1 & 1 & 1 \\ 10 & 13 & 16 \end{vmatrix}$ (by property (*d*))

$= 0$ (by property (*c*)).

Example 5. *Prove that* $\begin{vmatrix} (a+b-c-d)(ab-cd) & ab+cd & 1 \\ (a+c-d-b)(ac-db) & ac+db & 1 \\ (a+d-b-c)(ad-bc) & ad+bc & 1 \end{vmatrix} = 0.$ (O.)

It is easily verified that each of the expressions

$(a+b+c+d)(ab+cd) - (a+b-c-d)(ab-cd),$
$(a+b+c+d)(ac+db) - (a+c-d-b)(ac-db),$
$(a+b+c+d)(ad+bc) - (a+d-b-c)(ad-bc),$

is equal to $2(abc + abd + acd + bcd) = 2\Sigma$ (say). Hence by multiplying the elements of the second column by $a + b + c + d$ and subtracting those of the first column, all the elements of the new second column will be 2Σ. Removing the common factor 2Σ, the elements of the transformed second column will all be unity and the column will be identical with the third column. Hence the determinant vanishes.

Example 6. *Prove that* $(a-c)^2 + (b-d)^2$ *is a factor of the determinant*

$$\begin{vmatrix} a & b & c & d \\ d & a & b & c \\ c & d & a & b \\ b & c & d & a \end{vmatrix}$$

and find the two real linear factors. (O.)

7] EXERCISES 183

If the value of the given determinant is denoted by Δ,

$$\Delta = \begin{vmatrix} a-c & b-d & c & d \\ d-b & a-c & b & c \\ c-a & d-b & a & b \\ b-d & c-a & d & a \end{vmatrix} \quad (c_1' = c_1 - c_3, \ c_2' = c_2 - c_4)$$

$$= \begin{vmatrix} 0 & 0 & c+a & d+b \\ 0 & 0 & b+d & c+a \\ c-a & d-b & a & b \\ b-d & c-a & d & a \end{vmatrix} \quad (r_1' = r_1 + r_3, \ r_2' = r_2 + r_4)$$

$$= (c+a) \begin{vmatrix} 0 & 0 & c+a \\ c-a & d-b & b \\ b-d & c-a & a \end{vmatrix} - (d+b) \begin{vmatrix} 0 & 0 & b+d \\ c-a & d-b & a \\ b-d & c-a & d \end{vmatrix}$$

$$= (c+a)^2\{(c-a)^2 + (d-b)^2\} - (d+b)^2\{(c-a)^2 + (d-b)^2\}$$
$$= \{(a-c)^2 + (b-d)^2\}\{(a+c)^2 - (b+d)^2\}$$
$$= \{(a-c)^2 + (b-d)^2\}(a+c+b+d)(a+c-b-d),$$

which gives the required results.

EXERCISES 7 (a)

1. Evaluate the second order determinants:

   (a) $\begin{vmatrix} 6 & 2 \\ 9 & 7 \end{vmatrix}$, (b) $\begin{vmatrix} a+c & a \\ b+d & b \end{vmatrix}$, (c) $\begin{vmatrix} a+ib & c+id \\ -c+id & a-ib \end{vmatrix}$.

2. Evaluate the determinants:

   (a) $\begin{vmatrix} 1 & 1 & 1 \\ 4 & 5 & 6 \\ 8 & 9 & 10 \end{vmatrix}$, (b) $\begin{vmatrix} 1 & 1 & 1 \\ 1 & 1+a & 1 \\ 1 & 1 & 1+b \end{vmatrix}$.

3. Show that $\begin{vmatrix} b+c & c & b \\ c & a+c & a \\ b & a & a+b \end{vmatrix} = 4abc.$

4. Evaluate the determinant

   $$\begin{vmatrix} 1 & 1 & 1 \\ \tan A & \tan B & \tan C \\ \sin 2A & \sin 2B & \sin 2C \end{vmatrix},$$

   A, B, C being the angles of a triangle. (O.)

5. Express the determinant

   $$\begin{vmatrix} a^2 + ax + x^2 & a^2 + ay + y^2 & a^2 + az + z^2 \\ b^2 + bx + x^2 & b^2 + by + y^2 & b^2 + bz + z^2 \\ c^2 + cx + x^2 & c^2 + cy + y^2 & c^2 + cz + z^2 \end{vmatrix}$$

   as a product of linear factors. (O.)

6. Show that $\begin{vmatrix} 1 & 1 & 1 & 1 \\ 1 & 2 & 3 & 4 \\ 1 & 3 & 6 & 10 \\ 1 & 4 & 10 & 19 \end{vmatrix} = 0.$

7.4. Determinantal equations

Sometimes an equation is expressed as the vanishing of a determinant. The roots of such equations can be found by expanding the given determinant (using such of the properties of § 7.3 as may be appropriate), equating the expansion to zero and then solving the resulting equation in one of the usual ways. Two worked examples follow.

Example 7. *If no two of a, b, c are equal and*

$$\begin{vmatrix} 1 & bc + ax & a^2 \\ 1 & ca + bx & b^2 \\ 1 & ab + cx & c^2 \end{vmatrix} = 0,$$

prove that $x = a + b + c$. (N.U.)

Subtracting the first row from each of the others, the equation can be written

$$\begin{vmatrix} 1 & bc + ax & a^2 \\ 0 & ca - bc + (b - a)x & b^2 - a^2 \\ 0 & ab - bc + (c - a)x & c^2 - a^2 \end{vmatrix} = 0$$

The non-vanishing factors $b - a$, $c - a$ can now be removed from the second and third rows to give

$$\begin{vmatrix} 1 & bc + ax & a^2 \\ 0 & x - c & a + b \\ 0 & x - b & a + c \end{vmatrix} = 0$$

and, expansion by the first column leads to

$$(a + c)(x - c) - (a + b)(x - b) = 0.$$

Hence, $\quad (c - b)x = c(a + c) - b(a + b)$
$\quad\quad\quad\quad\quad\quad = (c - b)(a + b + c).$

so that $x = a + b + c$.

Example 8. *By means of the equation* $(x + b)(x + c) - f^2 = 0$, *prove that the equation in* x

$$\begin{vmatrix} x + a & h & g \\ h & x + b & f \\ g & f & x + c \end{vmatrix} = 0$$

has three real roots which are separated by the two roots of the first equation. (It may be assumed that a, b, c, f, g, h are all real and different from zero.) (C.)

Expanding the determinant, the equation in x can be written

$$\Delta(x) \equiv (x + a)(x + b)(x + c) - f^2(x + a) - g^2(x + b) - h^2(x + c) + 2fgh = 0.$$

The equation $(x + b)(x + c) - f^2 = 0$ can be written

$$x^2 + (b + c)x + bc - f^2 = 0$$

and its two roots, α and β (say), are given by

$$\alpha = -\tfrac{1}{2}(b + c + \mu), \ \beta = -\tfrac{1}{2}(b + c - \mu),$$

where $\quad \mu^2 = (b + c)^2 - 4(bc - f^2) = (b - c)^2 + 4f^2.$

Suppose $b > c$, then $\mu > b - c > 0$. Also

$2\alpha = -b - c - \mu < -b - c - (b - c) = -2b$, so that $\alpha < -b$.
$2\beta = -b - c + \mu > -b - c + b - c = -2c$, so that $\beta > -c$.

Hence, $a < -b < -c < \beta$. It is clear that $\Delta(x)$ is large and negative for large negative values of x and large and positive for large positive values of x. Also, since $(a + b)(a + c) - f^2 = 0$,
$$\Delta(a) = 2fgh - g^2(a + b) - h^2(a + c).$$
But $a + b < 0$ and $a + c < 0$ so that it is possible to write
$$a + b = -p^2, \ a + c = -q^2$$
and, as $(a + b)(a + c) = f^2$, $f = \pm pq$. Hence
$$\Delta(a) = \pm 2pqgh + g^2p^2 + h^2q^2 > 0.$$
Similarly it can be proved that $\Delta(\beta) < 0$. Hence, tabulating values of x and the signs of $\Delta(x)$,

x	$\Delta(x)$
Large negative	−
a	+
β	−
Large positive	+

which is sufficient to establish the required results

7.5. Differentiation of a determinant

The usual result (see, for example, *Advanced Level Pure Mathematics*, equation (8.5), page 134) for the differential coefficient of the product of n factors $a, b, c, \ldots, k$, all of which are functions of x, is

$$\frac{d}{dx}(abc \ldots k) = (bcd \ldots k)\frac{da}{dx} + (acd \ldots k)\frac{db}{dx} + \ldots + (abc \ldots j)\frac{dk}{dx}.$$

If, then, Δ is the nth order determinant $(a_1 b_2 c_3 \ldots k_n)$,

$$\frac{d\Delta}{dx} = \left(\frac{da_1}{dx} b_2 c_3 \ldots k_n\right) + \left(a_1 \frac{db_2}{dx} c_3 \ldots k_n\right) + \ldots + \left(a_1 b_2 c_3 \ldots \frac{dk_n}{dx}\right).$$
(7.13)

Thus $d\Delta/dx$ is the sum of n determinants each of the nth order. For the particular case of a third order determinant, the result (7.13) written out in full is

$$\frac{d}{dx}\begin{vmatrix} a_1 & b_1 & c_1 \\ a_2 & b_2 & c_2 \\ a_3 & b_3 & c_3 \end{vmatrix} = \begin{vmatrix} da_1/dx & b_1 & c_1 \\ da_2/dx & b_2 & c_2 \\ da_3/dx & b_3 & c_3 \end{vmatrix} + \begin{vmatrix} a_1 & db_1/dx & c_1 \\ a_2 & db_2/dx & c_2 \\ a_3 & db_3/dx & c_3 \end{vmatrix} + \begin{vmatrix} a_1 & b_1 & dc_1/dx \\ a_2 & b_2 & dc_2/dx \\ a_3 & b_3 & dc_3/dx \end{vmatrix},$$

and the reader should have no difficulty in interpreting (7.13) for other values of n. There is, of course, a corresponding formula for $d\Delta/dx$ in which rows are differentiated instead of columns.

Example 9. *If a and b are independent of x and*
$$\Delta = \begin{vmatrix} 1 & x & x^2 \\ 1 & a & a^2 \\ 1 & b & b^2 \end{vmatrix},$$

find $d\Delta/dx$.

Differentiating by columns,

$$\frac{d\Delta}{dx} = \begin{vmatrix} 0 & x & x^2 \\ 0 & a & a^2 \\ 0 & b & b^2 \end{vmatrix} + \begin{vmatrix} 1 & 1 & x^2 \\ 1 & 0 & a^2 \\ 1 & 0 & b^2 \end{vmatrix} + \begin{vmatrix} 1 & x & 2x \\ 1 & a & 0 \\ 1 & b & 0 \end{vmatrix}$$
$$= -b^2 + a^2 + 2x(b - a)$$
$$= (a - b)(a + b - 2x).$$

Alternatively, differentiating by rows,

$$\frac{d\Delta}{dx} = \begin{vmatrix} 0 & 1 & 2x \\ 1 & a & a^2 \\ 1 & b & b^2 \end{vmatrix} + \begin{vmatrix} 1 & x & x^2 \\ 0 & 0 & 0 \\ 1 & b & b^2 \end{vmatrix} + \begin{vmatrix} 1 & x & x^2 \\ 1 & a & a^2 \\ 0 & 0 & 0 \end{vmatrix}$$
$$= -b^2 + a^2 + 2x(b - a) = (a - b)(a + b - 2x).$$

7.6. Further examples of the evaluation of determinants

In this section some of the artifices which are useful in the evaluation of determinants are illustrated by means of worked examples. One useful method is to relate some particular nth order determinants with those of the same form but of lower orders and hence to obtain a difference equation whose solution gives an expression for the original determinant (Example 10). The method of induction is useful in certain instances (Example 11) and another useful artifice is that used in Example 12.

Example 10. *If Δ_n denotes the determinant*

$$\begin{vmatrix} \lambda & 1 & 0 & 0 & .. & 0 & 0 \\ 1 & \lambda & 1 & 0 & .. & 0 & 0 \\ 0 & 1 & \lambda & 1 & .. & 0 & 0 \\ .. & .. & .. & .. & .. & .. & .. \\ 0 & 0 & 0 & 0 & .. & \lambda & 1 \\ 0 & 0 & 0 & 0 & .. & 1 & \lambda \end{vmatrix}$$

with n rows and columns (where the elements in the principal diagonal are all λ, the elements adjacent to the principal diagonal are all unity and all the other elements vanish), prove that

$$\Delta_n = \lambda \Delta_{n-1} - \Delta_{n-2}.$$

Deduce that if $\lambda = 2 \cos \theta$, the value of the determinant Δ_n is $\sin(n+1)\theta$ cosec θ. (C.)

Expanding the determinant by its first row

$$\Delta_n = \lambda \alpha - \beta,$$

where α and β are given in terms of determinants of $n - 1$ rows and columns by

$$\alpha = \begin{vmatrix} \lambda & 1 & 0 & .. & 0 & 0 \\ 1 & \lambda & 1 & .. & 0 & 0 \\ .. & .. & .. & .. & .. & .. \\ .. & .. & .. & .. & \lambda & 1 \\ 0 & 0 & 0 & .. & 1 & \lambda \end{vmatrix}, \beta = \begin{vmatrix} 1 & 1 & 0 & .. & 0 & 0 \\ 0 & \lambda & 1 & .. & 0 & 0 \\ .. & .. & .. & .. & .. & .. \\ .. & .. & .. & .. & \lambda & 1 \\ 0 & 0 & 0 & .. & 1 & \lambda \end{vmatrix}$$

The determinant α is of precisely the same form as Δ_n but contains one less row and one less column, so that $\alpha = \Delta_{n-1}$. Expanding the determinant

β by its first column, it can be expressed as a determinant of precisely the same form as Δ_n but with two less rows and columns, so that $\beta = \Delta_{n-2}$. Hence
$$\Delta_n = \lambda \Delta_{n-1} - \Delta_{n-2}.$$
By the method of § 2.5 of Chapter 2, the solution of this difference equation is, when $\lambda = 2 \cos \theta$, given by
$$\Delta_n = A x_1^n + B x_2^n$$
where x_1, x_2 are the roots of the auxiliary equation
$$x^2 - 2(\cos \theta)x + 1 = 0,$$
and A, B are constants independent of n. The roots of the auxiliary quadratic equation are
$$\cos \theta \pm \{\cos^2 \theta - 1\}^{1/2}, \text{ or, } \cos \theta \pm i \sin \theta,$$
so that
$$\Delta_n = A(\cos \theta + i \sin \theta)^n + B(\cos \theta - i \sin \theta)^n$$
$$= (A + B) \cos n\theta + i(A - B) \sin n\theta,$$
when use is made of De Moivre's theorem. Since $\Delta_1 = \lambda = 2 \cos \theta$ and
$$\Delta_2 = \begin{vmatrix} \lambda & 1 \\ 1 & \lambda \end{vmatrix} = \lambda^2 - 1 = 4 \cos^2 \theta - 1,$$
the equations determining A and B are
$$2 \cos \theta = (A + B) \cos \theta + i(A - B) \sin \theta,$$
$$4 \cos^2 \theta - 1 = (A + B) \cos 2\theta + i(A - B) \sin 2\theta.$$
These simultaneous equations are easily solved to give
$$A + B = 1, \ i(A - B) = \cot \theta.$$
Hence,
$$\Delta_n = \cos n\theta + \cot \theta \sin n\theta$$
$$= \sin (n + 1)\theta \operatorname{cosec} \theta.$$

Example 11. *In a determinant of order n, every element in the principal diagonal is equal to x and every other element is equal to y. Show that the determinant has $(x - y)^{n-1}$ as a factor and find an expression for the determinant.* (O.C.)

If Δ_n denotes the determinant of order n, the determinant Δ_{n+1}, of order $n + 1$, can be written, by subtracting the $(n + 1)$th row from the first row, in the form

$$\Delta_{n+1} = \begin{vmatrix} x-y & 0 & 0 & 0 & \cdots & \cdots & 0 & y-x \\ y & x & y & y & \cdots & \cdots & y & y \\ y & y & x & y & \cdots & \cdots & y & y \\ \cdots & \cdots & \cdots & \cdots & \cdots & \cdots & \cdots & \cdots \\ y & y & y & y & \cdots & \cdots & x & y \\ y & y & y & y & \cdots & \cdots & y & x \end{vmatrix}$$

Expansion by the first row gives

$$\Delta_{n+1} = (x - y)\Delta_n + (-1)^n(y - x) \begin{vmatrix} y & x & y & y & \cdots & \cdots & y \\ y & y & x & y & \cdots & \cdots & y \\ \cdots & \cdots & \cdots & \cdots & \cdots & \cdots & \cdots \\ y & y & y & y & \cdots & \cdots & x \\ y & y & y & y & \cdots & \cdots & y \end{vmatrix}$$

$$= (x - y)\Delta_n + (-1)^n(y - x) \begin{vmatrix} 0 & x-y & 0 & 0 & \cdots & \cdots & 0 \\ 0 & 0 & x-y & 0 & \cdots & \cdots & 0 \\ \cdots & \cdots & \cdots & \cdots & \cdots & \cdots & \cdots \\ 0 & 0 & 0 & 0 & \cdots & \cdots & x-y \\ y & y & y & y & \cdots & \cdots & y \end{vmatrix}$$

subtracting the nth row from all the others. Expanding this determinant by its first column, its value is

$$(-1)^{n-1}y \begin{vmatrix} x-y & 0 & 0 & \cdots & \cdots & 0 \\ 0 & x-y & 0 & \cdots & \cdots & 0 \\ \cdots & \cdots & \cdots & \cdots & \cdots & \cdots \\ 0 & 0 & 0 & \cdots & \cdots & x-y \end{vmatrix} = (-1)^{n-1}y(x-y)^{n-1}.$$

Hence, $\quad \Delta_{n+1} = (x-y)\Delta_n + y(x-y)^n$,

so that if Δ_n has a factor $(x-y)^{n-1}$, Δ_{n+1} has a factor $(x-y)^n$. But $\Delta_2 = x^2 - y^2 = (x-y)(x+y)$ and the required result is true when $n = 2$. Hence the general result follows by induction. Δ_n is of degree n in x and y and, as it contains $(x-y)^{n-1}$ as a factor, it follows that

$$\Delta_n = (x-y)^{n-1}(a_n x + b_n y),$$

where a_n, b_n depend on n but not on x and y. Substituting in the result $\Delta_{n+1} = (x-y)\Delta_n + y(x-y)^n$, dividing by $(x-y)^n$ and equating the coefficients of x and y,

$$a_{n+1} = a_n, \quad b_{n+1} = 1 + b_n.$$

Since $\Delta_1 = x$, $a_1 = 1$, $b_1 = 0$ and it follows that

$$a_n = 1, \quad b_n = 1 + 1 + 1 + \ldots \text{ to } (n-1) \text{ terms} = n-1,$$

and hence $\quad \Delta_n = (x-y)^{n-1}\{x + (n-1)y\}.$

Example 12. *Show that if x is added to all the elements of any determinant, the resulting determinant has the value $A + Bx$, where A and B are independent of x.*

Show that for the determinant

$$\begin{vmatrix} a & b & b & b \\ p & a & b & b \\ q & r & a & b \\ s & t & u & a \end{vmatrix},$$

$A = \Delta$ and $bB = \Delta - (a-b)^4$ where Δ is the value of the determinant, and hence evaluate

$$\begin{vmatrix} a & b & b & b \\ c & a & b & b \\ c & c & a & b \\ c & c & c & a \end{vmatrix}. \qquad \text{(C.)}$$

If x is added to all the elements of a determinant Δ and the first row of the resulting determinant Δ' is subtracted from all the others, the elements of the first row will be all of the form $a + x$ and those of all the others will be independent of x. Expansion by the first row shows that the value of Δ' will be $A + Bx$ where A, B are independent of x.

Let $\quad \Delta = \begin{vmatrix} a & b & b & b \\ p & a & b & b \\ q & r & a & b \\ s & t & u & a \end{vmatrix}, \Delta' = \begin{vmatrix} a+x & b+x & b+x & b+x \\ p+x & a+x & b+x & b+x \\ q+x & r+x & a+x & b+x \\ s+x & t+x & u+x & a+x \end{vmatrix},$

so that $\Delta' = A + Bx$. Putting $x = 0$, A is the value of Δ' when $x = 0$, i.e. $A = \Delta$.

EVALUATION OF DETERMINANTS

Subtracting the first row from all the others, Δ' can be written

$$\Delta' = \begin{vmatrix} a+x & b+x & b+x & b+x \\ p-a & a-b & 0 & 0 \\ q-a & r-b & a-b & 0 \\ s-a & t-b & u-b & a-b \end{vmatrix}$$

$$= x \begin{vmatrix} 1+\dfrac{a}{x} & 1+\dfrac{b}{x} & 1+\dfrac{b}{x} & 1+\dfrac{b}{x} \\ p-a & a-b & 0 & 0 \\ q-a & r-b & a-b & 0 \\ s-a & t-b & u-b & a-b \end{vmatrix}$$

But $\Delta' = A + Bx = x\left(\dfrac{A}{x} + B\right)$ so that B is the value of the above determinant when $1/x$ is made zero in its first row. Hence

$$B = \begin{vmatrix} 1 & 1 & 1 & 1 \\ p-a & a-b & 0 & 0 \\ q-a & r-b & a-b & 0 \\ s-a & t-b & u-b & a-b \end{vmatrix}$$

$$= \dfrac{1}{b} \begin{vmatrix} b & b & b & b \\ p-a & a-b & 0 & 0 \\ q-a & r-b & a-b & 0 \\ s-a & t-b & u-b & a-b \end{vmatrix} = \dfrac{1}{b} \begin{vmatrix} a-(a-b) & b & b & b \\ p-(a-b) & a & b & b \\ q-(a-b) & r & a & b \\ s-(a-b) & t & u & a \end{vmatrix},$$

by adding the first row to all the others and writing the first element b of the principal diagonal in the form $a - (a - b)$. Expanding by the first column,

$$bB = \begin{vmatrix} a & b & b & b \\ p & a & b & b \\ q & r & a & b \\ s & t & u & a \end{vmatrix} - (a-b) \begin{vmatrix} 1 & b & b & b \\ 1 & a & b & b \\ 1 & r & a & b \\ 1 & t & u & a \end{vmatrix}$$

$$= \Delta - (a-b) \begin{vmatrix} 1 & b & b & b \\ 0 & a-b & 0 & 0 \\ 0 & r-b & a-b & 0 \\ 0 & t-b & u-b & a-b \end{vmatrix}$$

$$= \Delta - (a-b)^4.$$

Let $\Delta = \begin{vmatrix} a & b & b & b \\ c & a & b & b \\ c & c & a & b \\ c & c & c & a \end{vmatrix}$, $\Delta' = \begin{vmatrix} a+x & b+x & b+x & b+x \\ c+x & a+x & b+x & b+x \\ c+x & c+x & a+x & b+x \\ c+x & c+x & c+x & a+x \end{vmatrix}$,

so that, by the results obtained above,

$$\Delta' = A + Bx = \Delta + \{\Delta - (a-b)^4\}\dfrac{x}{b}.$$

Putting $x = -c$,

$$\Delta - \{\Delta - (a-b)^4\}\dfrac{c}{b} = \begin{vmatrix} a-c & b-c & b-c & b-c \\ 0 & a-c & b-c & b-c \\ 0 & 0 & a-c & b-c \\ 0 & 0 & 0 & a-c \end{vmatrix}$$

$$= (a-c)^4.$$

Hence $(b-c)\Delta = b(a-c)^4 - c(a-b)^4,$

giving $$\Delta = \dfrac{b(a-c)^4 - c(a-b)^4}{b-c}$$

EXERCISES 7 (b)

1. Solve the equation $\begin{vmatrix} x & 4 & -1 \\ 2 & x & 5 \\ 1 & 10 & x \end{vmatrix} = 0.$

2. Solve the following equation for x, given that $a^2 \neq 1$,
$$\begin{vmatrix} x^3 & a^2+1 & 1 \\ a^3 & x^2+1 & 1 \\ 1 & x^2+a^2 & 1 \end{vmatrix} = 0.$$
(O.C.)

3. Show that the derivative of the determinant
$$\begin{vmatrix} x & 1 & 1 & 1 & 1 \\ 1 & 0 & 1 & 1 & 1 \\ 1 & 1 & x & 1 & 1 \\ 1 & 1 & 1 & 0 & 1 \\ 1 & 1 & 1 & 1 & x \end{vmatrix}$$
with respect to x is the sum of three determinants each of which has $x - 1$ for a factor.
Hence, or otherwise, show that the value of the given determinant is $-(x - 1)^2(x - 4)$. (C.)

4. In a determinant Δ_n of n rows and columns, every element of the principal diagonal is 2, every element next to an element of the principal diagonal is 1 and every other element is 0. Prove that
$$\Delta_n - 2\Delta_{n-1} + \Delta_{n-2} = 0$$
and evaluate Δ_n. (O.)

5. Prove that the nth order determinant
$$\begin{vmatrix} x & a & a & a & .. & .. & a \\ a & x & 0 & 0 & .. & .. & 0 \\ a & 0 & x & 0 & .. & .. & .. \\ .. & .. & .. & .. & .. & .. & .. \\ a & 0 & 0 & 0 & .. & .. & x \end{vmatrix} = x^n - (n-1)a^2 x^{n-2}$$
(O.C.)

6. Prove that the value of
$$\begin{vmatrix} a_1 + \lambda & b_1 + 2\lambda & c_1 + 3\lambda \\ a_2 + \lambda & b_2 + 2\lambda & c_2 + 3\lambda \\ 1 + \lambda & 1 + 2\lambda & 1 + 3\lambda \end{vmatrix}$$
is of the form $A + B\lambda$ where A and B are independent of λ and find A and B.

7.7. Minors and cofactors

The expansions of the determinant (7.9) in terms of the minors $a_1, \beta_1, \ldots, \varkappa_1$ of its first row and in terms of the minors $a_1, a_2, \ldots, a_n$ of its first column have been given in (7.10), (7.12) respectively as
$$\Delta = (a_1 b_2 c_3 \ldots k_n) = a_1 a_1 - b_1 \beta_1 + c_1 \gamma_1 - \ldots + (-1)^{n-1} k_1 \varkappa_1$$
$$\Delta = (a_1 b_2 c_3 \ldots k_n) = a_1 a_1 - a_2 a_2 + a_3 a_3 - \ldots + (-1)^{n-1} a_n a_n.$$

MINORS AND COFACTORS

In terms of the minors of the second row and second column, the expansions would be, respectively,

$$\Delta = -a_2\alpha_2 + b_2\beta_2 - c_2\gamma_2 + \ldots + (-1)^n k_2\varkappa_2,$$
$$\Delta = -b_1\beta_1 + b_2\beta_2 - b_3\beta_3 + \ldots + (-1)^n b_n\beta_n,$$

with similar results when the determinant is expanded by other rows or columns. The presence of the plus and minus signs in these expressions is inconvenient and some simplification is obtained by the introduction of *cofactors* ('signed' minors).

The cofactor of the element in the rth row and sth column of a determinant is defined as $(-1)^{r+s}$ *times the minor of that element.* If s_r is the element of the rth row and sth column, it is convenient to denote its cofactor by S_r. For example, in the case of the determinant of (7.9),

$$A_1 = (-1)^{1+1}a_1 = a_1, \quad A_2 = (-1)^{2+1}a_2 = -a_2, \ldots$$
$$B_1 = (-1)^{1+2}\beta_1 = -\beta_1, \quad B_2 = (-1)^{2+2}\beta_2 = \beta_2, \ldots$$
$$C_1 = (-1)^{1+3}\gamma_1 = \gamma_1, \quad C_2 = (-1)^{2+3}\gamma_2 = -\gamma_2, \ldots$$

and so on. In terms of cofactors, the expansion of the determinant Δ of (7.9) by the rth row is given by

$$\Delta = a_r A_r + b_r B_r + c_r C_r + \ldots + k_r K_r,$$

and such a relation holds for $r = 1, 2, 3, \ldots, n$. If the elements in the sth column of the determinant are denoted by $s_1, s_2, s_3, \ldots, s_n$, the expansion by this column is

$$\Delta = s_1 S_1 + s_2 S_2 + s_3 S_3 + \ldots + s_n S_n,$$

this type of relation holding for all the n letters $a, b, c, \ldots, k$.

If the elements $a_r, b_r, c_r, \ldots, k_r$ of the rth row are identical with the elements $a_t, b_t, c_t, \ldots, k_t$ of any *other* (say the tth) row, the determinant vanishes and hence

$$a_t A_r + b_t B_r + c_t C_r + \ldots + k_t K_r = 0, \ (r \neq t).$$

Similarly if the elements $s_1, s_2, s_3, \ldots, s_n$ of the sth column are identical with the elements $t_1, t_2, t_3, \ldots, t_n$ of any *other* (say the tth) column, the determinant again vanishes and therefore

$$t_1 S_1 + t_2 S_2 + t_3 S_3 + \ldots + t_n S_n = 0.$$

The above results will be found to be of considerable importance in all work involving determinants. They can be memorised as follows:

(i) *the sum of the products of the elements of any row (or column) and their OWN cofactors is equal to the value of the determinant;* (7.14)

(ii) *the sum of the products of the elements of any row (or column) and the corresponding cofactors of any OTHER row (or column) is zero.* (7.15)

Examples of their use will be found in the remaining part of this chapter.

Example 13. *If Δ denotes the determinant*

$$\begin{vmatrix} a & h & g \\ h & b & f \\ g & f & c \end{vmatrix}$$

and the cofactors of $a, h, \ldots$ are denoted by $A, H, \ldots$, show that
$$aG^2 + bF^2 + cC^2 + 2fFC + 2gGC + 2hFG = C\Delta.$$

The relations (7.14), (7.15) applied to this determinant give
$$gG + fF + cC = \Delta,$$
$$hG + bF + fC = 0,$$
$$aG + hF + gC = 0.$$

Multiplying these relations by C, F and G respectively and adding, the required result follows immediately.

7.8. Multiplication of determinants

The product of two determinants of the nth order can be shown to be a determinant of the same order whose elements are, of course, functions of the elements of the original determinants. For example, consider the determinant Δ given by

$$\Delta \equiv \begin{vmatrix} a_1 & b_1 & 0 & 0 \\ a_2 & b_2 & 0 & 0 \\ -1 & 0 & x_1 & x_2 \\ 0 & -1 & y_1 & y_2 \end{vmatrix}. \qquad (7.16)$$

Expanding by the first row,

$$\Delta = a_1 \begin{vmatrix} b_2 & 0 & 0 \\ 0 & x_1 & x_2 \\ -1 & y_1 & y_2 \end{vmatrix} - b_1 \begin{vmatrix} a_2 & 0 & 0 \\ -1 & x_1 & x_2 \\ 0 & y_1 & y_2 \end{vmatrix}$$

$$= a_1 b_2 \begin{vmatrix} x_1 & x_2 \\ y_1 & y_2 \end{vmatrix} - a_2 b_1 \begin{vmatrix} x_1 & x_2 \\ y_1 & y_2 \end{vmatrix}$$

$$= (a_1 b_2 - a_2 b_1) \begin{vmatrix} x_1 & x_2 \\ y_1 & y_2 \end{vmatrix}$$

$$= \begin{vmatrix} a_1 & b_1 \\ a_2 & b_2 \end{vmatrix} \times \begin{vmatrix} x_1 & x_2 \\ y_1 & y_2 \end{vmatrix} = \begin{vmatrix} a_1 & b_1 \\ a_2 & b_2 \end{vmatrix} \times \begin{vmatrix} x_1 & y_1 \\ x_2 & y_2 \end{vmatrix}.$$
$$(7.17)$$

But, from (7.16)

$$\Delta = \begin{vmatrix} a_1 & b_1 & a_1 x_1 + b_1 y_1 & a_1 x_2 + b_1 y_2 \\ a_2 & b_2 & a_2 x_1 + b_2 y_1 & a_2 x_2 + b_2 y_2 \\ -1 & 0 & 0 & 0 \\ 0 & -1 & 0 & 0 \end{vmatrix} \quad \begin{pmatrix} c_3' = c_3 + x_1 c_1 + y_1 c_2, \\ c_4' = c_4 + x_2 c_1 + y_2 c_2 \end{pmatrix},$$

$$= - \begin{vmatrix} a_1 & a_1 x_1 + b_1 y_1 & a_1 x_2 + b_1 y_2 \\ a_2 & a_2 x_1 + b_2 y_1 & a_2 x_2 + b_2 y_2 \\ -1 & 0 & 0 \end{vmatrix} \quad \text{(expanding by the fourth row),}$$

$$= \begin{vmatrix} a_1 x_1 + b_1 y_1 & a_1 x_2 + b_1 y_2 \\ a_2 x_1 + b_2 y_1 & a_2 x_2 + b_2 y_2 \end{vmatrix}.$$

7] MULTIPLICATION OF DETERMINANTS

Equating this value of Δ with that obtained in (7.17),
$$\begin{vmatrix} a_1 & b_1 \\ a_2 & b_2 \end{vmatrix} \times \begin{vmatrix} x_1 & y_1 \\ x_2 & y_2 \end{vmatrix} = \begin{vmatrix} a_1x_1 + b_1y_1 & a_1x_2 + b_1y_2 \\ a_2x_1 + b_2y_1 & a_2x_2 + b_2y_2 \end{vmatrix}. \quad (7.18)$$

In a similar way, starting with the determinant
$$\begin{vmatrix} a_1 & b_1 & c_1 & 0 & 0 & 0 \\ a_2 & b_2 & c_2 & 0 & 0 & 0 \\ a_3 & b_3 & c_3 & 0 & 0 & 0 \\ -1 & 0 & 0 & x_1 & x_2 & x_3 \\ 0 & -1 & 0 & y_1 & y_2 & y_3 \\ 0 & 0 & -1 & z_1 & z_2 & z_3 \end{vmatrix},$$

it can be shown that
$$\begin{vmatrix} a_1 & b_1 & c_1 \\ a_2 & b_2 & c_2 \\ a_3 & b_3 & c_3 \end{vmatrix} \times \begin{vmatrix} x_1 & y_1 & z_1 \\ x_2 & y_2 & z_2 \\ x_3 & y_3 & z_3 \end{vmatrix}$$
$$= \begin{vmatrix} a_1x_1 + b_1y_1 + c_1z_1 & a_1x_2 + b_1y_2 + c_1z_2 & a_1x_3 + b_1y_3 + c_1z_3 \\ a_2x_1 + b_2y_1 + c_2z_1 & a_2x_2 + b_2y_2 + c_2z_2 & a_2x_3 + b_2y_3 + c_2z_3 \\ a_3x_1 + b_3y_1 + c_3z_1 & a_3x_2 + b_3y_2 + c_3z_2 & a_3x_3 + b_3y_3 + c_3z_3 \end{vmatrix}.$$
(7.19)

These results can be generalised to apply to the product of two determinants, each of the nth order. The form of the elements in the 'product' determinant should be studied carefully as the results can be used with great advantage in working particular examples. Some instances will be found in what follows.

Example 14. *Evaluate as the product of factors the determinant*
$$\begin{vmatrix} (a-x)^2 & (a-y)^2 & (a-z)^2 \\ (b-x)^2 & (b-y)^2 & (b-z)^2 \\ (c-x)^2 & (c-y)^2 & (c-z)^2 \end{vmatrix}. \quad (O)$$

The given determinant is, by (7.19), equal to the product of the two determinants
$$\begin{vmatrix} a^2 & a & 1 \\ b^2 & b & 1 \\ c^2 & c & 1 \end{vmatrix} \text{ and } \begin{vmatrix} 1 & -2x & x^2 \\ 1 & -2y & y^2 \\ 1 & -2z & z^2 \end{vmatrix}.$$

By arguments similar to those used in Example 1 of this chapter, these determinants are respectively equal to
$$-(a-b)(b-c)(c-a) \text{ and } -2(x-y)(y-z)(z-x).$$
Hence the value of the given determinant is
$$2(a-b)(b-c)(c-a)(x-y)(y-z)(z-x).$$

Example 15. *If* $\Delta_1 = \begin{vmatrix} 1 & \alpha & \alpha^2 \\ 1 & \beta & \beta^2 \\ 1 & \gamma & \gamma^2 \end{vmatrix}$, $\Delta_2 = \begin{vmatrix} 1 & \alpha^2 & \alpha^3 \\ 1 & \beta^2 & \beta^3 \\ 1 & \gamma^2 & \gamma^3 \end{vmatrix}$, use the notation $S_r = \alpha^r + \beta^r + \gamma^r$ to enable you to write Δ_1^2 and $\Delta_1\Delta_2$ as determinants of order 3.

By considering the form of these determinants, write down the two determinant factors of

$$\begin{vmatrix} S_0 & S_3 & S_4 \\ S_1 & S_4 & S_5 \\ S_2 & S_5 & S_6 \end{vmatrix}.$$ (O.)

Interchanging rows and columns, and using (7.19)

$$\Delta_1^2 = \begin{vmatrix} 1 & 1 & 1 \\ \alpha & \beta & \gamma \\ \alpha^2 & \beta^2 & \gamma^2 \end{vmatrix} \times \begin{vmatrix} 1 & 1 & 1 \\ \alpha & \beta & \gamma \\ \alpha^2 & \beta^2 & \gamma^2 \end{vmatrix}$$

$$= \begin{vmatrix} 1+1+1 & \alpha+\beta+\gamma & \alpha^2+\beta^2+\gamma^2 \\ \alpha+\beta+\gamma & \alpha^2+\beta^2+\gamma^2 & \alpha^3+\beta^3+\gamma^3 \\ \alpha^2+\beta^2+\gamma^2 & \alpha^3+\beta^3+\gamma^3 & \alpha^4+\beta^4+\gamma^4 \end{vmatrix} = \begin{vmatrix} S_0 & S_1 & S_2 \\ S_1 & S_2 & S_3 \\ S_2 & S_3 & S_4 \end{vmatrix}.$$

$$\Delta_1\Delta_2 = \begin{vmatrix} 1 & 1 & 1 \\ \alpha & \beta & \gamma \\ \alpha^2 & \beta^2 & \gamma^2 \end{vmatrix} \times \begin{vmatrix} 1 & 1 & 1 \\ \alpha^2 & \beta^2 & \gamma^2 \\ \alpha^3 & \beta^3 & \gamma^3 \end{vmatrix}$$

$$= \begin{vmatrix} 1+1+1 & \alpha^2+\beta^2+\gamma^2 & \alpha^3+\beta^3+\gamma^3 \\ \alpha+\beta+\gamma & \alpha^3+\beta^3+\gamma^3 & \alpha^4+\beta^4+\gamma^4 \\ \alpha^2+\beta^2+\gamma^2 & \alpha^4+\beta^4+\gamma^4 & \alpha^5+\beta^5+\gamma^5 \end{vmatrix} = \begin{vmatrix} S_0 & S_2 & S_3 \\ S_1 & S_3 & S_4 \\ S_2 & S_4 & S_5 \end{vmatrix}.$$

$$\begin{vmatrix} S_0 & S_3 & S_4 \\ S_1 & S_4 & S_5 \\ S_2 & S_5 & S_6 \end{vmatrix} = \begin{vmatrix} 1+1+1 & \alpha^3+\beta^3+\gamma^3 & \alpha^4+\beta^4+\gamma^4 \\ \alpha+\beta+\gamma & \alpha^4+\beta^4+\gamma^4 & \alpha^5+\beta^5+\gamma^5 \\ \alpha^2+\beta^2+\gamma^2 & \alpha^5+\beta^5+\gamma^5 & \alpha^6+\beta^6+\gamma^6 \end{vmatrix}$$

$$= \begin{vmatrix} 1 & 1 & 1 \\ \alpha & \beta & \gamma \\ \alpha^2 & \beta^2 & \gamma^2 \end{vmatrix} \times \begin{vmatrix} 1 & 1 & 1 \\ \alpha^3 & \beta^3 & \gamma^3 \\ \alpha^4 & \beta^4 & \gamma^4 \end{vmatrix}.$$

Example 16. If $A, H, \ldots, C$ denote the cofactors of $a, h, \ldots, c$ in the determinant

$$\Delta = \begin{vmatrix} a & h & g \\ h & b & f \\ g & f & c \end{vmatrix},$$

show that (i) $\begin{vmatrix} A & H & G \\ H & B & F \\ G & F & C \end{vmatrix} = \Delta^2$, (ii) $BC - F^2 = a\Delta$.

It may be assumed that $\Delta \neq 0$.

(i) $\Delta \times \begin{vmatrix} A & H & G \\ H & B & F \\ G & F & C \end{vmatrix} = \begin{vmatrix} a & h & g \\ h & b & f \\ g & f & c \end{vmatrix} \times \begin{vmatrix} A & H & G \\ H & B & F \\ G & F & C \end{vmatrix}$

$$= \begin{vmatrix} aA+hH+gG & aH+hB+gF & aG+hF+gC \\ hA+bH+fG & hH+bB+fF & hG+bF+fC \\ gA+fH+cG & gH+fB+cF & gG+fF+cC \end{vmatrix}$$

$$= \begin{vmatrix} \Delta & 0 & 0 \\ 0 & \Delta & 0 \\ 0 & 0 & \Delta \end{vmatrix}, \text{ using (7.14), (7.15)},$$

$$= \Delta^3.$$

Division by Δ then gives the required result.

MULTIPLICATION OF DETERMINANTS

(ii) $\Delta(BC - F^2) = \begin{vmatrix} a & h & g \\ h & b & f \\ g & f & c \end{vmatrix} \times \begin{vmatrix} 1 & 0 & 0 \\ H & B & F \\ G & F & C \end{vmatrix}$

$= \begin{vmatrix} a & aH + hB + gF & aG + hF + gC \\ h & hH + bB + fF & hG + bF + fC \\ g & gH + fB + cF & gG + fF + cC \end{vmatrix}$

$= \begin{vmatrix} a & 0 & 0 \\ h & \Delta & 0 \\ g & 0 & \Delta \end{vmatrix}$, using (7.14), (7.15)

$= a\Delta^2$.

Again, division by Δ leads to the desired result.

The reader should note the following definitions:

(i) a determinant is said to be *symmetrical* if the elements of any row are identical with those of the corresponding column. Thus the determinant

$$\begin{vmatrix} a & h & g \\ h & b & f \\ g & f & c \end{vmatrix}$$

of Example 16 above is symmetrical;

(ii) a determinant is said to be *skew* if the element in the rth row and sth column is equal in magnitude but opposite in sign to the element in the sth row and rth column, for all values of r and s: an example is

$$\begin{vmatrix} a & h & g & l \\ -h & b & f & m \\ -g & -f & c & n \\ -l & -m & -n & p \end{vmatrix};$$

(iii) a skew determinant in which all the elements of the principal diagonal are zero is said to be *skew-symmetrical*;

(iv) a determinant whose elements are the cofactors of the elements of a given determinant Δ is called the *reciprocal* determinant of Δ: for example,

$$\begin{vmatrix} A & H & G \\ H & B & F \\ G & F & C \end{vmatrix}$$

is the reciprocal of the determinant

$$\begin{vmatrix} a & h & g \\ h & b & f \\ g & f & c \end{vmatrix}$$

of Example 16.

EXERCISES 7 (c)

1. If Δ denotes the determinant
$$\begin{vmatrix} a & h & g & x \\ h & b & f & y \\ g & f & c & z \\ x & y & z & 0 \end{vmatrix}$$
and $A, H, \ldots, C$ are the cofactors of $a, h, \ldots, c$ in the determinant
$$\begin{vmatrix} a & h & g \\ h & b & f \\ g & f & c \end{vmatrix},$$
show that
$$Ax^2 + By^2 + Cz^2 + 2Fyz + 2Gzx + 2Hxy = -\Delta.$$

2. Express the product of the two determinants
$$\begin{vmatrix} a+ib & c+id \\ -c+id & a-ib \end{vmatrix}, \quad \begin{vmatrix} x+iy & z+iu \\ -z+iu & x-iy \end{vmatrix}$$
as a determinant of order 2.

Hence, or otherwise, show that the product of two sums each of four squares can be expressed as the sum of four squares.

3. Show that
$$\begin{vmatrix} a & c & b \\ b & a & c \\ c & b & a \end{vmatrix} = a^3 + b^3 + c^3 - 3abc.$$

Hence show that the product
$$(x^3 + y^3 + z^3 - 3xyz)(a^3 + b^3 + c^3 - 3abc)$$
may be expressed in the form
$$A^3 + B^3 + C^3 - 3ABC. \tag{C.}$$

4. The sum $\alpha^r + \beta^r + \gamma^r$ is represented by S_r. Prove that
$$\begin{vmatrix} 1 & 1 & 1 \\ \alpha & \beta & \gamma \\ \alpha^2 & \beta^2 & \gamma^2 \end{vmatrix}^2 = \begin{vmatrix} S_0 & S_1 & S_2 \\ S_1 & S_2 & S_3 \\ S_2 & S_3 & S_4 \end{vmatrix}.$$

Express
$$\begin{vmatrix} S_0 & S_2 & S_1 \\ S_2 & S_4 & S_6 \\ S_3 & S_5 & S_7 \end{vmatrix}$$
as the product of two determinants. Hence, or otherwise, show that its value is
$$(\alpha\beta + \beta\gamma + \gamma\alpha)\Pi(\beta - \gamma)\Pi(\alpha^2 - \beta^2),$$
where $\Pi(\beta - \gamma) = (\beta - \gamma)(\gamma - \alpha)(\alpha - \beta)$ and similarly for $\Pi(\beta^2 - \gamma^2)$.
(O.C.)

5. Show that
$$\begin{vmatrix} x & c & -b \\ -c & x & a \\ b & -a & x \end{vmatrix} \times \begin{vmatrix} a^2 + x^2 & ab + cx & ca - bx \\ ab - cx & b^2 + x^2 & bc + ax \\ ca + bx & bc - ax & c^2 + x^2 \end{vmatrix}$$
$= x^3(x^2 + a^2 + b^2 + c^2)^3.$

6. By expressing the determinant
$$\begin{vmatrix} 2 & a+b & a^2+b^2 \\ a+b & a^2+b^2 & a^3+b^3 \\ 1 & c & c^2 \end{vmatrix}$$
as the product of two determinants, show that its value is $(a-b)^2(c-a)(c-b)$.

7.9. The solution of simultaneous equations

Consider the n linear equations in the n variables $x, y, z, \ldots, w$,

$$\left.\begin{aligned} a_1x + b_1y + c_1z + \ldots + k_1w &= l_1, \\ a_2x + b_2y + c_2z + \ldots + k_2w &= l_2, \\ \cdots\cdots\cdots\cdots\cdots\cdots\cdots\cdots\cdots\cdots\cdots\cdots & \\ a_nx + b_ny + c_nz + \ldots + k_nw &= l_n. \end{aligned}\right\} \quad (7.20)$$

Let Δ denote the determinant of the nth order $(a_1b_2c_3 \ldots k_n)$ formed by the coefficients on the left-hand sides of the above equations and let $\Delta_1, \Delta_2, \Delta_3, \ldots, \Delta_n$ denote the nth order determinants $(l_1b_2c_3 \ldots k_n), (a_1l_2c_3 \ldots k_n), (a_1b_2l_3 \ldots k_n), \ldots, (a_1b_2c_3 \ldots l_n)$ respectively, these determinants being formed from Δ by replacing respectively the elements of the first, second, third, ..., nth columns by new elements $l_1, l_2, l_3, \ldots, l_n$.

Let $A_1, A_2, A_3, \ldots, A_n$ be the cofactors of $a_1, a_2, a_3, \ldots, a_n$ in the determinant Δ. Then multiplying the first of equations (7.20) by A_1, the second by A_2, the third by $A_3, \ldots$, the nth by A_n and adding

$(a_1A_1 + a_2A_2 + \ldots + a_nA_n)x + (b_1A_1 + b_2A_2 + \ldots + b_nA_n)y + \ldots + (k_1A_1 + k_2A_2 + \ldots + k_nA_n)w = l_1A_1 + l_2A_2 + \ldots + l_nA_n.$

By (7.14), (7.15) the coefficient of x in this equation is equal to Δ, the coefficients of $y, z, \ldots, w$ all vanish and the term on the right is equal to Δ_1. Hence

$$\Delta x = \Delta_1$$

and, if $\Delta \neq 0$, $x = \Delta_1/\Delta$. In a similar way it can be shown that $y = \Delta_2/\Delta, z = \Delta_3/\Delta, \ldots, w = \Delta_n/\Delta$, and the solution of the simultaneous equations can therefore be expressed by means of determinants.

It may perhaps be helpful to write out in full the solution for a particular case. Thus the solution of the three linear equations

$$\left.\begin{aligned} a_1x + b_1y + c_1z &= l_1, \\ a_2x + b_2y + c_2z &= l_2, \\ a_3x + b_3y + c_3z &= l_3, \end{aligned}\right\} \quad (7.21)$$

is conveniently written in the form

$$\frac{x}{\begin{vmatrix} l_1 & b_1 & c_1 \\ l_2 & b_2 & c_2 \\ l_3 & b_3 & c_3 \end{vmatrix}} = \frac{y}{\begin{vmatrix} a_1 & l_1 & c_1 \\ a_2 & l_2 & c_2 \\ a_3 & l_3 & c_3 \end{vmatrix}} = \frac{z}{\begin{vmatrix} a_1 & b_1 & l_1 \\ a_2 & b_2 & l_2 \\ a_3 & b_3 & l_3 \end{vmatrix}} = \frac{1}{\begin{vmatrix} a_1 & b_1 & c_1 \\ a_2 & b_2 & c_2 \\ a_3 & b_3 & c_3 \end{vmatrix}},$$

(7.22)

provided that $\begin{vmatrix} a_1 & b_1 & c_1 \\ a_2 & b_2 & c_2 \\ a_3 & b_3 & c_3 \end{vmatrix} \neq 0.$

If $\Delta = 0$ and $\Delta_1, \Delta_2, \ldots, \Delta_n$ are not all zero, the general solution

$$x = \Delta_1/\Delta, \, y = \Delta_2/\Delta, \ldots, w = \Delta_n/\Delta,$$

of the n simultaneous equations shows that the equations possess a solution only if infinite values of some or all the variables are admitted. All that can be asserted in this case is that the equations do not admit a finite solution. A particular case in which $\Delta = 0$ and all of $\Delta_1, \Delta_2, \ldots, \Delta_n$ also vanish is discussed in § 7.10.

Example 17. *Use the method of this section to solve the equations*

$$\left. \begin{array}{r} x + y + z = 5, \\ 2x - y + z = 7, \\ 3x + y - 5z = 13. \end{array} \right\}$$

Here

$$\Delta = \begin{vmatrix} 1 & 1 & 1 \\ 2 & -1 & 1 \\ 3 & 1 & -5 \end{vmatrix} = \begin{vmatrix} 1 & 0 & 0 \\ 2 & -3 & -1 \\ 3 & -2 & -8 \end{vmatrix} \quad (c_2' = c_2 - c_1, \, c_3' = c_3 - c_1)$$

$$= 24 - 2 = 22;$$

$$\Delta_1 = \begin{vmatrix} 5 & 1 & 1 \\ 7 & -1 & 1 \\ 13 & 1 & -5 \end{vmatrix} = \begin{vmatrix} 12 & 0 & 2 \\ 7 & -1 & 1 \\ 20 & 0 & -4 \end{vmatrix} \quad (r_1' = r_1 + r_2, \, r_3' = r_3 + r_2)$$

$$= 48 + 40 = 88;$$

$$\Delta_2 = \begin{vmatrix} 1 & 5 & 1 \\ 2 & 7 & 1 \\ 3 & 13 & -5 \end{vmatrix} = \begin{vmatrix} 1 & 0 & 0 \\ 2 & -3 & -1 \\ 3 & -2 & -8 \end{vmatrix} \quad (c_2' = c_2 - 5c_1, \, c_3' = c_3 - c_1)$$

$$= 24 - 2 = 22;$$

$$\Delta_3 = \begin{vmatrix} 1 & 1 & 5 \\ 2 & -1 & 7 \\ 3 & 1 & 13 \end{vmatrix} = \begin{vmatrix} 3 & 0 & 12 \\ 2 & -1 & 7 \\ 5 & 0 & 20 \end{vmatrix} \quad (r_1' = r_1 + r_2, \, r_3' = r_3 + r_2)$$

$$= -60 + 60 = 0.$$

Hence,

$$x = \Delta_1/\Delta = 88/22 = 4, \, y = \Delta_2/\Delta = 22/22 = 1, \, z = \Delta_3/\Delta = 0.$$

7.10. The consistency of sets of simultaneous equations

If, in equations (7.20), $l_1 = l_2 = \ldots = l_n = 0$, the equations

$$\left.\begin{array}{l} a_1x + b_1y + c_1z + \ldots + j_1v + k_1w = 0, \\ a_2x + b_2y + c_2z + \ldots + j_2v + k_2w = 0, \\ \cdots\cdots\cdots\cdots\cdots\cdots\cdots\cdots\cdots\cdots, \\ a_nx + b_ny + c_nz + \ldots + j_nv + k_nw = 0, \end{array}\right\} \quad (7.23)$$

form a set of n homogeneous equations in effectively $n - 1$ variables (the ratios of any $n - 1$ of the quantities $x, y, z, \ldots$ to the remaining one). These equations are consistent, in the sense of § 7.1, if the values of these $n - 1$ ratios found from any $n - 1$ of the above equations satisfy the remaining equation (the trivial case in which $x, y, z, \ldots, w$ are all zero being ignored).

The second, third, ..., nth of equations (7.23) can be solved by the method of § 7.9 to give

$$(a_2b_3c_4\ldots j_n)x = -(k_2b_3c_4\ldots j_n)w, \; (a_2b_3c_4\ldots j_n)y = -(a_2k_3c_4\ldots j_n)w,$$
$$(a_2b_3c_4\ldots j_n)z = -(a_2b_3k_4\ldots j_n)w, \text{ etc.},$$

and, if these values are substituted in the first of (7.23), there results, after division by $-w$ and multiplication by $(a_2b_3c_4\ldots j_n)$,

$$a_1(k_2b_3c_4\ldots j_n) + b_1(a_2k_3c_4\ldots j_n) + c_1(a_2b_3k_4\ldots j_n) + \\ \ldots - k_1(a_2b_3c_4\ldots j_n) = 0.$$

By moving respectively the first, second, ..., $(n-2)$th columns so that they become the $(n-1)$th column of the first $n-2$ of the determinants on the left-hand side of this relation, it can be written

$$a_1(b_2c_3\ldots k_n) - b_1(a_2c_3\ldots k_n) + c_1(a_2b_3\ldots k_n) + \\ \ldots + (-1)^{n-1}k_1(a_2b_3\ldots j_n) = 0.$$

The left-hand side of this relation is, of course, the nth order determinant $(a_1b_2c_3\ldots k_n)$ and the condition for the consistency of the homogeneous equations (7.23) can therefore be written

$$\begin{vmatrix} a_1 & b_1 & c_1 & \cdots & \cdots & j_1 & k_1 \\ a_2 & b_2 & c_2 & \cdots & \cdots & j_2 & k_2 \\ \cdots & \cdots & \cdots & \cdots & \cdots & \cdots & \cdots \\ \cdots & \cdots & \cdots & \cdots & \cdots & \cdots & \cdots \\ a_n & b_n & c_n & \cdots & \cdots & j_n & k_n \end{vmatrix} = 0. \quad (7.24)$$

An important example of this result occurs in coordinate geometry. The three straight lines $a_1x + b_1y + c_1 = 0$, $a_2x + b_2y + c_2 = 0$ and $a_3x + b_3y + c_3 = 0$ are concurrent if these equations are satisfied simultaneously by the coordinates (x, y) of the point of concurrence.

By taking $n = 3$, $z = 1$ in equations (7.23), the result (7.24) gives the necessary condition for concurrence of the lines in the form

$$\begin{vmatrix} a_1 & b_1 & c_1 \\ a_2 & b_2 & c_2 \\ a_3 & b_3 & c_3 \end{vmatrix} = 0.$$

Example 18. *Prove that $\lambda = 3$ is the only real value of λ for which the equations $-\lambda x + y + 2z = 0$, $x + \lambda y + 3z = 0$ and $x + 3y + \lambda z = 0$, where x, y, z are not all zero, are consistent.*

By (7.24), the condition for consistency is

$$\begin{vmatrix} -\lambda & 1 & 2 \\ 1 & \lambda & 3 \\ 1 & 3 & \lambda \end{vmatrix} = 0.$$

Expanding the determinant, this gives $\lambda^3 - 6\lambda - 9 = 0$. This cubic in λ is clearly satisfied when $\lambda = 3$ and, removing the factor $(\lambda - 3)$, by the roots of the quadratic $\lambda^2 + 3\lambda + 3 = 0$. The roots of this quadratic being complex, the required result follows.

7.11. Elimination

It has been shown in § 7.10 that, when n equations are satisfied by $n - 1$ variables, it is possible by solving $n - 1$ of the given equations and substituting the values so obtained in the remaining equation, to obtain a relation between the coefficients in the various equations which must hold if the equations are consistent. The process of obtaining this condition for consistency is termed the *elimination* of the variables, and the consistency condition itself is called the *eliminant* of the given equations. It is often possible to employ determinantal notation to avoid the labour of solving all but one of the equations and substituting in the remaining equation in other types of equation than those discussed in § 7.10, and some examples follow.

Example 19 *Eliminate x between the two equations*
$$ax^3 + bx^2 + cx + d = 0, \quad Ax^2 + Bx + C = 0.$$

By multiplying the first equation by x and the second by x^2 and x in succession, the following five equations are obtained,

$$\begin{aligned} ax^3 + bx^2 + cx + d &= 0, \\ ax^4 + bx^3 + cx^2 + dx &= 0, \\ Ax^2 + Bx + C &= 0, \\ Ax^3 + Bx^2 + Cx &= 0, \\ Ax^4 + Bx^3 + Cx^2 &= 0. \end{aligned}$$

If x^4, x^3, x^2 and x are regarded as four distinct variables, the required eliminant is, by § 7.10,

$$\begin{vmatrix} 0 & a & b & c & d \\ a & b & c & d & 0 \\ 0 & 0 & A & B & C \\ 0 & A & B & C & 0 \\ A & B & C & 0 & 0 \end{vmatrix} = 0.$$

Example 20. *Show that the equations* $ax^2 + bx + c = 0$, $x^3 = \lambda$ *have a common root if*

$$\begin{vmatrix} a & b & c \\ b & c & a\lambda \\ c & a\lambda & b\lambda \end{vmatrix} = 0.$$

Multiplying the first equation successively by x, x^2 and writing $x^3 = \lambda$,

$$bx^2 + cx + a\lambda = 0,$$
$$cx^2 + a\lambda x + b\lambda = 0.$$

Eliminating x^2 and x between these two equations and $ax^2 + bx + c = 0$, the required result follows.

EXERCISES 7 (d)

1. Solve the simultaneous equations
$$x + y + z = 6,\ 2x + 3y - z = 5,\ x + v + 5z = 18.$$

2. Solve the equations
$$4x - 3y + 2z + 7 = 0,\ 6x + 2y - 3z = 33,\ 2x - 4y - z + 3 = 0.$$

3. Find for what values of a the following equations are consistent:
$$x + a^2 y + a = 0,$$
$$ax + y + a^2 = 0,$$
$$a^2 x + ay + 1 = 0.$$

4. Determine the values of λ so that the equations
$$5x - 2y - 6z = \lambda x,$$
$$2x - 3y - 4z = \lambda y,$$
$$x + y = \lambda z,$$
may be satisfied by values of x, y and z not all zero.

5. Express, by the vanishing of a fourth order determinant, the result of eliminating x from the two quadratic equations $ax^2 + bx + c = 0$, $Ax^2 + Bx + C = 0$.

6. If $ax^3 + bx^2 + cx + d = 0$, $Ax^2 + Bx + C = 0$ show that
$$(aB - bA)x^2 + (aC - cA)x - dA = 0,$$
and
$$(aC - cA)x^2 + (bC - cB - dA)x - dB = 0.$$

Hence show that the result of eliminating x between the original two equations can be written in the form

$$\begin{vmatrix} A & B & C \\ aB - bA & aC - cA & -dA \\ aC - cA & bC - cB - dA & -dB \end{vmatrix} = 0.$$

EXERCISES 7 (e)

1. Express as a product of linear factors the determinant
$$\begin{vmatrix} 4 & x+1 & x+1 \\ x+1 & (x+2)^2 & 1 \\ x+1 & 1 & (x+2)^2 \end{vmatrix}.$$
(N.U.)

2. Prove that
$$\begin{vmatrix} 2bc - a^2 & a^2 & a^2 \\ b^2 & 2ca - b^2 & b^2 \\ c^2 & c^2 & 2ab - c^2 \end{vmatrix}$$
is divisible by $abc(a + b + c)$ and express the determinant as a product of real factors. (O.)

3. Prove that
$$\begin{vmatrix} 1 & \cos C & \cos B \\ \cos C & 1 & \cos A \\ \cos B & \cos A & 1 \end{vmatrix} = 4\cos A \cos B \cos C$$
if A, B, C are the angles of a triangle. (O.)

4. Evaluate the determinant
$$\begin{vmatrix} \dfrac{1}{x+1} & \dfrac{1}{x+2} & \dfrac{1}{x+3} \\ \dfrac{1}{x+2} & \dfrac{1}{x+3} & \dfrac{1}{x+4} \\ \dfrac{1}{x+3} & \dfrac{1}{x+4} & \dfrac{1}{x+5} \end{vmatrix}$$
(O.)

5. Evaluate as a product of linear factors the determinant
$$\begin{vmatrix} a^3 & a^2 & a & 1 \\ b^3 & b^2 & b & 1 \\ c^3 & c^2 & c & 1 \\ 3c^2 & 2c & 1 & 0 \end{vmatrix}.$$
(O.)

6. Express
$$\begin{vmatrix} 1 & a & a^3 \\ 1 & b & b^3 \\ 1 & c & c^3 \end{vmatrix}$$
as a product of linear factors in a, b, c.
Hence, or otherwise, prove that, if $\alpha + \beta + \gamma = \pi$, then
$$\begin{vmatrix} 1 & \sin \alpha & \sin 3\alpha \\ 1 & \sin \beta & \sin 3\beta \\ 1 & \sin \gamma & \sin 3\gamma \end{vmatrix} = -16 \sin \alpha \sin \beta \sin \gamma \sin \tfrac{1}{2}(\beta - \gamma) \times$$
$$\times \sin \tfrac{1}{2}(\gamma - \alpha) \sin \tfrac{1}{2}(\alpha - \beta).$$
(C.)

7. Prove that
$$\begin{vmatrix} 1 & 1 & 1 \\ \cos \alpha & \cos \beta & \cos \gamma \\ \cos 2\alpha & \cos 2\beta & \cos 2\gamma \end{vmatrix} = 2(\cos \beta - \cos \gamma)(\cos \gamma - \cos \alpha) \times$$
$$\times (\cos \alpha - \cos \beta).$$

Prove also that
$$\begin{vmatrix} \cos \alpha & \cos \beta & \cos \gamma \\ \cos 2\alpha & \cos 2\beta & \cos 2\gamma \\ \cos 3\alpha & \cos 3\beta & \cos 3\gamma \end{vmatrix}$$
$$= 4(\cos \beta - \cos \gamma)(\cos \gamma - \cos \alpha)(\cos \alpha - \cos \beta) \times$$
$$\times (\cos \alpha + \cos \beta + \cos \gamma + 2 \cos \alpha \cos \beta \cos \gamma).$$
(N.U.)

EXERCISES

8. Prove that
$$\begin{vmatrix} 1 & \frac{2x}{x+y} & \frac{2x}{x+z} \\ \frac{2y}{x+y} & 1 & \frac{2y}{y+z} \\ \frac{2z}{x+z} & \frac{2z}{y+z} & 1 \end{vmatrix} = \left[\frac{(y-z)(z-x)(x-y)}{(y+z)(z+x)(x+y)}\right]^2$$
(O.C.)

9. Prove that
$$\begin{vmatrix} \sin x & \sin^3 x & \cos x \\ \sin y & \sin^3 y & \cos y \\ \sin z & \sin^3 z & \cos z \end{vmatrix}$$
$$= \sin(x+y+z)\sin(y-z)\sin(z-x)\sin(x-y). \quad \text{(O.)}$$

10. Expand
$$\begin{vmatrix} -1 & \cos\psi & \cos\phi \\ \cos\psi & -1 & \cos\theta \\ \cos\phi & \cos\theta & -1 \end{vmatrix}$$
and show that its value is $4\cos\sigma\cos(\sigma-\theta)\cos(\sigma-\phi)\cos(\sigma-\psi)$ where $\sigma = \frac{1}{2}(\theta+\phi+\psi)$.

Show that
$$x^2 + y^2 + z^2 - 2yz\cos\theta - 2zx\cos\phi + 2xy\cos(\theta+\phi)$$
is the product of two linear factors, and find these factors. (O.)

11. Solve the equation
$$\begin{vmatrix} x & 1 & 2 \\ 1 & x+2 & 3 \\ 2 & 3 & x+4 \end{vmatrix} = 0.$$
(O.)

12. Find t from the equation
$$\begin{vmatrix} t & a & b & c \\ c & t & a & b \\ b & c & t & a \\ a & b & c & t \end{vmatrix} = 0.$$

13. Prove that, if $\beta\gamma + \gamma\alpha + \alpha\beta = 0$, one root of the equation
$$\begin{vmatrix} x^2 + \alpha^2 & x+\alpha & \beta\gamma \\ x^2 + \beta^2 & x+\beta & \gamma\alpha \\ x^2 + \gamma^2 & x+\gamma & \alpha\beta \end{vmatrix} = 0$$
is $x = 0$ and find the other root. (O.)

14. If $\bar{a}, \bar{b}, \bar{c}$ are the complex conjugates of a, b, c respectively, and if p, q, r are real, show that the equation
$$\begin{vmatrix} a-x & b & p \\ c & q-x & b \\ r & \bar{c} & \bar{a}-x \end{vmatrix} = 0$$
has either three real roots or one real root and a pair of conjugate complex roots.

It is given that q is a root when $a = i, b = c = p = r = 1$. Find q and solve the equation completely. (C.)

15. Prove that the sum of the roots of the equation

$$\begin{vmatrix} x & h & g \\ h & x & f \\ g & f & x \end{vmatrix} = 0$$

is zero, and that the sum of the squares of the roots is $2(f^2 + g^2 + h^2)$.

Taking f, g, h to be real, and assuming that the roots are then all real, prove that no root exceeds

$$2\sqrt{\{\tfrac{1}{3}(f^2 + g^2 + h^2)\}}$$

in absolute value. (C.)

16. Prove that the determinant

$$\Delta = \begin{vmatrix} a & b & c & d \\ -d & a & b & c \\ -c & -d & a & b \\ -b & -c & -d & a \end{vmatrix}$$

has $(a + b\omega + c\omega^2 + d\omega^3)$ as a factor, where ω is any one of the roots of the equation $\omega^4 = -1$.

Hence, or otherwise, prove that the nth power of this determinant is expressible in the form

$$\Delta^n = \begin{vmatrix} A & B & C & D \\ -D & A & B & C \\ -C & -D & A & B \\ -B & -C & -D & A \end{vmatrix}$$

where A, B, C, D are functions of a, b, c, d (which should not be found explicitly). (O.C.)

17. Show that
$$\begin{vmatrix} a_1 + \alpha_1 & b_1 + \beta_1 \\ a_2 + \alpha_2 & b_2 + \beta_2 \end{vmatrix}$$
is a factor of the determinant

$$\begin{vmatrix} a_1 & b_1 & \beta_1 & \alpha_1 \\ a_2 & b_2 & \beta_2 & \alpha_2 \\ \alpha_2 & \beta_2 & b_2 & a_2 \\ \alpha_1 & \beta_1 & b_1 & a_1 \end{vmatrix}$$

and express the other factor as a determinant. (O.)

18. Prove that if u_n is the determinant of the nth order

$$\begin{vmatrix} x & a & a & \cdots & \cdots & a \\ b & x & a & \cdots & \cdots & a \\ b & b & x & \cdots & \cdots & a \\ \cdots & \cdots & \cdots & \cdots & \cdots & \cdots \\ b & b & b & \cdots & \cdots & x \end{vmatrix},$$

$u_n = (x - a)u_{n-1} + a(x - b)^{n-1} = (x - b)u_{n-1} + b(x - a)^{n-1}$, and evaluate u_n. (O.)

19. Show that

$$\begin{vmatrix} t_1+x & a+x & a+x & a+x \\ b+x & t_2+x & a+x & a+x \\ b+x & b+x & t_3+x & a+x \\ b+x & b+x & b+x & t_4+x \end{vmatrix} = A + Bx,$$

where $\quad A = \dfrac{af(b) - bf(a)}{a-b}, \; B = \dfrac{f(b)-f(a)}{a-b}$

and $\quad f(t) \equiv (t-t_1)(t-t_2)(t-t_3)(t-t_4)$.

Find also the values of A and B if $b = a$. (O.)

20. If $A_1, B_1, \ldots$ are the cofactors of $a_1, b_1, \ldots$ in

$$\Delta = \begin{vmatrix} a_1 & b_1 & c_1 \\ a_2 & b_2 & c_2 \\ a_3 & b_3 & c_3 \end{vmatrix}$$

and, if r, s, t denote the numbers 1, 2, 3 in any order, show that

$$A_r B_s C_t - a_r b_s c_t \Delta$$

has a value independent of this order.

Prove that this value may be written as

$$-\begin{vmatrix} b_1 c_1 & c_1 a_1 & a_1 b_1 \\ b_2 c_2 & c_2 a_2 & a_2 b_2 \\ b_3 c_3 & c_3 a_3 & a_3 b_3 \end{vmatrix}. \quad \text{(O.)}$$

21. If $\alpha, \beta, \gamma, \delta$ are the roots of $x^4 - px^3 + qx^2 - rx + s = 0$ and

$$\Delta \equiv \begin{vmatrix} \alpha^6 & \alpha^4 & \alpha^2 & 1 \\ \beta^6 & \beta^4 & \beta^2 & 1 \\ \gamma^6 & \gamma^4 & \gamma^2 & 1 \\ \delta^6 & \delta^4 & \delta^2 & 1 \end{vmatrix}, \; D \equiv \begin{vmatrix} \alpha^3 & \alpha^2 & \alpha & 1 \\ \beta^3 & \beta^2 & \beta & 1 \\ \gamma^3 & \gamma^2 & \gamma & 1 \\ \delta^3 & \delta^2 & \delta & 1 \end{vmatrix},$$

show that $\Delta = D(pqr - p^2 s - r^2)$ and evaluate both determinants as products of linear factors. (O.)

22. By factorising the determinant, or otherwise, show that

$$\begin{vmatrix} x & y & z & u \\ u & x & y & z \\ z & u & x & y \\ y & z & u & x \end{vmatrix} = (x^2 + z^2 - 2yu)^2 - (u^2 + y^2 - 2zx)^2.$$

Express

$$\{(x^2 + z^2 - 2yu)^2 - (u^2 + y^2 - 2zx)^2\} \times$$
$$\times \{(X^2 + Z^2 - 2YU)^2 - (U^2 + Y^2 - 2ZX)^2\}$$

in the form $(A^2 + C^2 - 2BD)^2 - (D^2 + B^2 - 2CA)^2$, giving explicit expressions for A, B, C, D in terms of X, Y, Z, U, x, y, z, u. (C.)

23. If $x + y + z = 0, \; (a-b)x + ay + (a+b)z = 0,$
$(a-b)^2 x + a^2 y + (a+b)^2 z = 0$ and x, y, z are not all zero, prove that $b = 0$.

24. In the system of equations $-ny + mz = a$, $nx - lz = b$, $-mx + ly = c$, $lx + my + nz = p$, l, m, n, a, b, c, p are given real numbers and l, m, n, are not all zero. Prove that $la + mb + nc = 0$.
(C.)

25. Use the equations $x^3 + px^2 + qx + r = 0$, $x^2 + bx + c = 0$ to express $(r + bc - pc)^2 - (cq - c^2 - br)(bp - b^2 - q + c)$ as a determinant of the fifth order.
(C.)

CHAPTER 8

FUNCTIONS OF A CONTINUOUS VARIABLE. GENERAL THEOREMS OF THE DIFFERENTIAL CALCULUS

8.1. Introduction

The purpose of this and the two following chapters is to fill in some of the gaps which occurred in the discussion of differentiation and integration given in Chapters 7–13 of *Advanced Level Pure Mathematics*. In that book, the foundations of the subject were very hastily laid in order that the reader should be quickly able to appreciate the value of the calculus as a tool. It is now intended both to strengthen the foundations and to extend some of the techniques and applications.

8.2. Limits of functions of a continuous variable

In Chapter 3 (§ 3.1 *et seq.*) of this book a short discussion was given of the limit of a function of the positive integral variable n—the limit of a function $f(x)$ of the continuous real variable x can be treated similarly. When the variable x is supposed to assume successively *all* values corresponding to *all* the points on the straight line of § 1.1 (Fig. 1), commencing at some definite point of the line and progressing always to the right, x is said to tend to *infinity* and this is denoted symbolically by $x \to \infty$. The reader should observe that the difference implied by the symbols $n \to \infty$ and $x \to \infty$ is that n proceeds to the right of the line of Fig. 1 by a series of leaps whereas x proceeds to the right quite smoothly—this distinction is sometimes emphasised by saying that x tends *continuously* to infinity.

Limiting the discussion to single-valued functions of the continuous real variable x, formal definitions of the behaviour of such functions when $x \to \infty$ (corresponding almost exactly to the behaviour of convergent, divergent and oscillating sequences) can be framed as follows:

(*a*) The function $f(x)$ is said to tend to *a limit l as x tends to infinity* if, when any positive number δ (however small) is chosen, a number $x_0(\delta)$ (depending on δ) can be found such that $|f(x) - l| < \delta$ whenever $x \geqslant x_0(\delta)$. In such a case it is usual to write

$$\lim_{x \to \infty} f(x) = l,$$

and, when there is no risk of ambiguity, lim. $f(x) = l$ or $f(x) \to l$.

(*b*) The function $f(x)$ is said *to tend to infinity with x* if, when any positive number Δ (however large) is chosen, a number $x_0(\Delta)$ (depending on $\varDelta$) can be found such that $f(x) > \varDelta$ whenever $x \geqslant x_0(\varDelta)$. Again, if there is no risk of ambiguity, such behaviour is indicated by the symbols $f(x) \to \infty$ and the reader should have no difficulty in framing a definition to cover the case $f(x) \to -\infty$.

(c) If the conditions of neither (a) nor (b) above are satisfied, then $f(x)$ is said *to oscillate as x tends to infinity*. If $|f(x)|$ is less than some positive constant K whenever $x \geqslant x_0$, $f(x)$ is said *to oscillate finitely*—if this condition is not satisfied $f(x)$ is said *to oscillate infinitely*.

Similar definitions can clearly be devised to describe the behaviour of $f(x)$ as x tends to negative infinity. Alternatively, by writing $x = -y$, the behaviour of $f(x)$ as $x \to -\infty$ can be described by the behaviour of $f(-y)$ $\{= \phi(y)$, say$\}$ as $y \to \infty$.

Other correspondences between functions of the continuous real variable and those of a positive integral variable occur in the notion of functions which increase or decrease steadily and in certain general theorems on limits. Thus, corresponding to the definitions of § 3.3, *a function $f(x)$ is said to increase steadily with x, or to be monotonic increasing, if $f(x_2) \geqslant f(x_1)$ whenever $x_2 > x_1$*. In certain cases this condition is only satisfied from some definite value x_0 of x onwards so that the $f(x_2) \geqslant f(x_1)$ holds only when $x_2 > x_1 \geqslant x_0$—such functions would be said to increase steadily when $x \geqslant x_0$. The theorem of § 3.3 remains true, the proof merely requiring the alteration of n to x and some obvious verbal changes—hence, *a function which increases steadily with x either tends to a limit or to positive infinity as $x \to \infty$*. If the possibility of equality is excluded from the definition of a monotonic increasing function, that is, if $f(x_2) > f(x_1)$ whenever $x_2 > x_1$, $f(x)$ is said to be monotonic increasing in the *strict sense* and it is left to the reader to provide corresponding definitions and a corresponding theorem for functions which decrease steadily. The theorems on sums, products and quotients of limits proved in § 3.4 also remain true, both enunciations and proofs requiring only verbal alterations. Hence, if $f(x)$ and $F(x)$ are functions of a continuous real variable x and if $f(x) \to l$, $F(x) \to L$ as $x \to \infty$, then

$$\left. \begin{array}{l} f(x) \pm F(x) \to l \pm L, \\ f(x).F(x) \to lL, \\ f(x)/F(x) \to l/L \text{ (provided } L \neq 0). \end{array} \right\} \qquad (8.1)$$

So far the behaviour of a function has been considered only when the variable tends to positive or negative infinity. Very similar considerations apply when the variable approaches a finite value a. Thus $f(x)$ is said to tend to a limit l as x tends to a if, when any positive number δ (however small) is chosen, a number $\varepsilon(\delta)$ (depending on δ) can be found such that $|f(x) - l| < \delta$ whenever $0 < |x - a| \leqslant \varepsilon(\delta)$, and it is usual in such a case to write

$$\lim_{x \to a} f(x) = l. \qquad (8.2)$$

If values of x are restricted to those greater than a (that is, if $a < x \leqslant a + \varepsilon(\delta)$), $f(x)$ is said to tend to a limit l when x approaches a *from the right*. This is often written as
$$\lim_{x \to a+0} f(x) = l,$$
with the similar notation
$$\lim_{x \to a-0} f(x) = l$$
when x approaches *a from the left*, and it is worth noting that $f(a + 0)$, $f(a - 0)$ are sometimes used to denote these two limits. The symbolic statement (8.2) is strictly equivalent to the two statements
$$\lim_{x \to a+0} f(x) = l \text{ and } \lim_{x \to a-0} f(x) = l.$$
The reader should be able to provide for himself definitions to cover the cases in which $f(x) \to \pm \infty$ as x tends to a through values greater or less than a. The special case in which $a = 0$ can be included in these and the previous definitions, the notations $x \to +0$, $x \to -0$ being used to denote x approaching zero through positive and negative values respectively. It is worth pointing out also that the limit theorems of (8.1) are true (with obvious slight changes) when x tends to a finite value a and that, if there is a number ε such that $f(x_2) \geqslant f(x_1)$ whenever $a + \varepsilon > x_2 > x_1 > a - \varepsilon$, then $f(x)$ *is said to increase steadily (or to be monotonic increasing) in the neighbourhood of* $x = a$, with, of course, a similar definition for a decreasing function.

Example 1. *Find the values of a, b, c and d if*
$$\lim_{x \to 0} \left\{ \frac{a \cos x + b \sin x + c e^x + d}{x^3} \right\} = 1. \quad \text{(C.)}$$

Using the series expansions of $\cos x$, $\sin x$ and e^x, the expression whose limit is to be found can be written
$$\frac{1}{x^3} \left\{ a\left(1 - \frac{x^2}{2!} + \frac{x^4}{4!} - \ldots\right) + b\left(x - \frac{x^3}{3!} + \frac{x^5}{5!} - \ldots\right) \right.$$
$$\left. + c\left(1 + x + \frac{x^2}{2!} + \frac{x^3}{3!} + \frac{x^4}{4!} + \ldots\right) + d \right\}$$
$$= \frac{a+c+d}{x^3} + \frac{b+c}{x^2} + \frac{c-a}{2x} + \frac{c-b}{6} + \left(\frac{c+a}{24}\right)x + \text{higher powers of } x.$$

If this is to tend to unity as $x \to 0$, it follows that
$$a + c + d = 0, b + c = 0, c - a = 0, c - b = 6.$$
These equations are easily solved to give $a = 3, b = -3, c = 3, d = -6$.

Example 2. *Find* $\lim_{x \to \infty} \left\{ \dfrac{x + \sqrt{(x^4 - x^2 + 1)}}{2x^2 + 1 + \sqrt{(x^4 + 1)}} \right\}.$ (C.)

Putting $x = 1/y$, the required limit is equivalent to
$$\lim_{y \to 0} \left\{ \frac{\frac{1}{y} + \sqrt{\left(\frac{1}{y^4} - \frac{1}{y^2} + 1\right)}}{\frac{2}{y^2} + 1 + \sqrt{\left(\frac{1}{y^4} + 1\right)}} \right\} = \lim_{y \to 0} \left\{ \frac{y + \sqrt{(1 - y^2 + y^4)}}{2 + y^2 + \sqrt{(1 + y^4)}} \right\} = \frac{1}{3}.$$

Example 3. *If $\phi(x) \to a$ as $x \to a$ and $f(y) \to \beta$ as $y \to a$ and $y = \phi(x)$ show that $f(y) \to \beta$ as $x \to a$.*

Evaluate
$$\lim_{x \to 1} \left(\frac{x^3 - 1}{x - 1}\right)^3.$$

Since $f(y) \to \beta$ as $y \to a$, corresponding to an arbitrary positive number δ, there corresponds a positive number ε such that
$$|f(y) - \beta| < \delta$$
whenever $0 < |y - a| \leqslant \varepsilon$. Again, since $\phi(x) \to a$ as $x \to a$, corresponding to ε, there exists a positive number ε_1 such that
$$|\phi(x) - a| < \varepsilon$$
whenever $0 < |x - a| \leqslant \varepsilon_1$. Hence $|f(y) - \beta| < \delta$ for all x satisfying $0 < |x - a| \leqslant \varepsilon_1$ and the required result follows.

Here,
$$y = \phi(x) = \frac{x^3 - 1}{x - 1} = x^2 + x + 1$$
and $\phi(x) \to 3$ as $x \to 1$. Also $f(y) = y^3$ so that $f(y) \to 3^3$ as $y \to 3$. Hence the required limit of y^3 as $x \to 1$ is 27.

8.3. Continuous functions of a real variable

The graphical representation of a function of a real variable gives a general idea of what is meant by continuity. Thus the graph of Fig. 28 is a 'continuous' curve and that of Fig. 29 has 'discon-

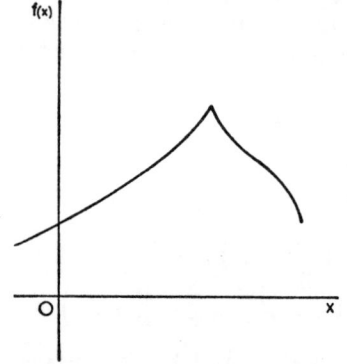

Fig. 28 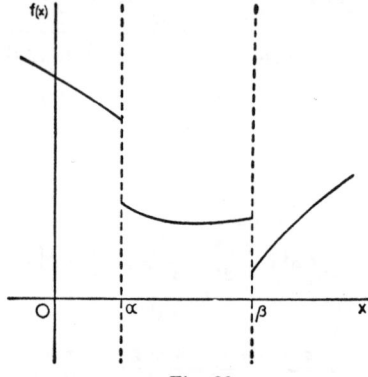   Fig. 29

tinuities' when $x = a$ and when $x = \beta$. It is natural to speak of functions possessing these types of graph as being continuous and discontinuous respectively, but more precise definitions are required in analysis.

Formally, *the function $f(x)$ is said to be continuous when $x = a$ if $f(x)$ tends to the same definite limit as x tends to a from either side and if this definite limit is equal to $f(a)$*: symbolically, $f(x)$ is continuous when $x = a$ if
$$\lim_{x \to a - 0} f(x) = f(a) = \lim_{x \to a + 0} f(x).$$

An alternative and equivalent definition is: *the function $f(x)$ is said to be continuous when $x = a$ if, when any positive number δ is chosen, a number $\varepsilon(\delta)$ can be found such that $|f(x) - f(a)| < \delta$ whenever $0 < |x - a| \leqslant \varepsilon(\delta)$*. If, when $x = a$, the above condition is not satisfied, the function is said to be *discontinuous* when $x = a$.

If a function $f(x)$ is continuous for *all* values of the variable throughout a certain interval it is said to be *continuous in that interval*. A function is *continuous everywhere* if it is continuous for *every* value of the variable. The end-values (or points) of an interval, for example, $x = a$, $x = \beta$ of the interval (a, β) of Fig. 29, are slightly different from the interior points of the interval—for if a is an interior point it is possible to find a 'neighbourhood' of a which lies entirely inside the given interval; this property is not, however, true for a or β. As far as the interval (a, β) is concerned, continuity is determined by 'one-sided' limits and $f(x)$ is continuous at $x = \beta$ if

$$\lim_{x \to \beta - 0} f(x) = f(\beta),$$

and it is continuous at $x = a$ if

$$\lim_{x \to a + 0} f(x) = f(a).$$

To allow for the special behaviour at the end-points of a finite interval, it is convenient to distinguish between intervals as follows:

the *open interval* (a, β) consists of all the points of the interval with the exception of the end-points, i.e. $a < x < \beta$;

the *closed interval* (a, β) consists of the open interval (a, β) together with the end-points, i.e. $a \leqslant x \leqslant \beta$;

the *half-open interval* (a, β) is one in which one of the end-points is included, i.e. $a < x \leqslant \beta$ or $a \leqslant x < \beta$.

It follows from the extensions of the limit theorems of (8.1) that

(a) the sum or difference of two continuous functions is continuous;
(b) the product of two continuous functions is continuous;
(c) the quotient of two continuous functions is continuous provided that the denominator does not vanish;

from the result of Example 3 of this chapter it can be deduced that

(d) a continuous function of a continuous function is itself continuous.

Immediate consequences are that the polynomial

$$P(x) \equiv a_0 x^n + a_1 x^{n-1} + a_2 x^{n-2} + \ldots + a_{n-1} x + a_n$$

is continuous for all finite values of x and that the rational function

$$R(x) = P(x)/Q(x),$$

where $P(x)$ and $Q(x)$ are polynomials, is continuous in an interval which does not contain values of x for which $Q(x)$ vanishes.

The notion of continuity with which this section opened suggests the following important property of continuous functions:

if a and b are two points in an interval of continuity of $f(x)$, then $f(x)$ will take at least once all values between $f(a)$ and $f(b)$ as x ranges from a to b.

Such a property is indeed true but a rigid proof is not attempted here. In this connection some further definitions are useful—$f(x)$ is said to be *bounded above* in the closed interval (α, β) if a constant A can be found such that $f(x) < A$ for all values of x in the interval; the function is similarly said to be *bounded below* if a constant B can be found such that $f(x) > B$ for all x in the interval. A function which is bounded both above and below is said to be *bounded*. The *upper bound* of $f(x)$ in the interval (α, β) is a number M such that $f(x) \leqslant M$ for all values of x in the interval and such that there exists a value of x in the interval for which $f(x) > M - \delta$ where δ is any arbitrary positive number, the *lower bound* m is similarly defined and the difference $M - m$ is called the *oscillation* of $f(x)$ in the interval. Two further important properties of continuous functions can be established but, again, a rigid discussion is not attempted here. The properties in question are:

(i) *if $f(x)$ is continuous in the closed interval (α, β), then $f(x)$ is bounded in the interval;*

(ii) *if $f(x)$ is continuous in the closed interval (α, β) and M, m are its upper and lower bounds, then $f(x)$ assumes the values M and m at least once in the interval.*

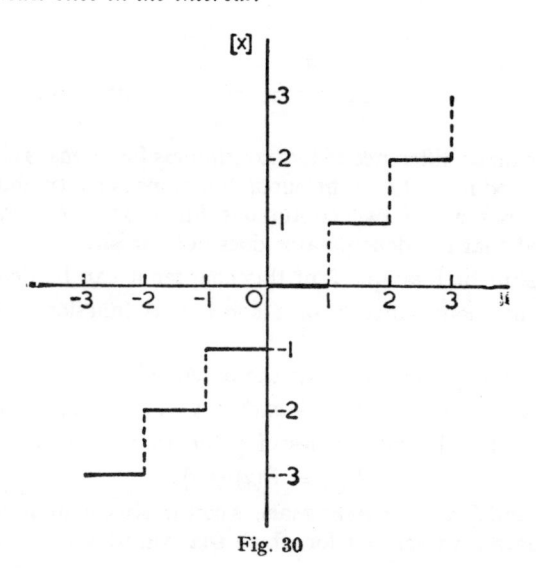

Fig. 30

Example 4. *The symbol $[x]$ denotes the greatest integer not exceeding x. Write down the values of $[2\frac{1}{2}]$, $[2]$, $[-2\frac{1}{2}]$, $[-2]$.*
Indicate the shapes of the graphs of

$$\text{(i) } [x], \qquad \text{(ii) } x - [x]. \qquad \text{(N.U.)}$$

$$[2\tfrac{1}{2}] = 2, \ [2] = 2, \ [-2\tfrac{1}{2}] = -3, \ [-2] = -2.$$

The graphs required are shown in Figs. 30, 31 and it should be noted that the left-hand end-points of the thick lines but not the right-hand ones belong to the curve.

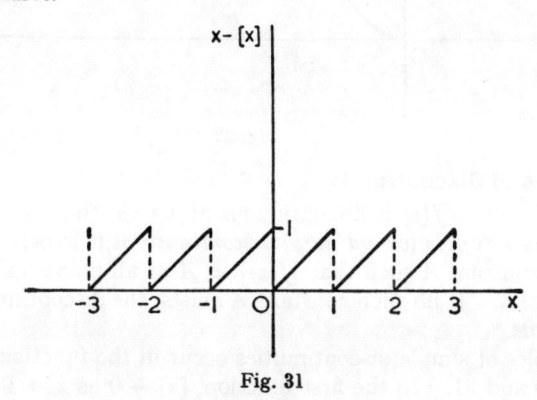

Fig. 31

Example 5. *The function $f(x)$ is defined for all x by means of the formula*

$$f(x) = \begin{cases} 0 & \text{for } x < 0, \\ a + b \cos px & \text{for } 0 < x < 1, \\ 1 & \text{for } x > 1. \end{cases}$$

Prove that values of a, b and p can be chosen so that the graph of $y = f(x)$ is continuous for all x and has, at every point, a continuous gradient which is never negative. Sketch the graph in this case. (N.U.)

Since $f(x) \to 0$ as $x \to -0$, continuity at $x = 0$ requires that
$a + b \cos px \to 0$ as $x \to +0$. Hence $a + b = 0$.
Since $f(x) \to 1$ as $x \to 1 + 0$, continuity at $x = 1$ requires that
$a + b \cos px \to 1$ as $x \to 1 - 0$. This gives $a + b \cos p = 1$. These two relations give

$$a = -b = \frac{1}{1 - \cos p},$$

so that, when $0 < x < 1$, $\quad f(x) = \dfrac{1 - \cos px}{1 - \cos p}.$

The gradient is given by $\quad f'(x) = \dfrac{p \sin px}{1 - \cos p}$

and continuity in gradient requires that $f'(x)$ vanishes as $x \to +0$ and as $x \to 1 - 0$. For this to occur, $p = (2n + 1)\pi$ where n is an integer or zero. When $n = 0$, $f'(x) = (\pi/2) \sin \pi x$ and this is never negative in the interval $(0, 1)$. A rough sketch of the graph is given in Fig. 32.

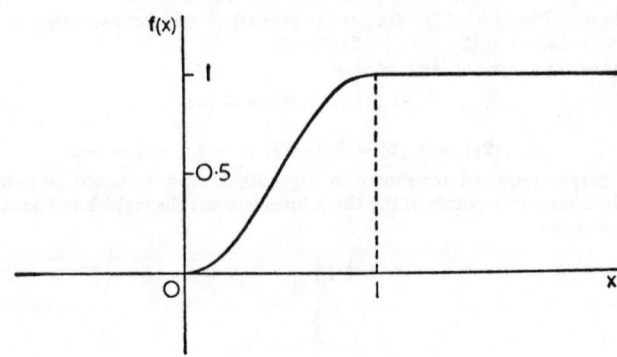

Fig. 32

8.4. Types of discontinuity

If the function $f(x)$ is discontinuous at $x = a$, the discontinuity is classified as a *simple (or ordinary) discontinuity* if it is possible to find a positive constant A such that $|f(x)| < A$ at all points in the neighbourhood of a. If no such constant A exists, the discontinuity is said to be *infinite*.

Examples of simple discontinuities occur in the functions depicted in Figs. 30 and 31. In the first function, $[x] \to 0$ as $x \to 1 - 0$ while $[x] \to 1$ as $x \to 1 + 0$; the two one-sided limits exist at $x = 1$ but they are not equal. There is inequality of the one-sided limits at all integer values of x in both the functions considered in Example 4.

If, with the notation of Example 4, $f(x) = [1 - x^2]$, then $f(x) \to 0$ as $x \to \pm 0$ but $f(0) = 1$. The one-sided limits at $x = 0$ are equal to each other but not to the value of the function itself at this value of x. The behaviour of this function at $x = 0$ provides another example of an ordinary discontinuity. Yet another example of an ordinary discontinuity occurs when one or both of the one-sided limits does not exist—for example, there is this type of behaviour in the function $\cos(1/x)$ at $x = 0$.

An infinite discontinuity in $f(x)$ occurs at $x = a$ when

(a) $f(x) \to \pm\infty$ as $x \to a$,

(b) $f(x) \to \pm\infty$ as $x \to a + 0$, $f(x) \to \mp\infty$ as $x \to a - 0$, unlike signs corresponding.

An example of each type is provided by (a) $1/x^4$ at $x = 0$, (b) $\tan x$ at $x = \pi/2$. A discontinuity which is neither simple nor infinite is said to be *oscillatory*—e.g. $\sin(1/x)$ at $x = 0$.

Example 6. *n is the positive integral variable, x the continuous real variable and y is defined by*

$$y = \lim_{n \to \infty} \left(\frac{x^n}{x^n + 1} \right).$$

Discuss the behaviour of y at $x = 1$.

Using the results of § 3.5 (Chapter 3), $y = 0$ when $|x| < 1$, $y = \frac{1}{2}$ when $x = 1$ and, when $x > 1$,

$$y = \lim_{n \to \infty} \left(\frac{1}{1 + (1/x)^n} \right) = 1.$$

Hence y possesses an ordinary discontinuity at $x = 1$.

8.5. Continuous functions of more than one variable

The foregoing notions of continuity and discontinuity may be extended to functions of more than one independent variable but their application gives rise to difficulties which are beyond the scope of the present work. All that will be attempted here is to give formal definitions for a function of two variables. Thus, the function $f(x, y)$ is said to tend to a limit l as x tends to a and y tends to b if, given any positive number δ (however small), it is possible to find $\varepsilon(\delta)$ (depending on δ) such that

$$|f(x, y) - l| < \delta$$

whenever $0 < |x - a| \leqslant \varepsilon(\delta)$ and $0 < |y - b| \leqslant \varepsilon(\delta)$. The function is said to be continuous at $x = a$, $y = b$ if $f(x, y) \to f(a, b)$ when $x \to a$, $y \to b$ in *any* manner.

The reader should observe that the continuity of $f(x, y)$ with respect to both variables x and y means that it is continuous with respect to x (or y) when any fixed value is given to y (or x). That the converse is not necessarily true can be seen from the following example.

Example 7. *Find the limits to which the function*

$$f(x, y) = \frac{x + (x + y)^2}{px + y + (x + y)^2}$$

tends when (a) *first x tends to zero and then y tends to zero,* (b) *first y tends to zero and then x tends to zero,* (c) *after putting $y = mx$, x tends to zero.* (C.)

(a) $\lim\limits_{x \to 0} f(x, y) = \dfrac{y^2}{y + y^2} = \dfrac{y}{1 + y}$.

$\lim\limits_{y \to 0} [\lim\limits_{x \to 0} f(x, y)] = \lim\limits_{y \to 0} \left(\dfrac{y}{1 + y} \right) = 0$.

(b) $\lim\limits_{y \to 0} f(x, y) = \dfrac{x + x^2}{px + x^2} = \dfrac{1 + x}{p + x}$.

$\lim\limits_{x \to 0} [\lim\limits_{y \to 0} f(x, y)] = \lim\limits_{x \to 0} \left(\dfrac{1 + x}{p + x} \right) = \dfrac{1}{p}$.

(c) $\lim\limits_{x \to 0} f(x, mx) = \lim\limits_{x \to 0} \left[\dfrac{x + (x + mx)^2}{px + mx + (x + mx)^2} \right]$

$= \lim\limits_{x \to 0} \left[\dfrac{1 + (1 + m)^2 x}{p + m + (1 + m)^2 x} \right] = \dfrac{1}{p + m}$.

EXERCISES 8 (a)

1. Evaluate

   (a) $\lim\limits_{x \to 0} \left(\dfrac{3x^3 + 6x^4 + 2x^5}{x^3} \right)$, (b) $\lim\limits_{x \to \infty} \left(\dfrac{3x^2 + 6x - 4}{5x^2 - 2x + 3} \right)$.

2. Find the limit as x tends to zero of y/x when
$$\frac{x^2}{a^2} - \frac{y^2}{b^2} = x^4.$$

3. Discuss the continuity of the function
$$\frac{x - p}{q - x}$$
in the closed interval (p, q).

4. Plot the graphs of the functions

   (a) $y = \lim\limits_{n \to \infty} \left(\dfrac{2x^n + 3}{5x^n + 7} \right)$, (b) $y = \lim\limits_{n \to \infty} \dfrac{2(x^{n+1} + 1)}{x^n + 1}$.

5. Show that a function which increases or decreases steadily in the neighbourhood of $x = a$ can have at most an ordinary discontinuity at $x = a$.

6. If $f(x, y) = 2xy/(x^2 + y^2)$, show that $f(x, y)$ is a continuous function of x and a continuous function of y. Show also that if x and y tend to zero along the straight line $y = mx$ the limit of $f(x, y)$ may have any value between ± 1 and hence that $f(x, y)$ is not a continuous function of x and y.

8.6. Differentiability

In *Advanced Level Pure Mathematics* the derivative (or derived function) of a function $f(x)$ was defined by

$$f'(x) = \lim_{\delta x \to 0} \left\{ \frac{f(x + \delta x) - f(x)}{\delta x} \right\}, \qquad (8.3)$$

and rules for finding the derivatives of certain standard functions, sums, products, quotients, etc., were set up. The implications of the definition (8.3) were not at that time considered in detail, the primary objective then being to proceed rapidly to applications of the calculus.

It should now be apparent that the definition (8.3) implies that one-sided limits of the expression in { } as $\delta x \to 0$ through positive and negative values both exist and that the two limits are equal to each other. In such circumstances the function is said to be *differentiable*. It follows that *if a function $f(x)$ is differentiable for all values of x in the closed interval (a, b), then $f(x)$ is necessarily continuous at every point of the interval.* The converse is not, however, true, as is easily demonstrated by the following example. The function whose graph consists of two straight lines making angles ϕ_1, ϕ_2 respectively with the x-axis

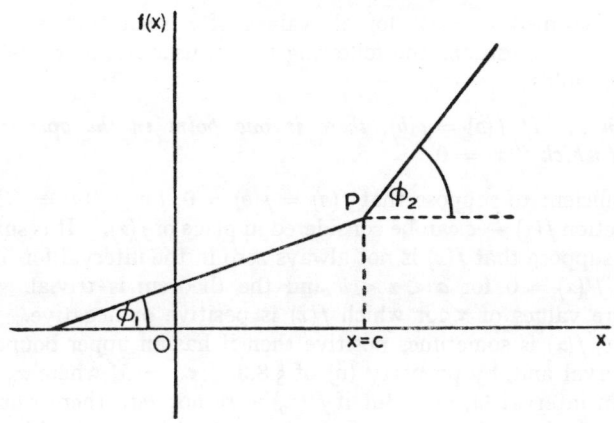

Fig. 33

(Fig. 33) is certainly continuous but its derivative does not exist at the point P where $x = c$. The expression $\{f(c + \delta x) - f(c)\}/\delta x$ here has the limit $\tan \phi_2$ as $\delta x \to 0$ by positive values and the limit $\tan \phi_1$ as $\delta x \to 0$ by negative values.

The rules for finding the derivatives of sums, products, quotients, etc., should, strictly speaking, be enunciated and developed more carefully than was done previously. Thus, for example, the rule for differentiating the sum of two functions should read—if two functions of x are differentiable for a given value of x then the derivative of the sum is the sum of the derivatives for this value of x; the strict proof of the rule would require an appeal to the theorem that the limit of a sum is equal to the sum of the limits. It is left as an exercise for the reader to state and prove, in this stricter sense, the other rules commonly used in the differential calculus.

8.7. Some important general theorems

Suppose that $f(x)$ has a positive derivative $f'(x_0)$ for some value x_0 of x. Then
$$\frac{f(x_0 + \delta x) - f(x_0)}{\delta x}$$
converges to a positive limit as $\delta x \to 0$. This can happen only if $\{f(x_0 + \delta x) - f(x_0)\}$ and δx have the same sign for sufficiently small values of x, a result which can be stated in the form:

Theorem 1. *If $f'(x_0) > 0$, $f(x)$ is strictly increasing at $x = x_0$,*

with a corresponding theorem (left as an exercise for the reader) in the case in which $f'(x_0) < 0$.

In what follows $f(x)$ is supposed continuous for $a \leqslant x \leqslant b$ and

$f'(x)$ is assumed to exist for all values of x such that $a < x < b$. Under these conditions the following result, usually known as *Rolle's theorem*, holds:

Theorem 2. *If $f(a) = f(b)$, there is one point in the open interval (a, b) at which $f'(x) = 0$.*

It is sufficient to suppose that $f(a) = f(b) = 0$, for if $f(a) = f(b) = c$, the function $f(x) - c$ can be considered in place of $f(x)$. It is sufficient also to suppose that $f(x)$ is not always zero in the interval for, in such a case, $f'(x) = 0$ for $a < x < b$ and the theorem is trivial. Hence there are values of x for which $f(x)$ is positive or negative. If, for example, $f(x)$ is sometimes positive then it has an upper bound M in the interval and, by property (ii) of § 8.3, $f(x_0) = M$ where x_0 lies in the open interval (a, b). But if $f'(x_0)$ were not zero there would, by Theorem 1 above, be values of x near x_0 at which $f(x) > M$ and this would contradict the definition of the upper bound M. Hence $f'(x_0) = 0$.

These two results can be combined into a third which is often useful in proving inequalities. The conditions on $f(x)$ are assumed to be as before and the result in question can be expressed as:

Theorem 3. *If $f'(x) > 0$ for every x in the open interval (a, b), then $f(x)$ is strictly increasing throughout the interval.*

If $a < x_1 < x_2 < b$ and $f(x_1) = f(x_2)$, then, by Theorem 2 above, there is a value of x between x_1 and x_2 for which $f'(x) = 0$ and this contradicts the assumption that $f'(x) > 0$. If $a < x_1 < x_2 < b$ and $f(x_1) > f(x_2)$, then, by Theorem 1 above, there is a value x' near to and greater than x_1 for which $f(x') > f(x_1) > f(x_2)$. It follows, since $f(x)$ is differentiable and therefore continuous, that there is a value x'' between x' and x_2 for which $f(x'') = f(x_1)$. By Theorem 2 there is therefore a value of x between x_1 and x'' for which $f'(x) = 0$ and again the assumption that $f'(x) > 0$ is violated. Hence $f(x_1) < f(x_2)$ for values x_1, x_2 of x such that $a < x_1 < x_2 < b$. The inequality just proved can be extended to cover the cases $x_1 = a$ and $x_2 = b$ as follows. From the above, $f(x) < f(x')$ if $a < x < x' < b$ so that $f(x)$ is strictly decreasing as x approaches a from the right. It follows that $f(a) = \lim_{x \to a+0} f(x) < f(x')$ and it can be shown similarly that $f(x') > f(b)$.

A useful corollary of this theorem is given in the result—*if $f'(x) > 0$ throughout the interval (a, b) and if also $f(a) \geqslant 0$, then $f(x)$ is positive throughout the interval*, with a similar result when $f'(x) < 0$.

Example 8. *Differentiate $x \sin x - \tfrac{1}{2} \sin^2 x$ with respect to x and from your result prove that, for $0 < x < \pi/2$,*

$$0 < x \sin x - \tfrac{1}{2} \sin^2 x < \tfrac{1}{2}(\pi - 1). \qquad \text{(O.C.)}$$

If $f(x) = x \sin x - \frac{1}{2} \sin^2 x$, $f'(x) = \sin x(1 - \cos x) + x \cos x$. For $0 < x < \pi/2$, $0 < \sin x < 1$, $0 < \cos x < 1$ so that $f'(x) > 0$, Theorem 3 gives
$$f(0) < f(x) < f(\pi/2),$$
i.e.
$$0 < x \sin x - \frac{1}{2} \sin^2 x < \frac{\pi}{2} - \frac{1}{2}.$$

Example 9. *If $x > 0$, prove that*
$$x - \frac{x^2}{2} < \log_e (1 + x) < x - \frac{x^2}{2} + \frac{x^3}{3}. \tag{C.}$$

Let $f(x) = \log_e (1 + x) - x + \frac{x^2}{2}$ so that
$$f'(x) = \frac{1}{1 + x} - 1 + x = \frac{x^2}{1 + x} > 0.$$
Hence $f(x) > f(0)$ giving $\log_e (1 + x) - x + \frac{x^2}{2} > 0$ and the first inequality is established.

Let $F(x) = x - \frac{x^2}{2} + \frac{x^3}{3} - \log_e (1 + x)$ so that
$$F'(x) = 1 - x + x^2 - \frac{1}{1 + x} = \frac{x^3}{1 + x} > 0.$$
Hence $F(x) > F(0)$ giving $x - \frac{x^2}{2} + \frac{x^3}{3} - \log_e (1 + x) > 0$ and the complete result follows.

8.8. Mean value theorems

Other important general results can be established from the preceding theorems. Theorems 4 and 5 below belong to this category and are usually known as *mean value theorems*.

Theorem 4.* *If $f(x)$ is continuous in the closed interval (a, b) and differentiable in the open interval, there is a value ξ of x between a and b such that $f(b) - f(a) = (b - a)f'(\xi)$.*

The function
$$f(b) - f(x) - \frac{b - x}{b - a}\{f(b) - f(a)\}$$
vanishes when $x = a$ and $x = b$ and hence, by Rolle's theorem, there is a value ξ of x in the open interval for which its derivative vanishes.

Hence
$$-f'(\xi) + \frac{1}{b - a}\{f(b) - f(a)\} = 0$$
and the required result has been established.

By writing $b = a + h$, $\xi = a + \theta h$ where $0 < \theta < 1$, an alternative and often useful form of the theorem is expressed by
$$f(a + h) = f(a) + hf'(a + \theta h). \tag{8.4}$$

* This theorem is often called the *first* mean value theorem

Theorem 5. *If $f(x)$, $F(x)$ both satisfy the conditions of Theorem 4 and if* (i) $F(b) \neq F(a)$, (ii) $f'(x)$, $F'(x)$ *never vanish for the same value of x, there is a value ξ of x between a and b such that*

$$\frac{f(b) - f(a)}{F(b) - F(a)} = \frac{f'(\xi)}{F'(\xi)}.$$

This theorem is due to Cauchy and is a generalisation of Theorem 4 to which it reduces in the special case in which $f(x) = x$. It is often known as *Cauchy's* mean value theorem. The proof is very similar to that of the previous theorem and runs as follows.

The function $\quad f(b) - f(x) - \dfrac{f(b) - f(a)}{F(b) - F(a)}\{F(b) - F(x)\}$

vanishes when $x = a$ and when $x = b$. Its derivative therefore vanishes for some value ξ in the interval so that

$$-f'(\xi) + \frac{f(b) - f(a)}{F(b) - F(a)}F'(\xi) = 0.$$

$F'(\xi)$ cannot vanish, for otherwise so would $f'(\xi)$ and this would contradict assumption (ii). Hence $F'(\xi) \neq 0$ and the required result follows by transposing $f'(\xi)$ and dividing by $F'(\xi)$.

Example 10. *If $f'(x) = 0$ for all values of x in the interval $a < x < b$, show that $f(x)$ is constant over this interval.*

If x_1 is any value of x such that $a < x_1 < b$, Theorem 4 gives

$$f(x_1) - f(a) = (x_1 - a)f'(\xi)$$

where $a < \xi < x_1$. Hence, since $f'(\xi) = 0$, $f(x_1) = f(a)$ and, as x_1 is arbitrary, $f(x)$ has therefore the constant value $f(a)$ over the interval.

EXERCISES 8 (b)

1. Prove that $\left(1 - \dfrac{\theta^2}{2}\right)\cos\theta + \theta\sin\theta$ increases as θ increases so long as θ is positive and less than π. (C.)

2. Prove that, if $x > 0$,
$$x - \frac{x^3}{3} < \tan^{-1}x < x - \frac{x^3}{3} + \frac{x^5}{5}.\qquad\text{(C.)}$$

3. Prove that $e^{-x} - 1 + x$ is never negative.
   If $f(x) = (e^x - 1)/x$ for $x \neq 0$ and $f(0) = 1$, prove that $f(x)$ is an increasing function of x for all x. Sketch the graph of $f(x)$. (N.U.)

4. If $0 < x < \pi/2$, establish the inequalities
   (i) $\tan x > x$, \quad (ii) $1 < \dfrac{x}{\sin x} < \dfrac{\pi}{2}$. \hfill (C.)

5. Find the value of ξ from the first mean value theorem (Theorem 4) when $f(x) = x^3$, $a = 1$, $b = 2$.

6. If $f(x)$ and $F(x)$ satisfy the conditions of Cauchy's mean value theorem (Theorem 5), show that

$$\frac{f(\xi) - f(a)}{F(b) - F(\xi)} = \frac{f'(\xi)}{F'(\xi)} \text{ where } a < \xi < b.$$

8.9. Repeated differentiation

In *Advanced Level Pure Mathematics* (§ 8.12) only the evaluation of the first few derivatives of given functions was attempted. Here some examples are given in which the nth derivatives can be found.

Since

$$\left(\frac{d}{dx}\right)^n (x^m) = m(m-1) \ldots (m-n+1)x^{m-n},$$

it is possible to find the nth derivative of any polynomial, and since

$$\left(\frac{d}{dx}\right)^n \frac{1}{(x-a)^m} = \frac{(-1)^n m(m+1) \ldots (m+n-1)}{(x-a)^{m+n}},$$

it is possible to write down the nth derivative of any rational function expressed as a sum of partial fractions. Another useful artifice is that used in Example 12 below.

Example 11. *Find the nth derivative of the function*

$$\{x(x+1)(x+2)(x+3)(x+4)\}^{-1}. \quad \text{(N.U.)}$$

If $f(x)$ denotes the given function, the usual rules for partial fractions give

$$f(x) \equiv \frac{1}{24x} - \frac{1}{6(x+1)} + \frac{1}{4(x+2)} - \frac{1}{6(x+3)} + \frac{1}{24(x+4)}.$$

The required nth derivative is therefore

$$(-1)^n (n)! \left[\frac{1}{24x^{n+1}} - \frac{1}{6(x+1)^{n+1}} + \frac{1}{4(x+2)^{n+1}} - \frac{1}{6(x+3)^{n+1}} + \frac{1}{24(x+4)^{n+1}} \right]$$

Example 12. *Prove that* $\dfrac{d^n}{dx^n}(e^x \sin x) = 2^{\frac{1}{2}n} e^x \sin\left(x + \dfrac{n\pi}{4}\right)$

Show that, when x is zero, the value of

$$\frac{d^n}{dx^n} \left\{ (e^x + e^{-x}) \sin x \right\}$$

is zero if n is even and $(-1)^m 2^{\frac{1}{2}(n+1)}$ if n is odd, m being the greatest integer in $n\pi/4$. (C.)

$$\frac{d}{dx}(e^x \sin x) = e^x(\sin x + \cos x) = \sqrt{2}.e^x\left(\frac{1}{\sqrt{2}} \sin x + \frac{1}{\sqrt{2}} \cos x\right)$$

$$= 2^{\frac{1}{2}} e^x \sin\left(x + \frac{\pi}{4}\right),$$

establishing the result for $n = 1$.

Assuming the result for the nth derivative,

$$\frac{d^n}{dx^n}(e^x \sin x) = \frac{d}{dx}\left\{2^{\frac{1}{2}n}e^x \sin\left(x + \frac{n\pi}{4}\right)\right\}$$

$$= 2^{\frac{1}{2}n}\left\{e^x \sin\left(x + \frac{n\pi}{4}\right) + e^x \cos\left(x + \frac{n\pi}{4}\right)\right\}$$

$$= 2^{\frac{1}{2}(n+1)} e^x \left\{\frac{1}{\sqrt{2}} \sin\left(x + \frac{n\pi}{4}\right) + \frac{1}{\sqrt{2}} \cos\left(x + \frac{n\pi}{4}\right)\right\}$$

$$= 2^{\frac{1}{2}(n+1)} e^x \sin\left\{x + \frac{(n+1)\pi}{4}\right\},$$

and the required result follows by induction.

Working as above, it will be found that

$$\frac{d^n}{dx^n}(e^{-x} \sin x) = (-1)^n 2^{\frac{1}{2}n} e^{-x} \sin\left(x - \frac{n\pi}{4}\right),$$

so that, when $x = 0$,

$$\frac{d^n}{dx^n}\left\{(e^x + e^{-x}) \sin x\right\} = 2^{\frac{1}{2}n}\left\{\sin\frac{n\pi}{4} + (-1)^n \sin\left(-\frac{n\pi}{4}\right)\right\},$$

from which the required result follows.

8.10. Leibnitz' theorem

If the first n derivatives of two functions u and v can be found, the nth derivative of the product uv can be obtained from the following theorem due to *Leibnitz*. In stating and proving this theorem it is convenient to use suffixes to denote differentiations so that, for example, u_n denotes the nth derivative of u. The theorem to be proved is

$$(uv)_n = u_n v + {}^nC_1 u_{n-1} v_1 + {}^nC_2 u_{n-2} v_2 + \ldots + {}^nC_r u_{n-r} v_r + \ldots + u v_n, \tag{8.5}$$

${}^nC_r$ denoting, as usual, the coefficient of x^r in the expansion of $(1 + x)^n$.

To prove the theorem, it is clear that

$$(uv)_1 = u_1 v + u v_1,$$

so that the result is true for $n = 1$. Assuming the result to be true for the nth derivative, one more differentiation gives

$$(uv)_{n+1} = u_{n+1} v + u_n v_1 + {}^nC_1(u_n v_1 + u_{n-1} v_2) + {}^nC_2(u_{n-1} v_2 + u_{n-2} v_3) + \ldots + {}^nC_r(u_{n-r+1} v_r + u_{n-r} v_{r+1}) + \ldots + u_1 v_n + u v_{n+1}$$

$$= u_{n+1} v + (1 + {}^nC_1) u_n v_1 + ({}^nC_1 + {}^nC_2) u_{n-1} v_2 + \ldots + ({}^nC_{r-1} + {}^nC_r) u_{n-r+1} v_r + \ldots + u v_{n+1},$$

and since, ${}^{n+1}C_r = {}^nC_r + {}^nC_{r-1}$, it follows that the result is true for the $(n + 1)$th derivative. The truth of (8.5) then follows by the principle of mathematical induction.

Example 13. *Find $d^n y/dx^n$ when $y = x^2 e^{ax}$.*

Take $u = e^{ax}$, $v = x^2$ so that $u_n = a^n e^{ax}$, $v_1 = 2x$, $v_2 = 2$, $v_3 = v_4 = \ldots = 0$. Hence Leibnitz' theorem gives

$$\frac{d^n y}{dx^n} = a^n e^{ax} \cdot x^2 + n a^{n-1} e^{ax} \cdot 2x + \frac{n(n-1)}{2} a^{n-2} e^{ax} \cdot 2$$
$$= a^{n-2} e^{ax} \{a^2 x^2 + 2nax + n(n-1)\}.$$

Example 14. *Prove that the polynomial $X_n = \dfrac{d^n}{dx^n}(x^2 - 1)^n$ satisfies the equation*

$$(1 - x^2)\frac{d^2 X_n}{dx^2} - 2x \frac{dX_n}{dx} + n(n+1) X_n = 0. \qquad \text{(C.)}$$

Let $u = (x^2 - 1)^n$ so that, by logarithmic differentiation,

$$\frac{u_1}{u} = \frac{2nx}{x^2 - 1}$$

and hence, $(1 - x^2) u_1 + 2nxu = 0$. Differentiating again

$$(1 - x^2) u_2 + 2(n-1) x u_1 + 2nu = 0.$$

Using Leibnitz' theorem to differentiate n times

$$(1 - x^2) u_{n+2} + n u_{n+1}(-2x) + \frac{n(n-1)}{2} u_n(-2) +$$
$$+ 2(n-1)\{x u_{n+1} + n u_n\} + 2n u_n = 0,$$

giving $(1 - x^2) u_{n+2} - 2x u_{n+1} + n(n+1) u_n = 0$.

Since $u_n = X_n$, the required result follows.

EXERCISES 8 (c)

1. Differentiate n times with respect to x

    (i) $\dfrac{x}{x^2 - a^2}$,   (ii) $\dfrac{x^2}{(x^2 - a^2)^2}$. (O.)

2. Differentiate n times with respect to x the function $e^{ax} \cos bx$.

3. Prove that, if $y = e^{-x^2}$,

$$\frac{d^{n+2} y}{dx^{n+2}} + 2x \frac{d^{n+1} y}{dx^{n+1}} + 2(n+1) \frac{d^n y}{dx^n} = 0.$$

Show also that, if $\left(\dfrac{d}{dx}\right)^n e^{-x^2} = e^{-x^2} f(x)$, then

$$\frac{d^2 f}{dx^2} - 2x \frac{df}{dx} + 2nf = 0. \qquad \text{(O.)}$$

4. Prove that, if $y = x^2 \cos x$, then

$$x^2 \frac{d^2 y}{dx^2} - 4x \frac{dy}{dx} + (x^2 + 6) y = 0.$$

Deduce that, when $x = 0$,

$$(n-2)(n-3) \frac{d^n y}{dx^n} + n(n-1) \frac{d^{n-2} y}{dx^{n-2}} = 0. \qquad \text{(C.)}$$

5. Show that if $y = (x - a)^n(x - b)^n$,

$$(x - a)(x - b)\frac{d^2y}{dx^2} - (n - 1)(2x - a - b)\frac{dy}{dx} - 2ny = 0,$$

and that, if $u = \dfrac{d^n y}{dx^n}$,

$$(x - a)(x - b)\frac{d^2u}{dx^2} + (2x - a - b)\frac{du}{dx} - n(n + 1)u = 0. \quad \text{(O.)}$$

6. If $y = x^n \log_e x$, show that

$$x^2 \frac{d^2y}{dx^2} - (2n - 1)x\frac{dy}{dx} + n^2 y = 0,$$

and that

$$x^2 \frac{d^{p+2}y}{dx^{p+2}} + (2p - 2n + 1)x\frac{d^{p+1}y}{dx^{p+1}} + (p - n)^2 \frac{d^p y}{dx^p} = 0. \quad \text{(C.)}$$

8.11. The general mean value theorem

The first mean value theorem (Theorem 4, § 8.8) is a special case of a more general result which is often known as *Taylor's theorem*. This asserts that *if the $(n - 1)$th derivative $f^{(n-1)}(x)$ of $f(x)$ is continuous in the closed interval (a, b) and if the nth derivative $f^{(n)}(x)$ exists in the open interval*, then

$$f(b) = f(a) + (b - a)f'(a) + \frac{(b - a)^2}{(2)!}f''(a) +$$

$$\ldots + \frac{(b - a)^{n-1}}{(n - 1)!}f^{(n-1)}(a) + \frac{(b - a)^n}{(n)!}f^{(n)}(\xi), \quad (8.6)$$

where $a < \xi < b$.

The proof lies in the application of Rolle's theorem to the function

$$\phi_n(x) - \left(\frac{b - x}{b - a}\right)^n \phi_n(a)$$

where

$$\phi_n(x) = f(b) - f(x) - (b - x)f'(x) - \ldots - \frac{(b - x)^{n-1}}{(n - 1)!}f^{(n-1)}(x).$$

This function vanishes for both $x = a$ and $x = b$ and, after a little reduction, its derivative is

$$\frac{n(b - x)^{n-1}}{(b - a)^n}\left\{\phi_n(a) - \frac{(b - a)^n}{(n)!}f^{(n)}(x)\right\}.$$

By Theorem 2 (§ 8.7) this vanishes for some value ξ of x such that $a < \xi < b$ and the result (8.6) follows on substitution for $\phi_n(a)$. By writing $b = a + h$, $\xi = a + \theta_n h$, the result (8.6) can be transposed into the equivalent, and sometimes more useful form

$$f(a+h) = f(a) + hf'(a) + \frac{h^2}{(2)!}f''(a) +$$
$$\ldots + \frac{h^{n-1}}{(n-1)!}f^{(n-1)}(a) + \frac{h^n}{(n)!}f^{(n)}(a+\theta_n h), \quad (8.7)$$
where $0 < \theta_n < 1$.

8.12. Taylor's and Maclaurin's series

If $f(x)$ is a function possessing derivatives of *all* orders in the interval $(a - \varepsilon, a + \varepsilon)$ surrounding the point $x = a$ and if $|h| < \varepsilon$, it follows from (8.7) that

$$f(a+h) - S_n = R_n, \quad (8.8)$$

where $\quad S_n = \sum_{r=0}^{n-1} \frac{h^r}{(r)!} f^{(r)}(a) \text{ and } R_n = \frac{h^n}{(n)!} f^{(n)}(a+\theta_n h), \quad (8.9)$

θ_n lying between 0 and 1. R_n is the *remainder*[*] when $f(a+h)$ is approximated to by the series defining S_n. If R_n tends to zero as n tends to infinity, it follows that

$$f(a+h) = \lim_{n \to \infty} S_n = f(a) + hf'(a) + \frac{h^2}{(2)!}f''(a) + \ldots \quad (8.10)$$

and this expression for $f(a+h)$ is known as *Taylor's series*. When $a = 0$, the formula (8.10) becomes

$$f(h) = f(0) + hf'(0) + \frac{h^2}{(2)!}f''(0) + \ldots, \quad (8.11)$$

a result due to Maclaurin. The reader should note that to establish the validity of Taylor's or Maclaurin's series in particular cases requires a detailed consideration of the behaviour of the remainder R_n as $n \to \infty$; it is not sufficient that all the derivatives exist.

As an example, take $f(x) = \sin x$. Then clearly $f(x)$ has derivatives of all orders for all values of x. Also $|f^{(n)}(x)| \leq 1$ for all values of x and n since $|f^{(n)}(x)|$ is $|\sin x|$ or $|\cos x|$ according as whether n is even or odd. Hence

$$|R_n| \leq \frac{h^n}{(n)!}$$

and this, by § 3.5 (b), tends to zero as n tends to infinity whatever the value of h. Hence, by (8.10),

$$\sin(x+h) = \sin x + h\cos x - \frac{h^2}{(2)!}\sin x - \frac{h^3}{(3)!}\cos x + \ldots$$

for all values of x and h and, in particular, putting $x = 0$,

$$\sin h = h - \frac{h^3}{(3)!} + \frac{h^5}{(5)!} - \ldots$$

---

[*] The function R_n defined by (8.9) is known as *Lagrange's form of the remainder*. There are other forms not discussed here.

for all values of h. It can be shown in exactly the same way that

$$\cos h = 1 - \frac{h^2}{(2)!} + \frac{h^4}{(4)!} - \cdots$$

and the results anticipated in §§ 5.12, 6.8 have been established.

Example 15. *Given that* $y = \sin^{-1} x$, *prove that*

$$(1-x^2)\frac{d^2y}{dx^2} - x\frac{dy}{dx} = 0.$$

Hence prove that

$$(1-x^2)y_{n+2} - (2n+1)xy_{n+1} - n^2y_n = 0,$$

where y_n *denotes* $d^n y/dx^n$. *Expand* y *as a Maclaurin series as far as the term in* x^5. (O.C.)

If $y = \sin^{-1} x$, $\quad \dfrac{dy}{dx} = \dfrac{1}{\sqrt{(1-x^2)}}, \; \dfrac{d^2y}{dx^2} = \dfrac{x}{(1-x^2)^{3/2}}.$

Hence, $\quad (1-x^2)\dfrac{d^2y}{dx^2} - \dfrac{x}{\sqrt{(1-x^2)}} = 0,$

giving, on substitution for dy/dx,

$$(1-x^2)\frac{d^2y}{dx^2} - x\frac{dy}{dx} = 0.$$

Differentiating n times by Leibnitz' theorem,

$$(1-x^2)y_{n+2} + n(-2x)y_{n+1} + \frac{n(n-1)}{2}(-2)y_n - xy_{n+1} - n(1)y_n = 0,$$

which reduces to

$$(1-x^2)y_{n+2} - (2n+1)xy_{n+1} - n^2y_n = 0.$$

When $x = 0$, $y = 0$, $y_1 = 1$, $y_2 = 0$ and, from the last result,

$$y_{n+2} = n^2 y_n.$$

Hence, putting $n = 1, 2, 3$ in turn,

$$y_3 = y_1 = 1, \; y_4 = 4y_2 = 0, \; y_5 = 9y_3 = 9.$$

If y_n now denotes the nth derivative of y with respect to x evaluated at $x = 0$, (8.11) gives

$$y = y_0 + xy_1 + \frac{x^2}{(2)!}y_2 + \frac{x^3}{(3)!}y_3 + \frac{x^4}{(4)!}y_4 + \frac{x^5}{(5)!}y_5 + \cdots$$

$$= x + \frac{x^3}{6} + \frac{3x^5}{40} + \cdots.$$

Example 16. *If* $y\sqrt{(1+x^2)} = \log_e\{x + \sqrt{(1+x^2)}\}$, *prove that*

$$(1+x^2)\frac{dy}{dx} + xy = 1.$$

Assuming that y *can be expanded in a series*

$$a_0 + a_1 x + \ldots + a_n x^n + \ldots$$

prove that $\quad a_{2m} = 0, \; a_{2m+1} = (-1)^m \dfrac{2.4 \ldots 2m}{3.5 \ldots (2m+1)}.$ (C.)

Differentiating the given relation and using the notation y_n to denote the nth derivative of y,

$$y_1\sqrt{(1+x^2)} + \frac{yx}{\sqrt{(1+x^2)}} = \frac{1 + \frac{x}{\sqrt{(1+x^2)}}}{x + \sqrt{(1+x^2)}} = \frac{1}{\sqrt{(1+x^2)}},$$

from which it follows that
$$(1+x^2)y_1 + xy = 1.$$
Differentiating this relation n times, using Leibnitz' theorem,
$$(1+x^2)y_{n+1} + n(2x)y_n + \frac{n(n-1)}{2}(2)y_{n-1} + xy_n + n(1)y_{n-1} = 0,$$
giving $\quad(1+x^2)y_{n+1} + (2n+1)xy_n + n^2 y_{n-1} = 0.$

When $x=0$, $y=0$, $y_1=1$ (from the first result proved above) and $y_{n+1} = -n^2 y_{n-1}$. It follows that $y_2 = 0$, $y_3 = -2^2 y_1 = -2^2$, $y_4 = -3^2 y_2 = 0$, $y_5 = -4^2 y_3 = 4^2 \cdot 2^2$ and so on.

Hence, when $x = 0$,
$$y_{2m} = 0, \quad y_{2m+1} = (-1)^m (2m)^2 (2m-2)^2 \ldots 4^2 \cdot 2^2.$$
Comparing the series $\Sigma a_n x^n$ with the Maclaurin series for y, $a_n = \frac{y_n}{(n)!}$,
so that $\quad a_{2m} = 0, \quad a_{2m+1} = (-1)^m \dfrac{2 \cdot 4 \ldots 2m}{3 \cdot 5 \ldots (2m+1)}.$

EXERCISES 8 (d)

1. Prove that, if $f''(x)$ exists,
$$f(a+h) = f(a) + hf'(a) + \tfrac{1}{2}h^2 f''(a + \theta h),$$
where $0 < \theta < 1$.
Hence, or otherwise, prove that
$$\lim_{h \to 0} \left\{ \frac{f(a+2h) - 2f(a+h) + f(a)}{h^2} \right\} = f''(a)$$
provided $f''(x)$ is a continuous function of x.

2. If $\log_e y = xy$, find the values of dy/dx and d^2y/dx^2 when $x=0$ and hence show that the Maclaurin expansion of y in powers of x is
$$y = 1 + x + \tfrac{3}{2}x^2 + \ldots.$$
Find also the expansion of x in powers of $(y-1)$ up to the term in $(y-1)^2$. (L.U.)

3. Prove that if $y = \log_e \cos x$, then
$$\frac{d^3 y}{dx^3} + 2\frac{d^2 y}{dx^2}\frac{dy}{dx} = 0.$$
Hence, or otherwise, obtain the Maclaurin expansion of $\log_e \cos x$ as far as the term in x^4.
Deduce the approximate relation
$$\log_e 2 = \frac{\pi^2}{16}\left(1 + \frac{\pi^2}{96}\right). \quad\text{(O.C.)}$$

228   MATHEMATICAL ANALYSIS   [8

4. If $y = (1 + x)^2 \log_e (1 + x)$, evaluate dy/dx and d^2y/dx^2 and prove that, if $n > 2$, then
$$\frac{d^n y}{dx^n} = (-1)^{n-1} \frac{2 \cdot (n-3)!}{(1+x)^{n-2}}.$$
Prove that the expansion of y as far as the term in x^4 is
$$y = x + \tfrac{3}{2}x^2 + \tfrac{1}{3}x^3 - \tfrac{1}{12}x^4. \qquad \text{(O.C.)}$$

5. Show that, if
$$e^x \sin x = a_0 + \frac{a_1}{(1)!}x + \frac{a_2}{(2)!}x^2 + \ldots + \frac{a_n}{(n)!}x^n + \ldots$$
then $a_{4n} = 0$, $a_{4n+1} = (-1)^n 4^n$ and determine a_{4n+2} and a_{4n+3}. (C.)

6. If $y = (\sin^{-1} x)^2$ and y_n denotes the value of $d^n y/dx^n$ when $x = 0$, prove that $y_{n+2} = n^2 y_n$ and find the expansion of y in ascending powers of x giving the general term. (O.)

8.13 Indeterminate forms

If two functions $f(x)$ and $F(x)$ both vanish for some value a of x, the function $f(x)/F(x)$ is said to take the *indeterminate form* $0/0$ when $x = a$. It often happens, however, that
$$\lim_{x \to a} \left\{ \frac{f(x)}{F(x)} \right\}$$
is definite and methods for its evaluation are given below.

(i) *Let $f(a) = F(a) = 0$ and let $f'(a)$ and $F'(a)$ both exist and have a definite ratio.*

Then,
$$\frac{f(x)}{F(x)} = \frac{f(x) - f(a)}{F(x) - F(a)} = \frac{\{f(x) - f(a)\}/(x-a)}{\{F(x) - F(a)\}/(x-a)},$$
so that
$$\lim_{x \to a} \left\{ \frac{f(x)}{F(x)} \right\} = \frac{f'(a)}{F'(a)}. \qquad (8.12)$$

(ii) *Let $f(a) = F(a) = 0$, let $f'(x)$, $F'(x)$ exist near (but not necessarily at) $x = a$, and let $f'(x)/F'(x)$ tend to a limit l as $x \to a$.*

By Cauchy's mean value theorem (Theorem 5),
$$\frac{f(x)}{F(x)} = \frac{f(x) - f(a)}{F(x) - F(a)} = \frac{f'(\xi)}{F'(\xi)},$$
where $a < \xi < x$. Hence, assuming the existence of the limits in question,
$$\lim_{x \to a+0} \left\{ \frac{f(x)}{F(x)} \right\} = \lim_{\xi \to a+0} \left\{ \frac{f'(\xi)}{F'(\xi)} \right\} = l.$$
Here, it has been assumed that $x > a$; if $x < a$
$$\frac{f(x)}{F(x)} = \frac{f(a) - f(x)}{F(a) - F(x)} = \frac{f'(\xi')}{F'(\xi')},$$

where $x < \xi' < a$ and, in this case

$$\lim_{x \to a-0} \left\{ \frac{f(x)}{F(x)} \right\} = \lim_{\xi' \to a-0} \left\{ \frac{f'(\xi')}{F'(\xi')} \right\} = l.$$

Hence, if $f'(x)/F'(x)$ possesses a unique finite limit as $x \to a$,

$$\lim_{x \to a} \left\{ \frac{f(x)}{F(x)} \right\} = \lim_{x \to a} \left\{ \frac{f'(x)}{F'(x)} \right\}. \tag{8.13}$$

If, however, $f'(x)/F'(x)$ possesses a one-sided limit only, then the corresponding one-sided limit of $f(x)/F(x)$ exists and these two limits are equal.

(iii) *Let $f(a) = F(a) = 0$ and let both $f(x)/F(x)$ and $f'(x)/F'(x)$ take the indeterminate form $0/0$ when $x = a$.*

As in (ii)

$$\lim_{x \to a} \left\{ \frac{f(x)}{F(x)} \right\} = \lim_{\xi \to a} \left\{ \frac{f'(\xi)}{F'(\xi)} \right\},$$

provided the limit on the right exists. If $f'(a) = F'(a) = 0$ but the ratio $f''(a)/F''(a)$ is definite then, by (i) above, the limit on the right exists and is equal to $f''(a)/F''(a)$.

Hence $$\lim_{x \to a} \left\{ \frac{f(x)}{F(x)} \right\} = \lim_{x \to a} \left\{ \frac{f''(x)}{F''(x)} \right\}.$$

In general, if the mth derivatives of $f(x)$, $F(x)$ vanish for $m < n$, repeated applications of the above result gives

$$\lim_{x \to a} \left\{ \frac{f(x)}{F(x)} \right\} = \lim_{x \to a} \left\{ \frac{f^{(n)}(x)}{F^{(n)}(x)} \right\}, \tag{8.14}$$

whenever the limit on the right exists.

Example 17. *Evaluate the limit of $(1 - \cos \theta)/\theta$ as $\theta \to 0$ and prove that the limit of*

$$\frac{\cos^2 \pi x}{e^{2x} - 2ex}$$

as x approaches the value $1/2$ is $\pi^2/2e$. (C.)

Both the functions in question take the indeterminate form $0/0$. For the first function, the derivatives of numerator and denominator being respectively $\sin \theta$ and unity,

$$\lim_{\theta \to 0} \left(\frac{1 - \cos \theta}{\theta} \right) = \lim_{\theta \to 0} \left(\frac{\sin \theta}{1} \right) = 0.$$

For the second function

$$\lim_{x \to 1/2} \left\{ \frac{\cos^2 \pi x}{e^{2x} - 2ex} \right\} = \lim_{x \to 1/2} \left\{ \frac{-2\pi \cos \pi x \sin \pi x}{2e^{2x} - 2e} \right\}$$

and the function on the right again takes the indeterminate form $0/0$ at $x = 1/2$. Differentiating numerator and denominator once more, the required limit

$$= \lim_{x \to 1/2} \left\{ \frac{2\pi^2 \sin^2 \pi x - 2\pi^2 \cos^2 \pi x}{4e^{2x}} \right\} = \frac{\pi^2}{2e}.$$

There are many variations of the preceding rules for the evaluation of indeterminate forms. Most of these can be reduced to the foregoing by a simple substitution. For example, limits when x tends to infinity may be reduced, by means of the substitution $y = 1/x$, to limits in which y tends to zero.

Again if, when $x = a$, $f(x)/F(x)$ takes the form ∞/∞ instead of $0/0$, the same rules can be applied. To show this, let x' and x be two values close enough to a to ensure that $f'(x)$ and $F'(x)$ exist in the interval (x', x) and that there are no values in this interval at which they simultaneously vanish. Then Cauchy's mean value theorem gives, if $x' < \xi < x$,

$$\frac{f(x) - f(x')}{F(x) - F(x')} = \frac{f'(\xi)}{F'(\xi)}.$$

This can be written

$$\frac{f(x)}{F(x)} \left\{ \frac{1 - f(x')/f(x)}{1 - F(x')/F(x)} \right\} = \frac{f'(\xi)}{F'(\xi)},$$

giving

$$\frac{f(x)}{F(x)} = \frac{f'(\xi)}{F'(\xi)} \left\{ \frac{1 - F(x')/F(x)}{1 - f(x')/f(x)} \right\}.$$

If $\lim_{x \to a} \{f'(x)/F'(x)\} = l$ and if x' and x are taken close enough to a, $f'(\xi)/F'(\xi)$ will be as near as we please to l. If also x' is kept fixed and x be made to approach a, the expression in $\{\ \}$ tends to unity and it follows that

$$\lim_{x \to a} \left\{ \frac{f(x)}{F(x)} \right\} = \lim_{x \to a} \left\{ \frac{f'(x)}{F'(x)} \right\}. \tag{8.15}$$

Example 18. *Find the limit as x tends to a of*

$$\frac{\log_e \sin (x - a)}{\log_e \tan (x - a)}.$$

This function takes the indeterminate form ∞/∞ at $x = a$, and the result (8.15) gives

$$\lim_{x \to a} \left\{ \frac{\log_e \sin (x - a)}{\log_e \tan (x - a)} \right\} = \lim_{x \to a} \left\{ \frac{\cot (x - a)}{\cot (x - a) \sec^2 (x - a)} \right\}$$
$$= \lim_{x \to a} \cos^2 (x - a) = 1.$$

Among other indeterminate forms are those which can be expressed symbolically* as $0 \times \infty$, $\infty - \infty$ and 0^0. These can be dealt with as follows.

---

* For brevity, it is convenient to speak, for example, of the form $0 \times \infty$. Stated fully, this indicates that two functions $f(x)$, $F(x)$ are such that $f(x) \to 0$ and $F(x) \to \infty$ when $x \to a$ and that possibly $\lim_{x \to a} \{f(x)F(x)\}$ may be finite and definite.

INDETERMINATE FORMS

(a) *Form* $0 \times \infty$. Suppose it is required to evaluate the limit when $x \to a$ (if it exists) of $f(x).F(x)$ given that $f(x) \to 0$, $F(x) \to \infty$ as $x \to a$. By writing

$$f(x).F(x) = \frac{f(x)}{1/F(x)}$$

the indeterminate form is 0/0 and the previous method applies.

(b) *Form* $\infty - \infty$. If the limit to be evaluated is that of $f(x) - F(x)$ when $f(x) \to \infty$ and $F(x) \to \infty$ as $x \to a$, by writing

$$f(x) - F(x) = \frac{\left(\dfrac{1}{F(x)} - \dfrac{1}{f(x)}\right)}{1/(f(x)F(x))}$$

the form is again reduced to 0/0.

(c) *Form* 0^0. If $f(x) \to 0$ and $F(x) \to 0$ and the function to be considered is $[f(x)]^{F(x)}$ the procedure is as follows.

Let
$$y = [f(x)]^{F(x)}$$
so that
$$\log_e y = F(x) \log_e f(x)$$
$$= \frac{F(x)}{1/\log_e f(x)}$$

and this is of the form 0/0. If $\log_e y$ tends to a limit l, since e^x is a continuous function of x,

$$\lim. y = e^l.$$

Limits involving the indeterminate forms ∞^0 and 1^∞ can be dealt with similarly.

Example 19. *Evaluate the following limits:*

(i) $\lim\limits_{x \to 0} \left(\dfrac{1}{x} - \cot x\right)$, (ii) $\lim\limits_{x \to 0} x^x$, (iii) $\lim\limits_{x \to \pi/2} (\sin x)^{\tan x}$.

(i) $\lim\limits_{x \to 0} \left(\dfrac{1}{x} - \cot x\right)$ is of the form $\infty - \infty$ and it can be written as

$$\lim_{x \to 0} \left(\frac{\tan x - x}{x \tan x}\right) \qquad \left[\text{form } \frac{0}{0}\right]$$

$$= \lim_{x \to 0} \left(\frac{\sec^2 x - 1}{x \sec^2 x + \tan x}\right) \qquad \left[\text{form } \frac{0}{0}\right]$$

$$= \lim_{x \to 0} \left(\frac{2 \sec^2 x \tan x}{\sec^2 x + 2x \sec^2 x \tan x + \sec^2 x}\right) = 0.$$

(ii) Let $y = x^x$ so that y takes the form 0^0 when $x = 0$.
Since $\log_e y = x \log_e x$,

$$\lim_{x \to 0} (\log_e y) = \lim_{x \to 0} (x \log_e x) \qquad [\text{form } 0 \times \infty]$$

$$= \lim_{x \to 0} \left(\frac{\log_e x}{1/x}\right) \qquad \left[\text{form } \frac{\infty}{\infty}\right]$$

$$= \lim_{x \to 0} \left(\frac{1/x}{-1/x^2}\right) = \lim_{x \to 0} (-x) = 0.$$

Hence $\lim\limits_{x \to 0} y = e^0 = 1$. The reader should observe that in changing the indeterminate form $0 \times \infty$ to ∞/∞ by writing $x \log_e x$ as $(\log_e x) \div (1/x)$, a definite result has been obtained by using (8.15). If, however, the form is changed to $0/0$ by writing $x \log_e x$ as $x \div (1/\log_e x)$, no progress results from repeated application of (8.14).

(iii) If $y = (\sin x)^{\tan x}$, y takes the form 1^∞ at $x = \pi/2$.

However,

$$\lim_{x \to \pi/2} (\log_e y) = \lim_{x \to \pi/2} (\tan x \log_e \sin x) \qquad [\text{form } \infty \times 0]$$

$$= \lim_{x \to \pi/2} \left(\frac{\log_e \sin x}{\cot x} \right) \qquad \left[\text{form } \frac{0}{0} \right]$$

$$= \lim_{x \to \pi/2} \left(\frac{\cot x}{-\operatorname{cosec}^2 x} \right)$$

$$= \lim_{x \to \pi/2} (-\sin x \cos x) = 0.$$

Thus $\lim\limits_{x \to \pi/2} y = e^0 = 1$.

EXERCISES 8 (e)

1. Find the limit, as x tends to zero, of

$$\frac{x \cos x - \sin x}{x^3}. \qquad \text{(C.)}$$

2. Evaluate the following limits:

   (i) $\lim\limits_{x \to a} \left(\dfrac{x^3 - a^3}{x^2 - a^2} \right)$, (ii) $\lim\limits_{x \to 0} \left(\dfrac{\tan x - x}{x - \sin x} \right)$. (O.C.)

3. Find the limits, as x tends to zero, of

$$\frac{\sin kx}{\tan x}, \quad \frac{\sin (\tan^{-1} px)}{\tan (\sin^{-1} qx)}, \quad \frac{(e^{nx} - 1)^2}{\sec x - 1}. \qquad \text{(O.C.)}$$

4. Find the limit as x tends to zero of

$$\frac{\log_e (1 + x \sin x)}{\cos x - 1}. \qquad \text{(C.)}$$

5. Show that $\lim\limits_{x \to \pi/2} \left\{ \dfrac{2x}{\pi} \sec x - \tan x \right\} = -\dfrac{2}{\pi}.$

6. Show that, as x tends to zero, both

$$(1 + \sin x)^{\cot x} \text{ and } (1 + \tan x)^{\operatorname{cosec} x}$$

tend to e.

EXERCISES 8 (f)

1. Evaluate

   (i) $\lim\limits_{x \to 0} \dfrac{1 - \sqrt{(1 - x^2)}}{1 - \cos 2x}$, (ii) $\lim\limits_{x \to 0} \dfrac{(1 - e^x) \tan x}{x \log_e (1 + x)}$. (C.)

2. (a) What value must be assessed to the function $(x^3 + 8)/(x + 2)$ when $x = -2$ so that it may be everywhere continuous?
   (b) Draw the graph of $\tan^{-1}(1/x)$ and discuss any discontinuity this function may possess.

3. A function $f(x)$ is defined as follows:
$$f(x) = (x^2 - a^2)/a, \quad 0 < x < a,$$
$$f(x) = 0, \quad x = a,$$
$$f(x) = a(x^2 - a^2)/x^2, \quad x > a.$$
Show that $f(x)$ is continuous at $x = a$ and sketch its graph. Discuss the continuity of $f'(x)$ at $x = a$.

4. Prove that, if $0 < x < \pi/2$,
   (i) $0 < x - \sin x \cos x < \pi/2$,
   (ii) $0 < x \sin x - \dfrac{1}{6}\sin^3 x < \dfrac{\pi}{2} - \dfrac{1}{6}$. (O.C.)

5. Prove that
   (i) If x is not zero then $\log_e \cosh x < \tfrac{1}{2}x^2$;
   (ii) $\theta - \sin\theta < a^3\left(\dfrac{\theta}{a} - \sin\dfrac{\theta}{a}\right)$ if $a > 1$ and $0 < \theta < \pi$. (O.)

6. If $x > 0$, prove that $\sinh^{-1} x < x < \sinh x$ and that $x^{-1}\sinh x$ increases with x.
   Prove also that $x^2 < \sinh x \sinh^{-1} x$. (O.)

7. Prove, by differentiation or otherwise, that if x is positive
$$x > \frac{5 \sin x}{4 + \cos x}.$$ (N.U.)

8. By differentiation, or otherwise, prove that, if n is positive and $x > 1$, then
$$n(x^{2n+2} - 1) > (n + 1)x(x^{2n} - 1).$$ (C.)

9. Prove that, if $0 < \theta < 1$,
$$\theta + \frac{\theta^3}{3} < \tan\theta < \theta + \frac{2\theta^3}{3}.$$ (C.)

10. Prove that, if y_r denotes $d^r y/dx^r$, then
$$\frac{d^2 x}{dy^2} = -\frac{y_2}{y_1^3}, \quad \frac{d^3 x}{dy^3} = \frac{3y_2^2 - y_1 y_3}{y_1^5}.$$
Show that, if $y_2^2 = \tfrac{1}{3} y_1 y_3$, then
$$y = a \pm \sqrt{(bx + c)} \text{ or } y = ax + b$$
where a, b, c are constants. (O.)

11. Prove that, if $x = a \cot\theta$,
$$\frac{d^n}{dx^n}\left(\frac{a^2}{x^2 + a^2}\right) = \left(-\frac{1}{a}\right)^n (n)! \sin(n + 1)\theta \sin^{n+1}\theta.$$ (C.)

12. Prove, by induction or otherwise, the following results for $n \geqslant 1$,

(i) $\dfrac{d^n}{dx^n}(x^n \log_e x) = (n)!\left(\log_e x + 1 + \dfrac{1}{2} + \ldots + \dfrac{1}{n}\right);$

(ii) $\dfrac{d^n}{dx^n}\{\log_e (x^2 + 1)\} = 2(-1)^{n-1}(n-1)!\ \sin^n \theta\ \cos n\theta,$ where $\cot \theta = x.$ (O.)

13. Prove that the result of differentiating the equation
$$(1 + x^2)\dfrac{dy}{dx} - 2x = 0$$
$(n + 1)$ times with respect to $x(n \geqslant 1)$ is
$$(1 + x^2)y_{n+2} + 2(n + 1)xy_{n+1} + n(n + 1)y_n = 0,$$
where y_n denotes $d^n y/dx^n$.
Hence verify that, if k is a positive integer and
$$z = \dfrac{d^k}{dx^k} \log_e (1 + x^2),$$
then z is a solution of the equation
$$(1 + x^2)\dfrac{d^2 z}{dx^2} + 2(k + 1)x\dfrac{dz}{dx} + k(k + 1)z = 0.$$ (O.C.)

14. Prove that the value of the $(n + 1)$th differential coefficient of $\{\log_e (1 - x)\}^2$ at $x = 0$ is
$$2(n)!\left(1 + \dfrac{1}{2} + \dfrac{1}{3} + \ldots + \dfrac{1}{n}\right).$$ (O.C.)

15. Show that $\dfrac{d^n}{dx^n}(e^{x \cot \alpha} \cos x) = \dfrac{e^{x \cot \alpha} \cos (x + n\alpha)}{\sin^n \alpha},$
and hence find the expansion of $e^{x \cot \alpha} \cos x$ in powers of x.
Deduce that
$$\cos x \cosh x = 1 - \dfrac{2^2 x^4}{(4)!} + \dfrac{2^4 x^8}{(8)!} - \dfrac{2^6 x^{12}}{(12)!} + \ldots$$ (C.)

16. If u and v are two functions of x and D^r denotes d^r/dx^r, prove that
$$D^3\left(\dfrac{u}{v}\right) = -\dfrac{1}{v^4}\begin{vmatrix} u & v & 0 & 0 \\ Du & Dv & v & 0 \\ D^2u & D^2v & 2Dv & v \\ D^3u & D^3v & 3D^2v & 3Dv \end{vmatrix}.$$ (O.)

17. If a function y of the variable x always satisfies the equation
$$(1 - x^2)\dfrac{dy}{dx} = 1 + xy$$
and if y and x are zero together, prove that
$$y = x + \dfrac{2}{3}x^3 + \dfrac{2.4}{3.5}x^5 + \ldots + \dfrac{2.4 \ldots 2n}{3.5 \ldots 2n+1}x^{2n+1} + \ldots$$ (O.)

18. If $\sqrt{(1+x^2)} \log_e \{x + \sqrt{(1+x^2)}\} = \sum\limits_{r=0}^{\infty} \dfrac{a_r x^r}{(r)!}$, prove that

   (i) $a_{2n} = 0$,   (ii) $a_1 = 1, a_3 = 2$,
   (iii) $a_{2n+1} = (-1)^{n-1} 2^{2n-1}(n)!(n-1)!(n > 1)$.

   For what values of x is the series convergent? (O.)

19. By use of Maclaurin's series, or otherwise, prove that
$$\sin x \sinh x = \frac{2x^2}{(2)!} - \frac{2^3 x^6}{(6)!} + \frac{2^5 x^{10}}{(10)!} - \ldots$$ (C.)

20. Prove that, if $y = x^m e^{-x}$ and D denotes differentiation with respect to x,
$$xD^{m+1}y + xD^m y + mD^{m-1}y = 0.$$
Prove that, if $D^n(x^n e^{-x}) = e^{-x} L_n(x)$, $L_n(x)$ is a polynomial of degree n and satisfies the recurrence formula
$$L_{n+1}(x) = (2n + 1 - x) L_n(x) - n^2 L_{n-1}(x).$$ (O.)

21. A set of functions $J_n(x)$, $n = 0, \pm 1, \pm 2, \ldots$, satisfy the following equations
$$J_{n-1}(x) + J_{n+1}(x) = \frac{2n}{x} J_n(x),$$
$$J_{n-1}(x) - J_{n+1}(x) = 2\frac{d}{dx} J_n(x).$$
Show that $\left(\dfrac{1}{x}\dfrac{d}{dx}\right)^m \{x^n J_n(x)\} = x^{n-m} J_{n-m}(x)$

and $\left(\dfrac{1}{x}\dfrac{d}{dx}\right)^m \{x^{-n} J_n(x)\} = (-1)^m x^{-n-m} J_{n+m}(x).$

Also prove that
$$x^2 \frac{d^2 J_n(x)}{dx^2} + x\frac{dJ_n(x)}{dx} + (x^2 - n^2) J_n(x) = 0.$$ (C.)

22. If $f'(x)$ is negative when $a < x < b$, zero when $x = b$ and positive when $b < x < c$, prove that $f(b) < f(\xi)$, ξ being any value of x other than b in the interval $a < x < c$. (O.)

23. Find the limit of $\dfrac{2^x - 2^{-x}}{3^x - 3^{-x}}$

   (i) as $x \to 0$,   (ii) as $x \to 1$,   (iii) as $x \to \infty$. (N.U.)

24. Show that
$$\lim_{x \to 0} \left\{ \frac{\cos(\sin x) + \sin(1 - \cos x) - 1}{x^4} \right\} = \frac{1}{6}.$$ (C.)

25. Find the limits of

   (i) $\dfrac{1 + \log_e x - x^x}{1 + \log_e x - x}$ as $x \to 1$,

   (ii) $\dfrac{\sin(\sin x) - \tan(\tan x)}{x^3}$ as $x \to 0$. (O.)

CHAPTER 9

PARTIAL DIFFERENTIATION

9.1. Partial derivatives of functions of more than one variable

In what has preceded, attention has been chiefly directed to functions of a single variable. The present chapter is devoted to a discussion of the differential coefficients of functions of several variables. For the sake of brevity, the discussion is mainly restricted to functions of two variables only, but much of the work is applicable also to functions of three or more variables.

Suppose that $f(x, y)$ is a function of two real variables x and y and suppose also that the limits

$$\lim_{\delta x \to 0} \left\{ \frac{f(x + \delta x, y) - f(x, y)}{\delta x} \right\}, \lim_{\delta y \to 0} \left\{ \frac{f(x, y + \delta y) - f(x, y)}{\delta y} \right\}$$

both exist for all the values of x and y under discussion. Then for such values of x and y, these limits are termed the *partial derivatives* (or *partial differential coefficients*) of f with respect to x and y respectively and are denoted by

$$\frac{\partial f}{\partial x}, \frac{\partial f}{\partial y} \text{ or } f'_x, f'_y.$$

The reader should note that these definitions presuppose the independence of x and y and that the partial derivatives $\partial f/\partial x$, $\partial f/\partial y$ are obtained by the usual rules of the differential calculus treating y and x respectively as constants for this purpose.

Example 1. *The variables r, θ, x, y are connected by the equations*
$$r = \sqrt{(x^2 + y^2)}, \theta = \tan^{-1}(y/x).$$
Determine the partial differential coefficients

$$\frac{\partial r}{\partial x}, \frac{\partial r}{\partial y}, \frac{\partial \theta}{\partial x}, \frac{\partial \theta}{\partial y}$$

and verify that
$$\frac{\partial r}{\partial y} \Big/ \frac{\partial r}{\partial x} = -\frac{\partial \theta}{\partial x} \Big/ \frac{\partial \theta}{\partial y}. \tag{O.C.}$$

Since $r = \sqrt{(x^2 + y^2)}$, $\partial r/\partial x$ is obtained by differentiating $\sqrt{(x^2 + y^2)}$ with respect to x treating y as a constant, so that

$$\frac{\partial r}{\partial x} = \frac{x}{\sqrt{(x^2 + y^2)}}.$$

$\partial r/\partial y$ is obtained by differentiating $\sqrt{(x^2 + y^2)}$ with respect to y treating x as a constant, giving

$$\frac{\partial r}{\partial y} = \frac{y}{\sqrt{(x^2 + y^2)}}.$$

In a similar way,
$$\frac{\partial \theta}{\partial x} = \frac{-y/x^2}{1+(y/x)^2} = -\frac{y}{x^2+y^2},$$
$$\frac{\partial \theta}{\partial y} = \frac{1/x}{1+(y/x)^2} = \frac{x}{x^2+y^2}.$$

From these results
$$\frac{\partial r}{\partial y}\bigg/\frac{\partial r}{\partial x} = \frac{y}{x} = -\frac{\partial \theta}{\partial x}\bigg/\frac{\partial \theta}{\partial y}.$$

The reader is warned that there are major differences in statements involving differentiability and continuity between functions of a single variable and those of several variables. Thus, for example, if a function of one variable is differentiable in a closed interval then the function is necessarily continuous in that interval (see § 8.6). The mere existence of $\partial f/\partial x$ and $\partial f/\partial y$ is not, however, sufficient to ensure the continuity of $f(x, y)$ but it is not proposed to pursue such matters in the present discussion which is intended to serve only as a brief introduction to partial differentiation.

9.2. Higher partial derivatives

In general the partial derivatives $\partial f/\partial x$, $\partial f/\partial y$ will themselves be functions of x and y and these functions may well possess partial derivatives with respect to both x and y. The four functions so obtained, viz.

$$\frac{\partial}{\partial x}\left(\frac{\partial f}{\partial x}\right), \frac{\partial}{\partial y}\left(\frac{\partial f}{\partial x}\right), \frac{\partial}{\partial x}\left(\frac{\partial f}{\partial y}\right) \text{ and } \frac{\partial}{\partial y}\left(\frac{\partial f}{\partial y}\right)$$

are called the second partial derivatives and are usually written

$$\frac{\partial^2 f}{\partial x^2}, \frac{\partial^2 f}{\partial y \partial x}, \frac{\partial^2 f}{\partial x \partial y} \text{ and } \frac{\partial^2 f}{\partial y^2}$$

or
$$f''_{xx}, f''_{yx}, f''_{xy} \text{ and } f''_{yy}.$$

Higher derivatives (if they exist) can be expressed in a similar notation and, in general,
$$\frac{\partial^{m+n} f}{\partial x^m \partial y^n}$$
denotes the result of differentiating a function $f(x, y)$ n times with respect to y treating x as a constant and then differentiating this result m times with respect to x treating y as a constant. The extension of the notation to functions of more than two variables should cause no difficulty to the reader.

Example 2. *Find the four second partial derivatives of sin (xy).*

If $f = \sin(xy)$
$$\frac{\partial f}{\partial x} = y \cos(xy), \frac{\partial f}{\partial y} = x \cos(xy).$$

$$\frac{\partial^2 f}{\partial x^2} = \frac{\partial}{\partial x}\{y\cos(xy)\} = -y^2\sin(xy).$$

$$\frac{\partial^2 f}{\partial y\partial x} = \frac{\partial}{\partial y}\{y\cos(xy)\} = \cos(xy) - xy\sin(xy).$$

$$\frac{\partial^2 f}{\partial x\partial y} = \frac{\partial}{\partial x}\{x\cos(xy)\} = \cos(xy) - xy\sin(xy).$$

$$\frac{\partial^2 f}{\partial y^2} = \frac{\partial}{\partial y}\{x\cos(xy)\} = -x^2\sin(xy).$$

In the above example it should be noted that

$$\frac{\partial^2 f}{\partial y\partial x} = \frac{\partial^2 f}{\partial x\partial y}$$

and that the equality of these two partial derivatives holds under quite general conditions on the function $f(x, y)$ involved. It can be shown in fact (but this is not attempted here) that the above two partial derivatives are equal for given values of x and y if one of them exists in the neighbourhood of the given values and if also it is continuous there.

EXERCISES 9 (a)

1. Find the value of $\dfrac{\partial}{\partial x}(xe^y + ye^x)$ when $x = 0, y = 1$. (O.C.)

2. Prove that, if u is a function of the variables x, y and z, which is given by
$$u = a(x - y) + 2b(x + y) + abz + c,$$
then u satisfies the equation
$$\left(\frac{\partial u}{\partial x}\right)^2 - \left(\frac{\partial u}{\partial y}\right)^2 = 8\frac{\partial u}{\partial z},$$
whatever the values of the constants a, b, c. (O.C.)

3. Prove that if $V = (1 - 2xy + y^2)^{-1/2}$
then
$$x\frac{\partial V}{\partial x} - y\frac{\partial V}{\partial y} = y^2 V^3,$$
and
$$\frac{\partial}{\partial x}\left\{(1 - x^2)\frac{\partial V}{\partial x}\right\} + \frac{\partial}{\partial y}\left\{y^2\frac{\partial V}{\partial y}\right\} = 0.$$ (O.C.)

4. Prove that, if $V = a_0 x^3 + a_1 x^2 y + a_2 xy^2$, where a_0, a_1 and a_2 are constants, then
$$x\frac{\partial V}{\partial x} + y\frac{\partial V}{\partial y} = 3V.$$
Prove also that
$$\frac{\partial}{\partial x}\left(x^2\frac{\partial V}{\partial x}\right) + \frac{\partial}{\partial y}\left(y^2\frac{\partial V}{\partial y}\right) - 8V$$
is independent of y. (O.C.)

DIFFERENTIAL COEFFICIENT

5. Prove that $V = f(\alpha x + \beta y + \gamma z)$ satisfies the equation
$$\frac{\partial^2 V}{\partial x^2} + \frac{\partial^2 V}{\partial y^2} + \frac{\partial^2 V}{\partial z^2} = 0$$
whatever the function f, if the constants α, β, γ satisfy a certain relation. (O.)

6. Variables u and v are defined in terms of x and y by the equations
$$u = x \sin x \cosh y - y \cos x \sinh y,$$
$$v = y \sin x \cosh y + x \cos x \sinh y.$$
Write down the values of
$$\frac{\partial u}{\partial x}, \frac{\partial v}{\partial y}, \frac{\partial u}{\partial y}, \frac{\partial v}{\partial x}.$$

Prove that (i) $2\dfrac{\partial}{\partial x}(uv) = \dfrac{\partial}{\partial y}(v^2 - u^2)$,

(ii) $\dfrac{\partial^2 u}{\partial x^2} + \dfrac{\partial^2 u}{\partial y^2} = \dfrac{\partial^2 v}{\partial x^2} + \dfrac{\partial^2 v}{\partial y^2}.$ (O.C.)

9.3. The differential coefficient of a function of two functions

Suppose that $x = f(t)$, $y = g(t)$ where f and g are functions of the single variable t with continuous differential coefficients $f'(t)$ and $g'(t)$. Suppose also that $\phi(x, y)$ is a function of the two variables x and y and that $\partial \phi/\partial x$, $\partial \phi/\partial y$ are continuous functions of both the variables for all values of x and y under discussion. Then the function of two functions $\phi\{f(t), g(t)\}$ is a function of the single variable t and, by definition,

$$\frac{d\phi}{dt} = \lim_{\delta t \to 0} \left[\frac{\phi\{f(t + \delta t), g(t + \delta t)\} - \phi\{f(t), g(t)\}}{\delta t} \right]$$

$$= \lim_{\delta t \to 0} \left[\frac{\phi(x + \delta x, y + \delta y) - \phi(x, y)}{\delta t} \right],$$

where it has been supposed that when t changes to $t + \delta t$, x and y change respectively to $x + \delta x$ and $y + \delta y$. This can be written

$$\frac{d\phi}{dt} = \lim_{\delta t \to 0} \left[\frac{\phi(x + \delta x, y + \delta y) - \phi(x, y + \delta y)}{\delta x} \cdot \frac{\delta x}{\delta t} + \frac{\phi(x, y + \delta y) - \phi(x, y)}{\delta y} \cdot \frac{\delta y}{\delta t} \right]$$

and, by the mean value theorem (§ 8.8), if θ and θ' each lie between 0 and 1,

$$\frac{\phi(x + \delta x, y + \delta y) - \phi(x, y + \delta y)}{\delta x} = \phi'_x(x + \theta \delta x, y + \delta y),$$

$$\frac{\phi(x, y + \delta y) - \phi(x, y)}{\delta y} = \phi'_y(x, y + \theta' \delta y).$$

But as δt tends to zero, both δx and δy tend to zero and $\delta x/\delta t$, $\delta y/\delta t$ tend respectively to dx/dt and dy/dt, so that

$$\frac{d\phi}{dt} = \phi'_x(x, y)\frac{dx}{dt} + \phi'_y(x, y)\frac{dy}{dt},$$

or, expressed in the alternative notation for partial derivatives,

$$\frac{d\phi}{dt} = \frac{\partial \phi}{\partial x}\frac{dx}{dt} + \frac{\partial \phi}{\partial y}\frac{dy}{dt}. \tag{9.1}$$

The reader should compare this result with the formula

$$\frac{d\phi}{dt} = \frac{d\phi}{dx}\frac{dx}{dt}$$

for the differential coefficient of a function of a single function.

9.4. The mean value theorem for a function of two variables

Writing $a = x$, $h = \delta x$ in the mean value theorem given by equation (8.4) of Chapter 8,

$$f(x + \delta x) = f(x) + \delta x f'(x + \theta \delta x),$$

where $0 < \theta < 1$. If $y = f(x)$, this can be written in the form

$$\delta y = f(x + \delta x) - f(x) = f'(x + \theta \delta x)\delta x, \tag{9.2}$$

and it is proposed in this section to obtain the corresponding formula for a function $z = f(x, y)$ of two variables.

Suppose that z increases by δz when x, y increase respectively by δx and δy so that

$$\delta z = f(x + \delta x, y + \delta y) - f(x, y).$$

If $f(x + t\delta x, y + t\delta y)$ is regarded as some function $\phi(t)$ of t, then using the mean value theorem for a function of a single variable,

$$\delta z = f(x + \delta x, y + \delta y) - f(x, y) = \phi(1) - \phi(0) = \phi'(\theta), \tag{9.3}$$

where $0 < \theta < 1$. But, by (9.1),

$$\phi'(t) = \frac{d}{dt}[f(x + t\delta x, y + t\delta y)]$$
$$= f'_x(x + t\delta x, y + t\delta y)\delta x + f'_y(x + t\delta x, y + t\delta y)\delta y;$$

writing $t = \theta$ and substituting in (9.3),

$$\delta z = f'_x(x + \theta \delta x, y + \theta \delta y)\delta x + f'_y(x + \theta \delta x, y + \theta \delta y)\delta y, \tag{9.4}$$

and this is the formula required.

9.5. Differentials

As f'_x and f'_y are supposed to be continuous functions of x and y, equation (9.4) can be written

$$\delta z = \{f'_x(x, y) + \varepsilon\}\delta x + \{f'_y(x, y) + \eta\}\delta y,$$

9]   DIFFERENTIALS   241

where ε and η tend to zero as δx and δy tend to zero. The result (9.4) can be expressed in the form
$$\delta z \simeq f'_x \delta x + f'_y \delta y, \qquad (9.5)$$
an approximation which has its counterpart
$$\delta y \simeq f'(x) \delta x$$
for a function $y = f(x)$ of a single variable.

So far, no meaning has been assigned to the symbol dy standing by itself. It is now defined, in the case of a function of a single variable $y = f(x)$, by $dy = f'(x)\delta x$. Writing $y = x$, so that $f(x) = x$ and $f'(x) = 1$, it follows that $dx = \delta x$. dy and dx defined in this way are known as *differentials* and the definition gives $dy = f'(x)dx$. Division by dx leads to
$$\frac{dy}{dx} = f'(x) \qquad (9.6)$$
and here dy/dx denotes the quotient of the differentials dy, dx and not, as previously, the differential coefficient of y with respect to x. The symbol dy/dx takes on, therefore, a second meaning but the result (9.6) remains valid in both cases.

In the case of functions of more than one variable, differentials can be defined similarly. Thus, when $z = f(x, y)$, the differential dz is defined by
$$dz = f'_x \delta x + f'_y \delta y.$$
By putting $z = x$, $f(x, y) = x$, $f'_x = 1$, $f'_y = 0$ and this gives $dx = \delta x$. Similarly by writing $z = y$, $dy = \delta y$ and the definition can be written
$$dz = f'_x dx + f'_y dy; \qquad (9.7)$$
this is an exact equation corresponding to the approximation of (9.5).

The reader should note that if $z = f(x, y)$ and if x and y are not independent but functions of a single variable t, the result (9.1) gives
$$\frac{dz}{dt} = f'_x \frac{dx}{dt} + f'_y \frac{dy}{dt}.$$
Multiplying this result by the differential dt, and, since in the notation of differentials,
$$dz = \frac{dz}{dt}dt, \; dx = \frac{dx}{dt}dt, \; dy = \frac{dy}{dt}dt$$
it follows that
$$dz = f'_x dx + f'_y dy$$
so that (9.7) remains true whether x and y are independent or not.

Example 3. *The volume V of a right circular cone is given in terms of its semi-vertical angle α and the radius r of its base by the formula*
$$V = \tfrac{1}{3}\pi r^3 \cot \alpha.$$
The radius r and the angle α are found by measurement to be 6 m and 45°, but these are liable to errors of $\pm$ 0.05 m and $\pm \tfrac{1}{2}°$ respectively. Find (to

one place of decimals) the greatest percentage error which can occur in the calculated value of the volume. (O.C.)

The given formula for V can be written in the form
$$\log V = \log (\pi/3) + 3 \log r + \log \cot \alpha,$$
and (9.7) gives
$$d(\log V) = \frac{\partial}{\partial r}(\log V)\, dr + \frac{\partial}{\partial \alpha}(\log V)\, d\alpha.$$

Since
$$\frac{\partial}{\partial \alpha}(\log \cot \alpha) = -\frac{\operatorname{cosec}^2 \alpha}{\cot \alpha} = -\frac{1}{\sin \alpha \cos \alpha},$$

this gives
$$\frac{dV}{V} = 3\frac{dr}{r} - \frac{d\alpha}{\sin \alpha \cos \alpha}.$$

Now, $r = 6$, $\alpha = 45°$, $dr = \pm 0.05$, $d\alpha = \pm \tfrac{1}{2}° = \pm 0.00873$ radians and the greatest value of dV occurs when dr and $d\alpha$ have unlike signs. Hence
$$\frac{dV}{V} = \pm \left(3 \times \frac{0.05}{6} + \frac{0.00873}{\sin 45° \cos 45°}\right)$$
$$= \pm (0.025 + 0.01746) = \pm 0.04246.$$

The percentage error in V is $100\, dV/V$ and this (to one place of decimals) is 4·2.

9.6. Differentiation of implicit functions

Suppose that two variables y and x are connected, not by an explicit relation $y = f(x)$ but by an implicit one $F(x, y) = 0$. Since F is always zero, $dF = 0$ and, assuming that $F(x, y)$ is differentiable
$$\frac{\partial F}{\partial x}dx + \frac{\partial F}{\partial y}dy = 0.$$

If y is considered as a function of the independent variable x, chosen so as to satisfy $F(x, y) = 0$, this relation gives, provided that $\partial F/\partial y \neq 0$,
$$\frac{dy}{dx} = -\frac{\partial F/\partial x}{\partial F/\partial y}. \tag{9.8}$$

For the present work it is assumed that $F(x, y)$ is such that
$$\frac{\partial^2 F}{\partial y \partial x} = \frac{\partial^2 F}{\partial x \partial y}$$

(see § 9.2), and for the subsequent part of this chapter, the notation
$$\left.\begin{array}{c} p = \dfrac{\partial F}{\partial x},\ q = \dfrac{\partial F}{\partial y}, \\[4pt] r = \dfrac{\partial^2 F}{\partial x^2},\ s = \dfrac{\partial^2 F}{\partial x \partial y},\ t = \dfrac{\partial^2 F}{\partial y^2} \end{array}\right\} \tag{9.9}$$

will be found convenient. In this notation, (9.8) can be written

$$\frac{dy}{dx} = -\frac{p}{q}, \tag{9.10}$$

and the second derivative is therefore given by

$$\frac{d^2y}{dx^2} = -\frac{q\dfrac{dp}{dx} - p\dfrac{dq}{dx}}{q^2}.$$

But $\quad\dfrac{dp}{dx} = \dfrac{\partial p}{\partial x} + \dfrac{\partial p}{\partial y}\dfrac{dy}{dx} = r + s\left(-\dfrac{p}{q}\right) = \dfrac{qr - ps}{q},$

and $\quad\dfrac{dq}{dx} = \dfrac{\partial q}{\partial x} + \dfrac{\partial q}{\partial y}\dfrac{dy}{dx} = s + t\left(-\dfrac{p}{q}\right) = \dfrac{qs - pt}{q},$

so that, after a little reduction,

$$\frac{d^2y}{dx^2} = -\frac{q^2r - 2pqs + p^2t}{q^3}. \tag{9.11}$$

The higher derivatives can be found in a similar way but the results are necessarily more complicated.

A similar argument can be applied to functions of more than two variables but it is necessary to be clear as to which of the several variables are to be regarded as independent. Thus, if $F(x, y, z) = 0$,

$$\frac{\partial F}{\partial x}dx + \frac{\partial F}{\partial y}dy + \frac{\partial F}{\partial z}dz = 0,$$

and, if $\partial F/\partial z \neq 0$, the partial derivatives of z with respect to the independent variables x and y are given by

$$\frac{\partial z}{\partial x} = -\frac{\partial F/\partial x}{\partial F/\partial z}, \quad \frac{\partial z}{\partial y} = -\frac{\partial F/\partial y}{\partial F/\partial z}.$$

Example 4. *If* $(x/a)^2 + (y/b)^2 = 1$, *show that*

$$\frac{dy}{dx} = -\frac{b^2x}{a^2y} \text{ and } \frac{d^2y}{dx^2} = -\frac{b^4}{a^2y^3}.$$

Here $F(x, y) \equiv (x/a)^2 + (y/b)^2 - 1$ and, in the notation of (9.9),

$$p = 2x/a^2,\ q = 2y/b^2,$$
$$r = 2/a^2,\ s = 0,\ t = 2/b^2.$$

Hence, from (9.1) and (9.11),

$$\frac{dy}{dx} = -\frac{2x/a^2}{2y/b^2} = -\frac{b^2x}{a^2y},$$

$$\frac{d^2y}{dx^2} = -\frac{(2y/b^2)^2(2/a^2) + (2x/a^2)^2(2/b^2)}{(2y/b^2)^3}$$

$$= -\frac{b^4}{a^2y^3}\left(\frac{x^2}{a^2} + \frac{y^2}{b^2}\right) = -\frac{b^4}{a^2y^3},$$

as $(x/a)^2 + (y/b)^2 = 1$.

EXERCISES 9 (b)

1. Measurements of the sides b, c and of the angle A of a plane triangle ABC are known to be correct to one per cent. The angle A is measured as $45°$. Find the greatest percentage error in the area (Δ) of the triangle if it is calculated from the formula
$$\Delta = \tfrac{1}{2}bc \sin A.$$

2. The area of a plane triangle ABC is computed from the side a and the angles B and C. Show that the error $(d\Delta)$ in the computed area (Δ) is given by
$$\frac{d\Delta}{\Delta} = 2\frac{da}{a} + \frac{\sin^2 B \cdot dC + \sin^2 C \cdot dB}{\sin A \sin B \sin C},$$
where da, dB, dC are the errors in a, B, C.

3. V is a function of r and h defined by the formula $V = \pi r^2 h$. Prove that

   (i) $r\dfrac{\partial V}{\partial r} + 2h\dfrac{\partial V}{\partial h} = 4V$;   (ii) $\dfrac{dV}{V} = 2\dfrac{dr}{r} + \dfrac{dh}{h}$.

   Deduce from (ii) that, if r and h are connected by the relation
$$2\pi rh + 2\pi r^2 = A,$$
where A is a constant, then V has a maximum value when
$$r = \sqrt{\left(\frac{A}{6\pi}\right)}. \tag{O.C.}$$

4. (a) If x and y are connected by the implicit relation
$$e^x + e^y = 2xy,$$
find the value of dy/dx.

   (b) Find $\partial z/\partial x$ when
$$\left(\frac{x}{a}\right)^n + \left(\frac{y}{b}\right)^n + \left(\frac{z}{c}\right)^n = 1.$$

5. If $x^3 + y^3 - 3xy = 0$, find dy/dx and d^2y/dx^2.

6. If $z = f(x, y)$ and, by means of this relation, y is expressed as a function of x and z in the form $y = F(x, z)$, prove that
$$\frac{\partial F}{\partial x} + \frac{\partial F}{\partial z} \cdot \frac{\partial f}{\partial x} = 0 \text{ and } \frac{\partial f}{\partial x} + \frac{\partial f}{\partial y} \cdot \frac{\partial F}{\partial x} = 0.$$

9.7. Change of variables

It is important in mathematical physics to be able to express the partial derivatives of a function of two or more independent variables in terms of the derivatives of the function with respect to other variables to which the first are related. Thus, if V is a function of

… CHANGE OF VARIABLES …

x and y (Cartesian coordinates) it is often useful to be able to express $\partial V/\partial x$, $\partial V/\partial y$ and higher derivatives in terms of $\partial V/\partial r$, $\partial V/\partial \theta$, etc., where r and θ (polar coordinates) are related to x, y by the relations $x = r \cos \theta, y = r \sin \theta$. The method of effecting such transformations is through relations derived from (9.7) and should be clear from the following examples.

Example 5. *If $x = r \cos \theta$, $y = r \sin \theta$, and if f is a function of x and y, prove that*

$$\frac{\partial f}{\partial x} = \cos \theta \, \frac{\partial f}{\partial r} - \frac{\sin \theta}{r} \frac{\partial f}{\partial \theta},$$

$$\frac{\partial f}{\partial y} = \sin \theta \, \frac{\partial f}{\partial r} + \frac{\cos \theta}{r} \frac{\partial f}{\partial \theta}.$$

Show further that

$$\frac{\partial^2 f}{\partial x^2} + \frac{\partial^2 f}{\partial y^2} = \frac{\partial^2 f}{\partial r^2} + \frac{1}{r} \frac{\partial f}{\partial r} + \frac{1}{r^2} \frac{\partial^2 f}{\partial \theta^2}.$$ (O.C.)

The relation

$$df = \frac{\partial f}{\partial x} dx + \frac{\partial f}{\partial y} dy$$

yields

$$\left. \begin{array}{l} \dfrac{\partial f}{\partial r} = \dfrac{\partial f}{\partial x} \dfrac{\partial x}{\partial r} + \dfrac{\partial f}{\partial y} \dfrac{\partial y}{\partial r}, \\[6pt] \dfrac{\partial f}{\partial \theta} = \dfrac{\partial f}{\partial x} \dfrac{\partial x}{\partial \theta} + \dfrac{\partial f}{\partial y} \dfrac{\partial y}{\partial \theta}. \end{array} \right\}$$

But $x = r \cos \theta$, $y = r \sin \theta$ so that

$$\frac{\partial x}{\partial r} = \cos \theta, \; \frac{\partial y}{\partial r} = \sin \theta, \; \frac{\partial x}{\partial \theta} = -r \sin \theta, \; \frac{\partial y}{\partial \theta} = r \cos \theta,$$

and substitution of these gives

$$\left. \begin{array}{l} \dfrac{\partial f}{\partial r} = \cos \theta \, \dfrac{\partial f}{\partial x} + \sin \theta \, \dfrac{\partial f}{\partial y}, \\[6pt] \dfrac{\partial f}{\partial \theta} = -r \sin \theta \, \dfrac{\partial f}{\partial x} + r \cos \theta \, \dfrac{\partial f}{\partial y}. \end{array} \right\}$$

Multiplying these relations respectively by $\cos \theta$, $\dfrac{1}{r} \sin \theta$ and subtracting leads to

$$\frac{\partial f}{\partial x} = \cos \theta \, \frac{\partial f}{\partial r} - \frac{\sin \theta}{r} \frac{\partial f}{\partial \theta};$$

in a similar way (multiplying by $\sin \theta$, $\dfrac{1}{r} \cos \theta$ and adding)

$$\frac{\partial f}{\partial y} = \sin \theta \, \frac{\partial f}{\partial r} + \frac{\cos \theta}{r} \frac{\partial f}{\partial \theta}.$$

The first of these latter expressions shows that the operations

$$\frac{\partial}{\partial x} \text{ and } \cos \theta \, \frac{\partial}{\partial r} - \frac{\sin \theta}{r} \frac{\partial}{\partial \theta}$$

are equivalent. Hence

$$\frac{\partial^2 f}{\partial x^2} = \frac{\partial}{\partial x}\left(\frac{\partial f}{\partial x}\right) = \left(\cos\theta\frac{\partial}{\partial r} - \frac{\sin\theta}{r}\frac{\partial}{\partial \theta}\right)\left(\cos\theta\frac{\partial f}{\partial r} - \frac{\sin\theta}{r}\frac{\partial f}{\partial \theta}\right)$$

$$= \cos^2\theta\frac{\partial^2 f}{\partial r^2} + \frac{\cos\theta\sin\theta}{r^2}\frac{\partial f}{\partial \theta} - \frac{\cos\theta\sin\theta}{r}\frac{\partial^2 f}{\partial r\partial\theta} + \frac{\sin^2\theta}{r}\frac{\partial f}{\partial r} -$$
$$- \frac{\sin\theta\cos\theta}{r}\frac{\partial^2 f}{\partial r\partial\theta} + \frac{\sin\theta\cos\theta}{r^2}\frac{\partial f}{\partial \theta} + \frac{\sin^2\theta}{r^2}\frac{\partial^2 f}{\partial \theta^2}$$

$$= \cos^2\theta\frac{\partial^2 f}{\partial r^2} - \frac{2\sin\theta\cos\theta}{r}\frac{\partial^2 f}{\partial r\partial\theta} + \frac{\sin^2\theta}{r}\frac{\partial f}{\partial r} + \frac{\sin^2\theta}{r^2}\frac{\partial^2 f}{\partial \theta^2} +$$
$$+ \frac{2\sin\theta\cos\theta}{r^2}\frac{\partial f}{\partial \theta}.$$

In a similar way,

$$\frac{\partial^2 f}{\partial y^2} = \frac{\partial}{\partial y}\left(\frac{\partial f}{\partial y}\right) = \left(\sin\theta\frac{\partial}{\partial r} + \frac{\cos\theta}{r}\frac{\partial}{\partial \theta}\right)\left(\sin\theta\frac{\partial f}{\partial r} + \frac{\cos\theta}{r}\frac{\partial f}{\partial \theta}\right)$$

$$= \sin^2\theta\frac{\partial^2 f}{\partial r^2} + \frac{2\sin\theta\cos\theta}{r}\frac{\partial^2 f}{\partial r\partial\theta} + \frac{\cos^2\theta}{r}\frac{\partial f}{\partial r} + \frac{\cos^2\theta}{r^2}\frac{\partial^2 f}{\partial \theta^2} -$$
$$- \frac{2\sin\theta\cos\theta}{r^2}\frac{\partial f}{\partial \theta}.$$

By addition it follows that

$$\frac{\partial^2 f}{\partial x^2} + \frac{\partial^2 f}{\partial y^2} = \frac{\partial^2 f}{\partial r^2} + \frac{1}{r}\frac{\partial f}{\partial r} + \frac{1}{r^2}\frac{\partial^2 f}{\partial \theta^2}.$$

Example 6. *f is a function of x, y, z and $x = vw/u$, $y = wu/v$, $z = uv/w$. Prove that*

$$u\frac{\partial f}{\partial u} + v\frac{\partial f}{\partial v} + w\frac{\partial f}{\partial w} = x\frac{\partial f}{\partial x} + y\frac{\partial f}{\partial y} + z\frac{\partial f}{\partial z}. \tag{O.}$$

Since
$$\frac{\partial f}{\partial u} = \frac{\partial f}{\partial x}\frac{\partial x}{\partial u} + \frac{\partial f}{\partial y}\frac{\partial y}{\partial u} + \frac{\partial f}{\partial z}\frac{\partial z}{\partial u}$$

and since
$$\frac{\partial x}{\partial u} = -\frac{vw}{u^2}, \quad \frac{\partial y}{\partial u} = \frac{w}{v}, \quad \frac{\partial z}{\partial u} = \frac{v}{w},$$

it follows that
$$\frac{\partial f}{\partial u} = -\frac{vw}{u^2}\frac{\partial f}{\partial x} + \frac{w}{v}\frac{\partial f}{\partial y} + \frac{v}{w}\frac{\partial f}{\partial z}.$$

Similarly,
$$\frac{\partial f}{\partial v} = \frac{w}{u}\frac{\partial f}{\partial x} - \frac{wu}{v^2}\frac{\partial f}{\partial y} + \frac{u}{w}\frac{\partial f}{\partial z},$$

$$\frac{\partial f}{\partial w} = \frac{v}{u}\frac{\partial f}{\partial x} + \frac{u}{v}\frac{\partial f}{\partial y} - \frac{uv}{w^2}\frac{\partial f}{\partial z}.$$

The required result follows by multiplying these three relations by u, v and w respectively and adding.

Example 7. *If $u = \log(x^2 + y^2)$ and $v = y/x$, show that*

$$x\frac{\partial \phi}{\partial x} + y\frac{\partial \phi}{\partial y} = 2\frac{\partial \phi}{\partial u} \text{ and } -y\frac{\partial \phi}{\partial x} + x\frac{\partial \phi}{\partial y} = (1+v^2)\frac{\partial \phi}{\partial v},$$

where ϕ is any function of x and y.

Hence show that if $y\frac{\partial \phi}{\partial x} = x\frac{\partial \phi}{\partial y}$ and $\phi(x, y) = x + \frac{1}{x}$ when $y = 0$,

then
$$\phi(x, y) = \frac{x^2 + y^2 + 1}{(x^2 + y^2)^{1/2}}. \tag{O.}$$

Since
$$\frac{\partial \phi}{\partial x} = \frac{\partial \phi}{\partial u}\frac{\partial u}{\partial x} + \frac{\partial \phi}{\partial v}\frac{\partial v}{\partial x}, \quad \frac{\partial \phi}{\partial y} = \frac{\partial \phi}{\partial u}\frac{\partial u}{\partial y} + \frac{\partial \phi}{\partial v}\frac{\partial v}{\partial y},$$

and since
$$\frac{\partial u}{\partial x} = \frac{2x}{x^2+y^2}, \quad \frac{\partial u}{\partial y} = \frac{2y}{x^2+y^2}, \quad \frac{\partial v}{\partial x} = -\frac{y}{x^2}, \quad \frac{\partial v}{\partial y} = \frac{1}{x},$$

it follows that
$$\frac{\partial \phi}{\partial x} = \frac{2x}{x^2+y^2}\frac{\partial \phi}{\partial u} - \frac{y}{x^2}\frac{\partial \phi}{\partial v},$$
$$\frac{\partial \phi}{\partial y} = \frac{2y}{x^2+y^2}\frac{\partial \phi}{\partial u} + \frac{1}{x}\frac{\partial \phi}{\partial v}.$$

The first of the required results follows by multiplying these relations by x and y respectively and adding.

The second result is obtained in a similar way (by multiplying respectively by $-y$ and x and adding) when use is made of the relation
$$\frac{y^2 + x^2}{x^2} = 1 + v^2.$$

If $y\dfrac{\partial \phi}{\partial x} = x\dfrac{\partial \phi}{\partial y}$, the result just obtained shows that $\dfrac{\partial \phi}{\partial v} = 0$ and therefore that ϕ is a function of u only. It follows that ϕ is a function of x and y in the combination $x^2 + y^2$, and since it reduces to $x + 1/x$ when $y = 0$,
$$\phi = (x^2 + y^2)^{1/2} + \frac{1}{(x^2+y^2)^{1/2}} = \frac{x^2 + y^2 + 1}{(x^2+y^2)^{1/2}}.$$

EXERCISES 9 (c)

1. If z is a function of x and y, if x, y are related to u, v by $x = uv$, $y = 1/v$, and if
$$2xy\frac{\partial z}{\partial x} + 2(1-y^2)\frac{\partial z}{\partial y} + x^2yz = 0,$$

show that
$$2uv\frac{\partial z}{\partial u} + 2(1-v^2)\frac{\partial z}{\partial v} + u^2vz = 0.$$

2. If $x = r\cosh\theta$, $y = r\sinh\theta$, express
$$\frac{\partial r}{\partial x}, \frac{\partial r}{\partial y}, \frac{\partial \theta}{\partial x}, \frac{\partial \theta}{\partial y}$$
in terms of r and θ.

If V is a function of x and y, prove that
$$\frac{\partial^2 V}{\partial x^2} - \frac{\partial^2 V}{\partial y^2} = \frac{\partial^2 V}{\partial r^2} + \frac{1}{r}\frac{\partial V}{\partial r} - \frac{1}{r^2}\frac{\partial^2 V}{\partial \theta^2}. \quad \text{(O.)}$$

3. If u is a function of x and y and if the variables x and y are changed to r and θ, where
$$x = r\sec\theta, \quad y = r\tan\theta,$$
show that
$$\frac{\partial^2 u}{\partial x^2} - \frac{\partial^2 u}{\partial y^2} = \frac{\partial^2 u}{\partial r^2} + \frac{1}{r}\frac{\partial u}{\partial r} - \frac{\cos^2\theta}{r^2}\frac{\partial^2 u}{\partial \theta^2} + \frac{\sin\theta\cos\theta}{r^2}\frac{\partial u}{\partial \theta}. \quad \text{(C.)}$$

4. The variables x, y in $f(x, y)$ are changed to ξ, η by the substitutions
$$x = \tfrac{1}{2}(\xi^2 - \eta^2), \quad y = \xi\eta.$$
Prove that
$$\xi\frac{\partial f}{\partial \xi} + \eta\frac{\partial f}{\partial \eta} = 2\left(x\frac{\partial f}{\partial x} + y\frac{\partial f}{\partial y}\right),$$
$$\frac{\partial^2 f}{\partial \xi^2} + \frac{\partial^2 f}{\partial \eta^2} = (\xi^2 + \eta^2)\left(\frac{\partial^2 f}{\partial x^2} + \frac{\partial^2 f}{\partial y^2}\right). \tag{C.}$$

5. If V is a function of two independent variables x and y, and if $u = x^3 - 3xy^2$, $v = 3x^2y - y^3$, prove that
$$3\frac{\partial V}{\partial u} = \frac{x^2 - y^2}{(x^2 + y^2)^2}\frac{\partial V}{\partial x} - \frac{2xy}{(x^2 + y^2)^2}\frac{\partial V}{\partial y}.$$
Show further that $3\left(u\dfrac{\partial V}{\partial u} + v\dfrac{\partial V}{\partial v}\right) = x\dfrac{\partial V}{\partial x} + y\dfrac{\partial V}{\partial y}.$

6. Transform the equation $\dfrac{\partial^2 V}{\partial x^2} + \dfrac{\partial^2 V}{\partial y^2} = 0$ into an equation in ξ and η where $\xi = x + iy$, $\eta = x - iy$, and hence show that the only homogeneous polynomial V of degree n which satisfies the equation is $A(x + iy)^n + B(x - iy)^n$, where A and B are constants. (O.)

9.8. Euler's theorems on homogeneous functions

A *homogeneous function of degree n* of several variables $x, y, z, \ldots$ is one which can be written in the form
$$x^n F\left(\frac{y}{x}, \frac{z}{x}, \ldots\right).$$
Writing $Y = y/x$, $Z = z/x$, etc., if u is such a function
$$u = x^n F(Y, Z, \ldots),$$
and, since $\partial Y/\partial x = -y/x^2$, $\partial Z/\partial x = -z/x^2$, etc.,
$$\begin{aligned}\frac{\partial u}{\partial x} &= nx^{n-1}F + x^n\frac{\partial F}{\partial x} \\ &= nx^{n-1}F + x^n\left\{\frac{\partial F}{\partial Y}\frac{\partial Y}{\partial x} + \frac{\partial F}{\partial Z}\frac{\partial Z}{\partial x} + \ldots\right\} \\ &= nx^{n-1}F + x^n\left\{-\frac{y}{x^2}\frac{\partial F}{\partial Y} - \frac{z}{x^2}\frac{\partial F}{\partial Z} - \ldots\right\} \\ &= nx^{n-1}F - x^{n-2}\left\{y\frac{\partial F}{\partial Y} + z\frac{\partial F}{\partial Z} + \ldots\right\}.\end{aligned}$$
Also, since $\partial Y/\partial y = 1/x$, $\partial Z/\partial y = 0$, etc.,
$$\frac{\partial u}{\partial y} = x^n\frac{\partial F}{\partial y} = x^n\frac{\partial F}{\partial Y}\frac{\partial Y}{\partial y} = x^{n-1}\frac{\partial F}{\partial Y},$$
and, similarly
$$\frac{\partial u}{\partial z} = x^{n-1}\frac{\partial F}{\partial Z}, \text{ etc.}$$

EULER'S THEOREMS

Multiplying these results by x, y, z, etc., and adding,

$$x\frac{\partial u}{\partial x} + y\frac{\partial u}{\partial y} + z\frac{\partial u}{\partial z} + \ldots = nx^n F = nu, \qquad (9.12)$$

a result originally due to Euler.

General theorems of a similar character but involving the partial derivatives of higher orders than the first were also proved by Euler. Here only the special case involving the second partial derivatives of a homogeneous function u of degree n in two variables x and y will be discussed. In this case, the derivatives $\partial u/\partial x$, $\partial u/\partial y$ will be homogeneous functions of degree $(n-1)$ and, applying (9.12) to these functions,

$$\left(x\frac{\partial}{\partial x} + y\frac{\partial}{\partial y}\right)\frac{\partial u}{\partial x} = (n-1)\frac{\partial u}{\partial x},$$

and

$$\left(x\frac{\partial}{\partial x} + y\frac{\partial}{\partial y}\right)\frac{\partial u}{\partial y} = (n-1)\frac{\partial u}{\partial y}.$$

Multiplying by x and y respectively and adding,

$$x^2\frac{\partial^2 u}{\partial x^2} + 2xy\frac{\partial^2 u}{\partial x \partial y} + y^2\frac{\partial^2 u}{\partial y^2} = (n-1)\left(x\frac{\partial u}{\partial x} + y\frac{\partial u}{\partial y}\right).$$

Since u is homogeneous and of degree n,

$$x\frac{\partial u}{\partial x} + y\frac{\partial u}{\partial y} = nu,$$

so that, finally,

$$x^2\frac{\partial^2 u}{\partial x^2} + 2xy\frac{\partial^2 u}{\partial x \partial y} + y^2\frac{\partial^2 u}{\partial y^2} = n(n-1)u. \qquad (9.13)$$

Example 8. *If V is a homogeneous function of x, y, z of degree n and if*

$$\frac{\partial^2 V}{\partial x^2} + \frac{\partial^2 V}{\partial y^2} + \frac{\partial^2 V}{\partial z^2} = 0,$$

prove that $\left(\dfrac{\partial^2}{\partial x^2} + \dfrac{\partial^2}{\partial y^2} + \dfrac{\partial^2}{\partial z^2}\right)\{(x^2 + y^2 + z^2)\}V = 2(2n+3)V.$ (O.)

$$\frac{\partial}{\partial x}\{(x^2 + y^2 + z^2)V\} = 2xV + (x^2 + y^2 + z^2)\frac{\partial V}{\partial x}.$$

$$\frac{\partial^2}{\partial x^2}\{(x^2 + y^2 + z^2)V\} = 2V + 4x\frac{\partial V}{\partial x} + (x^2 + y^2 + z^2)\frac{\partial^2 V}{\partial x^2}.$$

From this and the two similar expressions for the second partial derivatives with respect to y and z, there results

$$\left(\frac{\partial^2}{\partial x^2} + \frac{\partial^2}{\partial y^2} + \frac{\partial^2}{\partial z^2}\right)\{(x^2 + y^2 + z^2)V\}$$

$$= 6V + 4\left(x\frac{\partial V}{\partial x} + y\frac{\partial V}{\partial y} + z\frac{\partial V}{\partial z}\right) + (x^2 + y^2 + z^2)\left(\frac{\partial^2 V}{\partial x^2} + \frac{\partial^2 V}{\partial y^2} + \frac{\partial^2 V}{\partial z^2}\right)$$

$$= 6V + 4nV = 2(2n+3)V,$$

when use is made of the relations

$$x\frac{\partial V}{\partial x} + y\frac{\partial V}{\partial y} + z\frac{\partial V}{\partial z} = nV,$$

$$\frac{\partial^2 V}{\partial x^2} + \frac{\partial^2 V}{\partial y^2} + \frac{\partial^2 V}{\partial z^2} = 0.$$

9.9. Exact differentials

If $Pdx + Qdy$ be an exact differential of some function z of x and y, then by comparing the given expression with

$$dz = \frac{\partial z}{\partial x}dx + \frac{\partial z}{\partial y}dy,$$

it follows that $\qquad \dfrac{\partial z}{\partial x} = P, \dfrac{\partial z}{\partial y} = Q.$

Assuming that z is such that

$$\frac{\partial^2 z}{\partial y \partial x} = \frac{\partial^2 z}{\partial x \partial y},$$

the necessary condition that $Pdx + Qdy$ should be an exact differential can be written in the form

$$\frac{\partial P}{\partial y} = \frac{\partial Q}{\partial x}. \tag{9.14}$$

Example 9. *If $Pdx + Qdy + Rdz$ can be made a perfect differential of some function of x, y, z by multiplying each term by a common factor, show that*

$$P\left(\frac{\partial Q}{\partial z} - \frac{\partial R}{\partial y}\right) + Q\left(\frac{\partial R}{\partial x} - \frac{\partial P}{\partial z}\right) + R\left(\frac{\partial P}{\partial y} - \frac{\partial Q}{\partial x}\right) = 0.$$

If u is the function of x, y, z of which, after multiplication by a common factor λ, $Pdx + Qdy + Rdz$ is the differential, it follows as in § 9.9 that

$$\frac{\partial u}{\partial x} = \lambda P, \ \frac{\partial u}{\partial y} = \lambda Q, \ \frac{\partial u}{\partial z} = \lambda R.$$

Hence, assuming $\qquad \dfrac{\partial^2 u}{\partial y \partial z} = \dfrac{\partial^2 u}{\partial z \partial y},$

$$\frac{\partial}{\partial y}(\lambda R) = \frac{\partial^2 u}{\partial y \partial z} = \frac{\partial^2 u}{\partial z \partial y} = \frac{\partial}{\partial z}(\lambda Q).$$

This gives $\qquad \lambda\left(\dfrac{\partial R}{\partial y} - \dfrac{\partial Q}{\partial z}\right) + R\dfrac{\partial \lambda}{\partial y} - Q\dfrac{\partial \lambda}{\partial z} = 0.$

Similarly, $\qquad \lambda\left(\dfrac{\partial P}{\partial z} - \dfrac{\partial R}{\partial x}\right) + P\dfrac{\partial \lambda}{\partial z} - R\dfrac{\partial \lambda}{\partial x} = 0,$

and $\qquad \lambda\left(\dfrac{\partial Q}{\partial x} - \dfrac{\partial P}{\partial y}\right) + Q\dfrac{\partial \lambda}{\partial x} - P\dfrac{\partial \lambda}{\partial y} = 0.$

The required result follows by multiplying respectively by P, Q, R and adding.

EXERCISES 9 (d)

1. If $u = \phi(H_n)$ where H_n is a homogeneous function of degree n in $x, y, \ldots$ and if this relation is solved for H_n to give $H_n = F(u)$, show that
$$x\frac{\partial u}{\partial x} + y\frac{\partial u}{\partial y} + \ldots = n\frac{F(u)}{F'(u)}.$$
Deduce the values of $x\dfrac{\partial u}{\partial x} + y\dfrac{\partial u}{\partial y}$ when

   (i) $u = \sin^{-1}\left(\dfrac{\sqrt{x} - \sqrt{y}}{\sqrt{x} + \sqrt{y}}\right),$ (ii) $u = \tan^{-1}\left(\dfrac{x^3 + y^3}{x + y}\right).$

2. Prove that, if f is a homogeneous polynomial in x and y of degree n and suffixes denote partial differentiations, then

   (i) $xf'_x + yf'_y = nf,$
   (ii) $xf''_{xx} + yf''_{xy} = (n-1)f'_x,$
   (iii) $(n-1)\begin{vmatrix} f''_{xx} & f''_{xy} & f'_x \\ f''_{xy} & f''_{xy} & f'_y \\ f'_x & f'_y & 0 \end{vmatrix} = nf(f''^2_{xy} - f''_{xx}f''_{yy}).$ (C.)

3. If z is a function of x/y, show that
$$x^r\frac{\partial^r z}{\partial x^r} + {}^rC_1 x^{r-1}y\frac{\partial^r z}{\partial y \partial x^{r-1}} + \ldots + {}^rC_s x^{r-s}y^s \frac{\partial^r z}{(\partial y)^s(\partial x)^{r-s}} + \ldots + y^r\frac{\partial^r z}{\partial y^r} = 0. \quad \text{(C.)}$$

4. If $z = \dfrac{y}{x}f(x+y)$ and suffixes denote partial differentiations, show that
$$xz'_x + yz'_y = \frac{y}{x}(x+y)f'(x+y),$$
$$x^2 z''_{xx} + 2xyz''_{xy} + y^2 z''_{yy} = \frac{y}{x}(x+y)^2 f''(x+y),$$
in which $f'(t)$ stands for df/dt and so on. (C.)

5. Show that $(x^2 - 2xy + y)dx + (y^2 - x^2 + x)dy$ is an exact differential and that the expression $ydx + 2xdy$ can be made into an exact differential by multiplication by $1/xy$.

6. If P, Q, R are homogeneous polynomials in x, y, z, all of the same degree and if, for all x, y, z,
$$P\left(\frac{\partial Q}{\partial z} - \frac{\partial R}{\partial y}\right) + Q\left(\frac{\partial R}{\partial x} - \frac{\partial P}{\partial z}\right) + R\left(\frac{\partial P}{\partial y} - \frac{\partial Q}{\partial x}\right) = 0,$$
show that $\dfrac{\partial}{\partial y}\left(\dfrac{P}{xP + yQ + zR}\right) = \dfrac{\partial}{\partial x}\left(\dfrac{Q}{xP + yQ + zR}\right).$ (C.)

9.10. Taylor's theorem for a function of two variables

Results analogous to those of §§ 8.11, 8.12 (which relate to a function of a single variable) can be obtained for functions of several independent variables. To fix ideas, only the case of a function of two variables will be considered here, but the discussion is easily extended when more variables are involved.

Let $F(x,y)$ be such that its partial derivatives up to order n are continuous inside a circular region whose centre has coordinates (a, b) and which contains the point $(a + h, b + k)$. Let

$$x = a + ht, y = b + kt, (0 \leqslant t \leqslant 1),$$

then $\qquad F(x, y) = F(a + ht, b + kt) = \phi(t)$ (say).

By (9.1), $\qquad \dfrac{d\phi}{dt} = \dfrac{\partial F}{\partial x}\dfrac{dx}{dt} + \dfrac{\partial F}{\partial y}\dfrac{dy}{dt},$

and, since $dx/dt = h$, $dy/dt = k$, this can be written if the notation (9.9) is used,

$$\phi'(t) = hp + kq.$$

In a similar way, $\quad \phi''(t) = \left(h\dfrac{\partial}{\partial x} + k\dfrac{\partial}{\partial y}\right)(hp + kq)$

$$= h^2 r + 2hks + k^2 t$$

and similar expressions can be derived for the higher derivatives of $\phi(t)$.

From (8.7) with $a = 0, h = t, f = \phi,$

$$\phi(t) = \phi(0) + t\phi'(0) + \dfrac{t^2}{(2)!}\phi''(0) + \ldots + \dfrac{t^n}{(n)!}\phi^{(n)}(\theta_n t),$$

where $0 < \theta_n < 1$. Writing $t = 1$ and substituting for $\phi(1)$, $\phi(0)$, $\phi'(0)$, $\phi''(0)$, etc., this gives

$$F(a + h, b + k) = F(a, b) + hp + kq + \dfrac{1}{(2)!}(h^2 r + 2hks + k^2 t) + \ldots,$$

(9.15)

the partial derivatives $p, q, r, s, t, \ldots$ (i.e. $\partial F/\partial x$, $\partial F/\partial y$, $\partial^2 F/\partial x^2$, $\partial^2 F/\partial x \partial y$, $\partial^2 F/\partial y^2$, ...) being evaluated when $x = a, y = b$. The remainder after n terms in (9.15) is

$$\dfrac{1}{(n)!}\phi^{(n)}(\theta_n), (0 < \theta_n < 1),$$

and this can, if required, be expressed in terms of the partial derivatives of order n of F. By analogy with (8.7), the result (9.15) is often known as *Taylor's theorem* for a function of two independent variables. An analogy to Maclaurin's theorem follows by writing $a = b = 0$.

9.11. Maxima and minima—two independent variables

Let $F(x, y)$ be a function of two independent variables x, y which satisfies the conditions of § 9.10. The function $F(x, y)$ is said to be *stationary* when $x = a, y = b$ if $F(a + h, b + k) - F(a, b)$ preserves an invariable sign for all sufficiently small independent values of h and k.

By (9.15)

$$F(a + h, b + k) - F(a, b) = hp + kq + \frac{1}{(2)!}(h^2 r + 2hks + k^2 t) + \ldots$$
(9.16)

and, by taking h and k sufficiently small, the first degree terms $hp + kq$ can be made to govern the sign of the right-hand side and therefore also that of $F(a + h, b + k) - F(a, b)$. Hence a *necessary* condition for $F(x, y)$ to assume a stationary value at $x = a, y = b$ is that

$$hp + kq = 0,$$

and, as h and k are independent, it follows that $p = q = 0$. *Thus stationary values of $F(x, y)$ can be found by equating $\partial F/\partial x$ and $\partial F/\partial y$ to zero and solving the resulting simultaneous algebraical equations.*

The value $F(a, b)$ is said to be a *maximum* or *minimum* according as $F(a + h, b + k)$ is less or greater than $F(a, b)$ for all sufficiently small values of h and k. In both cases $F(a, b)$ will be stationary and, from (9.16),

$$F(a + h, b + k) - F(a, b) = \frac{1}{(2)!}(h^2 r + 2hks + k^2 t) + R,$$

where R consists of terms of the third and higher powers of h and k. By taking h and k sufficiently small, the sign of the expression on the left is governed by the sign of the expression

$$h^2 r + 2hks + k^2 t$$

and $F(a, b)$ will be a maximum or minimum according as this expression is negative or positive. Now

$$h^2 r + 2hks + k^2 t = \frac{1}{r}\{(hr + ks)^2 + (rt - s^2)k^2\},$$

and the expression takes the same sign as r for all h, k if $rt > s^2$. If this inequality is satisfied, it follows that r and t will be of the same sign and $F(a, b)$ will be a maximum or minimum according as r, t are both negative or both positive. The case in which $rt \leqslant s^2$ requires further investigation but this will not be undertaken here.

To sum up: $F(x, y)$ is stationary for values of x and y given by

$$\frac{\partial F}{\partial x} = 0, \frac{\partial F}{\partial y} = 0. \qquad (9.17)$$

If $x = a$, $y = b$ is a solution of this pair of equations, $F(a, b)$ is either a maximum or a minimum provided that, at $x = a$, $y = b$,

$$\frac{\partial^2 F}{\partial x^2} \cdot \frac{\partial^2 F}{\partial y^2} > \left(\frac{\partial^2 F}{\partial x \partial y}\right)^2; \qquad (9.18)$$

further, if (9.18) is satisfied, $F(a, b)$ is *a maximum or a minimum* according as

$$\frac{\partial^2 F}{\partial x^2} \text{ and } \frac{\partial^2 F}{\partial y^2}$$

evaluated at $x = a$, $y = b$ are *both negative or both positive*.

Example 10. *Show that the function* $xy + \dfrac{1}{x} + \dfrac{1}{y}$ *has a minimum value when* $x = y = 1$.

Here $F(x, y) = xy + \dfrac{1}{x} + \dfrac{1}{y}$ and

$$\frac{\partial F}{\partial x} = y - \frac{1}{x^2}, \frac{\partial F}{\partial y} = x - \frac{1}{y^2},$$

$$\frac{\partial^2 F}{\partial x^2} = \frac{2}{x^3}, \frac{\partial^2 F}{\partial x \partial y} = 1, \frac{\partial^2 F}{\partial y^2} = \frac{2}{y^3}.$$

Both first partial derivatives vanish when $x = y = 1$ and, for these values of x and y,

$$\frac{\partial^2 F}{\partial x^2} = 2, \frac{\partial^2 F}{\partial x \partial y} = 1, \frac{\partial^2 F}{\partial y^2} = 2.$$

Here the condition (9.18) is satisfied and, $\partial^2 F/\partial x^2$, $\partial^2 F/\partial y^2$ being positive, it follows that $F(1, 1)$ is a minimum.

Example 11. *A long rectangular piece of metal of width 2a is made into a trough by bending up the sides to form equal angles with the base. Find the amount to be bent up and the angle of inclination of the sides that will render the carrying capacity a maximum value.*

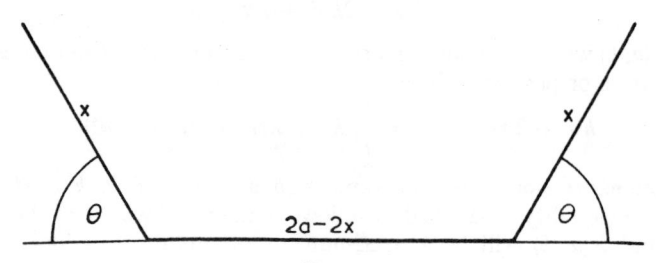

Fig. 34

Let the amount bent up be x and the angle of inclination of the sides be θ (Fig. 34). Then clearly the width of the base is $2(a - x)$, the height of the trough is $x \sin \theta$ and the width of its top is $2(a - x + x \cos \theta)$. The mean width is $2(a - x) + x \cos \theta$ and, since the carrying capacity is proportional

to the cross-sectional area, values of x and θ are sought which render the function
$$F(x, \theta) = x \sin \theta \{2(a - x) + x \cos \theta\}$$
a maximum.
After a little reduction, it is found that
$$\frac{\partial F}{\partial x} = 2 \sin \theta(a - 2x + x \cos \theta),$$
$$\frac{\partial F}{\partial \theta} = x\{2(a - x) \cos \theta + x(\cos^2 \theta - \sin^2 \theta)\}.$$

Stationary values of F occur when these two expressions vanish and, since $x = 0$, $\theta = 0$ clearly give a minimum value to F, the equations for solution are
$$a - 2x + x \cos \theta = 0,$$
$$2(a - x) \cos \theta + x(\cos^2 \theta - \sin^2 \theta) = 0.$$
Substitution of $x = a/(2 - \cos \theta)$ from the first of these equations in the second leads to $\cos \theta = \frac{1}{2}$ or $\theta = 60°$. The resulting value of x is $2a/3$ and it is left as an exercise for the reader to check that, for these values of x and θ,
$$\frac{\partial^2 F}{\partial x^2} \cdot \frac{\partial^2 F}{\partial \theta^2} > \left(\frac{\partial^2 F}{\partial x \partial \theta}\right)^2, \quad \frac{\partial^2 F}{\partial x^2} < 0, \quad \frac{\partial^2 F}{\partial \theta^2} < 0,$$
and thus that the carrying capacity is a maximum.

EXERCISES 9 (e)

1. Show that, if a and b are real,
$$e^{ax} \cos by = 1 + ax + \frac{a^2x^2 - b^2y^2}{(2)!} + \cdots \infty$$

2. Find the stationary values of xyz when $z = 1 + x + y$.

3. Find values of x and y for which
   (i) $x^3y^2 - x^4y^2 - x^3y^3$ and (ii) $\sin x \sin y \sin (x + y)$ have maximum values.

4. Show that the cheapest rectangular tank, to be made of sheet metal and open at the top and to contain a given volume, is one whose base is square and whose depth is half the length of a side of the base.

5. P is a variable point in the plane of a triangle ABC. Show that $PA^2 + PB^2 + PC^2$ is least when P lies at the centroid of the triangle.

6. Find the stationary values of $(2ax - x^2)(2by - y^2)$ where a and b are both positive and discuss if they are maximum or minimum values.

EXERCISES 9 (f)

1. Find the value of a if $V = x^3 + axy^2$ satisfies the equation
$$\frac{\partial^2 V}{\partial x^2} + \frac{\partial^2 V}{\partial y^2} = 0.$$

If the function V above satisfies this equation, if $U = r^n V$ and $r^2 = x^2 + y^2$, show that

$$\frac{\partial^2 U}{\partial x^2} + \frac{\partial^2 U}{\partial y^2} = n(n+6) r^{n-2} V.$$

2. If $\phi = \log_x y$, find $\partial \phi / \partial x$ and $\partial \phi / \partial y$ in terms of x and y. (O.)

3. Given that $x = u + e^{-v} \cos u$, $y = v + e^{-v} \sin u$, prove that

$$\frac{\partial u}{\partial x} \text{ is not equal to } 1 \Big/ \left(\frac{\partial x}{\partial u} \right).$$

Prove also that $\dfrac{\partial u}{\partial x} = \dfrac{\partial v}{\partial y}$. (O.)

4. If $\psi(u)$, where $u = \log_e (x^2 - y^2)$, satisfies the equation

$$(x^2 - y^2) \left(\frac{\partial^2 \psi}{\partial x^2} - \frac{\partial^2 \psi}{\partial y^2} \right) + 4\psi = 0,$$

show that $\dfrac{d^2 \psi}{du^2} + \psi = 0.$ (O.)

5. If $z = x\phi(x+y) + x^2 \psi(x+y)$, prove that

$$x^2 \left(\frac{\partial^2 z}{\partial x^2} - 2 \frac{\partial^2 z}{\partial x \partial y} + \frac{\partial^2 z}{\partial y^2} \right) - 2x \left(\frac{\partial z}{\partial x} - \frac{\partial z}{\partial y} \right) + 2z = 0. \quad \text{(O.)}$$

6. The length of the hypotenuse of a right-angled triangle is calculated from the length of its other two sides. These are measured as 17 and 23 centimetres respectively with a possible error of one-tenth of a centimetre in each. Find the possible error in the calculated length of the hypotenuse.

7. The time of oscillation (T) and the length (l) of a simple pendulum are related by the formula

$$T = 2\pi \sqrt{\left(\frac{l}{g} \right)},$$

where g is the acceleration due to gravity. If this relation is used to calculate the value of g from measurements of T and l and if positive errors of $\frac{1}{2}$ per cent. are made in both these measurements, find the resulting percentage error in the value of g.

8. If the circumradius R and the area Δ of a triangle ABC are regarded as functions of b, c, A, prove that

$$\frac{\partial R}{\partial A} = 4R \sin A \frac{\partial R}{\partial b} \frac{\partial R}{\partial c},$$

$$\frac{\partial R}{\partial b} \frac{\partial \Delta}{\partial c} + \frac{\partial R}{\partial c} \frac{\partial \Delta}{\partial b} = \frac{1}{2} R \sin A. \quad \text{(C.)}$$

9. Find dy/dx if $y^x + x^y = (x+y)^{x+y}$.

10. If $\sin^2 x + \sin^2 y = 2 \cos x \cos y$, show that

$$\frac{dy}{dx} = -\sin x \operatorname{cosec} y.$$

11. If z is defined implicitly as a function of x and y by the relation $3y = 3xz + z^3$, show that
$$\frac{\partial^2 z}{\partial x^2} + x\frac{\partial^2 z}{\partial y^2} = 0.$$

12. Given that $x = u/(u^2 + v^2)$, $y = v/(u^2 + v^2)$ and that f is a function of x and y, show that
$$(x^2 + y^2)\left(\frac{\partial^2 f}{\partial x^2} + \frac{\partial^2 f}{\partial y^2}\right) = (u^2 + v^2)\left(\frac{\partial^2 f}{\partial u^2} + \frac{\partial^2 f}{\partial v^2}\right). \quad \text{(O.C.)}$$

13. ϕ is a function of the two variables u and v which are themselves given as functions of x and y by the equations
$$u = x \cosh x \cos y - y \sinh x \sin y,$$
$$v = x \sinh x \sin y + y \cosh x \cos y.$$
Obtain $\dfrac{\partial \phi}{\partial x}$ and $\dfrac{\partial \phi}{\partial y}$ in terms of $\dfrac{\partial \phi}{\partial u}$, $\dfrac{\partial \phi}{\partial v}$, x and y.

If $\phi = u^2 + v^2$, show that
$$\phi\left(\frac{\partial^2 \phi}{\partial x^2} + \frac{\partial^2 \phi}{\partial y^2}\right) = \left(\frac{\partial \phi}{\partial x}\right)^2 + \left(\frac{\partial \phi}{\partial y}\right)^2 = 4\phi\left\{\left(\frac{\partial u}{\partial x}\right)^2 + \left(\frac{\partial u}{\partial y}\right)^2\right\}. \quad \text{(O.)}$$

14. The variables x, y and u, v are connected by the relations $x = ue^v$, $y = ue^{-v}$. Prove that, if V is a function of the variables x and y,
$$u\frac{\partial V}{\partial u} = x\frac{\partial V}{\partial x} + y\frac{\partial V}{\partial y}, \quad \frac{\partial V}{\partial v} = x\frac{\partial V}{\partial x} - y\frac{\partial V}{\partial y},$$
$$u\frac{\partial^2 V}{\partial u \partial v} - \frac{\partial V}{\partial v} = x^2\frac{\partial^2 V}{\partial x^2} - y^2\frac{\partial^2 V}{\partial y^2}. \quad \text{(O.)}$$

15. If $u = x^2 - y^2$, $v = 2xy$ and V is a function of x and y, prove that
$$(x^2 + y^2)\left\{\left(\frac{\partial V}{\partial x}\right)^2 + \left(\frac{\partial V}{\partial y}\right)^2\right\} = 4(u^2 + v^2)\left\{\left(\frac{\partial V}{\partial u}\right)^2 + \left(\frac{\partial V}{\partial v}\right)^2\right\},$$
and $\quad (x^2 + y^2)\left(\dfrac{\partial^2 V}{\partial x^2} + \dfrac{\partial^2 V}{\partial y^2}\right) = 4(u^2 + v^2)\left(\dfrac{\partial^2 V}{\partial u^2} + \dfrac{\partial^2 V}{\partial v^2}\right). \quad \text{(O.)}$

16. If $ux = vy = x^2 + y^2$, and ϕ is a function of x and y (or u and v), prove that
$$(x^2 + y^2)^2\left(\frac{\partial^2 \phi}{\partial x^2} + \frac{\partial^2 \phi}{\partial y^2}\right) = u^4\frac{\partial^2 \phi}{\partial u^2} + v^4\frac{\partial^2 \phi}{\partial v^2} + 2\left(u^3\frac{\partial \phi}{\partial u} + v^3\frac{\partial \phi}{\partial v}\right). \quad \text{(O.)}$$

17. Show that if V is a function of x, y, z, then
$$\left(\frac{\partial}{\partial x} + i\frac{\partial}{\partial y}\right)\left(\frac{\partial}{\partial x} - i\frac{\partial}{\partial y}\right)V = \frac{\partial^2 V}{\partial x^2} + \frac{\partial^2 V}{\partial y^2}.$$

If V is converted into a function of r, θ, ϕ by putting $x = r \sin \theta \cos \phi$, $y = r \sin \theta \sin \phi$, $z = r \cos \theta$, express
$$\frac{\partial V}{\partial x} + i\frac{\partial V}{\partial y}, \quad \frac{\partial V}{\partial x} - i\frac{\partial V}{\partial y}, \quad \frac{\partial V}{\partial z},$$

in terms of $r, \theta, \phi, \dfrac{\partial V}{\partial r}, \dfrac{\partial V}{\partial \theta}, \dfrac{\partial V}{\partial \phi}$; hence, or otherwise, show that

$$\dfrac{\partial^2 V}{\partial x^2} + \dfrac{\partial^2 V}{\partial y^2} + \dfrac{\partial^2 V}{\partial z^2}$$
$$= \dfrac{\partial^2 V}{\partial r^2} + \dfrac{2}{r}\dfrac{\partial V}{\partial r} + \dfrac{1}{r^2}\dfrac{\partial^2 V}{\partial \theta^2} + \dfrac{\cot\theta}{r^2}\dfrac{\partial V}{\partial \theta} + \dfrac{1}{r^2 \sin^2\theta}\dfrac{\partial^2 V}{\partial \phi^2}. \quad \text{(O.)}$$

18. If x and y are functions of ξ and η, and

$$a = \left(\dfrac{\partial x}{\partial \xi}\right)^2 + \left(\dfrac{\partial y}{\partial \xi}\right)^2,\ b = \left(\dfrac{\partial x}{\partial \eta}\right)^2 + \left(\dfrac{\partial y}{\partial \eta}\right)^2,\ h = \dfrac{\partial x}{\partial \xi}\dfrac{\partial x}{\partial \eta} + \dfrac{\partial y}{\partial \xi}\dfrac{\partial y}{\partial \eta},$$

and if
$$H = \dfrac{\partial x}{\partial \xi}\dfrac{\partial y}{\partial \eta} - \dfrac{\partial x}{\partial \eta}\dfrac{\partial y}{\partial \xi},$$

show that
$$H\dfrac{\partial U}{\partial x} = \dfrac{\partial U}{\partial \xi}\dfrac{\partial y}{\partial \eta} - \dfrac{\partial U}{\partial \eta}\dfrac{\partial y}{\partial \xi},$$
$$H\dfrac{\partial U}{\partial y} = -\dfrac{\partial U}{\partial \xi}\dfrac{\partial x}{\partial \eta} + \dfrac{\partial U}{\partial \eta}\dfrac{\partial x}{\partial \xi},$$

where U is any differentiable function of x, y.
Prove also that

$$\dfrac{\partial^2 U}{\partial x^2} + \dfrac{\partial^2 U}{\partial y^2} = \dfrac{1}{H}\dfrac{\partial}{\partial \xi}\left\{\dfrac{b\dfrac{\partial U}{\partial \xi} - h\dfrac{\partial U}{\partial \eta}}{H}\right\} - \dfrac{1}{H}\dfrac{\partial}{\partial \eta}\left\{\dfrac{h\dfrac{\partial U}{\partial \xi} - a\dfrac{\partial U}{\partial \eta}}{H}\right\}. \quad \text{(C.)}$$

19. If $\phi(x, y)$ is a real-valued function of the two real variables x and y and if
$$\dfrac{\partial^2 \phi}{\partial x^2} + \dfrac{\partial^2 \phi}{\partial y^2} = 0$$
throughout some region of the (x, y)-plane, then ϕ is said to be *harmonic* in that region.
   (i) Show that, if both ϕ and e^ϕ are harmonic for all (x, y), then ϕ must be a constant.
   (ii) Find all the harmonic functions which possess circular symmetry, i.e. find f such that $\phi \equiv f(x^2 + y^2)$ is harmonic everywhere save perhaps at the origin. (O.)

20. A function $f(x, t)$ satisfies the equation
$$k\dfrac{\partial^2 f}{\partial x^2} = \dfrac{\partial f}{\partial t}. \quad \text{(i)}$$
On transforming the independent variables from x, t to ξ, τ where
$$\xi = \dfrac{x}{\sqrt{(kt)}},\ \tau = t,$$
the function $f(x, t)$ is transformed into $\phi(\xi, \tau)$. Show that
$$\dfrac{\partial^2 \phi}{\partial \xi^2} + \dfrac{1}{2}\xi\dfrac{\partial \phi}{\partial \xi} = \tau\dfrac{\partial \phi}{\partial \tau}.$$
Find in the form of an integral the most general solution of (i) in the form $F[x/\sqrt{(kt)}]$. (C.)

21. If $x^2 + y^2 = (x + y)^z$, prove that
$$x\frac{\partial z}{\partial x} + y\frac{\partial z}{\partial y} = 0. \qquad \text{(O.)}$$

22. (i) If $u = \log_e v$ and v is a homogeneous function in x and y of degree n, prove that
$$x\frac{\partial u}{\partial x} + y\frac{\partial u}{\partial y} = n.$$

(ii) If V is any homogeneous function of x and y of the first degree, prove that
$$\frac{\partial^2 V}{\partial x^2}\bigg/y^2 = \frac{\partial^2 V}{\partial y^2}\bigg/x^2 = -\frac{\partial^2 V}{\partial x \partial y}\bigg/xy. \qquad \text{(O.)}$$

23. Show that if A, B, C are the angles of a plane triangle, the maximum value of $\cos A \cos B \cos C$ is $1/8$.

24. Prove that $Mdx + Ndy$ can be made into an exact differential by multiplication by a function independent of y if
$$N\frac{\partial}{\partial y}\left(\frac{\partial N}{\partial x} - \frac{\partial M}{\partial y}\right) - \frac{\partial N}{\partial y}\left(\frac{\partial N}{\partial x} - \frac{\partial M}{\partial y}\right) = 0.$$
Verify that this condition is satisfied when $M = x^2 - y^2 + y$, $N = (1 + x)(x + 2y)$.

25. If $x = \phi e^\theta$, $v = \theta e^\phi$, prove that
$$\frac{\partial x}{\partial \theta}\frac{\partial \theta}{\partial x} = \frac{\partial y}{\partial \phi}\frac{\partial \phi}{\partial y} = 1 - \frac{\partial x}{\partial \phi}\frac{\partial \phi}{\partial x} = 1 - \frac{\partial v}{\partial \theta}\frac{\partial v}{\partial y}.$$

CHAPTER 10

MORE ADVANCED METHODS OF INTEGRATION. FURTHER THEOREMS IN THE INTEGRAL CALCULUS

10.1. Introduction

In *Advanced Level Pure Mathematics* (Chapters 10–13), the ideas of the integral calculus were introduced, the more elementary methods of integration discussed and some applications to geometry and mechanics were explained. It is the purpose of the present chapter to take the reader a little further in the techniques of the integral calculus and, in Chapter 11, to give some further applications.

As a starting-point, it is assumed that the following indefinite integrals (a, b are constants and the arbitrary constant of integration is to be added in all cases) have been covered by previous work:

$$\int (ax+b)^n dx = \frac{(ax+b)^{n+1}}{(n+1)a} (n \neq -1); \int \frac{dx}{ax+b} = \frac{1}{a} \log_e (ax+b);$$

$$\int \cos ax\, dx = \frac{1}{a} \sin ax; \qquad \int \sin ax\, dx = -\frac{1}{a} \cos ax;$$

$$\int \sec^2 ax\, dx = \frac{1}{a} \tan ax; \qquad \int \operatorname{cosec}^2 ax\, dx = -\frac{1}{a} \cot ax;$$

$$\int e^{ax} dx = \frac{1}{a} e^{ax}; \qquad \int \frac{f'(x)}{f(x)} dx = \log_e f(x);$$

$$\int \tan ax\, dx = -\frac{1}{a} \log_e \cos ax; \qquad \int \cot ax\, dx = \frac{1}{a} \log_e \sin ax;$$

$$\int \operatorname{cosec} ax\, dx = \frac{1}{a} \log_e \tan \frac{1}{2} ax; \qquad \int \sec ax\, dx = \frac{1}{a} \log_e \tan \left(\frac{\pi}{4} + \frac{ax}{2}\right);$$

$$\int \frac{dx}{\sqrt{(a^2 - x^2)}} = \sin^{-1}\left(\frac{x}{a}\right) (x < a); \int \frac{dx}{a^2 + x^2} = \frac{1}{a} \tan^{-1}\left(\frac{x}{a}\right);$$

$$\int \frac{dx}{a^2 - x^2} = \frac{1}{2a} \log_e \left(\frac{a+x}{a-x}\right) (x < a);$$

$$\int \frac{dx}{x^2 - a^2} = \frac{1}{2a} \log_e \left(\frac{x-a}{x+a}\right) (x > a);$$

$$\int u \frac{dv}{dx} dx = uv - \int v \frac{du}{dx} dx \text{ (integration by parts)}.$$

The reader should also recall that rational algebraical fractions can often be made to depend on the integrals

EXAMPLES OF INTEGRATION

$$\int \frac{dx}{x+a}, \int \frac{dx}{x^2+a^2}$$

by first resolving the integrand into partial fractions.

Once again it is emphasised that a certain maturity of judgment is often required to effect an integration and the student is again advised to work through as many exercises as possible. A few illustrative examples and a fairly extensive set of exercises follow; these involve no new principles but all have been taken from examination papers at Scholarship level.

Example 1. *Evaluate* $\int \left(\frac{1}{x} + \frac{1}{x^2}\right) \log_e x \, dx.$ (C.)

In the rule for integration by parts, put

$$u = \log_e x, \frac{dv}{dx} = \frac{1}{x} + \frac{1}{x^2},$$

so that

$$\frac{du}{dx} = \frac{1}{x}, v = \log_e x - \frac{1}{x}.$$

Hence

$$\int \left(\frac{1}{x} + \frac{1}{x^2}\right) \log_e x \, dx = \log_e x \left(\log_e x - \frac{1}{x}\right) - \int \left(\log_e x - \frac{1}{x}\right) \frac{1}{x} dx$$

$$= (\log_e x)^2 - \frac{\log_e x}{x} - \frac{1}{x} - \int \frac{\log_e x}{x} dx.$$

To find $\int x^{-1} \log_e x \, dx$, put $u = \log_e x$, $dv/dx = x^{-1}$ so that $du/dx = x^{-1}$, $v = \log_e x$ and

$$\int \frac{\log_e x}{x} dx = (\log_e x)^2 - \int \frac{\log_e x}{x} dx,$$

giving

$$\int \frac{\log_e x}{x} dx = \frac{1}{2}(\log_e x)^2.$$

Substituting the value of this integral and adding the constant of integration, the required result is

$$\int \left(\frac{1}{x} + \frac{1}{x^2}\right) \log_e x \, dx = \frac{1}{2}(\log_e x)^2 - \frac{1}{x}(\log_e x + 1) + C.$$

Example 2. *Evaluate* $\int \frac{(x^2+1) dx}{x^4 + x^2 + 1}.$ (O.)

Dividing numerator and denominator by x^2, the integrand can be written

$$\frac{1 + \frac{1}{x^2}}{x^2 + 1 + \frac{1}{x^2}} = \frac{1 + \frac{1}{x^2}}{\left(x - \frac{1}{x}\right)^2 + 3}.$$

Writing $x - x^{-1} = t$, so that $(1 + x^{-2})dx = dt$, the integral is equivalent to

$$\int \frac{dt}{t^2 + 3} = \frac{1}{\sqrt{3}} \tan^{-1}\left(\frac{t}{\sqrt{3}}\right) + C = \frac{1}{\sqrt{3}} \tan^{-1}\left(\frac{x^2 - 1}{x\sqrt{3}}\right) + C.$$

Alternatively, since $x^4 + x^2 + 1 = (x^2 + x + 1)(x^2 - x + 1)$, the integrand can be expressed in partial fractions. Thus

$$\int \frac{(x^2+1)dx}{x^4+x^2+1} = \frac{1}{2}\int \left\{ \frac{1}{x^2+x+1} + \frac{1}{x^2-x+1} \right\} dx$$

$$= \frac{1}{2}\int \left\{ \frac{1}{\frac{3}{4}+\left(x+\frac{1}{2}\right)^2} + \frac{1}{\frac{3}{4}+\left(x-\frac{1}{2}\right)^2} \right\} dx$$

$$= \frac{1}{\sqrt{3}}\left\{ \tan^{-1}\left(\frac{2x+1}{\sqrt{3}}\right) + \tan^{-1}\left(\frac{2x-1}{\sqrt{3}}\right) \right\} + C'.$$

It is left as an exercise to show that this is equivalent to the previous result.

Example 3. *Evaluate the definite integral* $\displaystyle\int_a^b \frac{dx}{x\sqrt{\{(x-a)(b-x)\}}}$ *where* $0 < a < b$.

(C.)

For integrals involving the factors $(x - a)$, $(b - x)$, the substitution $x = a \cos^2 t + b \sin^2 t$ is often useful. With this substitution, $dx = (-2a \cos t \sin t + 2b \sin t \cos t)dt = 2(b - a) \sin t \cos t\, dt$, $x - a = a(\cos^2 t - 1) + b \sin^2 t = (b - a) \sin^2 t$, $b - x = b(1 - \sin^2 t) - a \cos^2 t = (b - a) \cos^2 t$, so that

$$\frac{dx}{\sqrt{\{(x-a)(b-x)\}}} = 2dt.$$

When $x = a$, $t = 0$ and when $x = b$, $t = \pi/2$ so that the given integral

$$= \int_0^{\pi/2} \frac{2dt}{a \cos^2 t + b \sin^2 t} = \int_0^{\pi/2} \frac{2 \sec^2 t\, dt}{a + b \tan^2 t}$$

$$= \frac{2}{b}\int_0^{\pi/2} \frac{d(\tan t)}{\frac{a}{b} + \tan^2 t} = \frac{2}{\sqrt{(ab)}}\left[\tan^{-1}\left\{ \sqrt{\left(\frac{b}{a}\right)} \tan t \right\} \right]_0^{\pi/2} = \frac{\pi}{\sqrt{(ab)}}.$$

Example 4. *Prove that*

$$\int \tan x \tan 2x \tan 3x\, dx = \log_e \cos x + \tfrac{1}{2} \log_e \cos 2x - \tfrac{1}{3} \log_e \cos 3x + C. \quad (O.)$$

$$\tan 3x = \tan(x + 2x) = \frac{\tan x + \tan 2x}{1 - \tan x \tan 2x},$$

so that $\quad \tan 3x - \tan x \tan 2x \tan 3x = \tan x + \tan 2x$.

Hence the integrand is equivalent to $\tan 3x - \tan x - \tan 2x$ and the given integral

$$= \int (-\tan x - \tan 2x + \tan 3x)dx$$

$$= \log_e \cos x + \tfrac{1}{2} \log_e \cos 2x - \tfrac{1}{3} \log_e \cos 3x + C.$$

Example 5. *Evaluate* $\displaystyle\int \frac{dx}{a + \tan x}$, *a constant*. (C.)

If $\tan x = t$, $\sec^2 x\, dx = dt$ and

$$dx = \frac{dt}{\sec^2 x} = \frac{dt}{1 + \tan^2 x} = \frac{dt}{1 + t^2}.$$

Hence
$$\int \frac{dx}{a + \tan x} = \int \frac{dt}{(a+t)(1+t^2)}$$
$$= \frac{1}{1+a^2}\int\left\{\frac{1}{a+t} + \frac{a-t}{1+t^2}\right\}dt,$$

by the usual rules for partial fractions. This can be written
$$\int \frac{dx}{a+\tan x} = \frac{1}{1+a^2}\left\{\int \frac{dt}{a+t} + a\int \frac{dt}{1+t^2} - \frac{1}{2}\int \frac{d(t^2)}{1+t^2}\right\}$$
$$= \frac{1}{1+a^2}\left\{\log_e(a+t) + a\tan^{-1}t - \frac{1}{2}\log_e(1+t^2)\right\} + C$$
$$= \frac{1}{1+a^2}\left\{\log_e(a+\tan x) + ax - \frac{1}{2}\log_e(1+\tan^2 x)\right\} + C$$
$$= \frac{1}{1+a^2}\left\{ax + \log_e\left(\frac{a+\tan x}{\sec x}\right)\right\} + C$$
$$= \frac{1}{1+a^2}\left\{ax + \log_e(a\cos x + \sin x)\right\} + C.$$

EXERCISES 10 (a)

Integrate the following functions with respect to x:

1. $x^{-n} \log_e x$ $(n \neq 1)$. (C.)
2. $\log_e \{x + \sqrt{(x^2 + a^2)}\}$. (O.)
3. $x^m (\log_e x)^2 (m \neq -1)$. (C.)
4. $\dfrac{e^x(x^2 - x + 2)}{(x+1)^3}$. (O.)
5. $(x^2 + a^2)^{-3/2}$. (O.)
6. $\dfrac{x^3}{(x^2+a^2)(x^2+b^2)^2}$. (O.)
7. $\dfrac{1}{x^4 + a^4}$. (C.)
8. $\left(\dfrac{x-a}{b-x}\right)^{1/2}$. (O.)
9. $\dfrac{1}{x^2\sqrt{(x^2+1)}}$. (O.)
10. $\left(\dfrac{x}{1-x}\right)^{1/2}$. (O.C.)
11. $\tan x(1 + \sec x)$. (C.)
12. $\dfrac{x + \sin x}{1 + \cos x}$. (O.)
13. $\dfrac{1}{\sin x (1 + \cos x)}$. (O.)
14. $\dfrac{\cos^2 2x}{\sin^4 x \cos^2 x}$. (C.)

Evaluate the following definite integrals:

15. $\int_0^{\pi/2} x^3 \sin x\, dx$. (O.)
16. $\int_0^{\pi/4} \tan^3 x\, dx$. (C.)
17. $\int_0^{\pi/6} \sin x \tan x\, dx$. (L.U.)
18. $\int_0^{\pi/2} \dfrac{a\cos x + b\sin x}{\cos x + \sin x}dx$.
19. Evaluate $\int e^x\left(\dfrac{1+\sin x}{1+\cos x}\right)dx$ and deduce the value of
$$\int e^x\left(\frac{1-\sin x}{1-\cos x}\right)dx. \qquad \text{(O.)}$$

20. Prove that $\int_0^{\pi/3} (1 + \tan^6 \theta)d\theta = \frac{9}{5}\sqrt{3}$. (C.)

21. By the substitution $\tan x = t$, or otherwise, show that approximately
$$\int_0^{\pi/4} \frac{dx}{\cos^2 x + 3\sin^2 x} = 0{\cdot}6045.$$ (W.)

22. Show that $\int_0^{\pi/2} \sqrt{\left(\frac{1-\sin x}{1+\sin x}\right)}dx = \log_e 2$. (W.)

23. Evaluate $\int \frac{dx}{x^2 + x\sqrt{(x^2+1)}}$. (O.C.)

24. Integrate with respect to x
$$\frac{1 + (a+x)^{1/3}}{1 - (a+x)^{1/3}}.$$ (C.)

25. Evaluate $\int_0^1 \frac{\tan^{-1} x}{(1+x^2)^{3/2}}dx$. (N.U.)

26. Prove that
$$\int_0^a x^3 \log_e\{a + \sqrt{(a^2 - x^2)}\}dx = \frac{a^4}{48}(5 + 12 \log_e a).$$ (O.)

27. Integrate $\sin(x+a)\,\mathrm{cosec}\,(x+\beta)$ with respect to x. (O.)

28. Evaluate $\int_0^1 \frac{1 + x^{1/2}}{1 + x^{1/3}}dx$. (C.)

29. Show that if $a > b > c$ and $p > -1$, the integral
$$\int_a^b (a-x)^p(x-b)^p(x-c)^{-p-1}dx$$
transforms into itself by the substitution
$$xx' - c(x+x') = ab - c(a+b).$$ (O.)

30. Prove that, if $b > a > 0$, then
$$\int_a^b \frac{dx}{\{(x-a)(b-x)\}^{1/2}} = \pi.$$
Deduce, or prove otherwise, that if $a < \beta$, then
$$\int_a^\beta \frac{d\theta}{\{(e^\theta - e^a)(e^\beta - e^\theta)\}^{1/2}} = \pi e^{-\frac{1}{2}(a+\beta)}.$$ (O.)

10.2. Some integrals involving the hyperbolic functions

The definitions of the hyperbolic functions (§ 6.10) give
$$\int \sinh ax\,dx = \frac{1}{2}\int (e^{ax} - e^{-ax})dx = \frac{1}{2a}(e^{ax} + e^{-ax}) + C$$
$$= \frac{1}{a}\cosh ax + C, \qquad (10.1)$$

$$\int \cosh ax\,dx = \frac{1}{2}\int (e^{ax} + e^{-ax})dx = \frac{1}{2a}(e^{ax} - e^{-ax}) + C$$
$$= \frac{1}{a}\sinh ax + C. \tag{10.2}$$

It is easy to verify from the definitions of the functions that
$$\frac{d}{dx}(\sinh ax) = a\cosh ax,\quad \frac{d}{dx}(\cosh ax) = a\sinh ax,$$
so that
$$\frac{d}{dx}(\tanh ax) = \frac{d}{dx}\left(\frac{\sinh ax}{\cosh ax}\right) = a\left\{\frac{\cosh^2 ax - \sinh^2 ax}{\cosh^2 ax}\right\}$$
$$= a\,\text{sech}^2\,ax,$$
since $\cosh^2 ax - \sinh^2 ax = 1$. It follows that
$$\int \text{sech}^2\,ax\,dx = \frac{1}{a}\tanh ax + C, \tag{10.3}$$
and it can be shown similarly that
$$\int \text{cosech}^2\,ax\,dx = -\frac{1}{a}\coth ax + C. \tag{10.4}$$

Example 6. *Integrate cosech x with respect to x.*

$$\int \text{cosech}\,x\,dx = \int \frac{dx}{\sinh x} = 2\int \frac{dx}{e^x - e^{-x}}$$
$$= 2\int \frac{e^x}{e^{2x} - 1}dx = \int\left\{\frac{1}{e^x - 1} - \frac{1}{e^x + 1}\right\}e^x\,dx$$
$$= \log_e(e^x - 1) - \log_e(e^x + 1) + C$$
$$= \log_e\left(\frac{e^x - 1}{e^x + 1}\right) + C$$
$$= \log_e\left(\frac{e^{x/2} - e^{-x/2}}{e^{x/2} + e^{-x/2}}\right) + C$$
$$= \log_e \tanh \frac{x}{2} + C.$$

Just as the substitution $x = a\sin t$ (or $x = a\cos t$) was often useful when the integrand contained the terms $(a^2 - x^2)$, so the substitutions $x = a\sinh t$, $x = a\cosh t$ are often effective when the terms $(x^2 + a^2)$, $(x^2 - a^2)$ appear in the integrand. For example, putting $x = a\sinh t$, $dx/dt = a\cosh t$ and
$$\int \frac{dx}{\sqrt{(x^2 + a^2)}} = \int \frac{a\cosh t\,dt}{\sqrt{(a^2 \sinh^2 t + a^2)}}.$$
Since $\cosh^2 t - \sinh^2 t = 1$, the denominator can be written $a\cosh t$ so that
$$\int \frac{dx}{\sqrt{(x^2 + a^2)}} = \int dt = t + C = \sinh^{-1}\left(\frac{x}{a}\right) + C. \tag{10.5}$$

In the same way, putting $x = a \cosh t$, $dx/dt = a \sinh t$,

$$\int \frac{dx}{\sqrt{(x^2 - a^2)}} = \int \frac{a \sinh t \, dt}{\sqrt{(a^2 \cosh^2 t - a^2)}} = \int dt = t + C$$

$$= \cosh^{-1}\left(\frac{x}{a}\right) + C. \tag{10.6}$$

These two results should be compared with the similar result

$$\int \frac{dx}{\sqrt{(a^2 - x^2)}} = \sin^{-1}\left(\frac{x}{a}\right) + C = -\cos^{-1}\left(\frac{x}{a}\right) + C'. \tag{10.7}$$

The integrals (10.5), (10.6) can be expressed in terms of logarithmic functions by using the results (6.59), (6.58): thus,

$$\int \frac{dx}{\sqrt{(x^2 + a^2)}} = \sinh^{-1}\left(\frac{x}{a}\right) + C = \log_e\left\{\frac{x + \sqrt{(x^2 + a^2)}}{a}\right\} + C, \tag{10.8}$$

$$\int \frac{dx}{\sqrt{(x^2 - a^2)}} = \cosh^{-1}\left(\frac{x}{a}\right) + C = \log_e\left\{\frac{x + \sqrt{(x^2 - a^2)}}{a}\right\} + C. \tag{10.9}$$

The integrals of $\sqrt{(a^2 - x^2)}$, $\sqrt{(x^2 + a^2)}$, $\sqrt{(x^2 - a^2)}$ can be found in a similar way. Thus, to find the first integral, put $x = a \sin t$, $dx/dt = a \cos t$, so that

$$\int \sqrt{(a^2 - x^2)}\,dx = \int \sqrt{(a^2 - a^2 \sin^2 t)}\, a \cos t\, dt = a^2 \int \cos^2 t\, dt$$

$$= \frac{a^2}{2}\int (1 + \cos 2t)dt = \frac{a^2}{2}\left(t + \frac{\sin 2t}{2}\right) + C$$

$$= \frac{a^2}{2}(t + \sin t \cos t) + C$$

$$= \frac{a^2}{2}\sin^{-1}\left(\frac{x}{a}\right) + \frac{1}{2}x\sqrt{(a^2 - x^2)} + C. \tag{10.10}$$

For the second integral, put $x = a \sinh t$, $dx/dt = a \cosh t$, and

$$\int \sqrt{(x^2 + a^2)}\,dx = \int \sqrt{(a^2 \sinh^2 t + a^2)}\, a \cosh t\, dt = a^2 \int \cosh^2 t\, dt$$

$$= \frac{a^2}{2}\int (1 + \cosh 2t)dt = \frac{a^2}{2}\left(t + \frac{\sinh 2t}{2}\right) + C$$

$$= \frac{a^2}{2}(t + \sinh t \cosh t) + C$$

$$= \frac{a^2}{2}\sinh^{-1}\left(\frac{x}{a}\right) + \frac{1}{2}x\sqrt{(x^2 + a^2)} + C$$

$$= \frac{a^2}{2}\log_e\left\{\frac{x + \sqrt{(x^2 + a^2)}}{a}\right\} + \frac{1}{2}x\sqrt{(x^2 + a^2)} + C. \tag{10.11}$$

HYPERBOLIC FUNCTIONS

In a similar way, use of the substitution $x = a \cosh t$ leads to

$$\int \sqrt{(x^2 - a^2)}\,dx = -\frac{a^2}{2} \cosh^{-1}\left(\frac{x}{a}\right) + \frac{1}{2}x\sqrt{(x^2 - a^2)} + C$$

$$= -\frac{a^2}{2} \log_e \left\{ \frac{x + \sqrt{(x^2 - a^2)}}{a} \right\} + \frac{1}{2}x\sqrt{(x^2 - a^2)} + C. \tag{10.12}$$

Integrals of the form $\int R^{1/2}\,dx$ or $\int R^{-1/2}\,dx$ where $R = ax^2 + 2bx + c$ can be found by writing $R^{1/2}$ in one of the forms

$$a^{1/2}\left\{\left(x + \frac{b}{a}\right)^2 - \frac{b^2 - ac}{a^2}\right\}^{1/2} (a > 0),$$

or

$$(-a)^{1/2}\left\{\frac{b^2 - ac}{a^2} - \left(x + \frac{b}{a}\right)^2\right\}^{1/2} (a < 0),$$

and then using the appropriate standard integral. An integral of the form

$$\int \frac{(Ax + B)\,dx}{\{ax^2 + 2bx + c\}^{1/2}}$$

can be evaluated by first writing the numerator of the integrand in the form

$$\frac{A}{a}(ax + b) + B - \frac{Ab}{a}.$$

The integral then depends on the two integrals

$$\int \frac{(ax + b)\,dx}{\{ax^2 + 2bx + c\}^{1/2}} \quad \text{and} \quad \int \frac{dx}{\{ax^2 + 2bx + c\}^{1/2}};$$

the first of these is easily seen to be $\{ax^2 + 2bx + c\}^{1/2}$ and the second has already been discussed.

Some illustrative examples follow.

Example 7. *Integrate $\sqrt{(3x^2 + 4x - 7)}$ with respect to x.*

The required integral can be written in the form

$$\sqrt{3} \int \left\{ \left(x + \frac{2}{3}\right)^2 - \frac{25}{9} \right\}^{1/2} dx$$

and this, by (10.12), is

$$-\sqrt{3}\left(\frac{25}{18}\right) \log_e \left[\frac{\left(x + \frac{2}{3}\right) + \left\{\left(x + \frac{2}{3}\right)^2 - \frac{25}{9}\right\}^{1/2}}{(5/3)} \right] +$$

$$+ \frac{\sqrt{3}}{2}\left(x + \frac{2}{3}\right)\left\{\left(x + \frac{2}{3}\right)^2 - \frac{25}{9}\right\}^{1/2} + C.$$

After a little reduction, the final result is

$$\frac{1}{6}(3x + 2)\sqrt{(3x^2 + 4x - 7)} - \frac{25\sqrt{3}}{18} \log_e \left\{ \frac{3x + 2 + \sqrt{(9x^2 + 12x - 21)}}{5} \right\} + C.$$

Example 8. *Evaluate* $\int_{-1}^{1} \sqrt{\left(\dfrac{x+3}{x+1}\right)} dx$. (C.)

$$\int_{-1}^{1} \sqrt{\left(\dfrac{x+3}{x+1}\right)} dx = \int_{-1}^{1} \dfrac{x+3}{\sqrt{(x^2+4x+3)}} dx$$
$$= \int_{-1}^{1} \dfrac{\tfrac{1}{2}(2x+4)+1}{\sqrt{(x^2+4x+3)}} dx$$
$$= \dfrac{1}{2}\int_{-1}^{1} \dfrac{(2x+4)dx}{\sqrt{(x^2+4x+3)}} + \int_{-1}^{1} \dfrac{dx}{\sqrt{\{(x+2)^2-1\}}}$$
$$= \left[\sqrt{(x^2+4x+3)} + \log_e\left\{x+2+\sqrt{\{(x+2)^2-1\}}\right\}\right]_{-1}^{1}$$
$$= 2\sqrt{2} + \log_e(3+2\sqrt{2}).$$

Example 9. *Show that* $\int_0^{\pi/4} \sec^3\theta\, d\theta = \tfrac{1}{2}[\log_e(\sqrt{2}+1)+\sqrt{2}]$. (W.)

Writing $\tan\theta = t$ so that $\sec^2\theta\, d\theta = dt$,
$$\int_0^{\pi/4} \sec^3\theta\, d\theta = \int_0^1 \sqrt{(1+t^2)}\, dt$$
$$= \left[\tfrac{1}{2}\log_e\{t+\sqrt{(1+t^2)}\} + \tfrac{1}{2}t\sqrt{(1+t^2)}\right]_0^1$$
$$= \tfrac{1}{2}[\log_e(\sqrt{2}+1)+\sqrt{2}].$$

EXERCISES 10 (b)

1. Integrate $\cosh ax \cos bx$ with respect to x. (O.)

2. Prove that $\int_0^1 \dfrac{dx}{(1+e^x)(1+e^{-x})} = \dfrac{1}{2}\tanh\dfrac{1}{2}$. (N.U.)

3. Evaluate $\int_a^b \sqrt{\{(b-x)(x-a)\}}\,dx,\ 0 < a < b$. (C.)

4. Integrate with respect to x:
    (i) $\dfrac{1}{\sqrt{(x^2+2x+3)}}$,   (ii) $\dfrac{x^2+x+1}{\sqrt{(x^2+2x+3)}}$,
    (iii) $\dfrac{x^3+x^2+x+1}{\sqrt{(x^2+2x+3)}}$.

5. Evaluate $\int \dfrac{dx}{x+\sqrt{(x^2+1)}}$. (C.)

6. Integrate $x^{1/2}\sqrt{(1+x^3)}$ with respect to x. (C.)

10.3. The integration of $(a+b\cos x)^{-1}$ and similar functions

Since
$$a+b\cos x = a\left(\cos^2\dfrac{x}{2}+\sin^2\dfrac{x}{2}\right) + b\left(\cos^2\dfrac{x}{2}-\sin^2\dfrac{x}{2}\right)$$

INTEGRATION OF $(a + b\cos x)^{-1}$

$$= (a + b) \cos^2 \frac{x}{2} + (a - b) \sin^2 \frac{x}{2}$$

$$= (a - b) \cos^2 \frac{x}{2} \left(\frac{a + b}{a - b} + \tan^2 \frac{x}{2} \right),$$

it follows that

$$\int \frac{dx}{a + b \cos x} = \frac{2}{a - b} \int \frac{\frac{1}{2} \sec^2 \frac{x}{2} dx}{\frac{a + b}{a - b} + \tan^2 \frac{x}{2}}$$

$$= \frac{2}{a - b} \int \frac{d\left(\tan \frac{x}{2} \right)}{\frac{a + b}{a - b} + \tan^2 \frac{x}{2}}. \tag{10.13}$$

Since
$$\frac{a + b}{a - b} = \frac{a^2 - b^2}{(a - b)^2},$$

the first term in the denominator of the integrand is positive if $a^2 > b^2$ and, in this case,

$$\int \frac{dx}{a + b \cos x} = \frac{2}{a - b} \frac{1}{\sqrt{\left(\frac{a + b}{a - b} \right)}} \tan^{-1} \left\{ \frac{\tan \frac{x}{2}}{\sqrt{\left(\frac{a + b}{a - b} \right)}} \right\} + C$$

$$= \frac{2}{\sqrt{(a^2 - b^2)}} \tan^{-1} \left\{ \sqrt{\left(\frac{a - b}{a + b} \right)} \tan \frac{x}{2} \right\} + C. \tag{10.14}$$

The case in which $a^2 < b^2$ can be treated by rearranging (10.13) in the form

$$\int \frac{dx}{a + b \cos x} = \frac{2}{b - a} \int \frac{d\left(\tan \frac{x}{2} \right)}{\frac{b + a}{b - a} - \tan^2 \frac{x}{2}}.$$

$$= \frac{2}{b - a} \frac{1}{2\sqrt{\left(\frac{b + a}{b - a} \right)}} \log_e \left\{ \frac{\sqrt{\left(\frac{b + a}{b - a} \right)} + \tan \frac{x}{2}}{\sqrt{\left(\frac{b + a}{b - a} \right)} - \tan \frac{x}{2}} \right\} + C$$

$$= \frac{1}{\sqrt{(b^2 - a^2)}} \log_e \left\{ \frac{1 + \sqrt{\left(\frac{b - a}{b + a} \right)} \tan \frac{x}{2}}{1 - \sqrt{\left(\frac{b - a}{b + a} \right)} \tan \frac{x}{2}} \right\} + C$$

$$= \frac{2}{\sqrt{(b^2 - a^2)}} \tanh^{-1} \left\{ \sqrt{\left(\frac{b - a}{b + a} \right)} \tan \frac{x}{2} \right\} + C. \tag{10.15}$$

It remains to consider the case in which $a^2 = b^2$. The integral now takes one of the forms

$$\frac{1}{b}\int \frac{dx}{1 \pm \cos x},$$

i.e. $\quad \dfrac{1}{b}\int \dfrac{1}{2} \sec^2 \dfrac{x}{2} dx \quad$ or $\quad \dfrac{1}{b}\int \dfrac{1}{2} \operatorname{cosec}^2 \dfrac{x}{2} dx,$

and these are, respectively,

$$\frac{1}{b} \tan \frac{x}{2} + C \quad \text{and} \quad -\frac{1}{b} \cot \frac{x}{2} + C.$$

The integral with respect to x of

$$\frac{1}{a + b \cos x + c \sin x}$$

may be easily deduced, for

$$b \cos x + c \sin x = \sqrt{(b^2 + c^2)} \cos\left(x - \tan^{-1} \frac{c}{b}\right)$$

and the integral is, by this transformation, thrown into the form just considered. The integration of $(a + b \sin x)^{-1}$ can be effected in a similar way to that of $(a + b \cos x)^{-1}$ or it can be deduced from the previous results by writing $x = \dfrac{\pi}{2} + y$ so that

$$\int \frac{dx}{a + b \sin x} = \int \frac{dy}{a + b \cos y}.$$

The integration of $(a + b \cosh x)^{-1}$ and similar functions can also be treated similarly. Thus

$$\int \frac{dx}{a + b \cosh x} = \int \frac{dx}{a\left(\cosh^2 \dfrac{x}{2} - \sinh^2 \dfrac{x}{2}\right) + b\left(\cosh^2 \dfrac{x}{2} + \sinh^2 \dfrac{x}{2}\right)}$$

$$= \frac{2}{b-a} \int \frac{d\left(\tanh \dfrac{x}{2}\right)}{\dfrac{b+a}{b-a} + \tanh^2 \dfrac{x}{2}}.$$

This leads to the results

$$\int \frac{dx}{a + b \cosh x} = \frac{2}{\sqrt{(a^2 - b^2)}} \tanh^{-1}\left\{\sqrt{\left(\frac{a-b}{a+b}\right)} \tanh \frac{x}{2}\right\} (a^2 > b^2),$$
$$= \frac{2}{\sqrt{(b^2 - a^2)}} \tan^{-1}\left\{\sqrt{\left(\frac{b-a}{b+a}\right)} \tanh \frac{x}{2}\right\} (a^2 < b^2).$$

(10.16)

In working examples of these types of integral the reader is encouraged to work through the steps outlined above rather than to attempt to memorise formulae (10.14), (10.15), etc., and substitute numerical values for the constants.

Example 10. *Evaluate the definite integral* $\int_0^{\pi/2} \dfrac{dx}{2 + \cos x}$. (L.U.)

$$2 + \cos x = 2\left(\cos^2 \frac{x}{2} + \sin^2 \frac{x}{2}\right) + \cos^2 \frac{x}{2} - \sin^2 \frac{x}{2}$$

$$= 3\cos^2 \frac{x}{2} + \sin^2 \frac{x}{2} = \cos^2 \frac{x}{2}\left(3 + \tan^2 \frac{x}{2}\right).$$

Hence,

$$\int_0^{\pi/2} \frac{dx}{2 + \cos x} = 2\int_0^{\pi/2} \frac{\frac{1}{2}\sec^2 \frac{x}{2} dx}{3 + \tan^2 \frac{x}{2}} = 2\int_0^{\pi/2} \frac{d\left(\tan \frac{x}{2}\right)}{3 + \tan^2 \frac{x}{2}}$$

$$= \frac{2}{\sqrt{3}}\left[\tan^{-1}\left(\frac{1}{\sqrt{3}}\tan \frac{x}{2}\right)\right]_0^{\pi/2} = \frac{2}{\sqrt{3}}\tan^{-1}\left(\frac{1}{\sqrt{3}}\right)$$

$$= \frac{\pi}{3\sqrt{3}}.$$

Example 11. *Evaluate* $\int_0^{\pi/2} \dfrac{dx}{5 + 4\cos x + 3\sin x}$. (C.)

$$5 + 4\cos x + 3\sin x = 5\left(\cos^2 \frac{x}{2} + \sin^2 \frac{x}{2}\right) + 4\left(\cos^2 \frac{x}{2} - \sin^2 \frac{x}{2}\right) +$$
$$+ 6\sin \frac{x}{2}\cos \frac{x}{2}$$

$$= 9\cos^2 \frac{x}{2} + 6\sin \frac{x}{2}\cos \frac{x}{2} + \sin^2 \frac{x}{2}$$

$$= \left(3\cos \frac{x}{2} + \sin \frac{x}{2}\right)^2 = \cos^2 \frac{x}{2}\left(3 + \tan \frac{x}{2}\right)^2.$$

Hence,

$$\int_0^{\pi/2} \frac{dx}{5 + 4\cos x + 3\sin x} = 2\int_0^{\pi/2} \frac{\frac{1}{2}\sec^2 \frac{x}{2} dx}{\left(3 + \tan \frac{x}{2}\right)^2}$$

$$= 2\int_0^{\pi/2} \frac{d\left(\tan \frac{x}{2}\right)}{\left(3 + \tan \frac{x}{2}\right)^2}$$

$$= -2\left[\frac{1}{3 + \tan \frac{x}{2}}\right]_0^{\pi/2}$$

$$= -2\left(\frac{1}{4} - \frac{1}{3}\right) = \frac{1}{6}.$$

Example 12. *Evaluate* $\int_0^{\pi/2} \dfrac{dx}{1 + \sin a \sin x}$ *where* $0 < a < \dfrac{\pi}{2}$. (C.)

$$1 + \sin a \sin x = \cos^2 \frac{x}{2} + \sin^2 \frac{x}{2} + 2 \sin a \sin \frac{x}{2} \cos \frac{x}{2}$$
$$= \cos^2 \frac{x}{2}\left(1 + 2 \sin a \tan \frac{x}{2} + \tan^2 \frac{x}{2}\right).$$

Hence the given integral

$$= 2 \int_0^{\pi/2} \frac{d\left(\tan \frac{x}{2}\right)}{1 + 2 \sin a \tan \frac{x}{2} + \tan^2 \frac{x}{2}}$$

$$= 2 \int_0^{\pi/2} \frac{d\left(\tan \frac{x}{2}\right)}{1 - \sin^2 a + \left(\tan \frac{x}{2} + \sin a\right)^2} = 2 \int_0^{\pi/2} \frac{d\left(\tan \frac{x}{2}\right)}{\cos^2 a + \left(\tan \frac{x}{2} + \sin a\right)^2}$$

$$= \frac{2}{\cos a} \left[\tan^{-1}\left(\frac{\tan \frac{x}{2} + \sin a}{\cos a}\right) \right]_0^{\pi/2}$$

$$= \frac{2}{\cos a} \left\{ \tan^{-1}\left(\frac{1 + \sin a}{\cos a}\right) - \tan^{-1}(\tan a) \right\}.$$

Now,

$$\frac{1 + \sin a}{\cos a} = \frac{\cos^2 \frac{a}{2} + \sin^2 \frac{a}{2} + 2 \sin \frac{a}{2} \cos \frac{a}{2}}{\cos^2 \frac{a}{2} - \sin^2 \frac{a}{2}} = \frac{\left(\cos \frac{a}{2} + \sin \frac{a}{2}\right)^2}{\cos^2 \frac{a}{2} - \sin^2 \frac{a}{2}}$$

$$= \frac{\cos \frac{a}{2} + \sin \frac{a}{2}}{\cos \frac{a}{2} - \sin \frac{a}{2}} = \frac{1 + \tan \frac{a}{2}}{1 - \tan \frac{a}{2}} = \frac{\tan \frac{\pi}{4} + \tan \frac{a}{2}}{1 - \tan \frac{\pi}{4} \tan \frac{a}{2}}$$

$$= \tan\left(\frac{\pi}{4} + \frac{a}{2}\right).$$

Hence the value of the given integral reduces to

$$\frac{2}{\cos a}\left\{\frac{\pi}{4} + \frac{a}{2} - a\right\}, \text{ i.e. } \left(\frac{\pi}{2} - a\right) \sec a.$$

EXERCISES 10 (c)

1. Evaluate the definite integrals

    (i) $\int_0^{\pi/2} \dfrac{dx}{3 + 5 \cos x}$, (ii) $\int_0^\pi \dfrac{dx}{5 - 3 \cos x}$. (C.)

2. Evaluate

    (i) $\dfrac{1}{\pi}\int_0^\pi \dfrac{d\theta}{1 - \cos \theta \tanh a}$, (ii) $\dfrac{2}{\pi}\int_0^{\pi/2} \dfrac{d\theta}{1 - \cos^2 \theta \tanh^2 a}$. (O.)

3. Evaluate
$$\int \frac{d\theta}{(2 - \cos\theta)(3 - \cos\theta)}.$$ (O.)

4. Show that
$$\int \frac{\sqrt{2}\,dx}{2\sqrt{2} + \sin x + \cos x} = \frac{2}{\sqrt{3}} \tan^{-1}\left\{\frac{1}{\sqrt{3}} \tan\left(\frac{x}{2} - \frac{\pi}{8}\right)\right\} + C.$$

5. If $a^2 < 4bc$, show that
$$\int \frac{dx}{a + be^x + ce^{-x}} = \frac{2}{\sqrt{(4bc - a^2)}} \tan^{-1}\left\{\frac{2be^x + a}{\sqrt{(4bc - a^2)}}\right\} + C.$$

6. By means of the substitution
$$(1 + e\cos\theta)(1 - e\cos\phi) = 1 - e^2 \quad (e < 1),$$
transform the integral
$$I_n = \int (1 + e\cos\theta)^{-n} d\theta$$
into an integral involving ϕ.

Evaluate in terms of θ the integrals I_0, I_1 and I_2. (C.)

10.4. The integration of $1/(X\sqrt{Y})$ where X and Y are linear or quadratic functions

In this section the integral
$$\int \frac{dx}{X\sqrt{Y}}$$
where X and Y are linear or quadratic functions of x is discussed. The integration is straightforward when

(i) X and Y are both linear,
(ii) X is linear and Y quadratic,
(iii) X is quadratic and Y linear.

When X and Y are both quadratic, the integration can be performed but, except in a few special cases, it is not discussed here.

(i) *X and Y both linear.* Here the integral for discussion is
$$\int \frac{dx}{(ax + b)\sqrt{(Ax + B)}}$$
and an effective substitution is $y = \sqrt{(Ax + B)}$. With this substitution
$$dy = \frac{A\,dx}{2\sqrt{(Ax + B)}}$$
and
$$ax + b = \frac{a}{A}(y^2 - B) + b.$$

The integral therefore transforms into
$$2\int \frac{dy}{ay^2 - aB + bA}$$

and this is a standard form. The same substitution is effective also for the integral
$$\int \frac{f(x)dx}{(ax+b)\sqrt{(Ax+B)}},$$
where $f(x)$ is any rational integral algebraical function of x.

Example 13. *Evaluate* $\int_0^1 \frac{dx}{(1+x)\sqrt{(1+2x)}}.$ \hfill (N.U.)

Putting
$$y = \sqrt{(1+2x)},$$
$$dy = \frac{dx}{\sqrt{(1+2x)}}$$
and
$$1 + x = \tfrac{1}{2}(y^2+1).$$
When $x = 0, y = 1$ and when $x = 1, y = \sqrt{3}$, so that the integral transforms into
$$2\int_1^{\sqrt{3}} \frac{dy}{y^2+1} = 2\Big[\tan^{-1} y\Big]_1^{\sqrt{3}} = 2\Big[\frac{\pi}{3} - \frac{\pi}{4}\Big] = \frac{\pi}{6}.$$

(ii) *X linear, Y quadratic.* Here the general form of the integral is
$$\int \frac{dx}{(ax+b)\sqrt{(Ax^2+Bx+C)}}.$$
Putting $ax + b = 1/y$, logarithmic differentiation gives
$$\frac{adx}{ax+b} = -\frac{dy}{y},$$
and, after a little algebra,
$$Ax^2 + Bx + C = \frac{1}{a^2y^2}\{(b^2A - abB + a^2C)y^2 + (aB - 2bA)y + A\}^{1/2}.$$
The integral reduces therefore to the form
$$\int \frac{dy}{\sqrt{(py^2+qy+r)}}$$
where p, q and r are constants expressible in terms of a, b, A, B, C and such integrals have already been discussed. The integral
$$\int \frac{f(x)dx}{(ax+b)\sqrt{(Ax^2+Bx+C)}},$$
where $f(x)$ is any rational integral algebraical function of x can also be found. For, by ordinary division, $f(x)/(ax+b)$ can be expressed in the form
$$a_0x^n + a_1x^{n-1} + \ldots + a_{n-1}x + a_n + \frac{\lambda}{ax+b}$$
and the integral therefore reduces to a number of integrals of the form
$$\int \frac{x^r dx}{\sqrt{(Ax^2+Bx+C)}}$$
and one of the type discussed above.

Example 14. *Evaluate* $\int_0^1 \dfrac{dx}{(x+1)\sqrt{(x^2+1)}}$. (C.)

Putting $x + 1 = y^{-1}$, so that by logarithmic differentiation

$$\frac{dx}{x+1} = -\frac{dy}{y},$$

and

$$x^2 + 1 = \frac{2y^2 - 2y + 1}{y^2}.$$

When $x = 0$, $y = 1$ and when $x = 1$, $y = 1/2$, so that the integral becomes

$$\int_{\frac{1}{2}}^{1} \frac{dy}{\sqrt{(2y^2 - 2y + 1)}} = \frac{1}{\sqrt{2}} \int_{\frac{1}{2}}^{1} \frac{dy}{\sqrt{\left\{\left(y - \frac{1}{2}\right)^2 + \frac{1}{4}\right\}}}$$

$$= \frac{1}{\sqrt{2}} \left[\log_e \left\{ \frac{y - \frac{1}{2} + \sqrt{\left\{\left(y - \frac{1}{2}\right)^2 + \frac{1}{4}\right\}}}{\frac{1}{2}} \right\} \right]_{1/2}^{1} = \frac{1}{\sqrt{2}} \log_e (1 + \sqrt{2})$$

(iii) *X quadratic, Y linear.* The general form of the integral is now

$$\int \frac{dx}{(ax^2 + bx + c)\sqrt{(Ax + B)}}$$

and an effective substitution is $\sqrt{(Ax + B)} = y$. This leads to

$$\frac{dx}{\sqrt{(Ax+b)}} = \frac{2}{A} dy$$

and the expression $ax^2 + bx + c$ reduces to a quadratic in y^2. The integral is therefore transformed into

$$\int \frac{dy}{py^4 + qy^2 + r}$$

where p, q and r are constants depending on the constants in the original integral. This integral can be expressed, by the method of partial fractions, as the sum of two integrals of the form

$$\int \frac{(\lambda y + \mu) dy}{\alpha y^2 + \beta y + \gamma} \quad \text{and} \quad \int \frac{(\lambda' y + \mu') dy}{\alpha' y^2 + \beta' y + \gamma'}$$

and each of these integrals has already been discussed. The substitution $\sqrt{(Ax + B)} = y$ also permits evaluation of the integral

$$\int \frac{f(x) dx}{(ax^2 + bx + c)\sqrt{(Ax + B)}},$$

where $f(x)$ is a rational integral algebraical function, but the details are not pursued here.

Example 15. *Evaluate* $\int \frac{(2x-1)dx}{(x^2-x-2)\sqrt{(x+2)}}$. (C.)

Putting $\sqrt{(x+2)} = y$,

$$\frac{dx}{\sqrt{(x+2)}} = 2dy$$

and $x = y^2 - 2$. Hence

$$2x - 1 = 2y^2 - 5$$

and $x^2 - x - 2 = (y^2-2)^2 - (y^2-2) - 2 = y^4 - 5y^2 + 4$.

Hence the integral is equivalent to

$$2\int \frac{(2y^2-5)dy}{y^4-5y^2+4} = 2\int \left\{ \frac{1}{y^2-4} + \frac{1}{y^2-1} \right\} dy$$

$$= \frac{1}{2}\log_e \left(\frac{y-2}{y+2}\right) + \log_e \left(\frac{y-1}{y+1}\right) + C$$

$$= \log_e \left[\frac{\sqrt{\{\sqrt{(x+2)}-2\}\{\sqrt{(x+2)}-1\}}}{\sqrt{\{\sqrt{(x+2)}+2\}\{\sqrt{(x+2)}+1\}}} \right] + C$$

It is rather beyond the scope of the present work to deal fully with the general case in which both X and Y are quadratic. The effective substitution in this case is $\sqrt{(Y/X)} = y$ and the following example shows the method in a simple case.

Example 16. *Evaluate* $\int \frac{dx}{(1+x^2)\sqrt{(1-x^2)}}$. (O.)

Putting

$$\sqrt{\left(\frac{1-x^2}{1+x^2}\right)} = y,$$

it follows that

$$x^2 = \frac{1-y^2}{1+y^2}, \quad 1+x^2 = \frac{2}{1+y^2}, \quad 1-x^2 = \frac{2y^2}{1+y^2}.$$

It also follows, by logarithmic differentiation, that

$$2\frac{dx}{x} = \left(\frac{-2y}{1-y^2} - \frac{2y}{1+y^2}\right)dy,$$

giving

$$dx = \frac{-2y\,dy}{(1-y^2)^{1/2}(1+y^2)^{3/2}}.$$

After a little reduction, the integral is equivalent to

$$-\frac{1}{\sqrt{2}}\int \frac{dy}{\sqrt{(1-y^2)}} = -\frac{1}{\sqrt{2}}\sin^{-1} y + C = -\frac{1}{\sqrt{2}}\sin^{-1}\sqrt{\left(\frac{1-x^2}{1+x^2}\right)} + C.$$

EXERCISES 10 (d)

1. Integrate $(x+a)^{-1}(x+b)^{-1/2}$, where $a > b > 0$, with respect to x. (O.)

2. Evaluate $\int \frac{dx}{(x-a)\sqrt{(x^2+1)}}$. (C.)

3. If $0 < k < 1$, integrate with respect to t,

$$\frac{1}{(1-kt)\sqrt{(1-t^2)}}.$$ (C.)

4. Evaluate $\int_{1/2}^{X} \dfrac{dx}{x\sqrt{(5x^2 - 4x + 1)}}.$ (C.)

5. Evaluate $\int \dfrac{dx}{(x^2 + 2x + 2)\sqrt{(x+1)}}.$

6. Evaluate $\int \dfrac{dx}{(x^2 + b^2)\sqrt{(x^2 + a^2)}},\ a > b.$ (O.)

10.5. The integration of $e^{ax}\cos bx$ and $e^{ax}\sin bx$

The two integrals

$$C = \int e^{ax}\cos bx\, dx,\quad S = \int e^{ax}\sin bx\, dx$$

have important applications to many physical problems. They can be found by integrating by parts as follows:

$$C = \int e^{ax}\cos bx\, dx = e^{ax}\frac{\sin bx}{b} - \int a e^{ax}\frac{\sin bx}{b}dx$$

$$= \frac{e^{ax}\sin bx}{b} - \frac{a}{b}S,$$

$$S = \int e^{ax}\sin bx\, dx = -e^{ax}\frac{\cos bx}{b} + \int a e^{ax}\frac{\cos bx}{b}dx$$

$$= -\frac{e^{ax}\cos bx}{b} + \frac{a}{b}C.$$

Hence, apart from the arbitrary constant of integration,

$$bC + aS = e^{ax}\sin bx,$$
$$aC - bS = e^{ax}\cos bx.$$

Solving these equations for C and S,

$$\left.\begin{aligned}C = \int e^{ax}\cos bx\, dx &= \frac{e^{ax}(b\sin bx + a\cos bx)}{a^2 + b^2},\\ S = \int e^{ax}\sin bx\, dx &= \frac{e^{ax}(a\sin bx - b\cos bx)}{a^2 + b^2}.\end{aligned}\right\} \quad (10.17)$$

An alternative method of evaluating these integrals is to use the device employed in § 6.13 in the summation of certain trigonometrical series. Thus

$$C + iS = \int e^{ax}(\cos bx + i\sin bx)dx = \int e^{(a+ib)x}dx$$

$$= \frac{e^{(a+ib)x}}{a + ib} = \frac{e^{ax}(\cos bx + i\sin bx)}{a + ib}$$

$$= \frac{e^{ax}(a - ib)(\cos bx + i\sin bx)}{a^2 + b^2}$$

and the results (10.17) follow by equating real and imaginary parts.

10.6. Reduction formulae

Integrals which cannot be reduced to one of the standard integrals can, in some cases, be connected with another integral which can be so reduced. Examples have already occurred when using the method of integration by parts. Certain integrals involving powers of the variable or powers of functions of the variable can be related to integrals of the same form but containing *reduced* powers and such relations are called *reduction formulae*. Successive use of such formulae will often allow a given integral to be expressed in terms of a much simpler one.

Some of the more important formulae of this type are given below, others will be found in the exercises at the end of this section and at the end of this chapter.

(i) *Integral powers of tangent or cotangent.* If I_n denotes $\int \tan^n x\, dx$

$$I_n = \int \tan^n x\, dx = \int \tan^{n-2} x (\sec^2 x - 1)\, dx$$

$$= \int \tan^{n-2} x\, d(\tan x) - \int \tan^{n-2} x\, dx,$$

$$= \frac{\tan^{n-1} x}{n-1} - I_{n-2}. \qquad (10.18)$$

This formula relates I_n with I_{n-2}, and if n is a positive integer, successive use of it will ultimately relate I_n with either $\int \tan x\, dx$ or $\int \tan^2 x\, dx$.

Since, $\quad \int \tan x\, dx = \log_e \sec x,$

$$\int \tan^2 x\, dx = \int (\sec^2 x - 1)\, dx = \tan x - x,$$

any positive integral power of $\tan x$ can therefore be integrated.

If I'_n denotes $\int \cot^n x\, dx$, it can be shown similarly that

$$I'_n = -\frac{\cot^{n-1} x}{n-1} - I'_{n-2}, \qquad (10.19)$$

and hence the integration of positive integral powers of $\cot x$ is similarly possible.

(ii) *Integral powers of secant or cosecant.* Even positive integral powers of $\sec x$ are immediately integrable, since

$$\int \sec^{2n} x\,dx = \int \sec^{2n-2} x \sec^2 x\,dx$$

$$= \int (1 + \tan^2 x)^{n-1} d(\tan x)$$

$$= \tan x + {}^{n-1}C_1 \frac{\tan^3 x}{3} + {}^{n-1}C_2 \frac{\tan^5 x}{5} + \ldots + \frac{\tan^{2n-1} x}{2n-1}.$$

Odd positive integral powers of sec x can be integrated through a reduction formula by noting that

$$2n \sec^{2n+1} x - (2n-1) \sec^{2n-1} x = \frac{d}{dx}(\tan x \sec^{2n-1} x).$$

Hence, if I_{2n+1} denotes $\int \sec^{2n+1} x\,dx$,

$$I_{2n+1} = \frac{\tan x \sec^{2n-1} x}{2n} + \left(\frac{2n-1}{2n}\right) I_{2n-1}. \tag{10.20}$$

Successive use of this reduction formula enables I_{2n+1} to be related to the known integral $\int \sec x\,dx = \log_e \tan\left(\frac{\pi}{4} + \frac{x}{2}\right)$.

It is left as an exercise for the reader to obtain similar formulae relating to integrals of powers of cosec x.

(iii) *Integrals of the form* $\int \sin^m x \cos^n x\,dx$. Immediate integration of $\sin^m x \cos^n x$ is possible when either m or n is a positive odd integer.

For example,

$$\int \cos^2 x \sin^3 x\,dx = -\int \cos^2 x (1 - \cos^2 x)\,d(\cos x)$$

$$= -\tfrac{1}{3}\cos^3 x + \tfrac{1}{5}\cos^5 x + C.$$

When m and n are both positive even integers this method fails. Integration is possible by expressing $\sin^m x$, $\cos^n x$ in terms of sines and cosines of multiple angles and the device used in § 6.3 is useful in this connection. The method is, however, complicated when m or n is large and it is then preferable to use a reduction formula. Such a formula can be obtained by writing $P = \sin^{m+1} x \cos^{n-1} x$ so that

$$\frac{dP}{dx} = (m+1)\sin^m x \cos^n x - (n-1)\sin^{m+2} x \cos^{n-2} x$$

$$= (m+1)\sin^m x \cos^n x - (n-1)\sin^m x (1-\cos^2 x)\cos^{n-2} x$$

$$= (m+n)\sin^m x \cos^n x - (n-1)\sin^m x \cos^{n-2} x.$$

Hence, if $I_{m,n} = \int \sin^m x \cos^n x\,dx$,

$$(m+n)I_{m,n} = \sin^{m+1} x \cos^{n-1} x + (n-1)I_{m,n-2}. \tag{10.21}$$

By starting with $Q = \sin^{m-1} x \cos^{n+1} x$, it can be shown similarly that
$$(m+n)I_{m,n} = -\sin^{m-1} x \cos^{n+1} x + (m-1)I_{m-2,n}. \quad (10.22)$$
An alternative method of obtaining these two reduction formulae is to integrate $\sin^m x \cos^n x$ by parts and to make appropriate use of the identity $\cos^2 x + \sin^2 x = 1$.

The successive use of one of formulae (10.21), (10.22) will reduce the evaluation of $\int \sin^m x \cos^n x dx$ to that of one of the integrals $\int \sin^m x dx$, $\int \cos^n x dx$. If m and n are both positive even integers, $\int \sin^m x dx$ and $\int \cos^n x dx$ can be related to $\int 1.dx$ by successive use respectively of (10.22) with $n = 0$ and (10.21) with $m = 0$. By this means the integration of $\sin^m x \cos^n x$ can be completed.

Example 17. *Evaluate* $\int \sin^2 x \cos^6 x dx$

Writing $\sin x = s$, $\cos x = c$ and using (10.21) with $m = 2$, $n = 6$,
$$8 \int s^2 c^6 dx = s^3 c^5 + 5 \int s^2 c^4 dx.$$
Using (10.21) with $m = 2$, $n = 4$,
$$6 \int s^2 c^4 dx = s^3 c^3 + 3 \int s^2 c^2 dx,$$
and further use of the same formula with $m = 2$, $n = 2$ gives
$$4 \int s^2 c^2 dx = s^3 c + \int s^2 dx.$$
Eliminating $\int s^2 c^4 dx$, $\int s^2 c^2 dx$ from these relations
$$\int s^2 c^6 dx = \frac{1}{8} s^3 c^5 + \frac{5}{48} s^3 c^3 + \frac{5}{64} s^3 c + \frac{5}{64} \int s^2 dx.$$
The integral $\int s^2 dx$ can, of course, be found by writing $\sin^2 x = \frac{1}{2}(1 - \cos 2x)$.
Alternatively, carrying out the procedure outlined above and using (10.22) with $m = 2$, $n = 0$,
$$2 \int s^2 dx = -sc + \int 1.dx = x - sc + C.$$
The final result is therefore
$$\int s^2 c^6 dx = \frac{1}{8} s^3 c^5 + \frac{5}{48} s^3 c^3 + \frac{5}{64} s^3 c + \frac{5}{128}(x - sc) + C.$$

(iv) *Some important definite integrals.* Two important definite integrals which can be deduced from the reduction formulae (10.21), (10.22) are
$$\int_0^{\pi/2} \cos^n x dx \quad \text{and} \quad \int_0^{\pi/2} \sin^n x dx,$$
where n is a positive integer. Observing that, so long as $n > 1$,

REDUCTION FORMULAE

$\sin x \cos^{n-1} x$ vanishes both when $x = 0$ and when $x = \pi/2$, successive use of (10.21) with $m = 0$ gives

$$\int_0^{\pi/2} \cos^n x\, dx = \frac{n-1}{n} \int_0^{\pi/2} \cos^{n-2} x\, dx$$

$$= \frac{n-1}{n} \cdot \frac{n-3}{n-2} \int_0^{\pi/2} \cos^{n-4} x\, dx = \ldots.$$

Two cases arise according as n is even or odd. If n is even and equal to $2p$, the integral depends ultimately on $\int_0^{\pi/2} 1.dx$. The value of this integral being $\pi/2$,

$$\int_0^{\pi/2} \cos^{2p} x\, dx = \frac{2p-1}{2p} \cdot \frac{2p-3}{2p-2} \cdots \frac{5}{6} \cdot \frac{3}{4} \cdot \frac{1}{2} \cdot \frac{\pi}{2}.$$

If n is odd and equal to $2p + 1$, the ultimate integral to be evaluated is $\int_0^{\pi/2} \cos x\, dx$. Since the value of this integral is unity,

$$\int_0^{\pi/2} \cos^{2p+1} x\, dx = \frac{2p}{2p+1} \cdot \frac{2p-2}{2p-1} \cdots \frac{4}{5} \cdot \frac{2}{3}.$$

By writing $x = (\pi/2) - y$, it is clear that

$$\int_0^{\pi/2} \sin^n x\, dx = \int_{\pi/2}^0 \cos^n y(-dy) = \int_0^{\pi/2} \cos^n x\, dx,$$

and the final result, usually known as *Wallis' formula*, can be written

$$\int_0^{\pi/2} \cos^n x\, dx = \int_0^{\pi/2} \sin^n x\, dx = \frac{n-1}{n} \cdot \frac{n-3}{n-2} \cdots \frac{5}{6} \cdot \frac{3}{4} \cdot \frac{1}{2} \cdot \frac{\pi}{2}, \ (n \text{ even}),$$

$$= \frac{n-1}{n} \cdot \frac{n-3}{n-2} \cdots \frac{4}{5} \cdot \frac{2}{3}, \ (n \text{ odd}).$$

(10.23)

For the more general definite integral $\int_0^{\pi/2} \sin^m x \cos^n x\, dx$, where m and n are positive integers, (10.21) gives, since with $n > 1$, $\sin^{m+1} x \cos^{n-1} x$ vanishes at both the limits of integration,

$$\int_0^{\pi/2} \sin^m x \cos^n x\, dx = \frac{n-1}{m+n} \int_0^{\pi/2} \sin^m x \cos^{n-2} x\, dx. \quad (10.24)$$

Various cases again arise according as m and n are odd or even. If $m = 2p$, $n = 2q$, successive use of the above formula gives

$$\int_0^{\pi/2} \sin^{2p} x \cos^{2q} x\, dx = \frac{2q-1}{2p+2q} \cdot \frac{2q-3}{2p+2q-2} \cdots$$

$$\cdots \frac{3}{2p+4} \cdot \frac{1}{2p+2} \int_0^{\pi/2} \sin^{2p} x\, dx,$$

and substitution from (10.23) gives, after slight rearrangement,
$$\int_0^{\pi/2} \sin^{2p} x \cos^{2q} x\,dx = \frac{1.3.\ldots.(2p-1)1.3.\ldots.(2q-1)}{2.4.\ldots.2p(2p+2)(2p+4)\ldots(2p+2q)} \cdot \frac{\pi}{2}.$$

If $m = 2p$, $n = 2q + 1$, (10.24) gives
$$\int_0^{\pi/2} \sin^{2p} x \cos^{2q+1} x\,dx = \frac{2q}{2p+2q+1} \cdot \frac{2q-2}{2p+2q-1} \cdot \ldots$$
$$\ldots \cdot \frac{4}{2p+5} \cdot \frac{2}{2p+3} \int_0^{\pi/2} \sin^{2p} x \cos x\,dx.$$

Since
$$\int_0^{\pi/2} \sin^{2p} x \cos x\,dx = \left[\frac{\sin^{2p+1} x}{2p+1}\right]_0^{\pi/2} = \frac{1}{2p+1},$$
$$\int_0^{\pi/2} \sin^{2p} x \cos^{2q+1} x\,dx = \frac{2.4.\ldots.2q}{(2p+3)(2p+5)\ldots(2p+2q+1)} \cdot \frac{1}{(2p+1)}$$
$$= \frac{1.3.\ldots.(2p-1)2.4.\ldots.2q}{1.3.\ldots.(2p+1)(2p+3)(2p+5)\ldots(2p+2q+1)}.$$

Two similar formulae can be obtained when m is odd. All four results can be expressed in a single formula by means of the *Gamma function*. For present purposes this can be defined by
$$\Gamma(n+1) = n\Gamma(n),\ \Gamma(1) = 1,\ \Gamma(\tfrac{1}{2}) = \sqrt{\pi}. \qquad (10.25)$$

For all positive integral powers of m and n greater than unity, the value of the integral under discussion is then given by
$$\int_0^{\pi/2} \sin^m x \cos^n x\,dx = \frac{\Gamma\!\left(\dfrac{m+1}{2}\right)\Gamma\!\left(\dfrac{n+1}{2}\right)}{2\Gamma\!\left(\dfrac{m+n+2}{2}\right)}. \qquad (10.26)$$

Example 18. *Evaluate* $\int_0^{\pi/2} \sin^4 x \cos^6 x\,dx.$ (C.)

Using (10.26) the value of the given integral is
$$\frac{\Gamma\!\left(\dfrac{5}{2}\right)\Gamma\!\left(\dfrac{7}{2}\right)}{2\Gamma(6)}.$$

From (10.25)
$$\Gamma\!\left(\frac{5}{2}\right) = \frac{3}{2}\Gamma\!\left(\frac{3}{2}\right) = \frac{3}{2}\cdot\frac{1}{2}\Gamma\!\left(\frac{1}{2}\right) = \frac{3}{4}\sqrt{\pi},$$
$$\Gamma\!\left(\frac{7}{2}\right) = \frac{5}{2}\Gamma\!\left(\frac{5}{2}\right) = \frac{5}{2}\cdot\frac{3}{4}\sqrt{\pi} = \frac{15}{8}\sqrt{\pi},$$

and, when n is a positive integer, (10.25) shows that $\Gamma(n+1) = (n)!$, so that
$$\Gamma(6) = (5)! = 120.$$

Hence the required value is
$$\frac{\dfrac{3}{4}\sqrt{\pi} \times \dfrac{15}{8}\sqrt{\pi}}{2 \times 120} = \frac{3\pi}{512}.$$

(v) *The integral* $\int x^{m-1}(a+bx^n)^p dx$. If $a+bx^n = X$, the given integral is $\int x^{m-1}X^p dx$ and this can be related to any one of the following six integrals:

$$\int x^{m-1}X^{p-1}dx, \qquad \int x^{m-1}X^{p+1}dx,$$

$$\int x^{m-n-1}X^p dx, \qquad \int x^{m+n-1}X^p dx,$$

$$\int x^{m-n-1}X^{p+1}dx, \qquad \int x^{m+n-1}X^{p-1}dx.$$

To do this, let $P = x^{\lambda+1}X^{\mu+1}$ where λ and μ are the smaller indices of x and X respectively in the two expressions whose integrals are to be related. Find dP/dx and rearrange it as a linear function of the expressions whose integrals are to be related. The required reduction formula then follows on integration.

Example 19. *If* $I_n = \int_{-1}^{1}(1-x^2)^p dx$, *show that for* $p \geqslant 1$

$$I_p = \left(\frac{2p}{2p+1}\right)I_{p-1}.$$

Deduce that, when p is a positive integer,

$$I_p = \frac{2^{2p+1}(p!)^2}{(2p+1)!}. \tag{W.}$$

Take $P = x(1-x^2)^p$, so that

$$\frac{dP}{dx} = (1-x^2)^p - 2px^2(1-x^2)^{p-1}$$

$$= (2p+1)(1-x^2)^p - 2p(1-x^2)^{p-1}.$$

Hence,

$$x(1-x^2)^p = P = (2p+1)\int (1-x^2)^p dx - 2p\int (1-x^2)^{p-1}dx$$

and, since the left-hand side vanishes when $x = \pm 1$,

$$0 = (2p+1)\int_{-1}^{1}(1-x^2)^p dx - 2p\int_{-1}^{1}(1-x^2)^{p-1}dx.$$

Hence $\qquad I_p = \left(\dfrac{2p}{2p+1}\right)I_{p-1},$

and by successive use of this,

$$I_p = \frac{2p}{2p+1}\cdot\frac{2p-2}{2p-1}\cdot \ldots \cdot \frac{4}{5}\cdot\frac{2}{3}\cdot I_0.$$

Since $I_0 = \int_{-1}^{1} dx = 2$, this can be written

$$I_p = \frac{2^{2p+1}(p!)^2}{(2p+1)!}.$$

284 MATHEMATICAL ANALYSIS [10

(vi) *Some miscellaneous examples.* This section concludes with three further examples. Further examples will be found in Exercises 10 (*e*) and 10 (*h*).

Example 20. *Evaluate* $\int_0^1 (\log_e x)^n dx$, (*n* a positive integer). (O.)

Integrating by parts,

$$\int_0^1 (\log_e x)^n dx = \left[x(\log_e x)^n \right]_0^1 - \int_0^1 x \cdot n(\log_e x)^{n-1} \cdot \frac{1}{x} dx$$

$$= -n \int_0^1 (\log_e x)^{n-1} dx.$$

Since $\int_0^1 \log_e x \, dx = \left[x \log_e x \right]_0^1 - \int_0^1 x \cdot \frac{1}{x} dx = -1,$

successive use of the above reduction formula gives

$$\int_0^1 (\log_e x)^n dx = (-n)(-n+1) \ldots (-3)(-2)(-1) = (-1)^n(n)!$$

Example 21. *Find a reduction formula for* $\int x^n e^{x^2} dx$ *and show that*

$$\int_0^1 x^5 e^{x^2} dx = \frac{1}{2}(e - 2).$$ (C.)

Integrating by parts,

$$2 \int x^n e^{x^2} dx = \int x^{n-1} 2x e^{x^2} dx = x^{n-1} e^{x^2} - (n-1) \int x^{n-2} e^{x^2} dx,$$

so that the required reduction formula is

$$\int x^n e^{x^2} dx = \frac{1}{2} x^{n-1} e^{x^2} - \frac{1}{2}(n-1) \int x^{n-2} e^{x^2} dx.$$

Putting $n = 5$ and inserting the limits of integration,

$$\int_0^1 x^5 e^{x^2} dx = \left[\frac{x^4 e^{x^2}}{2} \right]_0^1 - 2 \int_0^1 x^3 e^{x^2} dx = \frac{1}{2} e - 2 \int_0^1 x^3 e^{x^2} dx.$$

But with $n = 3$, the reduction formula gives

$$2 \int_0^1 x^3 e^{x^2} dx = \left[x^2 e^{x^2} \right]_0^1 - 2 \int_0^1 x e^{x^2} dx = e - \left[e^{x^2} \right]_0^1 = 1.$$

Hence, $\int_0^1 x^5 e^{x^2} dx = \frac{1}{2} e - 1 = \frac{1}{2}(e - 2).$

Example 22. *Prove that*

$$\int \cos^m x \sin nx \, dx = -\frac{\cos^m x \cos nx}{m+n} + \frac{m}{m+n} \int \cos^{m-1} x \sin(n-1)x \, dx,$$

and deduce that

$$\int_0^{\pi/2} \cos^4 x \sin 6x \, dx = \frac{1}{5}.$$ (N.U.)

REDUCTION FORMULAE

Integrating by parts,

$$\int \cos^m x \sin nx\, dx = -\frac{\cos^m x \cos nx}{n} - \frac{m}{n}\int \cos^{m-1} x \sin x \cos nx\, dx$$

The left-hand side can be written as

$$\left(\frac{m+n}{n} - \frac{m}{n}\right)\int \cos^m x \sin nx\, dx$$
$$= \left(\frac{m+n}{n}\right)\int \cos^m x \sin nx\, dx - \frac{m}{n}\int \cos^{m-1} x \cos x \sin nx\, dx,$$

so that, after multiplication by n and a slight rearrangement

$$(m+n)\int \cos^m x \sin nx\, dx$$
$$= -\cos^m x \cos nx - m\int \cos^{m-1} x(\sin x \cos nx - \cos x \sin nx)\, dx$$
$$= -\cos^m x \cos nx + m\int \cos^{m-1} x \sin(n-1)x\, dx.$$

The required result then follows after division by $(m+n)$.
Inserting the limits of integration,

$$\int_0^{\pi/2} \cos^m x \sin nx\, dx$$
$$= \left[-\frac{\cos^m x \cos nx}{m+n}\right]_0^{\pi/2} + \frac{m}{m+n}\int_0^{\pi/2} \cos^{m-1} x \sin(n-1)x\, dx$$
$$= \frac{1}{m+n} + \frac{m}{m+n}\int_0^{\pi/2} \cos^{m-1} x \sin(n-1)x\, dx.$$

Taking appropriate values for m and n

$$\int_0^{\pi/2} \cos^4 x \sin 6x\, dx = \frac{1}{10} + \frac{4}{10}\int_0^{\pi/2} \cos^3 x \sin 5x\, dx,$$

$$\int_0^{\pi/2} \cos^3 x \sin 5x\, dx = \frac{1}{8} + \frac{3}{8}\int_0^{\pi/2} \cos^2 x \sin 4x\, dx,$$

$$\int_0^{\pi/2} \cos^2 x \sin 4x\, dx = \frac{1}{6} + \frac{2}{6}\int_0^{\pi/2} \cos x \sin 3x\, dx,$$

$$\int_0^{\pi/2} \cos x \sin 3x\, dx = \frac{1}{4} + \frac{1}{4}\int_0^{\pi/2} \sin 2x\, dx$$
$$= \frac{1}{4} - \frac{1}{8}\left[\cos 2x\right]_0^{\pi/2} = \frac{1}{2}.$$

Working backwards, it is easy to find successively that

$$\int_0^{\pi/2} \cos^2 x \sin 4x\, dx = \frac{1}{3}, \quad \int_0^{\pi/2} \cos^3 x \sin 5x\, dx = \frac{1}{4},$$

$$\int_0^{\pi/2} \cos^4 x \sin 6x\, dx = \frac{1}{5}.$$

EXERCISES 10 (e)

1. Prove that if $I_n = \int_0^{\pi/4} \tan^n \theta \, d\theta$,
$$I_n + I_{n-2} = \frac{1}{n-1} (n > 1).$$
Prove also that I_n lies between
$$\frac{1}{2(n-1)} \text{ and } \frac{1}{2(n+1)}, \text{ when } n > 1. \tag{O.}$$

2. Prove by means of a reduction formula that
$$\int_0^1 x^m (\log_e x)^n dx = \frac{(-1)^n n!}{(m+1)^{n+1}}. \tag{O.}$$

3. If $I_{m,n} = \int \sin^m x \cos^n x \, dx$, express $I_{m,n}$ in terms of $I_{m-2,n-2}$. (C.)

4. Prove that
$$(m+n)\int \cos^m x \cos nx \, dx - m\int \cos^{m-1} x \cos(n-1)x \, dx$$
$$= \cos^m x \sin nx.$$
Evaluate, when n is a positive integer
$$\int_0^{\pi/2} \cos^n x \cos nx \, dx. \tag{C.}$$

5. If for $q > 1$, $I_{p,q}$ denotes $\int_0^\pi e^{px} \sin^q x \, dx$, derive the reduction formula
$$(p^2 + q^2)I_{p,q} = q(q-1)I_{p,q-2}.$$
Hence show that for a positive even integral value of q
$$I_{p,q} = (q!)(e^{p\pi} - 1)/\{p(p^2 + 4)(p^2 + 16) \ldots (p^2 + q^2)\}. \tag{C.}$$

6. If n is a positive integer and if
$$I_n = \int_0^1 x^n (1-x)^{1/2} dx,$$
prove that $I_n = \left(\frac{2n}{2n+3}\right) I_{n-1}$,
and evaluate I_2. (L.I.C.)

7. If I_n denotes $\int_0^a x^n (a^2 - x^2)^{1/2} dx$, where $n > 1$, prove that
$$I_n = \left(\frac{n-1}{n+2}\right) a^2 I_{n-2}.$$
Hence, or otherwise, evaluate I_4. (C.)

8. By first finding the differential coefficient with respect to x of
$$P = \frac{\sin x}{(a + b \cos x)^{n-1}},$$
show that if $\quad U_n = \int_0^\pi \frac{dx}{(a + b \cos x)^n}(a > b),$
then
$$(n - 1)(a^2 - b^2)U_n - (2n - 3)aU_{n-1} + (n - 2)U_{n-2} = 0.$$
Hence, or otherwise, show that
$$U_2 = \frac{\pi a}{(a^2 - b^2)^{3/2}}. \qquad \text{(C.)}$$

10.7. Some general properties of the definite integral

There are certain general properties of the definite integral which are often useful in analysis. In establishing these for the definite integral $\int_a^b f(x)dx$, it is assumed that $f(x)$ is finite and continuous when $a \leqslant x \leqslant b$ and that $a < b$.

If $F(x)$ is a function such that
$$f(x) = \frac{d}{dx}F(x),$$
then
$$\int_a^b f(x)dx = F(b) - F(a) = -\{F(a) - F(b)\} = -\int_b^a f(x)dx. \quad (10.27)$$

It should be noted that the value of a definite integral is independent of the variable of integration for
$$\int_a^b f(y)dy = F(b) - F(a) = \int_a^b f(x)dx, \qquad (10.28)$$
and this fact is often of use in subsequent work.

Again,
$$\int_a^c f(x)dx + \int_c^b f(x)dx = F(c) - F(a) + F(b) - F(c)$$
$$= F(b) - F(a) = \int_a^b f(x)dx. \qquad (10.29)$$

It is more generally true and it can be shown similarly that
$$\int_a^c f(x)dx + \int_c^d f(x)dx + \ldots + \int_k^b f(x)dx = \int_a^b f(x)dx. \quad (10.30)$$

By writing $x = a + b - y$ in the definite integral $\int_a^b f(x)dx$, so that when $x = a$, $y = b$ and when $x = b$, $y = a$, it follows that, since $dx = -dy$,

$$\int_a^b f(x)dx = -\int_b^a f(a+b-y)dy = \int_a^b f(a+b-x)dx, \quad (10.31)$$

the last step following by (10.27) and (10.28). An important particular case occurs when $a = 0$, $b = a$, viz.,

$$\int_0^a f(x)dx = \int_0^a f(a-x)dx. \quad (10.32)$$

This relation is often helpful in evaluating certain trigonometrical integrals and two typical examples are given below.

Example 23. *Show that*

$$\int_0^\pi \frac{x \sin x}{1 + \cos^2 x}dx = \int_0^\pi \frac{(\pi - x) \sin x}{1 + \cos^2 x}dx = \frac{\pi^2}{4}. \quad \text{(O.C.)}$$

Taking $f(x) = (x \sin x)/(1 + \cos^2 x)$, $a = \pi$ in (10.32),

$$\int_0^\pi \frac{x \sin x}{1 + \cos^2 x}dx = \int_0^\pi \frac{(\pi - x) \sin (\pi - x)}{1 + \cos^2 (\pi - x)}dx = \int_0^\pi \frac{(\pi - x) \sin x}{1 + \cos^2 x}dx.$$

Hence,

$$2\int_0^\pi \frac{x \sin x}{1 + \cos^2 x}dx = \pi\int_0^\pi \frac{\sin x}{1 + \cos^2 x}dx = -\pi\int_0^\pi \frac{d(\cos x)}{1 + \cos^2 x}$$

$$= -\pi\Big[\tan^{-1}(\cos x)\Big]_0^\pi = -\pi\{\tan^{-1}(-1) - \tan^{-1}(1)\}$$

$$= -\pi\left(-\frac{\pi}{4} - \frac{\pi}{4}\right) = \frac{\pi^2}{2},$$

and the required result follows on division by 2.

Example 24. *Evaluate* $I = \int_0^{\pi/2} \log_e \sin x\, dx$.

From (10.32) with $a = \pi/2$,

$$I = \int_0^{\pi/2} \log_e \sin x\, dx = \int_0^{\pi/2} \log_e \sin\left(\frac{\pi}{2} - x\right)dx = \int_0^{\pi/2} \log_e \cos x\, dx.$$

Hence,

$$2I = \int_0^{\pi/2} \log_e \sin x\, dx + \int_0^{\pi/2} \log_e \cos x\, dx$$

$$= \int_0^{\pi/2} (\log_e \sin x + \log_e \cos x)dx = \int_0^{\pi/2} \log_e (\sin x \cos x)dx$$

$$= \int_0^{\pi/2} (\log_e \sin 2x - \log_e 2)dx$$

$$= \int_0^{\pi/2} \log_e \sin 2x\, dx - \frac{\pi}{2}\log_e 2. \quad (10.33)$$

By writing $2x = y$, it follows that

$$\int_0^{\pi/2} \log_e \sin 2x\, dx = \frac{1}{2}\int_0^\pi \log_e \sin y\, dy$$

$$= \frac{1}{2}\int_0^{\pi/2} \log_e \sin y\, dy + \frac{1}{2}\int_{\pi/2}^\pi \log_e \sin y\, dy.$$

By writing $y = \frac{\pi}{2} + z$ in the last integral on the right,

$$\int_{\pi/2}^{\pi} \log_e \sin y\, dy = \int_0^{\pi/2} \log_e \sin\left(\frac{\pi}{2} + z\right) dz = \int_0^{\pi/2} \log_e \cos z\, dz$$

$$= \int_0^{\pi/2} \log_e \sin y\, dy.$$

Hence $\quad\int_0^{\pi/2} \log_e \sin 2x\, dx = \int_0^{\pi/2} \log_e \sin y\, dy = I,$

and substitution in (10.33) gives

$$2I = I - \frac{\pi}{2} \log_e 2$$

so that $\quad I = \frac{\pi}{2} \log_e \frac{1}{2}.$

Now, from (10.29)

$$\int_0^{2a} f(x)\, dx = \int_0^a f(x)\, dx + \int_a^{2a} f(x)\, dx,$$

and, by writing $y = 2a - x$,

$$\int_a^{2a} f(x)\, dx = -\int_a^0 f(2a - y)\, dy = \int_0^a f(2a - x)\, dx.$$

Hence, $\quad\int_0^{2a} f(x)\, dx = \int_0^a f(x)\, dx + \int_0^a f(2a - x)\, dx. \qquad (10.34)$

If $f(x)$ is such that $f(2a - x) = f(x)$, it follows that

$$\int_0^{2a} f(x)\, dx = 2\int_0^a f(x)\, dx, \qquad (10.35)$$

while, if $f(2a - x) = -f(x)$,

$$\int_0^{2a} f(x)\, dx = 0. \qquad (10.36)$$

These last two results are often useful in evaluating trigonometrical integrals. For example, $\sin^n x = \sin^n (\pi - x)$ and hence (10.35) gives

$$\int_0^{\pi} \sin^n x\, dx = 2\int_0^{\pi/2} \sin^n x\, dx. \qquad (10.37)$$

Again, $\cos^{2n} x = \cos^{2n}(\pi - x)$ and $\cos^{2n+1} x = -\cos^{2n+1}(\pi - x)$ so that

$$\left. \begin{array}{l} \displaystyle\int_0^{\pi} \cos^{2n} x\, dx = 2\int_0^{\pi/2} \cos^{2n} x\, dx, \\[6pt] \displaystyle\int_0^{\pi} \cos^{2n+1} x\, dx = 0. \end{array} \right\} \qquad (10.38)$$

These results, together with Wallis' formulae (10.23) permit the numerical evaluation of the integrals $\displaystyle\int_0^{\pi} \sin^n x\, dx$ and $\displaystyle\int_0^{\pi} \cos^n x\, dx$.

From the definition of a definite integral as the limit of a sum it follows that *if $f(x) \geq 0$ when $a \leq x \leq b$, then*

$$\int_a^b f(x)dx \geq 0. \qquad (10.39)$$

If $H \leq f(x) \leq K$ when $a \leq x \leq b$, then by applying (10.39) to $f(x) - H$ and $K - f(x)$ respectively,

$$H(b-a) \leq \int_a^b f(x)dx \leq K(b-a). \qquad (10.40)$$

Taking H to be the least and K the greatest value of $f(x)$ in the interval (a, b), the integral is, by (10.40), equal to $\lambda(b-a)$ where λ lies between H and K. But as $f(x)$ is continuous there must be a value ξ for which $f(\xi) = \lambda$ (see § 8.3). Hence

$$\int_a^b f(x)dx = (b-a)f(\xi) \qquad (10.41)$$

where ξ lies between a and b; this result is often known as *the first mean value theorem for integrals*.

By applying (10.39) to the integrals $\int_a^b \{f(x) - H\}\phi(x)dx$, $\int_a^b \{K - f(x)\}\phi(x)dx$ and working similarly to the above, it is possible to establish the following *generalised mean value theorem for integrals: if $f(x)$, $\phi(x)$ are finite and continuous in the interval (a, b) and if $H \leq f(x) \leq K$ then, if $\phi(x)$ is positive*

$$H\int_a^b \phi(x)dx \leq \int_a^b f(x)\phi(x)dx \leq K\int_a^b \phi(x)dx, \qquad (10.42)$$

and

$$\int_a^b f(x)\phi(x)dx = f(\xi)\int_a^b \phi(x)dx \qquad (10.43)$$

where ξ lies between a and b. There is, of course, a corresponding result when $\phi(x)$ is negative.

A *second mean value theorem for integrals* can be established as follows. Let

$$\int_a^x f(t)dt = F(x)$$

so that $F'(x) = f(x)$ and $F(a) = 0$. Then, integrating by parts

$$\int_a^b f(x)\phi(x)dx = \int_a^b F'(x)\phi(x)dx$$
$$= \Big[F(x)\phi(x)\Big]_a^b - \int_a^b F(x)\phi'(x)dx$$
$$= F(b)\phi(b) - F(\xi)\int_a^b \phi'(x)dx, \ (a < \xi < b),$$

using (10.43) and assuming that $\phi(x)$ *has a differential coefficient of constant sign throughout the interval* (a, b). Since

$$\int_a^b \phi'(x)dx = \phi(b) - \phi(a),$$

this gives $\int_a^b f(x)\phi(x)dx = F(b)\phi(b) - \{\phi(b) - \phi(a)\}F(\xi).$

But $F(b) = \int_a^b f(x)dx,\ F(\xi) = \int_a^\xi f(x)dx,$ so that

$$\int_a^b f(x)\phi(x)dx = \phi(a)\int_a^\xi f(x)dx + \phi(b)\left\{\int_a^b f(x)dx - \int_a^\xi f(x)dx\right\}$$

$$= \phi(a)\int_a^\xi f(x)dx + \phi(b)\int_\xi^b f(x)dx, \qquad (10.44)$$

and this is the result required.

Example 25. *Defining* $\log_e t = \int_1^t \frac{du}{u}$, *prove that, for* $t > 0$, $\log_e t \leqslant t - 1$. (C.)

If $t > 1$, the first mean value theorem (10.41) gives

$$\log_e t = (t - 1)\frac{1}{\xi} \text{ where } 1 < \xi < t.$$

If $0 < t < 1$, the same theorem gives

$$\log_e t = (t - 1)\frac{1}{\xi} \text{ where } t < \xi < 1$$

and, in both cases, $\log_e t < t - 1$.

When $t = 1$, the definition shows that $\log_e t = 0$, so that for **all positive** t

$$\log_e t \leqslant t - 1.$$

EXERCISES 10 (*f*)

1. Use the result $\int_0^a f(x)dx = \int_0^a f(a - x)dx$ to evaluate

    (i) $\int_0^\pi \dfrac{xdx}{2 + \tan^2 x}$,

    (ii) $\int_0^\theta \log_e(1 + \tan\theta \tan x)dx,\ 0 < \theta < \dfrac{\pi}{2}.$ (C.)

2. If $f(x) \equiv f(a - x)$, prove that

$$2\int_0^a xf(x)dx = a\int_0^a f(x)dx.$$

Evaluate $\displaystyle\int_0^{\pi/2} \frac{x \sin x \cos x}{\sin^4 x + \cos^4 x}dx.$ (O.)

3. Prove that
$$\int_0^a f(x)dx = \frac{1}{2}\int_0^a \{f(x) + f(a-x)\}dx$$
and give a geometrical interpretation of this result.

Evaluate
$$\int_0^1 \frac{dx}{(x^2 - x + 1)(e^{2x-1} + 1)}.$$ (C.)

4. Show that
$$\int_0^{\pi/2} f(x)f\left(\frac{1}{2}\pi - x\right)dx = \frac{2}{\pi}\int_0^\pi f(x)dx$$
when $f(x) \equiv (\sin x)/x$. (O.)

5. If $f(x) \leqslant F(x)$ when $a \leqslant x \leqslant b$, prove that
$$\int_a^b f(x)dx \leqslant \int_a^b F(x)dx.$$

Prove that
$$\int_0^1 \frac{\sqrt{(1-x^2)}}{1+x^2}dx = \frac{\pi}{2}(\sqrt{2} - 1)$$
and hence prove that
$$\frac{\pi}{2}(\sqrt{2} - 1) < \int_0^1 \frac{\sqrt{(1-x^2)}}{1+x^4}dx < \frac{\pi}{4}.$$ (N.U.)

6. Given that $f_0(x) > 0$ for $x > 0$ and that
$$f_n(x) = \int_0^x f_{n-1}(u)du,$$
prove that
$$\frac{f_n(x)}{x^n} > \frac{f_{n+1}(x)}{x^{n+1}} \quad (n = 1, 2, 3, \ldots).$$

By repeated integration by parts, verify the formula
$$f_n(x) = \frac{1}{(n-1)!}\int_0^x (x-u)^{n-1}f_0(u)du,$$
for $n \geqslant 1$ and $x > 0$.

10.8. Infinite and improper integrals

Definite integrals in which the range of integration is infinite or in which the integrand becomes infinite at certain points in the range have not so far been considered in detail. Since any definite integral is the limit of a sum, a discussion of these types of integral involves the consideration of a double limit. This involves many difficulties and will not be fully gone into here. However, simple cases of these integrals occur in quite elementary problems and the discussion below should serve as a brief introduction.

(i) *Infinite integrals.* If $f(x)$ is continuous for $x \geqslant a$ and if
$$\lim_{b \to \infty} \left\{\int_a^b f(x)dx\right\}$$

exists and is equal to l, then the limit is written

$$\int_a^\infty f(x)dx$$

and is called an *infinite integral*; the integral is said to *converge* and to have the value l. On the other hand, if the above limit is infinite the integral is said to *diverge*; if neither of these alternatives occurs, the integral is said to *oscillate*.

As an example, consider the integral discussed in (10.17). Here

$$\int_0^X e^{ax} \cos bx\, dx = \left[\frac{e^{ax}(b \sin bx + a \cos bx)}{a^2 + b^2}\right]_0^X$$
$$= \frac{e^{aX}(b \sin bX + a \cos bX)}{a^2 + b^2} - \frac{a}{a^2 + b^2}. \quad (10.45)$$

If a is negative, e^{aX} tends to zero as X tends to infinity and the limit as $X \to \infty$ of $\int_0^X e^{ax} \cos bx\, dx$ is $-a/(a^2 + b^2)$. In this case, the integral converges and the result is written

$$\int_0^\infty e^{ax} \cos bx\, dx = -\frac{a}{a^2 + b^2} (a < 0). \quad (10.46)$$

Similarly
$$\int_0^\infty e^{ax} \sin bx\, dx = \frac{b}{a^2 + b^2} (a < 0). \quad (10.47)$$

When a is positive both integrals oscillate infinitely, for the first term on the right in (10.45) (and in the similar expression for the other integral) now tends to infinity as X tends to infinity.

Similar remarks apply to integrals in which the lower limit is minus infinity. Further, if both the infinite integrals $\int_a^\infty f(x)dx$ and $\int_{-\infty}^a f(x)dx$ converge, the integral $\int_{-\infty}^\infty f(x)dx$ is said to exist and its value is given by

$$\int_{-\infty}^\infty f(x)dx = \int_{-\infty}^a f(x)dx + \int_a^\infty f(x)dx,$$

for example,

$$\int_{-\infty}^0 \frac{dx}{1+x^2} = \int_0^\infty \frac{dx}{1+x^2} = \frac{1}{2}\int_{-\infty}^\infty \frac{dx}{1+x^2} = \frac{\pi}{2}.$$

(ii) *Improper integrals.* If $|f(x)|$ tends to infinity when $x \to a + 0$, the limit

$$\lim_{\delta \to +0} \left\{\int_{a+\delta}^b f(x)dx\right\}$$

may exist. In this case the limit is written

$$\int_a^b f(x)dx$$

and is called an *improper integral* (or an *infinite integral of the second kind*) and the integral is said to converge. Similar remarks apply to integrals in which the integrand becomes infinite at the upper limit. Thus, $\int_0^1 x^{-1/2}dx$ is an improper integral, since

$$\int_0^1 x^{-1/2}dx = \lim_{\delta \to +0}\left\{\int_\delta^1 x^{-1/2}dx\right\} = \lim_{\delta \to +0}\left[2x^{1/2}\right]_\delta^1 = \lim_{\delta \to +0}(2 - 2\delta^{1/2}) = 2.$$

On the other hand, $\int_0^1 x^{-2}dx$ does not exist (or diverges) for

$$\int_\delta^1 x^{-2}dx = \left[-x^{-1}\right]_\delta^1 = -1 + \frac{1}{\delta},$$

and this expression does not tend to a limit as $\delta \to +0$.

The above covers the cases in which the integrand is infinite at one of the limits of integration. It may well happen that $|f(x)| \to \infty$ as $x \to c$ where $a < c < b$ and the limit

$$\lim_{\delta \to +0}\left\{\int_a^{c-\delta} f(x)dx\right\} + \lim_{\delta' \to +0}\left\{\int_{c+\delta'}^b f(x)dx\right\}$$

may exist. If it does, the limit is again written as

$$\int_a^b f(x)dx$$

and this integral is again called an improper integral. It is possible, however, that the above limits do not exist for independent values of δ and δ' but that the limit

$$\lim_{\delta \to +0}\left\{\int_a^{c-\delta} f(x)dx + \int_{c+\delta}^b f(x)dx\right\}$$

does exist. In this case the limit is called *Cauchy's principal value* of $\int_a^b f(x)dx$ and this is often written $P\int_a^b f(x)dx$. The reader should have no difficulty in framing corresponding definitions to cover cases in which the integrand is infinite at more than one point in the range of integration. As an example of an improper integral of this type, consider $\int_{-1}^1 x^{-2/3}dx$. Since the integrand tends to infinity as x tends to zero, the limit to be discussed is

$$\lim_{\delta \to +0}\left\{\int_{-1}^{-\delta} x^{-2/3}dx\right\} + \lim_{\delta' \to +0}\left\{\int_{\delta'}^1 x^{-2/3}dx\right\}$$
$$= \lim_{\delta \to +0} 3(1 - \delta^{1/3}) + \lim_{\delta' \to +0} 3(1 - \delta'^{1/3}) = 6.$$

DIFFERENTIATION OF INTEGRALS

As a second example, consider $\int_0^2 \frac{dx}{1-x^2}dx$. Here the integrand tends to infinity as x tends to unity and the limit to be discussed is

$$\lim_{\delta \to +0}\left\{\int_0^{1-\delta} \frac{dx}{1-x^2}\right\} - \lim_{\delta' \to +0}\left\{\int_{1+\delta'}^2 \frac{dx}{x^2-1}\right\}$$

$$= \lim_{\delta \to +0}\left\{\frac{1}{2}\log_e\left(\frac{2-\delta}{\delta}\right)\right\} - \lim_{\delta' \to +0}\left\{\frac{1}{2}\log_e\frac{1}{3} - \frac{1}{2}\log_e\left(\frac{\delta'}{2+\delta'}\right)\right\}.$$

This limit exists only when $\delta = \delta'$ and it then becomes

$$\frac{1}{2}\log_e 3 + \lim_{\delta \to 0}\log_e\left(\frac{2-\delta}{2+\delta}\right) = \frac{1}{2}\log_e 3.$$

Hence $$P\int_0^2 \frac{dx}{1-x^2} = \frac{1}{2}\log_e 3.$$

10.9. Differentiation of definite integrals

The integrand of a definite integral often depends on variables other than the variable of integration. Suppose, for example, that $f(x, a)$ is a continuous function of two variables, then the definite integral of $f(x, a)$ with respect to x (assumed to exist) between given limits which may themselves contain a will be a function of a. Thus, let

$$F(a) = \int_a^b f(x, a)dx, \qquad (10.48)$$

and let $a + \delta a$, $b + \delta b$ be the values of a and b when a changes to $a + \delta a$. Then

$$F(a + \delta a) = \int_{a+\delta a}^{b+\delta b} f(x, a + \delta a)dx$$

and, by subtraction and some simple manipulation

$$F(a + \delta a) - F(a) = \int_a^b \{f(x, a + \delta a) - f(x, a)\}dx + $$
$$+ \int_b^{b+\delta b} f(x, a + \delta a)dx - \int_a^{a+\delta a} f(x, a + \delta a)dx. \qquad (10.49)$$

Since $f(x, a)$ is continuous with respect to δa, the first integral on the right-hand side of (10.49) can be written

$$\delta a \int_a^b \left\{f'_a(x, a) + \epsilon\right\}dx$$

where ϵ tends to zero with δa. Applying the first mean value theorem (10.41) to the other two integrals on the right of (10.49), these can be written as

$$\delta b . f(b + \theta \delta b, a + \delta a) \text{ and } \delta a . f(a + \theta' \delta a, a + \delta a)$$

where θ and θ' lie between 0 and 1. Hence (10.49) gives, after dividing by δa and proceeding to the limit as $\delta a \to 0$.

$$\frac{\partial}{\partial a}\int_a^b f(x, a)dx = \int_a^b \frac{\partial}{\partial a}f(x, a)dx + \frac{\partial b}{\partial a}\cdot f(b, a) - \frac{\partial a}{\partial a}\cdot f(a, a). \quad (10.50)$$

The last step in obtaining this general formula has, as the reader should have noticed, involved some further assumptions. Its validity depends on the continuity of $f(x, a)$, $f'_a(x, a)$ with respect to both x and a and also on the continuity of $\partial b/\partial a$ and $\partial a/\partial a$. An important special case arises when the limits a and b are independent of a, in which case

$$\frac{\partial}{\partial a}\int_a^b f(x, a)dx = \int_a^b \frac{\partial}{\partial a}f(x,a)dx. \quad (10.51)$$

These formulae are often of use in deducing values of definite integrals from those of other integrals. A typical example follows.

Example 26. Find the value of $\int_0^{\pi/2} \frac{d\theta}{x^2 \cos^2\theta + y^2 \sin^2\theta}$ and, by differentiation with respect to x and y, show that

$$\int_0^{\pi/2} \frac{d\theta}{(x^2 \cos^2\theta + y^2 \sin^2\theta)^2} = \frac{\pi}{4xy}\left(\frac{1}{x^2} + \frac{1}{y^2}\right). \quad \text{(O.)}$$

If the given integral is denoted by I,

$$I = \int_0^{\pi/2} \frac{d\theta}{x^2 \cos^2\theta + y^2 \sin^2\theta} = \int_0^{\pi/2} \frac{\sec^2\theta\, d\theta}{x^2 + y^2 \tan^2\theta}$$

$$= \frac{1}{y^2}\int_0^{\pi/2} \frac{d(\tan\theta)}{(x/y)^2 + \tan^2\theta} = \frac{1}{xy}\left[\tan^{-1}\left(\frac{y\tan\theta}{x}\right)\right]_0^{\pi/2} = \frac{\pi}{2xy}.$$

The limits of integration being independent of x, (10.51) gives

$$\frac{\partial I}{\partial x} = \int_0^{\pi/2} \frac{-2x\cos^2\theta\, d\theta}{(x^2\cos^2\theta + y^2\sin^2\theta)^2} = -2x\int_0^{\pi/2} \frac{\cos^2\theta\, d\theta}{(x^2\cos^2\theta + y^2\sin^2\theta)^2}.$$

But $I = \pi/(2xy)$, so that $\dfrac{\partial I}{\partial x} = -\dfrac{\pi}{2x^2 y}$,

and hence

$$\int_0^{\pi/2} \frac{\cos^2\theta\, d\theta}{(x^2\cos^2\theta + y^2\sin^2\theta)^2} = \frac{\pi}{4x^3 y}.$$

Similarly,

$$\int_0^{\pi/2} \frac{\sin^2\theta\, d\theta}{(x^2\cos^2\theta + y^2\sin^2\theta)^2} = \frac{\pi}{4xy^3}$$

and, by addition,

$$\int_0^{\pi/2} \frac{d\theta}{(x^2\cos^2\theta + y^2\sin^2\theta)^2} = \frac{\pi}{4xy}\left(\frac{1}{x^2} + \frac{1}{y^2}\right).$$

EXERCISES 10 (g)

1. Show that the improper integral $\int_{-1}^{1} x^{-1/3} dx$ is convergent and has the value zero.

2. If any of the following expressions are meaningless, explain why Evaluate each of the integrals which has a meaning:
$$\int_0^1 \frac{ax}{\sqrt{\{x(1-x)\}}}, \quad \int_{-4}^{-2} \frac{dx}{2x+1}, \quad \int_0^{3\pi/4} \tan\theta\, d\theta. \quad \text{(C.)}$$

3. Prove that $\int_{-1}^{1} x^{-2} dx$ is divergent and that the principal value of $\int_{-1}^{1} x^{-3} dx$ is zero.

4. Prove that $\int_{1}^{\infty} \frac{dx}{x^2 + 2x \cos a + 1} = \frac{a}{\sin a} \quad (0 < a < \pi).$ (C.)

5. Prove that $\int_0^{\pi/2} (\log_e \cos x) \cos x\, dx = \log_e 2 - 1.$ (O.)

6. Given that I_n denotes the integral
$$\int_0^{\pi/2} \frac{d\theta}{(x\cos^2\theta + y\sin^2\theta)^n}$$
in which x and y are both positive, prove that
$$\frac{\partial I_n}{\partial x} + \frac{\partial I_n}{\partial y} + n I_{n+1} = 0.$$
Show that $I_1 = \pi/\sqrt{(4xy)}$ and hence evaluate I_3. (O.C.)

7. If $I_{m,n} = \int_0^1 x^m (\log_e x)^n dx$, show that
$$\frac{\partial I_{m,n}}{\partial m} = I_{m,n+1}.$$
Show that $I_{m,0} = (m+1)^{-1}$ and deduce that
$$I_{m,n} = \frac{(-1)^n (n)!}{(m+1)^{n+1}}.$$

8. If $\pi y = \int_0^\pi \cos(x \sin\theta) d\theta$, show that
$$\frac{d^2 y}{dx^2} + \frac{1}{x}\frac{dy}{dx} + y = 0.$$

EXERCISES 10 (h)

1. If $I_r = \int_0^\infty \frac{x^r}{1+x^4} dx$, prove that $I_0 = I_2$ and $I_1 = \pi/4$.
By considering the value of $I_0 + I_2 - \sqrt{2} I_1$, or otherwise, prove that $I_2 = \pi/2\sqrt{2}.$ (O.)

2. Find $\displaystyle\int \frac{x^5 dx}{\sqrt{(a+bx^2)}}$. (C.)

3. Prove that $\displaystyle\int_1^2 \frac{(x^2+1)dx}{x^4+7x^2+1} = \frac{1}{3}\tan^{-1}\left(\frac{1}{2}\right)$. (C.)

4. Evaluate

   (i) $\displaystyle\int \sqrt{\left(\frac{a^2-x^2}{1-x^2}\right)} x\, dx,\ a>1$, (ii) $\displaystyle\int \log_e \left\{\frac{x(1+x^3)}{(1+x)(1+x^2)}\right\} dx$. (C.)

5. Show that $\displaystyle\int \frac{dx}{(x^2+a)^2 \sqrt{(x^2+b)}}$

   is rationalized by the substitution $1+bx^{-2}=y^2$ and evaluate the integral when $a=2$, $b=1$. (O.)

6. Evaluate $\displaystyle\int_1^2 \frac{17-24x+8x^2}{\{(2-x)(x-1)\}^{1/2}} dx$. (O.)

7. Evaluate $\displaystyle\int_0^\infty \frac{dx}{\{x+\sqrt{(1+x^2)}\}^n}$, where $n>1$. (O.C.)

8. Evaluate the definite integral

   $\displaystyle\int_0^{\pi/2} \sqrt{(\sin^6 \theta + \cos^6 \theta)} \sin 2\theta\, d\theta$. (O.)

9. Evaluate $\displaystyle\int_{\pi/4}^{\pi/2} \frac{dx}{\sin^2 x (\sin^2 x + 1)(\sin^2 x + 2)}$. (C.)

10. Prove that, if $a>1$

    $\displaystyle\int_1^\infty \frac{dx}{(1+x)\sqrt{(1+2ax+x^2)}} = \frac{1}{\sqrt{(2a-2)}} \sin^{-1}\sqrt{\left(\frac{a-1}{a+1}\right)}$.

    Show also that the limiting value of this result as a tends to unity is equal to the value of the integral when $a=1$. (O.)

11. Prove that $\displaystyle\int_0^\pi \frac{1-a\cos\theta}{1-2a\cos\theta+a^2} d\theta = \left.\begin{matrix}\pi \text{ if } |a|<1, \\ 0 \text{ if } |a|>1.\end{matrix}\right\}$ (N.U.)

12. (a) Evaluate

    $\displaystyle\int_0^a \frac{d\theta}{5+3\cos\theta}$ when (i) $a=\pi$. (ii) $a=2\pi$.

    (b) Evaluate $\displaystyle\int_0^\pi \frac{d\theta}{5+3\cos^2\theta}$. (N.U.)

13. Prove that, if $0<a<\pi$, then

    $\displaystyle\int_0^{\pi/2} \frac{d\theta}{1+\cos a \cos\theta} = \frac{a}{\sin a}$.

    What is the value of the integral when $\pi<a<2\pi$? (C.)

14. If $I_p = \int (x^2 + a)^p dx$, show that
$$(2p + 1)I_p - 2paI_{p-1} = x(x^2 + a)^p,$$
and hence, or otherwise, evaluate
$$\int_0^\infty \frac{dx}{(x^2 + a^2)^{3/2}}. \qquad \text{(C.)}$$

15. If $u_n = \int_0^{\pi/2} \sin^n x\, dx$ and n is a positive integer, prove that
$$nu_n u_{n-1} = \frac{1}{2}\pi,$$
and
$$0 < u_n < u_{n-1}.$$
Hence, or otherwise, prove that
$$nu_n^2 \to \frac{1}{2}\pi \text{ as } n \to \infty. \qquad \text{(C.)}$$

16. Obtain a formula of reduction for
$$I_{m,n} = \int \frac{\sin^m \theta}{\cos^n \theta} d\theta,$$
where m and n are positive and greater than 2.
Show that
$$\int_0^{\pi/4} \frac{\sin^6 \theta}{\cos^4 \theta} d\theta = \frac{5\pi}{8} - \frac{23}{12}. \qquad \text{(C.)}$$

17. Prove that if $I_m = \int_0^\infty x^m e^{-x} \sin x\, dx$ and m is a positive integer or zero, then
$$I_{m+2} - (m + 2)I_{m+1} + \tfrac{1}{2}(m + 2)(m + 1)I_m = 0.$$
Prove that this relation is satisfied by
$$I_m = 2^{-m}(m)!\{A(1 + i)^m + B(1 - i)^m\},$$
where A and B are arbitrary constants (which may be complex) and $i = \sqrt{(-1)}$.
Determine A and B by evaluating I_0 and I_1. (O.C.)

18. Prove that, if I_n denotes $\int_0^{\pi/2} \cos^n x \sin^3 x\, dx$,
$$I_n = \frac{2}{(n + 1)(n + 3)}.$$
For what values of n is this result true? Give reasons for your answer. Deduce that, when p is a positive integer
$$I_n + I_{n+2} + \ldots + I_{n+2p} = \frac{1}{n + 1} - \frac{1}{n + 2p + 3} \qquad \text{(O.C.)}$$

19. If $Q \equiv ax^2 + 2bx + c$, and
$$I_n = \int \frac{dx}{Q^{n+1}},$$
show, by differentiating $(Ax + B)/Q^n$ (where A, B are adjustable constants), or otherwise, that
$$2n(ac - b^2)I_n = \frac{ax+b}{Q^n} + (2n-1)aI_{n-1}.$$
Obtain a similar reduction formula for
$$J_n = \int \frac{x\,dx}{Q^{n+1}}.$$
Evaluate $\displaystyle\int_0^1 \frac{dx}{(x^2 - x + 1)^{1/2}}.$ \hfill (C.)

20. If $I_{m,n}$ denotes $\displaystyle\int_0^1 x^m(1-x)^n dx$ where $m > -1$ and $n > 0$, prove that
$$I_{m\,n} = \frac{n}{m+n+1} I_{m\,n-1},$$
and deduce that, if m and n are positive integers,
$$\frac{1}{m+1} - \frac{{}^nC_1}{m+2} + \frac{{}^nC_2}{m+3} - \cdots + \frac{(-1)^r {}^nC_r}{m+r+1} + \cdots$$
$$\cdots + \frac{(-1)^n}{m+n+1} = \frac{(m)!(n)!}{(m+n+1)!}. \quad \text{(C.)}$$

21. If
$$u_n = \int^\infty \frac{\theta\,d\theta}{\cosh^n \theta} \quad (n > 2),$$
prove that
$$u_n = \left(\frac{n-2}{n-1}\right) u_{n-2} - \frac{1}{(n-1)(n-2)}. \quad \text{(C.)}$$

22. Prove that for odd values of n,
$$\int_0^\pi \frac{\cos n\theta}{\cos \theta} d\theta = (-1)^{\frac{n-1}{2}} \pi.$$
If, for odd values of n, $I_n = \displaystyle\int_0^\pi \frac{\cos^2 n\theta}{\cos^2 \theta} d\theta$,
show that, $\quad I_n = I_{n-2} + 2\pi = n\pi.$ \hfill (C.)

23. If $y_r(x)$ satisfies the equation
$$\frac{d}{dx}\left\{(1-x^2)\frac{dy_r}{dx}\right\} + r(r+1)y_r = 0,$$
show that, if $m \neq n$, then
$$\int_{-1}^1 y_m(x) y_n(x) dx = 0. \quad \text{(C.)}$$

24. Prove that, if λ_r, λ_s are distinct roots of the equation
$$\lambda \cot \lambda l = C \text{ where } C \text{ is constant,}$$
$$\int_0^l \sin \lambda_r x \sin \lambda_s x\, dx = 0.$$
Given that
$$\phi(x) = \sum_{r=1}^n a_r \sin \lambda_r x,$$
where λ_r are the positive roots of the equation $\lambda \cot \lambda l = C$ arranged in ascending order and the a_r are constant, evaluate
$$\int_0^l \{\phi(x)\}^2 dx \qquad \text{(O.)}$$

25. Polynomials $f_0(x)$, $f_1(x)$, $f_2(x)$, ... are defined by the relation
$$f_n(x) = \frac{d^n}{dx^n}(x^2 - 1)^n$$
Prove that,
$$\int_{-1}^1 f_n(x) f_m(x)\, dx = 0$$
if $m \neq n$ and that
$$\int_{-1}^1 \{f_n(x)\}^2 dx = \frac{(n!)^2}{2n+1} 2^{2n+1}.$$
Show that, if $\phi(x)$ is any polynomial of degree m,
$$\phi(x) = \sum_{n=0}^m a_n f_n(x)$$
where
$$a_n = \frac{2n+1}{2^{2n+1}(n!)^2} \int_{-1}^1 \phi(x) f_n(x)\, dx. \qquad \text{(C.)}$$

26. If $f(x) = f(a - x)$, prove that
$$\int_0^a f(x)\, dx = 2 \int_0^{a/2} f(x)\, dx \text{ and } 2 \int_0^a x f(x)\, dx = a \int_0^a f(x)\, dx.$$
Evaluate
$$\int_0^\pi \frac{x \sin^3 x}{1 + \cos^2 x}\, dx. \qquad \text{(O.)}$$

27. Use the result $\int_0^a f(x)\, dx = \int_0^a f(a-x)\, dx$ and the substitution $x = \sin \theta$ to evaluate
$$\int_0^1 \frac{dx}{x + \sqrt{(1 - x^2)}}. \qquad \text{(C.)}$$

28. By means of the substitutions $x \pm x^{-1} = t$, or otherwise, effect the integrations
$$\int \frac{x^2 + 1}{(x^4 + x^2 + 1)^2} dx, \int \frac{x^2 - 1}{(x^4 + x^2 + 1)^2} dx, \int \frac{dx}{x^4 + x^2 + 1}. \qquad \text{(O.)}$$

29. Verify that
$$y = \int_0^x (x-t) f(t) dt$$
satisfies the differential equation
$$\frac{d^2y}{dx^2} = f(x)$$
with the initial conditions $y = dy/dx = 0$ when $x = 0$.

30. Prove that
$$\theta = \int_0^{\frac{x}{2a\sqrt{t}}} e^{-z^2} dz$$
satisfies the equation
$$\frac{\partial \theta}{\partial t} = a^2 \frac{\partial^2 \theta}{\partial x^2}$$
where a is a constant.

CHAPTER 11

SOME FURTHER GEOMETRICAL APPLICATIONS OF THE CALCULUS

11.1. Introduction

It is the purpose of this chapter to discuss some further geometrical applications of the differential and integral calculus; topics which have been treated already in *Advanced Level Pure Mathematics* will be extended and new topics introduced. To save space, where back references to *Advanced Level Pure Mathematics* are necessary, these will be indicated by the initials *A.L.P.M.* followed by a page number. While not pretending to be exhaustive, this chapter, together with the appropriate sections of the earlier book, should serve as an introduction to a study of plane curves.

11.2. Tangents and normals

The equation to the tangent at the point (x_1, y_1) to the curve $y = f(x)$ is, since the slope of the tangent is the value of the derivative at the point in question,

$$y - y_1 = \left(\frac{dy}{dx}\right)_{x=x_1}(x - x_1). \tag{11.1}$$

Since the normal at the same point is perpendicular to the tangent, its equation is

$$x - x_1 + \left(\frac{dy}{dx}\right)_{x=x_1}(y - y_1) = 0. \tag{11.2}$$

These equations are easily written down for curves whose equations are given in the explicit form $y = f(x)$. If, however, the curve is given by the implicit equation $f(x, y) = 0$, since

$$\frac{dy}{dx} = -\frac{\partial f}{\partial x}\bigg/\frac{\partial f}{\partial y},$$

more convenient forms for the tangent and normal are respectively

$$(x - x_1)\frac{\partial f}{\partial x} + (y - y_1)\frac{\partial f}{\partial y} = 0 \tag{11.3}$$

and

$$\frac{x - x_1}{\partial f/\partial x} = \frac{y - y_1}{\partial f/\partial y}, \tag{11.4}$$

where the partial derivatives are in each case evaluated at $x = x_1$, $y = y_1$.

Example 1. *Show that the curve*
$$y^4 - 4axy^2 + 3a^2x^2 - x^4 = 0$$
has tangents parallel to the axis of x at points for which $x^4 = 3a^4/4$.   (C.)

Here $f(x, y) \equiv y^4 - 4axy^2 + 3a^2x^2 - x^4$ and the curve has tangents parallel to the axis of x (i.e. with zero slope) where

$$\frac{\partial f}{\partial x} = -4ay^2 + 6a^2x - 4x^3 = 0.$$

This gives
$$y^2 = \frac{3ax}{2} - \frac{x^3}{a}$$

and substituting in the equation to the curve, written in the form $y^2(y^2 - 4ax) + 3a^2x^2 - x^4 = 0$,

$$\left(\frac{3ax}{2} - \frac{x^3}{a}\right)\left(-\frac{5ax}{2} - \frac{x^3}{a}\right) + 3a^2x^2 - x^4 = 0.$$

This reduces to
$$-\frac{3}{4}a^2x^2 + \frac{x^6}{a^2} = 0,$$

and this is satisfied when $x^4 = 3a^4/4$.

If the equation to the curve is given parametrically by the equations
$$x = f(t), y = g(t),$$
the slope of the tangent is given by
$$\frac{dy}{dx} = \frac{dy}{dt} \bigg/ \frac{dx}{dt} = \frac{g'(t)}{f'(t)}.$$

The equation to the tangent at the point 't' is therefore
$$y - g(t) = \frac{g'(t)}{f'(t)}\{x - f(t)\}$$

and this can be written
$$xg'(t) - yf'(t) = f(t)g'(t) - g(t)f'(t). \quad (11.5)$$

Similarly, the corresponding normal is
$$xf'(t) + yg'(t) = f(t)f'(t) + g(t)g'(t). \quad (11.6)$$

Example 2. *P is the point 't' on the curve $x = a(t - \sin t)$, $y = a(1 - \cos t)$. Find the equation to the normal to the curve at P and show that it passes through the point Q whose coordinates are $a(t + \sin t)$, $-a(1 - \cos t)$.*   (O.C.)

Here $f(t) = a(t - \sin t)$, $g(t) = a(1 - \cos t)$, $f'(t) = a(1 - \cos t)$, $g'(t) = a \sin t$. Equation (11.6) gives as the equation to the normal
$$xa(1 - \cos t) + ya \sin t = a(t - \sin t)a(1 - \cos t) + a(1 - \cos t)a \sin t;$$
this reduces to
$$x(1 - \cos t) + y \sin t = at(1 - \cos t)$$

and it is a simple matter to verify that this equation is satisfied by $x = a(t + \sin t)$, $y = -a(1 - \cos t)$.

TANGENTS AND NORMALS

It has been shown (*A.L.P.M.*, p. 229) that the length of arc (s) of the curve $y = f(x)$ measured from a point of abscissa a to one of abscissa x is given by

$$s = \int_a^x \sqrt{\left\{1 + \left(\frac{dy}{dx}\right)^2\right\}}\, dx.$$

It follows that
$$\frac{ds}{dx} = \sqrt{\left\{1 + \left(\frac{dy}{dx}\right)^2\right\}}. \tag{11.7}$$

If ψ is the angle (Fig. 35) between the tangent to the curve at the point (x, y) and the axis of x, $\tan \psi = dy/dx$ and hence

$$\frac{ds}{dx} = \sqrt{\{1 + \tan^2 \psi\}} = \sec \psi.$$

Thus $\cos \psi = dx/ds$ and $\sin \psi = \tan \psi \cos \psi = (dy/dx)(dx/ds) = dy/ds$. Hence in a given curve, x, y, s and the angle ψ are connected by the following three important relations

$$\sin \psi = \frac{dy}{ds}, \quad \cos \psi = \frac{dx}{ds}, \quad \tan \psi = \frac{dy}{dx}. \tag{11.8}$$

The relation between the length of arc (s) of a curve, measured from a given fixed point on the curve, and the angle (ψ) between the tangents at its extremities is called the *intrinsic equation* to the curve. If the equation to the curve is given in the form $y = f(x)$, if the axis of x is a tangent at the origin and if the length of arc is measured from the origin,
$$\tan \psi = f'(x)$$
and
$$\frac{ds}{dx} = \sqrt{[1 + \{f'(x)\}^2]}.$$

If s be determined by integration from the second relation and if x be then eliminated between the result of the integration and the above equation for $\tan \psi$, the required relation between s and ψ will be obtained. For a curve given parametrically by the equations $x = f(t)$, $y = g(t)$ the intrinsic equation can be determined in a similar way from the relations

$$\tan \psi = \frac{dy}{dt} \bigg/ \frac{dx}{dt} = \frac{g'(t)}{f'(t)},$$

$$\frac{ds}{dt} = \frac{dx}{dt}\frac{ds}{dx} = f'(t)[1 + \{g'(t)/f'(t)\}^2]^{1/2} = [\{f'(t)\}^2 + \{g'(t)\}^2]^{1/2}.$$

Example 3. *Find the intrinsic equation to the curve $y + 1 = \cosh x$.*

Here $\tan \psi = \sinh x$ and
$$\frac{ds}{dx} = \sqrt{\{1 + \sinh^2 x\}} = \cosh x.$$

Hence $s = \sinh x$, the constant of integration being chosen so that x and s vanish together.
The required intrinsic equation is therefore $s = \tan \psi$.

Relations can be found, similar to those of (11.8), giving the trigonometrical ratios of the angle ϕ between the tangent to a curve and the radius vector to its point of contact. Thus in Fig. 35, PT is the tangent at P to a curve given in polar coordinates by the equation

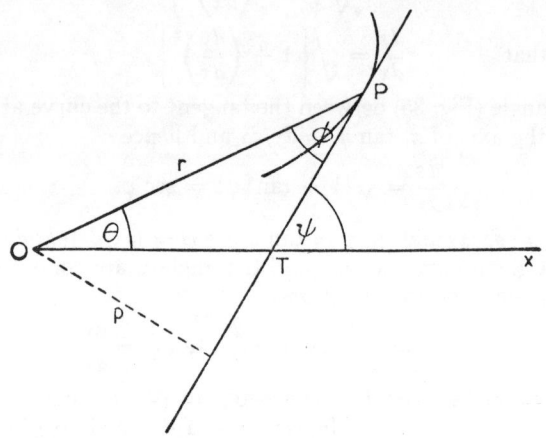

Fig. 35

$r = f(\theta)$: O is the origin or pole, Ox the axis of x, $OP(= r)$ is the radius vector and θ, ψ are respectively the angles made by OP and PT with Ox. It is clear from the diagram that $\phi = \psi - \theta$ and from the relations between cartesian and polar coordinates, $\sin \theta = y/r$, $\cos \theta = x/r$. Hence, using these relations and (11.8),

$$\sin \phi = \sin(\psi - \theta) = \sin \psi \cos \theta - \cos \psi \sin \theta$$
$$= \frac{dy}{ds}\frac{x}{r} - \frac{dx}{ds}\frac{y}{r} = \frac{x^2}{r}\frac{d}{ds}\left(\frac{y}{x}\right) = r \cos^2 \theta \frac{d}{ds}(\tan \theta) = r\frac{d\theta}{ds}, \quad (11.9)$$

$$\cos \phi = \cos(\psi - \theta) = \cos \psi \cos \theta + \sin \psi \sin \theta$$
$$= \frac{dx}{ds}\frac{x}{r} + \frac{dy}{ds}\frac{y}{r} = \frac{1}{2r}\frac{d}{ds}(x^2 + y^2) = \frac{1}{2r}\frac{d}{ds}(r^2) = \frac{dr}{ds}, \quad (11.10)$$

and, by division, $$\tan \phi = r\frac{d\theta}{dr}. \quad (11.11)$$

Since $\operatorname{cosec}^2 \phi = 1 + \cot^2 \phi$, (11.9) and (11.11) show that

$$\frac{1}{r^2}\left(\frac{ds}{d\theta}\right)^2 = 1 + \frac{1}{r^2}\left(\frac{dr}{d\theta}\right)^2,$$

giving $$\frac{ds}{d\theta} = \sqrt{\left\{r^2 + \left(\frac{dr}{d\theta}\right)^2\right\}} \quad (11.12)$$

as the formula corresponding to (11.7) when polar coordinates are involved.

TANGENTS AND NORMALS

If a perpendicular of length p is dropped from the pole O on to the tangent PT (Fig. 35) it is clear that

$$p = r \sin \phi = r^2 \frac{d\theta}{ds},$$

using (11.9). With the aid of (11.12) this can be written

$$\frac{1}{p^2} = \frac{1}{r^2} + \frac{1}{r^4}\left(\frac{dr}{d\theta}\right)^2. \tag{11.13}$$

For a curve whose polar equation is known, $dr/d\theta$ can, theoretically at least, be expressed in terms of r and (11.13) will yield a relation between p and r. A relation of this type is often a very convenient way of defining a curve: it is sometimes known as the (p, r) or *pedal* equation to the curve. Two examples follow: in the first, the pedal equation is deduced from the polar equation and in the second, it is obtained from the cartesian equation to the curve.

Example 4. *Find the (p, r) equations to the curves:*

(i) $r^2 = a^2 \sin 2\theta$,   (ii) $y^2 = 4a(x + a)$.

(i) From the equation to the curve,

$$r\frac{dr}{d\theta} = a^2 \cos 2\theta,$$

and substitution in (11.13) gives

$$\frac{1}{p^2} = \frac{1}{r^2} + \frac{a^4 \cos^2 2\theta}{r^6} = \frac{r^4 + a^4(1 - \sin^2 2\theta)}{r^6} = \frac{a^2}{r^4}$$

as $r^4 = a^4 \sin^2 2\theta$. Hence the required relation is $pa^2 = r^3$.

(ii) The equation to the tangent at (x_1, y_1) to the curve $y^2 = 4a(x + a)$ is

$$y - y_1 = \frac{2a}{y_1}(x - x_1),$$

or $\qquad 2ax - yy_1 + y_1^2 - 2ax_1 = 0.$

The perpendicular distance p from the origin on to this line is given by

$$p^2 = \frac{(y_1^2 - 2ax_1)^2}{4a^2 + y_1^2} = \frac{(2ax_1 + 4a^2)^2}{8a^2 + 4ax_1},$$

since $\qquad y_1^2 = 4a(x_1 + a).$

This reduces to $\qquad p^2 = a(x_1 + 2a) = ar,$

since $\qquad r^2 = x_1^2 + y_1^2 = x_1^2 + 4a(x_1 + a) = (x_1 + 2a)^2.$

EXERCISES 11 (a)

1. Show that the equations to the tangents to the curve
$$x^2y - ax^2 + a^2y = 0$$
at the points where $y = a/4$ are
$$y = \pm\frac{3\sqrt{3}}{8}x - \frac{a}{8},$$
and find the equations to the normals at these points.

2. Find the equations to the tangent and normal at the point 'θ' on the ellipse
$$x = a \cos \theta, \; y = b \sin \theta.$$

3. For a certain curve, the coordinates are given in terms of a parameter t by the equations $x = a(t + \sin t)$, $y = a(1 - \cos t)$.
Show that $s = 4a \sin \psi$ where s is the length of the arc of the curve measured from the origin and ψ is the inclination of the tangent to the line $y = 0$. (C.)

4. Prove that the curves $r = a(1 + \cos \theta)$, $r = b(1 - \cos \theta)$ cut at right angles. (O.)

5. The equation to a curve in polar coordinates is $r = a + bf(\theta)$ where b is small compared with a. Neglecting b^2/a^2, find the inclination of the tangent at the point 'θ' to the radius vector. (C.)

6. Find, in terms of r, the length of the perpendicular from the origin to the tangent at any point on the curve $r = a(1 - \cos \theta)$. Prove that the tangents to the curve at the points θ, $(\pi/3) + \theta$, $(2\pi/3) + \theta$ and $\pi + \theta$ form a rectangle. (C.)

11.3. Asymptotes

A line which meets a plane curve in two points both of which are at an infinite distance from the origin, but which is not itself altogether at infinity, is called an *asymptote* to the curve. The asymptotes of the hyperbola have already been found (*A.L.P.M.*, p. 330) and a similar method applies to the more general case in which the equation to the curve is a rational algebraical function of the nth degree.

If the equation to the curve is given by $f(x, y) = 0$, where $f(x, y)$ is a rational algebraical function of the nth degree, the abscissae of the points of intersection of the curve with the straight line $y = mx + c$ are given by
$$f(x, mx + c) = 0.$$
This equation is in general of the nth degree and, expressed in descending powers of x, can be written
$$a_0 x^n + a_1 x^{n-1} + a_2 x^{n-2} + \ldots + a_{n-1} x + a_n = 0, \quad (11.14)$$
or, as an equation in $1/x$,
$$a_0 + \frac{a_1}{x} + \frac{a_2}{x^2} + \ldots + \frac{a_{n-1}}{x^{n-1}} + \frac{a_n}{x^n} = 0,$$
where $a_0, a_1, \ldots, a_n$ are functions of m and c. This equation will have two zero roots, and the line $y = mx + c$ will therefore meet the curve in two points at infinity if m and c are chosen so that
$$a_0 = a_1 = 0.$$
It will be found that the equation $a_0 = 0$ contains m only and in a degree not greater than n and that the equation $a_1 = 0$ contains c only

in the first degree. Hence n (real or imaginary) values of m and c can be found and *a curve of the nth degree possesses n (real or imaginary) asymptotes.*

Example 5. *Find the asymptotes of the curve*
$$4x^3 - 4x^2y - xy^2 + y^3 - 4x^2 + 4xy = 0.$$

The abscissae of the points of intersection of the line $y = mx + c$ and the curve are given by
$$4x^3 - 4x^2(mx + c) - x(mx + c)^2 + (mx + c)^3 - 4x^2 + 4x(mx + c) = 0,$$
and values of m and c which make the line $y = mx + c$ an asymptote are given *by equating to zero the coefficients of the two highest powers of x* in this cubic equation. This gives
$$m^3 - m^2 - 4m + 4 = 0,$$
$$(3m^2 - 2m - 4)c + 4(m - 1) = 0.$$
The first relation gives $m = 1$, 2 and -2 and the corresponding values of c derived from the second relation are 0, -1 and 1. Hence the lines $y = x$, $y = \pm(2x - 1)$ are all asymptotes of the curve.

In the case of certain curves, some values of m which make the coefficient a_0 in equation (11.14) vanish also cause the coefficient a_1 of the next highest power of x to vanish identically. In such cases the line $y = mx + c$ intersects the curve at two points at infinity for any value of c. If c is now chosen so as to make the third coefficient a_2 equal to zero, each such value of m will be found to give rise to two values of c and hence to two *parallel asymptotes*, each of which meets the curve in three points at an infinite distance from the origin.

Example 6. *Find the asymptotes of the curve*
$$(x - y + 1)(x - y - 2)(x + y) = 8x - 1. \qquad (C.)$$

Writing $y = mx + c$, the abscissae of the points of intersection of the line and curve are given by
$$\{(1 - m)x + 1 - c\}\{(1 - m)x - 2 - c\}\{(1 + m)x + c\} = 8x - 1$$
Arranged as a cubic in x, this can be written
$$(1 - m)^2(1 + m)x^3 + \{(3c + 1)m^2 - 2cm - c - 1\}x^2 +$$
$$+ \{3c^2m + 2cm - c^2 - 2m - 10\}x + c(c - 1)(c + 2) + 1 = 0.$$
By equating to zero the coefficient of x^3, $m = -1$ or 1. With $m = -1$, the coefficient of x^2 reduces to $4c$ and hence $y = -x$ is an asymptote. With $m = 1$, the coefficient of x^2 vanishes identically and the coefficient of x becomes $2c^2 + 2c - 12$. This vanishes when $c = 2$ and -3, so that there are two parallel asymptotes $y = x + 2$ and $y = x - 3$.

Asymptotes parallel to the axis of y cannot be found by the foregoing methods for their slopes are not finite. However, *asymptotes which are parallel to either of the coordinate axes* can be found as follows. If the equation to the curve is

$$\begin{aligned}
a_0x^n + a_1x^{n-1}y + a_2x^{n-2}y^2 + \ldots + a_{n-1}xy^{n-1} + a_ny^n& \\
+ b_1x^{n-1} + b_2x^{n-2}y + \ldots \quad\quad\quad\quad + b_ny^{n-1}& \\
+ c_2x^{n-2} + \ldots \quad\quad\quad\quad\quad + c_ny^{n-2}& \\
+ \ldots \quad\quad\quad = 0,&
\end{aligned} \quad (11.15)$$

it can be rearranged in descending powers of x in the form

$$a_0x^n + (a_1y + b_1)x^{n-1} + \ldots = 0.$$

If $a_0 = 0$ and $a_1y + b_1 = 0$, the coefficients of the two highest powers of x in this equation vanish and the straight line $a_1y + b_1 = 0$ is therefore an asymptote. Similarly if $a_n = 0$, $a_{n-1}x + b_n = 0$ is an asymptote. Again if a_0, a_1 and b_1 all vanish and if

$$a_2y^2 + b_2y + c_2 = 0,$$

the two lines represented by this equation represent a pair of asymptotes, real or imaginary, parallel to the axis of x. These considerations show that, to find those asymptotes which are parallel to the coordinate axes, it is sufficient to *equate to zero the coefficients of the highest powers of x and y.*

Example 7. *Find the three asymptotes of the curve*

$$x^2y - xy^2 - x^2 + y^2 + x + y = 0$$

and prove that they are concurrent. (O.)

The coefficient of x^2 (the highest power of x) is $y - 1$ and the coefficient of y^2 (the highest power of y) is $-x + 1$ so that $y - 1 = 0$, $-x + 1 = 0$ are asymptotes. To find the third asymptote put $y = mx + c$, so that

$$x^2(mx + c) - x(mx + c)^2 - x^2 + (mx + c)^2 + x + y = 0.$$

This can be rearranged as

$$(m - m^2)x^3 + (c - 2cm - 1 + m^2)x^2 + \ldots = 0$$

and asymptotes are given by

$$m - m^2 = 0, \; c - 2cm - 1 + m^2 = 0.$$

These give $m = 0$, $c = 1$ (corresponding to the asymptote $y = 1$ already found) and $m = 1$, $c = 0$ giving the third asymptote $y = x$. The three asymptotes $y = 1$, $x = 1$, $y = x$ are clearly concurrent at the point (1, 1).

If a curve of the nth degree possesses n distinct asymptotes all of which pass through the origin, it follows by writing $y = mx$ in equation (11.15) and equating to zero the resulting coefficients of x^n and x^{n-1} that

$$a_0 + a_1m + a_2m^2 + \ldots + a_{n-1}m^{n-1} + a_nm^n = 0,$$
$$b_1 + b_2m + \ldots + b_nm^{n-1} = 0.$$

The first of these relations will give n values for the slopes of the various asymptotes and all n of these values can only satisfy the second relation if $b_1 = b_2 = \ldots = b_n = 0$. Hence if the equation to a curve of the nth degree is arranged in homogeneous sets of terms (of degree indicated by the suffix) in the form

$$u_n + u_{n-1} + u_{n-2} + \ldots = 0,$$

the curve will possess n distinct asymptotes all of which pass through the origin if all the terms forming u_{n-1} are absent.

Example 8. *Find, by change of origin, the condition that the asymptotes of the curve*

$$ax^3 + bx^2y + cxy^2 + dy^3 + ex^2 + fxy + gy^2 + hx + ky + l = 0$$

shall be concurrent. (O.)

Suppose the point of concurrence is (α, β), then the equation to the curve referred to axes OX, OY with this point as origin is

$$a(X+\alpha)^3 + b(X+\alpha)^2(Y+\beta) + c(X+\alpha)(Y+\beta)^2 + d(Y+\beta)^3 + e(X+\alpha)^2$$
$$+ f(X+\alpha)(Y+\beta) + g(Y+\beta)^2 + h(X+\alpha) + k(Y+\beta) + l = 0.$$

If the three asymptotes pass through the origin, the coefficients of the second degree terms (X^2, XY and Y^2) will all vanish: this gives

$$3a\alpha + b\beta + e = 0, \quad 2b\alpha + 2c\beta + f = 0, \quad c\alpha + 3d\beta + g = 0.$$

Eliminating α and β from these three relations, the required condition is

$$\begin{vmatrix} 3a & b & e \\ 2b & 2c & f \\ c & 3d & g \end{vmatrix} = 0.$$

EXERCISES 11 (b)

1. Find the real asymptote of the curve $y^3 - 2ax^2 + x^3 = 0$.

2. Show that the curve $y^2(x^2 + 1) = x^2(x^2 - 1)$ has two real asymptotes and find their equations.

3. Find the asymptotes of the curve

$$y = a + \frac{b}{(x-c)^2}.$$

4. The equation to a plane curve is

$$x^n \phi_n\left(\frac{y}{x}\right) + x^{n-1}\phi_{n-1}\left(\frac{y}{x}\right) + \ldots = 0,$$

where $\phi_n(y/x)$, $\phi_{n-1}(y/x)$, ... are rational algebraical functions of degrees n, $n-1$, ... If m satisfies the equation $\phi_n(m) = 0$, show that the line

$$y = mx - \frac{\phi_{n-1}(m)}{\phi_n'(m)}$$

is an asymptote.

Hence find the asymptotes of the curve $y^3 - x^2y + 2y^2 = 0$.

5. Find the asymptotes of the curve

$$x^3 + 2x^2 + x = xy^2 - 2xy + y.$$ (C.)

6. Find the asymptotes of the curve
$$xy(x^2 - y^2) + (x - y)(x^2 - y^2) - 2x^2 + 1 = 0$$
and show that their finite points of intersection with the curve lie on the circle $x^2 + y^2 = 1$. (O.)

11.4. Curvature

As an introduction to the subject of curvature, it is useful first to consider the 'order of contact' of two plane curves. For this purpose, suppose that $y = f(x)$, $y = g(x)$ are the equations to two curves which intersect at the point P whose abscissa is a. The curves are said to *touch* at the point P if they have the same tangent at this point: the necessary and sufficient conditions for this are clearly

$$f(a) = g(a), f'(a) = g'(a), \qquad (11.16)$$

and, in this case, the curves are said to have *first order contact* at P.

Looking at the matter in a slightly different way, let Q and R (Fig. 36) be points on the above curves $y = f(x)$, $y = g(x)$ respectively

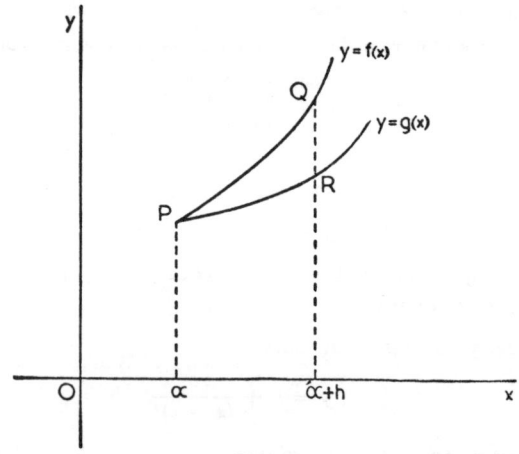

Fig. 36

with abscissae $a + h$. In view of the relations (11.16) and assuming the derivatives in question exist, the general mean value theorem (8.6) gives

$$QR = f(a + h) - g(a + h) = \frac{1}{2}h^2\{f''(a + \theta h) - g''(a + \theta h)\}$$

where $0 < \theta < 1$. Hence

$$\lim_{h \to 0} \left(\frac{QR}{h^2}\right) = \frac{1}{2}\{f''(a) - g''(a)\}$$

and, *when two curves touch at a point with abscissa* a, *the difference of their ordinates at a point with abscissa* $a + h$ *is at least of the second order of smallness in* h.

The degree of smallness of QR may be used as a sort of measure of the closeness of the contact between two curves and the above suggests (and the reader should have little difficulty in proving) that if the first n derivatives of $f(x)$, $g(x)$ exist and have equal values when $x = a$,

$$\lim_{h \to 0} \left(\frac{QR}{h^{n+1}} \right) = \frac{1}{(n+1)!} \{ f^{(n+1)}(a) - g^{(n+1)}(a) \},$$

and hence that QR is of the $(n+1)$th 'order of smallness'. These considerations lead to the following definition—if $f(a) = g(a)$, $f'(a) = g'(a)$, ..., $f^{(n)}(a) = g^{(n)}(a)$ but if $f^{(n+1)}(a) \neq g^{(n+1)}(a)$, the curves $y = f(x)$, $y = g(x)$ are said to have *contact of the nth order* at the point whose abscissa is a. As a simple example, take one of the curves ($y = g(x)$) as the straight line $y = mx + c$. Then $g''(x) = 0$ and the straight line has contact of the second order with the curve $y = f(x)$ at a point where $x = a$ if $f''(a) = 0$. Such a point being a point of inflexion (*A.L.P.M.*, p. 162), it follows that the tangent to a curve at a point of inflexion has second order contact with the curve (the tangent to the curve at any other point having first order contact).

The equation to the circle with centre $(\bar{x}, \bar{y})$ and radius ϱ is

$$(x - \bar{x})^2 + (y - \bar{y})^2 = \varrho^2, \tag{11.17}$$

and the first and second derivatives of y with respect to x are given by

$$x - \bar{x} + (y - \bar{y})\frac{dy}{dx} = 0, \tag{11.18}$$

$$1 + \left(\frac{dy}{dx}\right)^2 + (y - \bar{y})\frac{d^2y}{dx^2} = 0. \tag{11.19}$$

If this circle has *contact of the second order* at the point (x, y) with the curve $y = f(x)$, then y, dy/dx and d^2y/dx^2 will have the *same* values for the curve and the circle at the point whose abscissa is x. The circle is then called the *circle of curvature* of the curve at the point (x, y) and its centre and radius are called respectively the *centre and radius of curvature* of the curve at the point in question. Equation (11.19) gives

$$y - \bar{y} = -\frac{1 + \left(\dfrac{dy}{dx}\right)^2}{\dfrac{d^2y}{dx^2}}, \tag{11.20}$$

and it follows from (11.18) that

$$x - \bar{x} = \frac{\dfrac{dy}{dx}\left\{1 + \left(\dfrac{dy}{dx}\right)^2\right\}}{\dfrac{d^2y}{dx^2}}. \tag{11.21}$$

Substitution in (11.17) then gives for the radius of curvature (ϱ) of the curve $y = f(x)$ at the point (x, y)

$$\varrho = \frac{\left\{1 + \left(\frac{dy}{dx}\right)^2\right\}^{3/2}}{\frac{d^2y}{dx^2}}. \tag{11.22}$$

The coordinates $(\bar{x}, \bar{y})$ of the centre of curvature of the curve at the point (x, y) follow immediately from (11.21), (11.20) as

$$\left. \begin{aligned} \bar{x} &= x - \frac{\frac{dy}{dx}\left\{1 + \left(\frac{dy}{dx}\right)^2\right\}}{\frac{d^2y}{dx^2}}, \\ \bar{y} &= y + \frac{1 + \left(\frac{dy}{dx}\right)^2}{\frac{d^2y}{dx^2}}. \end{aligned} \right\} \tag{11.23}$$

The *measure of curvature* (or more simply the *curvature*) at a given point is defined as the reciprocal of the radius of curvature at the point in question. Since formula (11.22) involves a square root, the sign of ϱ remains to be settled. It is conventional to take the positive value of the root in the numerator of (11.22) so that the sign of ϱ is the same as the sign of d^2y/dx^2—this is equivalent to taking ϱ to be positive at points at which the curve lies above its tangent. It is worth noticing that, at a point of inflexion, d^2y/dx^2 vanishes and hence the radius of curvature is infinite (and the curvature zero) at such a point.

Example 9. *Prove that the coordinates $(\bar{x}, \bar{y})$ of the centre of curvature corresponding to any point (x, y) on the curve $ay^2 = x^3$ are*

$$\bar{x} = -\left(x + \frac{9x^2}{2a}\right), \ \bar{y} = 4y\left(1 + \frac{a}{3x}\right). \tag{O.}$$

Here $\quad y = \sqrt{\left(\frac{x^3}{a}\right)}, \ \frac{dy}{dx} = \frac{3}{2}\sqrt{\left(\frac{x}{a}\right)}, \ \frac{d^2y}{dx^2} = \frac{3}{4}\sqrt{\left(\frac{1}{ax}\right)}.$

Substitution in (11.23) gives

$$\bar{x} = x - \frac{\frac{3}{2}\sqrt{\left(\frac{x}{a}\right)}\left(1 + \frac{9x}{4a}\right)}{\frac{3}{4}\sqrt{\left(\frac{1}{ax}\right)}} = x - 2x\left(1 + \frac{9x}{4a}\right) = -\left(x + \frac{9x^2}{2a}\right),$$

$$\bar{y} = y + \frac{1 + \frac{9x}{4a}}{\frac{3}{4}\sqrt{\left(\frac{1}{ax}\right)}} = y + \frac{4}{3}\sqrt{(ax)} + 3\sqrt{\left(\frac{x^3}{a}\right)}$$

$$= y + \frac{4}{3}\frac{a}{x}\sqrt{\left(\frac{x^3}{a}\right)} + 3\sqrt{\left(\frac{x^3}{a}\right)} = 4y\left(1 + \frac{a}{3x}\right),$$

as $y = \sqrt{(x^3/a)}$.

11] CURVATURE 315

Example 10. *Prove that if ψ is the inclination to the $x-$ axis of the tangent at a point on the curve $y = a \log_e \sec (x/a)$, then the radius of curvature at this point is $a \sec \psi$.* (O.C.)

Here $y = -a \log_e \cos (x/a)$, $dy/dx = \tan (x/a)$, and
$$\frac{d^2y}{dx^2} = \frac{1}{a} \sec^2 \left(\frac{x}{a}\right).$$
Hence (11.22) gives, for the radius of curvature ϱ,
$$\varrho = \frac{\{1 + \tan^2(x/a)\}^{3/2}}{\frac{1}{a} \sec^2 \left(\frac{x}{a}\right)} = a \sec \left(\frac{x}{a}\right).$$
But $\tan \psi = dy/dx = \tan (x/a)$, so that $\psi = x/a$ and thus
$$\varrho = a \sec \psi.$$

The formula (11.22) is suitable for determining the radius of curvature at a point on a curve given by the explicit relation $y = f(x)$. If the equation to the curve is given in the parametric form $x = f(t)$, $y = g(t)$ and dots denote differentiation with respect to t,
$$\frac{dy}{dx} = \frac{dy}{dt} \Big/ \frac{dx}{dt} = \frac{\dot{y}}{\dot{x}},$$
$$\frac{d^2y}{dx^2} = \frac{d}{dx}\left(\frac{\dot{y}}{\dot{x}}\right) = \frac{d}{dt}\left(\frac{\dot{y}}{\dot{x}}\right)\frac{dt}{dx} = \frac{\dot{x}\ddot{y} - \ddot{x}\dot{y}}{\dot{x}^3}.$$
Substitution in (11.22) then gives
$$\varrho = \frac{(\dot{x}^2 + \dot{y}^2)^{3/2}}{\dot{x}\ddot{y} - \ddot{x}\dot{y}}. \tag{11.24}$$

If the equation to the curve is given in the implicit form $F(x, y) = 0$ and if
$$p = \frac{\partial F}{\partial x},\ q = \frac{\partial F}{\partial y},$$
$$r = \frac{\partial^2 F}{\partial x^2},\ s = \frac{\partial^2 F}{\partial x \partial y},\ t = \frac{\partial^2 F}{\partial y^2},$$
it follows from equations (9.10), (9.11) that
$$\frac{dy}{dx} = -\frac{p}{q},\ \frac{d^2y}{dx^2} = -\frac{q^2 r - 2pqs + p^2 t}{q^3}.$$
Substitution in (11.22) gives, as a convenient formula for ϱ,
$$\varrho = -\frac{(p^2 + q^2)^{3/2}}{q^2 r - 2pqs + p^2 t}. \tag{11.25}$$

Example 11. *Find the curvature of the curve*
$$2xy^2 + 2(x^2 - x + 2)y - (x^2 - 5x + 2) = 0$$
at the point (0, 1/2). (C.)

Here,

$$p = 2y^2 + 4xy - 2y - 2x + 5,\ q = 4xy + 2x^2 - 2x + 4,\ r = 4y - 2,$$
$$s = 4y + 4x - 2,\ t = 4x.$$

When $x = 0$, $y = 1/2$ these become

$$p = \frac{9}{2},\ q = 4,\ r = 0,\ s = 0,\ t = 0$$

and (11.25) shows that the curvature $(1/\varrho)$ is zero.

Yet another result (*Newton's formula*) can be obtained which is often useful when the equation to the curve is given in cartesian coordinates. This applies when the curve touches one of the coordinate axes at the origin. Suppose that the curve touches the axis of x at the origin and that the suffix 0 is used to denote values at this point. Then $y_0 = (dy/dx)_0 = 0$ and, from (11.22),

$$\frac{1}{\varrho_0} = \left(\frac{d^2y}{dx^2}\right)_0.$$

By Maclaurin's series (8.11), the equation to the curve can be written, using the above values for y and its first two derivatives, in the form

$$y = \frac{x^2}{(2)!}\frac{1}{\varrho_0} + \frac{x^3}{(3)!}\left(\frac{d^3y}{dx^3}\right)_0 + \ldots.$$

Dividing by x^2 and letting x tend to zero, the curvature at the origin is given by

$$\frac{1}{\varrho_0} = \lim_{x \to 0}\left(\frac{2y}{x^2}\right). \qquad (11.26)$$

Similarly, if the $y-$ axis is a tangent at the origin

$$\frac{1}{\varrho_0} = \lim_{x \to 0}\left(\frac{2x}{y^2}\right). \qquad (11.27)$$

Example 12. *Find the radius of curvature of the curve*

$$3y = x^2 + xy + y^4$$

at the origin. (C.)

The curve touches the $x-$ axis at the origin and its equation can be written

$$3 = \frac{x^2}{y} + x + y.$$

Hence as x (and therefore also y) tends to zero,

$$\lim_{x \to 0}\left(\frac{x^2}{y}\right) = 3.$$

But, by (11.26), the radius of curvature at the origin is given by

$$\varrho_0 = \lim_{x \to 0}\left(\frac{x^2}{2y}\right)$$

so the required value of the radius is 3/2.

CURVATURE

For a curve defined by its *intrinsic* equation (§ 11.2), a formula giving the radius of curvature ϱ in terms of s and ψ is required. This takes a very simple form and can be found as follows. Using (11.8),

$$\left\{1 + \left(\frac{dy}{dx}\right)^2\right\}^{3/2} = (1 + \tan^2 \psi)^{3/2} = \sec^3 \psi,$$

$$\frac{d^2y}{dx^2} = \frac{d}{dx}(\tan \psi) = \sec^2 \psi \frac{d\psi}{dx} = \sec^2 \psi \frac{d\psi}{ds}\frac{ds}{dx} = \sec^3 \psi \frac{d\psi}{ds}.$$

Dividing and using (11.22),

$$\varrho = \frac{ds}{d\psi}. \tag{11.28}$$

As a simple example, consider the curve $y + 1 = \cosh x$. By Example 3 of this chapter, its intrinsic equation is $s = \tan \psi$ and hence the radius of curvature at a point where the tangent makes an angle ψ with the $x-$ axis is given by

$$\varrho = \frac{ds}{d\psi} = \sec^2 \psi.$$

When the equation to a curve is expressed parametrically in terms of the length of arc s measured from a fixed point on the curve the following formulae are useful. By (11.8),

$$\cos \psi = \frac{dx}{ds}, \quad \sin \psi = \frac{dy}{ds}$$

so that

$$-\sin \psi \frac{d\psi}{ds} = \frac{d^2x}{ds^2}, \quad \cos \psi \frac{d\psi}{ds} = \frac{d^2y}{ds^2}.$$

Hence, by (11.28),

$$\frac{1}{\varrho} = -\frac{d^2x/ds^2}{dy/ds} = \frac{d^2y/ds^2}{dx/ds}, \tag{11.29}$$

and

$$\frac{1}{\varrho^2} = \left(\frac{d^2x}{ds^2}\right)^2 + \left(\frac{d^2y}{ds^2}\right)^2. \tag{11.30}$$

Example 13. *If a curve touches the axis of x at the origin and s is the arc measured from the origin to a point (x, y) on the curve, prove that*

$$x = s - \frac{s^3}{6\varrho^2}, \quad y = \frac{s^2}{2\varrho} - \frac{s^3}{6\varrho^2}\frac{d\varrho}{ds},$$

correct to terms of degree 3 in s, where ϱ is the radius of curvature of the curve at the origin and $d\varrho/ds$ is its rate of change there. (O.)

If x and y are expressed parametrically in terms of s, since x and y both vanish when $x = 0$, Maclaurin's series gives

$$x = s\left(\frac{dx}{ds}\right) + \frac{s^2}{(2)!}\left(\frac{d^2x}{ds^2}\right) + \frac{s^3}{(3)!}\left(\frac{d^3x}{ds^3}\right) + \cdots$$

$$y = s\left(\frac{dy}{ds}\right) + \frac{s^2}{(2)!}\left(\frac{d^2y}{ds^2}\right) + \frac{s^3}{(3)!}\left(\frac{d^3y}{ds^3}\right) + \cdots$$

all the derivatives being evaluated when $s = 0$

From (11.8), if ψ is the angle between the tangent to the curve and the $x-$ axis,

$$\frac{dx}{ds} = \cos\psi, \quad \frac{dy}{ds} = \sin\psi,$$

from which it follows, when use is made of (11.28), that

$$\frac{d^2x}{ds^2} = -\sin\psi\frac{d\psi}{ds} = -\frac{\sin\psi}{\varrho}, \quad \frac{d^2y}{ds^2} = \cos\psi\frac{d\psi}{ds} = \frac{\cos\psi}{\varrho};$$

a further differentiation gives

$$\frac{d^3x}{ds^3} = -\frac{\cos\psi}{\varrho}\frac{d\psi}{ds} + \frac{\sin\psi}{\varrho^2}\frac{d\varrho}{ds} = -\frac{\cos\psi}{\varrho^2} + \frac{\sin\psi}{\varrho^2}\frac{d\varrho}{ds},$$

and

$$\frac{d^3y}{ds^3} = -\frac{\sin\psi}{\varrho}\frac{d\psi}{ds} - \frac{\cos\psi}{\varrho^2}\frac{d\varrho}{ds} = -\frac{\sin\psi}{\varrho^2} - \frac{\cos\psi}{\varrho^2}\frac{d\varrho}{ds}.$$

Since the curve touches the axis of x at the origin, $\psi = 0$ at this point and the values of the derivatives at the origin are given by

$$\frac{dx}{ds} = 1, \quad \frac{dy}{ds} = 0, \quad \frac{d^2x}{ds^2} = 0, \quad \frac{d^2y}{ds^2} = \frac{1}{\varrho}, \quad \frac{d^3x}{ds^3} = -\frac{1}{\varrho^2}, \quad \frac{d^3y}{ds^3} = -\frac{1}{\varrho^2}\frac{d\varrho}{ds}.$$

The required result then follows by substituting these values in the expressions for x and y and neglecting terms in s^4 and above.

For a curve expressed in the (p, r) form yet another expression for ϱ is useful. Referring to Fig. 35, $p = r\sin\phi$ so that

$$\frac{dp}{dr} = \sin\phi + r\cos\phi\frac{d\phi}{dr}$$

$$= r\frac{d\theta}{ds} + r\frac{dr}{ds}\frac{d\phi}{dr},$$

when use is made of (11.9) and (11.10). This can be written

$$\frac{dp}{dr} = r\frac{d\theta}{ds} + r\frac{d\phi}{ds} = r\frac{d}{ds}(\theta + \phi) = r\frac{d\psi}{ds},$$

as, from Fig. 35, $\psi = \theta + \phi$. Since from (11.28) $\varrho = ds/d\psi$, the required formula relating ϱ, p and r is

$$\varrho = r\frac{dr}{dp}. \tag{11.31}$$

As an example, take the curve $r^2 = a^2\sin 2\theta$ for which the (p, r) equation (Example 4) is $pa^2 = r^3$. Here

$$a^2\frac{dp}{dr} = 3r^2$$

and the radius of curvature at the point whose radius vector is r is given by

$$\varrho = r\frac{dr}{dp} = \frac{ra^2}{3r^2} = \frac{a^2}{3r}.$$

Finally, it is important to derive a formula for ϱ when the equation to the curve is expressed in polar coordinates in the form $r = f(\theta)$. Starting with equation (11.13) in the form

$$p = \frac{r^2}{\sqrt{\left\{r^2 + \left(\dfrac{dr}{d\theta}\right)^2\right\}}}$$

it follows, after a little reduction, that

$$\frac{dp}{d\theta} = \frac{r\dfrac{dr}{d\theta}\left\{r^2 + 2\left(\dfrac{dr}{d\theta}\right)^2 - r\dfrac{d^2r}{d\theta^2}\right\}}{\left\{r^2 + \left(\dfrac{dr}{d\theta}\right)^2\right\}^{3/2}}$$

But from (11.31), $\quad \varrho = r\dfrac{dr}{dp} = r\left(\dfrac{dr/d\theta}{dp/d\theta}\right)$

and the required formula is

$$\varrho = \frac{\left\{r^2 + \left(\dfrac{dr}{d\theta}\right)^2\right\}^{3/2}}{r^2 + 2\left(\dfrac{dr}{d\theta}\right)^2 - r\dfrac{d^2r}{d\theta^2}}. \qquad (11.32)$$

Example 14. *Prove that the radius of curvature of the curve $r = a(1 - \cos\theta)$ is $(4a/3)\sin(\theta/2)$.* (C.)

Here, $\qquad \dfrac{dr}{d\theta} = a\sin\theta, \; \dfrac{d^2r}{d\theta^2} = a\cos\theta,$

$$r^2 + \left(\frac{dr}{d\theta}\right)^2 = a^2(1 - \cos\theta)^2 + a^2\sin^2\theta = 2a^2(1 - \cos\theta) = 4a^2\sin^2\frac{\theta}{2},$$

$$r^2 + 2\left(\frac{dr}{d\theta}\right)^2 - r\frac{d^2r}{d\theta^2} = a^2(1 - \cos\theta)^2 + 2a^2\sin^2\theta - a^2(1 - \cos\theta)\cos\theta$$

$$= 3a^2(1 - \cos\theta) = 6a^2\sin^2\frac{\theta}{2}.$$

Hence, from (11.32), $\quad \varrho = \dfrac{8a^3\sin^3(\theta/2)}{6a^2\sin^2(\theta/2)} = \dfrac{4a}{3}\sin\dfrac{\theta}{2}.$

EXERCISES 11 (c)

1. Find the length of the radius of curvature for the curve $xy = c^2$ at a point of intersection of the curve with the circle $x^2 + y^2 = 4c^2$. (C.)

2. Find the radius of curvature at the point P of the curve $y = c\cosh(x/c)$ for which $x = x_1$.
   Show also that the length of intercept on the normal to the curve between P and the line $y = 0$ is given numerically by the radius of curvature. (C.)

3. Show that all centres of curvature to the curve
$$y = \frac{\sin^2 x}{x^2}$$
at its points of contact with the $x-$ axis lie on a parabola. (O.)

4. A curve is given in parametric form by the equations
$$x = a(2\cos t + \cos 2t),\ y = a(2\sin t - \sin 2t).$$
Find the radius of curvature in terms of t. (O.)

5. The equation to a curve referred to the tangent and normal at the origin as axes is
$$2y = ax^2 + 2hxy + by^2.$$
Prove that $\quad a = \dfrac{1}{\varrho},\ h = -\dfrac{1}{3\varrho}\dfrac{d\varrho}{ds}$

where ϱ is the radius of curvature at the origin. (O.)

6. In the curve $y = a\log_e \sec(x/a)$, prove that $x = a\psi$ and $\varrho = a\sec\psi$, where ψ is the inclination of the tangent at any point to the $x-$ axis and ϱ is the radius of curvature at the point.

Prove also that $\quad a^2\left(\dfrac{d\varrho}{ds}\right)^2 = \varrho^2 - a^2.$ (C.)

7. Prove that, if k denotes the curvature $(d\psi/ds)$ and primes denote differentiation with respect to s,
$$\begin{vmatrix} x' & y' & 1 \\ x'' & y'' & 1 \\ x''' & y''' & 1 \end{vmatrix} = k^3 + k - k'.$$
(O.)

8. The equation of an 'equiangular spiral', in terms of polar coordinates (r, θ) is
$$r = ae^{\theta \cot \alpha}.$$
Show that its radius of curvature is $r\cosec\alpha$. (O.)

11.5. Double points

There exist points on some curves through which more than one branch of the curve passes: such points are called *multiple points*. If two branches pass through a point P (Fig. 37), P is called a *double point*; if three branches pass through P, it is known as a *triple point* and so on, a point through which r branches pass being referred to as *a multiple point of the rth order*. An inspection of Fig. 37 shows that there are two tangents to the curve at a double point P, one for each branch. A tangent to one branch intersects the curve in two points such as A and B which ultimately coincide at P, and this tangent also intersects the second branch of the curve at a third point C, ultimately also coinciding with P. A tangent at a double point therefore intersects the curve in *three ultimately coincident points*.

The two tangents at a double point may be real, imaginary or coincident. In the case of two real tangents, there are two real branches of the curve passing through the double point, the general shape of the curve is as shown in Fig. 37 and the double point is called a *node*. If the two tangents are imaginary, there are no real tangents in the immediate neighbourhood of the double point and the point is

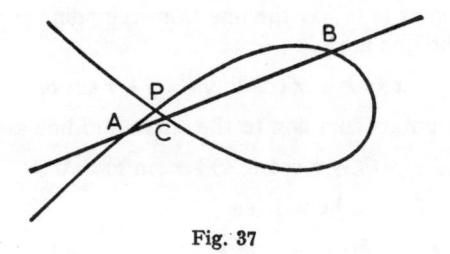

Fig. 37

simply an isolated one whose coordinates satisfy the equation to the curve; the point is in this case called a *conjugate point*. If the tangents at the double point are coincident, the two branches of the curve touch at the point in question; such a point is called a *cusp*. A cusp may be of two kinds as shown in Fig. 38. In Fig. 38 (a) the two branches PA, PB of the curve lie on opposite sides of the tangent at the cusp P and the cusp is said to be of the *first species*. Fig. 38 (b)

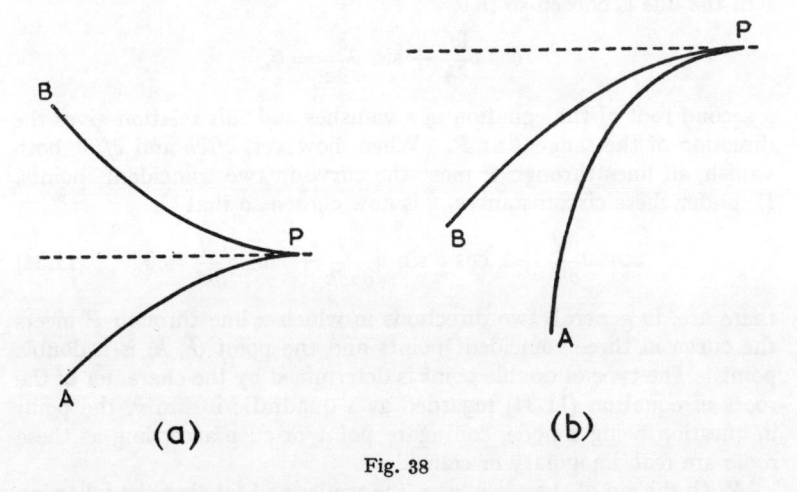

(a)   (b)

Fig. 38

shows a cusp of the *second species*, in which the branches PA, PB lie on the same side of the tangent at P.

Multiple points of higher orders than the second can be considered as combinations of double points—thus, a triple point may be regarded as a combination of three double points, each pair of the three branches forming a double point. Such points will not be further considered here.

The character of a point P with coordinates (h, k) on a curve $f(x, y) = 0$ can be examined by considering its intersection with a straight line through P. If the slope of this line is $\tan \theta$ and r is the distance of a point (x, y) on the line from the point P, the parametric equations of the line are

$$x = h + r \cos \theta, \quad y = k + r \sin \theta.$$

Values of r for points common to the curve and line are given by

$$f(h + r \cos \theta, k + r \sin \theta) = 0$$

and this, by (9.15), can be written

$$f(h, k) + r\left(\cos \theta \frac{\partial f}{\partial h} + \sin \theta \frac{\partial f}{\partial k}\right) +$$
$$+ \frac{r^2}{2}\left(\cos^2 \theta \frac{\partial^2 f}{\partial h^2} + 2 \cos \theta \sin \theta \frac{\partial^2 f}{\partial h \partial k} + \sin^2 \theta \frac{\partial^2 f}{\partial k^2}\right) + \ldots = 0,$$
(11.33)

where $\partial f/\partial h$, $\partial f/\partial k$, etc., denote the values of $\partial f/\partial x$, $\partial f/\partial y$, etc., at the point P. Since the point (h, k) lies on the curve, the first term $f(h, k)$ and therefore one root of this equation in r vanishes. If the direction θ of the line is chosen so that

$$\cos \theta \frac{\partial f}{\partial h} + \sin \theta \frac{\partial f}{\partial k} = 0,$$

a second root of the equation in r vanishes and this relation gives the direction of the tangent at P. When, however, $\partial f/\partial h$ and $\partial f/\partial k$ both vanish, all lines through P meet the curve in two coincident points. If, under these circumstances, θ is now chosen so that

$$\cos^2 \theta \frac{\partial^2 f}{\partial h^2} + 2 \cos \theta \sin \theta \frac{\partial^2 f}{\partial h \partial k} + \sin^2 \theta \frac{\partial^2 f}{\partial k^2} = 0 \qquad (11.34)$$

there are, in general, two directions in which a line through P meets the curve in three coincident points and the point (h, k) is a double point. The type of double point is determined by the character of the roots of equation (11.34) regarded as a quadratic in $\tan \theta$, the point in question being a node, conjugate point or cusp according as these roots are real, imaginary or coincident.

With the aid of these remarks, the reader will see that the following is the procedure for searching for and deciding the character of a double point on a curve $f(x, y) = 0$: find $\partial f/\partial x$ and $\partial f/\partial y$, equate each to zero and solve: test whether any of the values of x and y so obtained satisfy the equation to the curve: values of x and y which satisfy these requirements then give the coordinates of a double point and this is, in general,* a node, conjugate point or cusp according as $\left(\dfrac{\partial^2 f}{\partial x \partial y}\right)^2$ is greater than, less than or equal to $\dfrac{\partial^2 f}{\partial x^2}\cdot\dfrac{\partial^2 f}{\partial y^2}$.

Example 15. *Show that the curve*
$$64(x - \tfrac{1}{2}a)^3 = 27ay^2$$
has a cusp at the point $(a/2, 0)$.

Here
$$f(x, y) \equiv 64(x - \tfrac{1}{2}a)^3 - 27ay^2,$$
$$\frac{\partial f}{\partial x} = 192(x - \tfrac{1}{2}a)^2, \quad \frac{\partial f}{\partial y} = -54ay.$$

Hence $\partial f/\partial x$, $\partial f/\partial y$ both vanish when $x = a/2$, $y = 0$ and as these values of x and y satisfy the equation to the curve, the point $(a/2, 0)$ is a double point. Further
$$\frac{\partial^2 f}{\partial x^2} = 384(x - \tfrac{1}{2}a), \quad \frac{\partial^2 f}{\partial x \partial y} = 0, \quad \frac{\partial^2 f}{\partial y^2} = -54a,$$
and, when $x = a/2$,
$$\left(\frac{\partial^2 f}{\partial x \partial y}\right)^2 = \frac{\partial^2 f}{\partial x^2}\cdot\frac{\partial^2 f}{\partial y^2}.$$

Hence the point $(a/2, 0)$ is a cusp.

11.6. The nature of the origin

If a curve passes through the origin and if its equation is a rational algebraical function of x and y, it can be written in the form
$$(a_1 x + a_2 y) + (b_1 x^2 + b_2 xy + b_3 y^2) + \ldots$$
$$+ (k_1 x^n + k_2 x^{n-1} y + \ldots + k_{n+1} y^n) = 0. \qquad (11.35)$$

A line through the origin inclined at an angle θ to the axis of x will meet the curve in points whose distances from the origin are given by the roots of the equation in r obtained from (11.35) by writing $x = r\cos\theta$, $y = r\sin\theta$. This is

$$r(a_1 \cos\theta + a_2 \sin\theta) + r^2(b_1 \cos^2\theta + b_2 \cos\theta\sin\theta + b_3 \sin^2\theta) +$$
$$\ldots + r^n(k_1 \cos^n\theta + k_2 \cos^{n-1}\theta\sin\theta + \ldots + k_{n+1}\sin^n\theta) = 0.$$
$$(11.36)$$

If θ be chosen so that $a_1 \cos\theta + a_2 \sin\theta = 0$, two roots of (11.36) vanish and it is to be inferred that a line making an angle $\tan^{-1}(-a_1/a_2)$ with the x— axis cuts the curve in two coincident points at the origin and is therefore a tangent there. Hence the equation to the tangent at the origin is

$$a_1 x + a_2 y = 0$$

---

* When the conditions for a cusp hold, the curve is sometimes imaginary in the neighbourhood of the point in question and the point should then be classified as a conjugate point. Further investigation is therefore strictly necessary and further work is also required to decide the species of a cusp. Lack of space prevents these matters being pursued here.

and the reader should note that *if a curve passes through the origin, the terms of the first degree in x and y in its equation when equated to zero give the equation to the tangent at the origin.* If these terms are absent from the equation (i.e. if $a_1 = a_2 = 0$), it is generally possible to choose θ in (11.36) so that

$$b_1 \cos^2 \theta + b_2 \cos \theta \sin \theta + b_3 \sin^2 \theta = 0$$

and then three roots of (11.36) will vanish. In this case the pair of straight lines whose equation is

$$b_1 x^2 + b_2 xy + b_3 y^2 = 0 \qquad (11.37)$$

each cut the curve at the origin in three coincident points. There are therefore two branches of the curve intersecting at the origin, the origin is a double point and is a node, conjugate point or cusp according as the straight lines represented by (11.37) (the two tangents at the origin) are real, imaginary or coincident. In the same way if $a_1 = a_2 = b_1 = b_2 = b_3 = 0$, the origin will be a triple point and the equation of the tangents at the origin will be

$$c_1 x^3 + c_2 x^2 y + c_3 xy^2 + c_4 y^3 = 0$$

and so on.

Thus *the equation of the tangents at the origin can be written down by equating to zero the terms of the lowest degree in the equation to the curve.* From the resulting equation the character of the origin can be inferred by considering the reality, etc., of the tangents there.

Example 16. *Find the nature of the origin on the curve*
$$(x^2 + y^2)^2 = x^2 - 4y^2.$$

The equation to the curve can be written
$$x^4 + 2x^2 y^2 + y^4 - x^2 + 4y^2 = 0,$$
and the terms of the lowest degree are $-x^2 + 4y^2$. The tangents at the origin are therefore given by $x^2 - 4y^2 = 0$ or
$$x = \pm 2y.$$

Hence the curve has two real tangents at the origin and this point is therefore a node.

11.7. Envelopes

If x, y are rectangular coordinates and a is a parameter, the equation $f(x, y, a) = 0$ represents a curve and a different curve is, in general, obtained for each different (constant) value of a. When different values are assigned in this way to the parameter a, the equation $f(x, y, a) = 0$ is said to represent a *system* or *family* of curves. For instance, the equation $y^2 = 4a(x - a)$ represents a family of parabolas, each of which has its axis along the axis of x and its vertex at the point $(a, 0)$.

Two neighbouring curves of the family $f(x, y, a) = 0$, corresponding to values a, $a + \delta a$ of the parameter a, intersect in points whose coordinates are given by

$$f(x, y, a) = 0, \quad f(x, y, a + \delta a) = 0.$$

By the first mean value theorem, the second equation can be written

$$f(x, y, a + \delta a) = f(x, y, a) + \delta a \cdot \frac{\partial}{\partial a}\{f(x, y, a + \theta \delta a)\} = 0$$

where $0 < \theta < 1$ and the limiting values as $\delta a \to 0$ of the points of intersection will be given by

$$f(x, y, a) = 0, \quad \frac{\partial}{\partial a}\{f(x, y, a)\} = 0. \tag{11.38}$$

Such points lie on the curve $f(x, y, a) = 0$ and are referred to as *characteristic points* on the curve. The locus of such points is called the *envelope* of the family $f(x, y, a) = 0$ and the equations (11.38) when solved for x and y give the parametric equations of the envelope: alternatively, elimination of a from equations (11.38) leads to the cartesian equation to the envelope.

To take a simple example, consider the family of straight lines $x \cos \theta + y \sin \theta = a$, each of which is at a constant distance a from the origin. Here the parameter is θ and the envelope is given by eliminating θ between the equation to the line

$$x \cos \theta + y \sin \theta = a,$$

and the equation $\quad -x \sin \theta + y \cos \theta = 0,$

obtained from the equation to the line by differentiating with respect to the parameter θ. By eliminating θ between these two equations (by squaring and adding), the equation to the envelope is given in the form $x^2 + y^2 = a^2$ and is therefore a circle centre the origin and radius a. It is clear that any line of the family touches this circle and that two consecutive lines of the family are consecutive tangents to

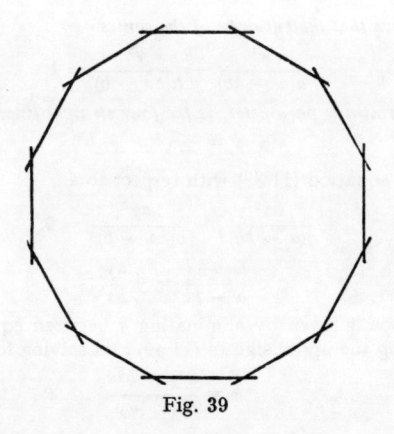

Fig. 39

the circle, their point of intersection tending to coincide with a point on the circle. If lines are drawn for values of θ which differ only slightly, it will be seen (Fig. 39) that the points of intersection and the parts of the tangents between them are scarcely indistinguishable from the envelope.

This is, in fact, a general property—*the envelope of a family of curves touches each curve of the family at its characteristic points.* For suppose the curves shown in Fig. 40 are those with parameters

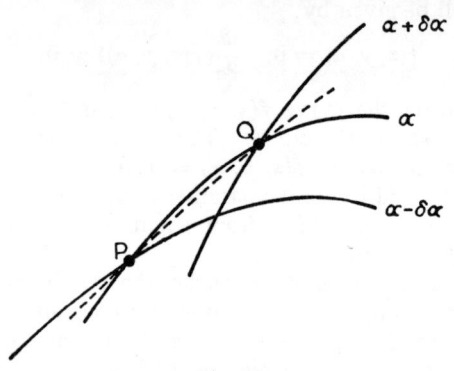

Fig. 40

$a - \delta a$, a and $a + \delta a$ and that P, Q are points of intersections of consecutive curves of the family. These points lie on a curve (shown dotted in the diagram) which ultimately becomes the envelope. As $\delta a \to 0$, P and Q approach each other and a line joining P to Q becomes both a tangent to the curve with parameter a and to the envelope. Thus the curve and envelope touch at a characteristic point.

Example 17. *Prove that the envelope of the conics*

$$\frac{x^2}{a(a - kb)} + \frac{y^2}{b(ka - b)} = 1, \tag{11.39}$$

where k is an arbitrary parameter, is the four straight lines

$$x \pm y = \pm \sqrt{(a^2 - b^2)}. \tag{C.}$$

Differentiating equation (11.39) with respect to k

$$\frac{bx^2}{a(a - kb)^2} - \frac{ay^2}{b(ka - b)^2} = 0,$$

from which
$$\frac{ka - b}{a - kb} = \pm \frac{ay}{bx}, \tag{11.40}$$

and the envelope is given by eliminating k between equations (11.39) and (11.40). Taking the upper sign in (11.40) and solving for k,

$$k = \frac{a^2 y + b^2 x}{ab(x + y)}$$

After substituting this value of k in (11.39) and a little reduction, there results $x^2 + 2xy + y^2 = a^2 - b^2$, so that
$$x + y = \pm \sqrt{(a^2 - b^2)}$$
forms part of the envelope. The other part, obtained similarly but with the lower sign in (11.40) gives
$$x - y = \pm \sqrt{(a^2 - b^2)}.$$

The reader should be warned that the result of eliminating the parameter a from the equations
$$f(x, y, a) = 0, \quad \frac{\partial}{\partial a}\{f(x, y, a)\} = 0$$
may contain loci other than the true envelope. It may include the loci of any double points which the curve may possess; this is illustrated in the following example but a general discussion on this point is beyond the scope of the present work.

Example 18. *Find the envelope of the curves* $a(y + a)^2 - x^3 = 0$.

Differentiating with respect to a,
$$(y + a)^2 + 2a(y + a) = 0$$
so that $y + a = 0$ or $y + 3a = 0$. Substituting in the equation to the curve, the envelope would appear to consist of the two lines $x = 0$ and $3x = -4^{1/3}y$. It will be found, however, that the given curve possesses a

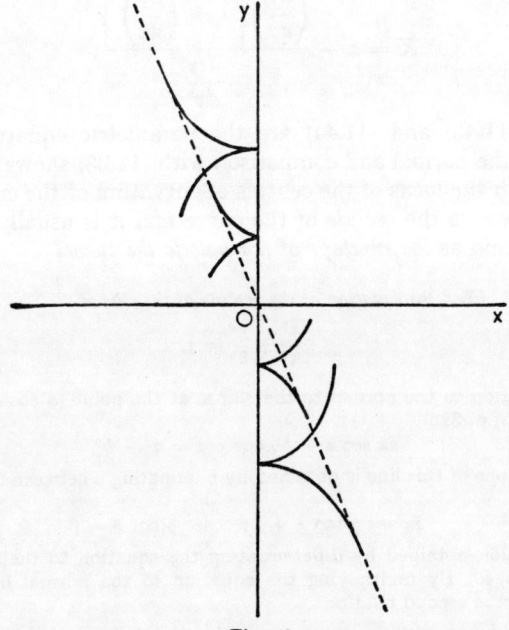

Fig. 41

cusp at the points $(0, -a)$, the tangent there being $y + a = 0$. A rough sketch of the curves for a few values of a shows that the line $x = 0$ is, in fact, a 'cusp-locus', while the line $3x = -4^{1/3}y$ is a true envelope (Fig. 41).

If X, Y are used to denote current coordinates, the equation to the normal to a curve $Y = f(X)$ at the point (x, y) is

$$X - x + \left(\frac{dy}{dx}\right)(Y - y) = 0. \tag{11.41}$$

Treating x as a parameter, the envelope of the normal is given by eliminating x and y between the equation to the curve, the equation to the normal (11.41) and the equation

$$-1 + \left(\frac{d^2y}{dx^2}\right)(Y - y) - \left(\frac{dy}{dx}\right)^2 = 0, \tag{11.42}$$

obtained by differentiating (11.41) with respect to x. Equation (11.42) gives

$$Y = y + \frac{1 + \left(\frac{dy}{dx}\right)^2}{\frac{d^2y}{dx^2}}, \tag{11.43}$$

and substitution in (11.41) then yields

$$X = x - \frac{\left(\frac{dy}{dx}\right)\left\{1 + \left(\frac{dy}{dx}\right)^2\right\}}{\frac{d^2y}{dx^2}}. \tag{11.44}$$

Equations (11.43) and (11.44) are the parametric equations to the envelope of the normal and comparison with (11.23) shows that this is identical with the locus of the centres of curvature of the curve. This locus is known as the *evolute* of the curve and it is usually most conveniently found as *the envelope of normals to the curve*.

Example 19. *Find the equation of the evolute of the ellipse*

$$\frac{x^2}{a^2} + \frac{y^2}{b^2} = 1. \tag{C.}$$

The equation to the normal to the ellipse at the point $(a \cos \phi, b \sin \phi)$ is $(A.L.P.M.,$ p. 322),

$$ax \sec \phi - by \csc \phi = a^2 - b^2.$$

The envelope of this line is obtained by eliminating ϕ between this equation and

$$ax \sec \phi \tan \phi + by \csc \phi \cot \phi = 0,$$

the equation obtained by differentiating the equation to the normal with respect to ϕ. By multiplying the equation to the normal by $\cot \phi$ and adding to the second relation

$$ax \sec \phi (\cot \phi + \tan \phi) = (a^2 - b^2) \cot \phi$$

giving, $$x = \left(\frac{a^2-b^2}{a}\right)\cos^3\phi.$$

In a similar way, $$y = -\left(\frac{a^2-b^2}{b}\right)\sin^3\phi,$$

and elimination of ϕ between these two relations leads to the equation of the evolute in the form
$$(ax)^{2/3} + (by)^{2/3} = (a^2 - b^2)^{2/3}.$$

EXERCISES 11 (d)

1. Show that there is a double point at $(0, -a)$ on the curve
$$x^2y^2 - (a+y)^2(b^2-y^2) = 0.$$
Show further that this point is a cusp if $b = a$.

2. Show that the origin is a node or conjugate point on the curve
$$y^2 = bx \sin(x/a)$$
according as a and b have like or unlike signs.

3. Find the equations to the tangents at the origin on the curve
$$x(x^2+y^2) = a(x^2-y^2)$$
and hence show that the origin is a node.

4. If a straight line moves so as to cut off distances (from the origin) a and b from the axes of x and y, related by the equation $a^n + b^n = c^n$ where c and n are constants, show that it envelops the curve
$$x^{\frac{n}{n+1}} + y^{\frac{n}{n+1}} = c^{\frac{n}{n+1}}. \tag{O.}$$

5. Find the point of contact of the line $x\cos\psi + y\sin\psi = p$ with its envelope, p being a function of ψ.
The tangent to a curve at P meets Ox at Q and Oy at R. The lines through Q and R parallel to Oy and Ox respectively meet at T.
Prove that the tangent at T to the locus of T and PT are equally inclined to the axes. (O.)

6. P is a variable point $(a\cos\theta, a\sin\theta)$ on the circle $x^2 + y^2 = a^2$ and PQ a line making an angle 2θ with the positive direction of the axis of x. Find the coordinates of the point at which PQ touches its envelope. (O.)

7. Show that the parametric equations of the evolute of the curve
$$x^{2/3} + y^{2/3} = a^{2/3}$$
are $x = \tfrac{1}{2}a(3\cos\theta - \cos 3\theta)$, $y = \tfrac{1}{2}a(3\sin\theta + \sin 3\theta)$. (O.)

8. The arc AP of a curve measured from a fixed point A is s and the radius of curvature at P is ϱ. Prove that the radius of curvature at the corresponding point of the evolute is $\varrho(d\varrho/ds)$. (O.)

11.8. Curve sketching

Some of the points to look for when making a rough sketch of a curve whose cartesian equation is given have been listed in the previous book (*A.L.P.M.*, p. 164). These are, briefly:

(i) symmetry about either of the coordinate axes or about the origin;
(ii) ranges of values of x (or y) for which y^2 (or x^2) is negative and hence for which there are no real points on the curve;
(iii) the points of intersection of the curve with the coordinate axes;
(iv) turning points and points of inflexion.

To these can now be added:
(v) find the asymptotes: both those parallel to the coordinate axes and the oblique ones (§ 11.3);
(vi) if the curve passes through the origin, determine its character there by equating to zero the terms of the lowest degree in its equation (§ 11.6): see also the paragraph following Example 21 below;
(vii) search for the position and character of any double points which the curve may possess (§ 11.5).

Two examples follow.

Example 20. *If a and b are positive constants, trace the curve*
$$y^2 = x^2\left(\frac{a+x}{b-x}\right).$$

In tracing the curve it is to be observed that:
(i) only even powers of y occur in the equation to the curve so that there is symmetry about the axis of x;
(ii) there are no real points on the curve when $x < -a$ or when $x > b$ for such values of x make y^2 negative;

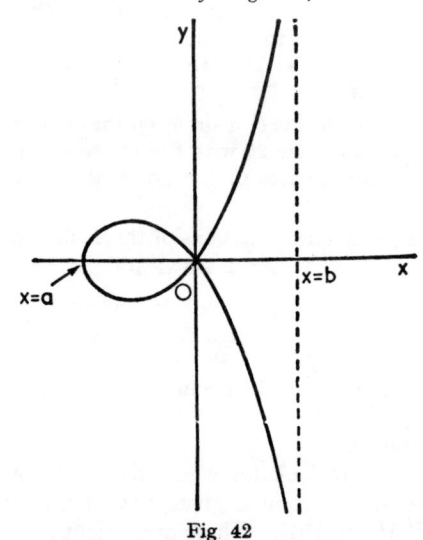

Fig 42

(iii) the curve intersects the x-axis ($y = 0$) when $x = 0$ and when $x = -a$;

(iv) by writing the equation to the curve in the form
$$y^2(b - x) - x^2(a + x) = 0$$
and equating to zero the coefficient of y^2, it can be inferred that the line $x = b$ is an asymptote: it is easily confirmed that this line is the only real asymptote;

(v) the tangents at the origin (given by equating to zero the terms of the lowest degree in the equation to the curve) are $by^2 - ax^2 = 0$, or $y = \pm x\sqrt{(b/a)}$, showing that the origin is a node.

A rough sketch of the curve is given in Fig. 42.

Example 21. *Trace the curve* $x^2(x - y) + y^2 = 0$. (C.)

It is to be observed that:

(i) there is no symmetry about the coordinate axes nor about the origin;
(ii) writing the equation to the curve in the form $y^2 - x^2y + x^3 = 0$, it follows that
$$2y = x^2 \pm \sqrt{(x^4 - 4x^3)},$$

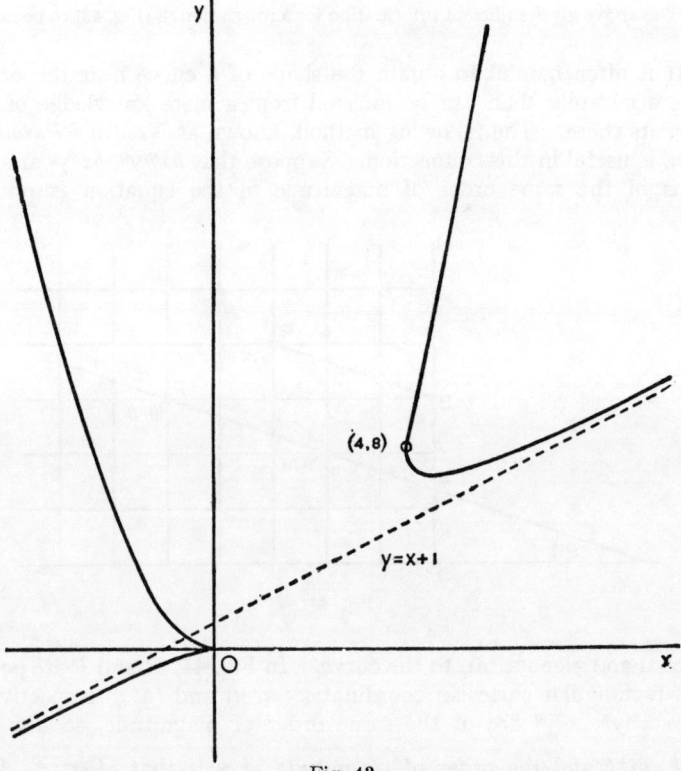

Fig. 43

and y is imaginary between $x = 0$ and $x = 4$: expanding the radical, it is easily found that

$$2y = x^2\left\{1 \pm \left(1 - \frac{2}{x} + \ldots\right)\right\},$$

so that for a given large value of x, there are two values of y which approximate respectively to $y = x^2$ and $y = x$;

(iii) the curve meets the coordinate axes only at the origin;

(iv) differentiating the equation to the curve and making a slight rearrangement,

$$(2y - x^2)\frac{dy}{dx} + 3x^2 - 2xy = 0,$$

so that dy/dx vanishes when $x = 0$ and when $3x = 2y$: thus the tangent is parallel to the axis of x at the origin and at the point of intersection of the curve with the line $3x = 2y$, and this is easily found to be the point $(9/2, 27/4)$: dy/dx is infinite when $2y = x^2$ and substitution in the equation to the curve shows that the tangent is perpendicular to the axis of x at the point $(4, 8)$;

(v) the only real asymptote is easily found to be the line $y = x + 1$;

(vi) the tangents at the origin are given by $y^2 = 0$, showing that the origin is a cusp.

The above gives sufficient information for a rough sketch (Fig. 43) to be made.

It is often helpful to obtain the shape of a curve near the origin more accurately than can be inferred from a mere knowledge of the tangents there. The following method, known as *Newton's Parallelogram*, is useful in this connection. Suppose that $ax^m y^n$, $bx^p y^q$ are two terms of the same order of magnitude in the equation (supposed

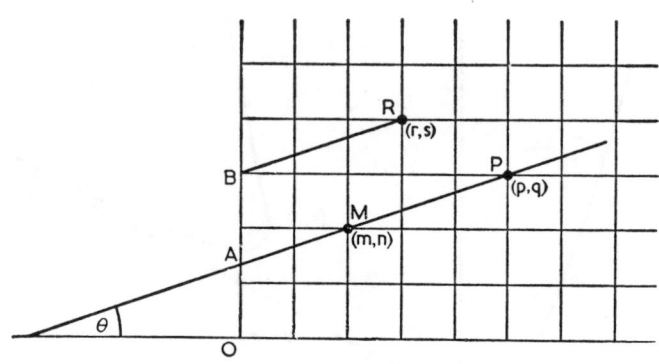

Fig. 44

rational and algebraical) to the curve. In Fig. 44, M and P are points with rectangular cartesian coordinates (m, n) and (p, q) respectively. Since $x^m y^n$, $x^p y^q$ are of the same order of magnitude, so also are x^{m-p}, y^{q-n} and the order of magnitude of x is that of $y^{\frac{q-n}{m-p}}$. It is easy to see from the diagram that the order of x is that of $y^{-\tan\theta}$ where θ is the angle made by the line PM with the horizontal axis. The order of the term $ax^m y^n$ is therefore that of $y^{n-m\tan\theta}$ and, since $OA = n - m\tan\theta$, its order is therefore measured by the intercept OA of Fig. 44. If R represents the position on the same diagram of any other term $cx^r y^s$ in the equation to the curve, the order of magnitude of this term is measured by OB, where B is the point in which a line through R parallel to PM meets the vertical axis of Fig. 44. If $OB > OA$, this term can therefore be rejected in comparison with the terms represented by M and P when tracing the curve near the origin. Similarly if $OB < OA$, the term can be rejected in comparison with the others when drawing the curve at great distances from the origin. These considerations give rise to the rule: if all the terms in the equation to a curve be represented as points on a diagram in the above manner and if when any two, say M and P, are selected, all the other points lie on the side of the line PM remote from the origin, then the terms represented by these points may be rejected when sketching the curve near the origin; if, on the other hand, M and P are selected so that all the other points lie nearer to the origin than PM, then the terms represented by these points may be rejected when sketching the curve far from the origin.

Example 22. *Trace the curve* $x^5 + y^5 - 5ax^2 y = 0$.

In Fig. 45, the terms x^5, y^5, $-5ax^2 y$ are represented respectively by the points A, B and C. The diagram shows at once that, near the origin, the curve may be represented by the first and third or by the second and third

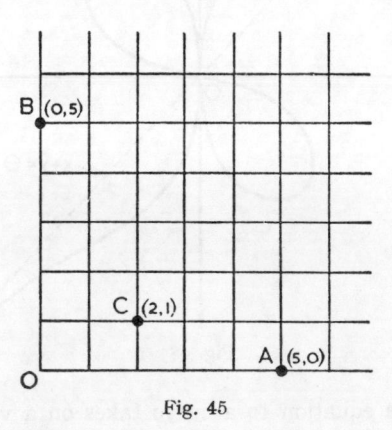

Fig. 45

terms, so that the form of the curve near the origin is that of the curves $x^3 = 5ay$ and $y^4 = 5ax^2$. The first of these is the well-known cubic curve shown in Fig. 46 and the second consists of the two parabolas $y^2 = \pm x\sqrt{(5a)}$ of Fig. 47. Far from the origin, Fig. 45 shows that the term $5ax^2 y$ may be

Fig. 46   Fig. 47

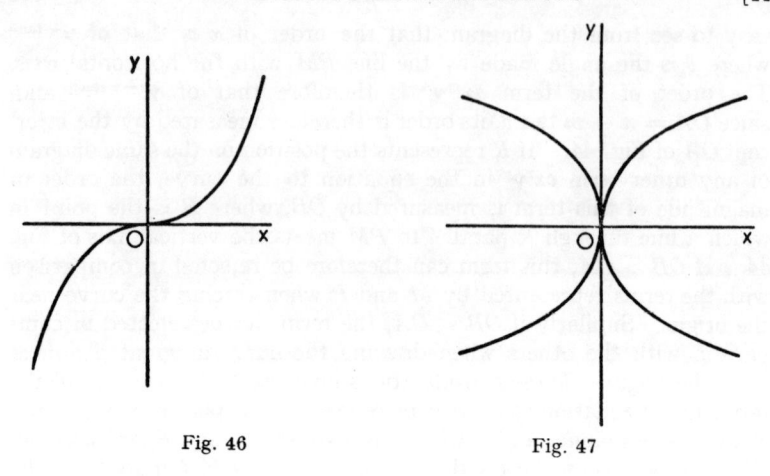

neglected and the factor $(x + y)$ of the remaining terms shows that the line $x + y = 0$ is an asymptote. A rough sketch of the curve (Fig. 48) can be made from these observations.

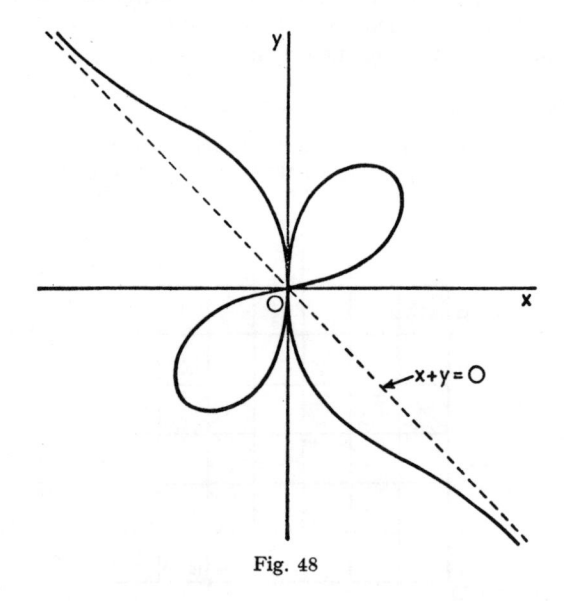

Fig. 48

Sometimes the equation to a curve takes on a very simple form when it is expressed in polar coordinates: this is especially true when the origin is a multiple point on the curve. A suggested procedure for building up information on which to base a rough sketch of a curve whose equation is given in the form $r = f(\theta)$ is as follows:

(i) examine for symmetry: if a change of sign of θ leaves the equation unaltered, there is symmetry about the initial line Ox: if only even powers of r occur, there is symmetry about the origin;

(ii) check from the given equation if there are any obvious limits between which r must lie;

(iii) if possible, make a very rough table showing how r varies with θ;

(iv) it is sometimes helpful to determine the direction of the tangent at given points by finding ϕ (equation (11.11));

(v) search for values of θ which make r very large: this will indicate the existence of asymptotes.

Example 23. *Trace the curve* $r^2 = a^2 \cos 2\theta$.

It is to be observed that:

(i) there is symmetry about the initial line and about the origin;

(ii) r is never greater than a: r^2 is negative when $(\pi/4) < \theta < (\pi/2)$ and no real points exist for such values of θ;

(iii)

θ	0	$\dfrac{\pi}{6}$	$\dfrac{\pi}{4}$
r	a	$\dfrac{a}{2}$	0

The above is sufficient to show that the curve consists of two similar loops as shown in Fig. 49.

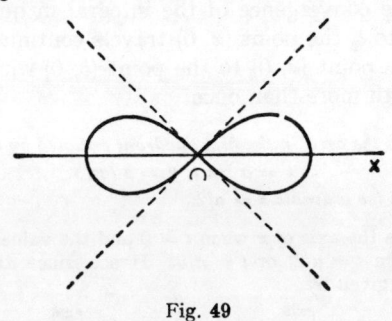

Fig. 49

EXERCISES 11 (e)

1. Sketch the curve $x^2(a^2 - x^2) = 8a^2y^2$. (O.)

2. Find the asymptote of the curve $y^2(2 - x) - x^3 = 0$ and give a rough sketch of the curve. (C.)

3. Draw a rough sketch of the curve $4ay^2 = (x - a)(x - 3a)^2$. (O.)

4. Draw a sketch of the curve
$$\frac{y^2}{c^2} = \frac{a^2 - x^2}{b^2 - x^2},$$
where a, b, c are positive and $a < b$. (C.)

5. Sketch the curve whose equation in polar coordinates is $r = a \cos 2\theta$. (C.)

6. Sketch the curve $r = a + b \cos \theta$, where a and b are positive,
(i) when $b > a$, (ii) when $b = a$, (iii) when $b < a$. (O.)

11.9. Some further formulae for plane areas

The plane area (A) bounded by the curve $y = f(x)$, the axis of x and ordinates at $x = a$, $x = b$ has already been established (*A.L.P.M.*, p. 179) as

$$A = \int_a^b f(x)\,dx, \tag{11.45}$$

and it is the purpose of this section to establish some further formulae for the area bounded by plane curves.

If the bounding curve is given by the parametric equations
$$x = f(t),\ y = g(t),$$
it follows immediately from (11.45) that the area bounded by the curve, the axis of x and ordinates at $x = a$, $x = b$ is given by

$$A = \int_{t_1}^{t_2} g(t) f'(t)\,dt, \tag{11.46}$$

where t_1, t_2 are the values of the parameter corresponding respectively to $x = a$, $x = b$. It should be noted that the validity of (11.46) requires, besides the convergence of the integral in question, that as t increases from t_1 to t_2 the point $(x, 0)$ travels continuously and in one direction from the point $(a, 0)$ to the point $(b, 0)$ without going over any part of its path more than once.

Example 24. *Find the area in the first quadrant enclosed by the curve*
$$x = a \sin t,\ y = a \tan t,$$
the axis of x and the ordinate $x = a/2$. (O.C.)

The curve meets the axis of x when $t = 0$ and the value of t when $x = a/2$ is given by $a \sin t = a/2$, or $t = \pi/6$. Hence, since $dx/dt = a \cos t$, the required area is given by

$$A = \int_0^{\pi/6} a \tan t \cdot a \cos t\,dt = a^2 \int_0^{\pi/6} \sin t\,dt$$
$$= a^2 \Big[-\cos t\Big]_0^{\pi/6} = \frac{a^2(2 - \sqrt{3})}{2}.$$

To obtain a formula giving the *sectorial area* bounded by a curve and two radius vectors, suppose (Fig. 50) that O is the origin, Ox the initial line and AB an arc of a curve whose equation in polar coordinates is $r = f(\theta)$. It is assumed that a given radius vector through O meets

PLANE AREAS

the arc AB in one point only and it is here assumed (but this restriction can be later removed) that the radius vector r of points on the curve does not decrease as θ increases. Suppose also that P is the point (r, θ) and Q the neighbouring point $(r + \delta r, \theta + \delta \theta)$ on the curve. With centre O, circular arcs PL, QM are drawn through P and Q to

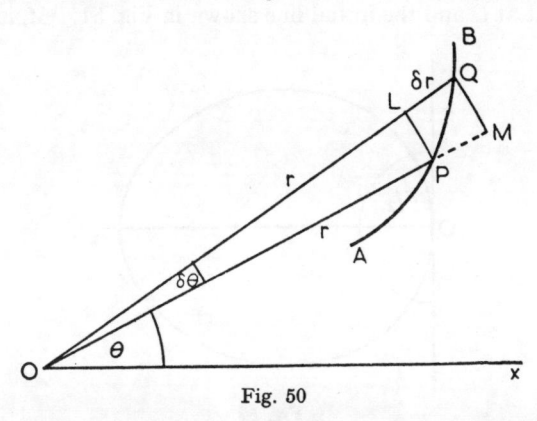

Fig. 50

meet OQ, OP respectively in L, M and it is clear from the figure that area $OPL \leqslant$ area $OPQ \leqslant$ area OQM, the signs of equality ocurring only when the curve AB is part of a circle centre O. If the area OPQ is denoted by δA, the usual formula for the area of a sector of a circle gives

$$\tfrac{1}{2}r^2\delta\theta \leqslant \delta A \leqslant \tfrac{1}{2}(r + \delta r)^2 \delta\theta.$$

It follows by the usual argument (see *A.L.P.M.*, p. 178) that

$$A = \lim_{\delta\theta \to 0} \Sigma \tfrac{1}{2} r^2 \delta\theta = \tfrac{1}{2} \int_\alpha^\beta r^2 d\theta, \tag{11.47}$$

A being the whole area included between radius vectors inclined at angles α, β to the initial line.

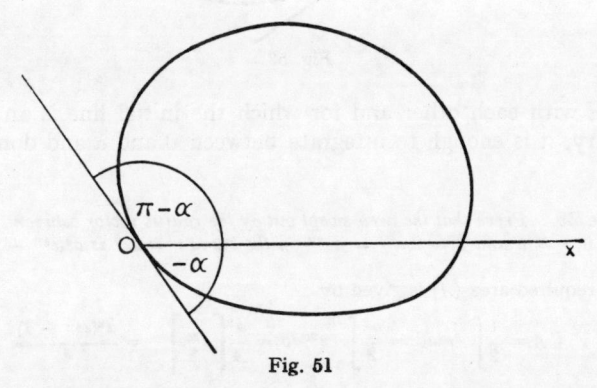

Fig. 51

338 MATHEMATICAL ANALYSIS [11

In using formula (11.47) it is worth noticing that if the curve is closed with the origin inside it, the limits of integration to be used in finding the whole area are 0, 2π. If the origin O is on the perimeter of a closed curve and if it is not a multiple point, the limits required in finding the total area are $-\alpha$ and $\pi - \alpha$ where α is the angle between the tangent at O and the initial line shown in Fig. 51. If, in addition,

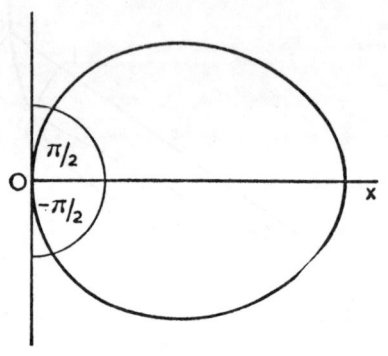

Fig. 52

the initial line is an axis of symmetry (Fig. 52), the limits become $-\pi/2, \pi/2$: it is sufficient in this case to integrate between 0 and $\pi/2$ and double the result. Finally (Fig. 53), in finding the whole area of a loop on which the origin O is a node at which the tangents are inclined at an

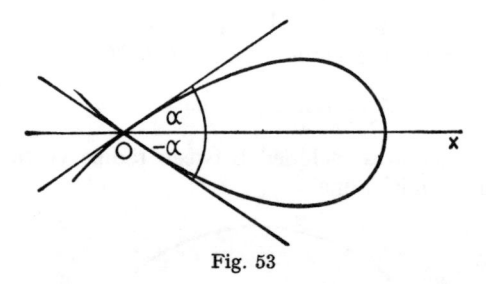

Fig. 53

angle 2α with each other and for which the initial line is an axis of symmetry, it is enough to integrate between 0 and α and double the result.

Example 25. *Prove that the area swept out by the radius vector between the points $\theta = 0$, $r = a$ and $\theta = 2\pi$, $r = ae^{2\pi}$ on the curve $r = ae^{\theta}$ is $a^2(e^{4\pi} - 1)/4$.*
(O.C.)

The required area (A) is given by

$$A = \frac{1}{2}\int_0^{2\pi} r^2 d\theta = \frac{a^2}{2}\int_0^{2\pi} e^{2\theta}d\theta = \frac{a^2}{4}\left[e^{2\theta}\right]_0^{2\pi} = \frac{a^2(e^{4\pi} - 1)}{4}.$$

Example 26. *Show that the area of the loop of the curve $r\cos\theta = a\cos 2\theta$ is*
$$a^2\left(2 - \frac{\pi}{2}\right). \tag{C.}$$

A rough sketch of the curve shows that the initial line is an axis of symmetry and that the origin is a node at which the two tangents are mutually perpendicular. Hence the required area

$$= \int_0^{\pi/4} r^2 d\theta = a^2 \int_0^{\pi/4} \frac{\cos^2 2\theta}{\cos^2\theta} d\theta = a^2 \int_0^{\pi/4} (2\cos\theta - \sec\theta)^2 d\theta$$

$$= a^2 \int_0^{\pi/4} (4\cos^2\theta - 4 + \sec^2\theta) d\theta = a^2 \int_0^{\pi/4} (2\cos 2\theta - 2 + \sec^2\theta) d\theta$$

$$= a^2\left[\sin 2\theta - 2\theta + \tan\theta\right]_0^{\pi/4} = a^2\left(2 - \frac{\pi}{2}\right).$$

When the bounding curve is defined by equations giving the cartesian coordinates of a point on it in terms of a parameter t, the sectorial area OAB of Fig. 50 is easily derived from formula (11.47). Since $x = r\cos\theta$, $y = r\sin\theta$, differentiation with respect to the parameter t gives

$$\frac{dx}{dt} = \frac{dr}{dt}\cos\theta - r\sin\theta\frac{d\theta}{dt}, \quad \frac{dy}{dt} = \frac{dr}{dt}\sin\theta + r\cos\theta\frac{d\theta}{dt}.$$

Multiplying by $y = r\sin\theta$, $x = r\cos\theta$ respectively and subtracting,

$$x\frac{dy}{dt} - y\frac{dx}{dt} = r^2\frac{d\theta}{dt}.$$

Hence the sectorial area OAB is given by

$$A = \frac{1}{2}\int_\alpha^\beta r^2 d\theta = \frac{1}{2}\int_{t_1}^{t_2}\left(x\frac{dy}{dt} - y\frac{dx}{dt}\right)dt, \tag{11.48}$$

where t_1, t_2 are the values of the parameter corresponding respectively to the points A and B.

Example 27. *A rod OA of length $2a$ turns about a fixed point O, and a rod AB of length a turns about A in the same plane with twice the angular velocity of OA. Initially OAB is a straight line. Find the whole area of the curve traced out by the point B.* (C.)

Taking the initial position of the rods as the axis of x and the angle turned through by OA at time t to be θ (Fig. 54), the angle between the rod AB and the x − axis at this instant is 2θ. It is clear from the diagram that the cartesian coordinates of the point B are given by

$$x = 2a\cos\theta + a\cos 2\theta, \quad y = 2a\sin\theta + a\sin 2\theta.$$

The curve traced out by B is clearly symmetrical with respect to the line Ox and, since O is an internal point, the whole area (A) of the curve is given by

$$A = \int_0^\pi \left(r\frac{dy}{d\theta} - y\frac{dx}{d\theta}\right)d\theta.$$

Fig. 54

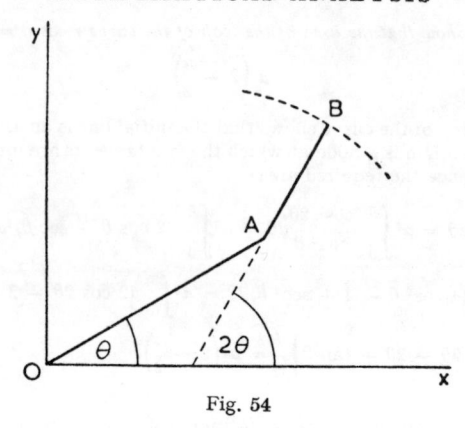

After a little reduction it is easily found that

$$x\frac{dy}{d\theta} - y\frac{dx}{d\theta} = 6a^2(1 + \cos\theta)$$

so that $\quad A = 6a^2 \int_0^\pi (1 + \cos\theta)d\theta = 6a^2\Big[\theta + \sin\theta\Big]_0^\pi = 6\pi a^2.$

EXERCISES 11 (f)

1. Sketch the curve $x = a\cos^3 t$, $y = a\sin^3 t$ and find its whole area. (O.C.)

2. Show that the curve $x = a\sin 2t$, $y = a\sin t$ consists of two equal loops and that the area of either is $4a^2/3$. (O.)

3. Find the area of the plane region whose boundary is given in polar coordinates by $r = a(1 + \cos\theta)$. (C.)

4. Find the area of the part of one loop of the curve $r = a\cos 2\theta$ that lies outside the circle $r = \sqrt{3}a/2$. (C.)

5. Sketch the curve $r^2(1 + \frac{1}{2}\tan^2\frac{1}{2}\theta) = a^2$ and find its area.
   (Negative values of r are to be ignored.) (O.C.)

6. Sketch the curve $\quad x = t^2 + 1,\ y = t(t^2 - 4)$.
   Show that it has a loop and find the area of this loop. (C.)

11.10. Further formulae for length of arc

The length (s) of arc of the curve $y = f(x)$ between two points on the curve at abscissae $x = a$, $x = b$ is (*A.L.P.M.*, p. 228)

$$s = \int_a^b \sqrt{\left\{1 + \left(\frac{dy}{dx}\right)^2\right\}}\,dx, \qquad (11.49)$$

and it is now proposed to derive similar formulae when the equation to the curve is given in other forms.

Since, $$\frac{ds}{dx} = \sqrt{\left\{1 + \left(\frac{dy}{dx}\right)^2\right\}}$$

and since, $$\frac{ds}{dt} = \frac{dx}{dt}\frac{ds}{dx}, \quad \frac{dy}{dx} = \frac{dy}{dt}\bigg/\frac{dx}{dt},$$

it follows that $$\frac{ds}{dt} = \sqrt{\left\{\left(\frac{dx}{dt}\right)^2 + \left(\frac{dy}{dt}\right)^2\right\}}.$$

Hence, when the equation to the curve is given parametrically by the equations $x = f(t)$, $y = g(t)$, a formula for the length of arc between two points on the curve at which the values of the parameter are t_1 and t_2 is

$$s = \int_{t_1}^{t_2} \sqrt{\left\{\left(\frac{dx}{dt}\right)^2 + \left(\frac{dy}{dt}\right)^2\right\}}\, dt. \tag{11.50}$$

Example 28. *Find the length of the whole perimeter of the closed curve*

$$x = a\cos^3 t, \quad y = a\sin^3 t. \tag{O.C.}$$

A rough sketch of the curve shows that it is symmetrical with respect to both of the coordinate axes and its whole length is therefore four times that in the first quadrant. Hence the required length

$$= 4\int_0^{\pi/2} \sqrt{\left\{\left(\frac{dx}{dt}\right)^2 + \left(\frac{dy}{dt}\right)^2\right\}}\, dt = 4a\int_0^{\pi/2} \sqrt{\{9\cos^4 t\sin^2 t + 9\sin^4 t\cos^2 t\}}\, dt$$

$$= 12a\int_0^{\pi/2} \sin t\cos t\, dt = 12a\left[\frac{\sin^2 t}{2}\right]_0^{\pi/2} = 6a.$$

Using (11.12), it follows that the length of arc of the curve whose equation in polar coordinates is $r = f(\theta)$ is given by

$$s = \int_{\theta_1}^{\theta_2} \sqrt{\left\{r^2 + \left(\frac{dr}{d\theta}\right)^2\right\}}\, d\theta \tag{11.51}$$

where θ_1, θ_2 are the θ coordinates of the end-points. If the equation to the curve is given in the inverse form $\theta = f(r)$, it is more convenient to use the formula (immediately derivable from (11.51))

$$s = \int_{r_1}^{r_2} \sqrt{\left\{1 + r^2\left(\frac{d\theta}{dr}\right)^2\right\}}\, dr, \tag{11.52}$$

where now r_1, r_2 are the r coordinates of the end-points.

Example 29. *Prove that the length of arc between the points $\theta = 0$, $r = a$, and $\theta = 2\pi$, $r = ae^{2\pi}$ on the curve $r = ae^\theta$ is $\sqrt{2}a(e^{2\pi} - 1)$.* (O.C.)

Here $dr/d\theta = ae^\theta$ so that $r^2 + \left(\dfrac{dr}{d\theta}\right)^2 = 2a^2 e^{2\theta}$,

and the required length is given by

$$\int_0^{2\pi} \sqrt{2}\,ae^\theta\, d\theta = \sqrt{2}\,a(e^{2\pi} - 1).$$

11.11. Volumes and surface areas of figures of revolution

The volume (V) and the surface area (S) of the solid formed by rotating through four right-angles the part of the curve $y = f(x)$ which lies between ordinates at $x = a$ and $x = b$ are given (*A.L.P.M.*, pp. 184, 231) by

$$V = \pi \int_a^b y^2 dx, \tag{11.53}$$

$$S = 2\pi \int_a^b y \sqrt{\left\{1 + \left(\frac{dy}{dx}\right)^2\right\}} dx. \tag{11.54}$$

Since $y = r \sin \theta$ and

$$\sqrt{\left\{1 + \left(\frac{dy}{dx}\right)^2\right\}} dx = ds = \sqrt{\left\{r^2 + \left(\frac{dr}{d\theta}\right)^2\right\}} d\theta,$$

the second formula gives for the surface area of the figure generated by rotating about the initial line the part of the polar curve $r = f(\theta)$ which lies between points for which $\theta = \alpha$, $\theta = \beta$,

$$S = 2\pi \int_\alpha^\beta r \sqrt{\left\{r^2 + \left(\frac{dr}{d\theta}\right)^2\right\}} \sin \theta \, d\theta. \tag{11.55}$$

Example 30. *The curve $r = a(1 + \cos \theta)$ is rotated through two right-angles about the initial line. Show that the area of the surface generated is $32\pi a^2/5$.* (L.I.C.)

The given curve is symmetrical about the initial line and the solid generated is equivalent to that formed by the rotation of the *upper* part of the curve through four right-angles. Here

$$r = a(1 + \cos \theta), \quad \frac{dr}{d\theta} = -a \sin \theta,$$

$$r^2 + \left(\frac{dr}{d\theta}\right)^2 = a^2(1 + \cos \theta)^2 + a^2 \sin^2 \theta = 2a^2(1 + \cos \theta) = 4a^2 \cos^2 \frac{\theta}{2}.$$

Hence the required surface area

$$= 4\pi a^2 \int_0^\pi (1 + \cos \theta) \sin \theta \cos \frac{\theta}{2} d\theta = 16\pi a^2 \int_0^\pi \cos^4 \frac{\theta}{2} \sin \frac{\theta}{2} d\theta$$

$$= 16\pi a^2 \left[-\frac{2}{5} \cos^5 \frac{\theta}{2}\right]_0^\pi = \frac{32\pi a^2}{5}.$$

The following interesting and useful theorems on figures of revolution are due to *Pappus* (about A.D. 300).

(i) *If a closed plane curve is rotated about an axis in its plane which does not intersect the curve, the surface area of the figure generated is equal to the length of the arc of the curve multiplied by the length of the path of the centre of mass of the arc.*

Suppose the axis of x is the axis of rotation and let y be an ordinate of the generating curve. The surface area generated is

$$2\pi \int y \sqrt{\left\{1 + \left(\frac{dy}{dx}\right)^2\right\}} dx = 2\pi \int y \, ds,$$

the integration being over the whole length of the generating curve. But the ordinate $\bar{y}$ of the centre of mass of the arc is given by

$$\bar{y} = \frac{\int y\,ds}{\int ds},$$

so that the surface area is $2\pi\bar{y}\int ds$ and this proves the theorem.

(ii) *If a closed plane curve is rotated about an axis in its plane which does not intersect the curve, the volume of the figure generated is equal to the area of the plane curve multiplied by the length of the path of the centre of mass of the area.*

Employing the previous notation, if δA be an element of the area of the generating curve, the volume generated in a complete revolution is

$$\lim_{\delta A \to 0} \Sigma(2\pi y \delta A).$$

But, if $\bar{y}$ is now the ordinate of the centre of mass of the area

$$\bar{y} = \lim_{\delta A \to 0} \left\{ \frac{\Sigma(y\delta A)}{\Sigma(\delta A)} \right\},$$

so that the volume generated is $2\pi\bar{y} \lim_{\delta A \to 0} \Sigma(\delta A)$ and the theorem is proved.

In the above the rotation of the generating curve has been taken to be through four right-angles but this restriction is clearly not essential. The theorems are useful both as stated and when used conversely—in the converse forms, the centre of mass of a plane arc (or of a plane area) may be found when the surface area (or the volume) generated by its revolution is known.

Example 31. *Show that the centre of mass of a semicircular arc of radius a is distant $2a/\pi$ from the diameter joining its two extremities.*

Here the length of arc of the generating curve is πa and the surface area of the figure generated is $4\pi a^2$. Hence if $\bar{y}$ is the required distance of the centre of mass from the bounding diameter, Pappus' first theorem gives

$$2\pi\bar{y} \times \pi a = 4\pi a^2,$$

so that $\bar{y} = 2a/\pi$.

Example 32. *Show that the volume of the solid formed by the rotation about the line $\theta = 0$ of the area bounded by the curve $r = f(\theta)$ and the lines $\theta = \alpha$ $\theta = \beta$ is*

$$\frac{2\pi}{3}\int_\alpha^\beta r^3 \sin\theta\,d\theta.$$

Hence find the volume of the solid formed by the revolution of the curve $r = a(1 + \cos\theta)$ about the line $\theta = 0$. (L.I.C.)

The element of area formed by radius vectors inclined at angles θ, $\theta + \delta\theta$ to the line $\theta = 0$ is, to the first order of small quantities, a triangle of area $(r^2\delta\theta)/2$ and its centre of mass is two-thirds along its median. Hence, neglecting small quantities of the first and higher orders, the distance of the centre of mass of the element from the axis of rotation is $(2r \sin \theta)/3$ and Pappus' second theorem gives for the volume of the solid formed by rotating the element

$$2\pi \times \frac{2}{3}r \sin \theta \times \frac{1}{2}r^2\delta\theta, \quad \text{or} \quad \frac{2\pi}{3}r^3 \sin \theta \delta\theta.$$

The whole volume is obtained by summing these elementary volumes for values of θ ranging from α to β and is therefore

$$\frac{2\pi}{3}\int_\alpha^\beta r^3 \sin \theta d\theta.$$

For the curve $r = a(1 + \cos \theta)$, the appropriate limits are $\theta = 0$, $\theta = \pi$ and the required volume

$$= \frac{2\pi}{3}\int_0^\pi a^3(1 + \cos \theta)^3 \sin \theta d\theta = \frac{2\pi a^3}{3}\left[-\frac{(1 + \cos \theta)^4}{4}\right]_0^\pi = \frac{8\pi a^3}{3}.$$

EXERCISES 11 (g)

1. Prove that, in the parabola $x = at^2$, $y = 2at$,

    $$s = a \log_e \cot \tfrac{1}{2}\psi + a \cot \psi \operatorname{cosec} \psi,$$

    where s is the length of an arc measured from the vertex and ψ is the angle the tangent makes with the axis. (O.)

2. Show that the whole length of the curve $r = a(1 - \cos \theta)$ is $8a$.

3. Sketch the curve whose parametric equations are

    $$x = a \sin t, \quad y = a \tan t.$$

    Prove that the volume generated by rotating the area in the first quadrant enclosed by the curve, the axis of x and the line $x = a/2$ through four right-angles about the axis of x is equal to

    $$\tfrac{1}{2}\pi a^3(\log_e 3 - 1). \quad (O.C.)$$

4. The arc of the parabola $r(1 + \cos \theta) = 2a$ between $\theta = 0$ and $\theta = \pi/2$ is rotated through four right-angles about the initial line. Show that the area of the curved surface generated is

    $$\frac{8\pi}{3}(\sqrt{8} - 1)a^2.$$

5. Prove that the centre of mass of a semicircular lamina of radius a is distant $4a/3\pi$ from the base of the semicircle.

6. Find the coordinates of the centre of mass of the smaller segment of the ellipse

    $$\frac{x^2}{a^2} + \frac{y^2}{b^2} = 1$$

    cut off by a latus rectum: and find the volume of the annular solid generated when this segment revolves about the minor axis of the ellipse. (O.)

EXERCISES 11 (h)

1. The curve $y = ax + bx^2$ passes through the points $(-0.2, 0.0167)$, and $(0.25, 0.026)$. Prove that the tangent at the origin makes an angle of approximately 34 seconds with the x—axis. (C.)

2. The intrinsic equation to a curve is $s = c \sin^n \psi$, where s is the distance along the curve from a fixed point O to the variable point P, ψ is the angle between the tangents to the curve at O and P, and n, c are constants. If the tangents to the curve at O and P intersect at T, prove that TP is proportional to s.
   Show also that, when $n = 2$, the cartesian equation to the curve is
   $$(3x)^{2/3} + (3y)^{2/3} = (2c)^{2/3}$$
   with a suitable choice of axes. (O.)

3. Starting from the cartesian equation to an ellipse, deduce its (p, r) equation with respect to a focus as pole. (O.)

4. Prove that the curve $(y - x)^2(y + x)(y + 2x) = a^2y^2$ has two parallel asymptotes. (O.)

5. Find the asymptotes of the curve $y^4 - y^2x^2 + 2y^3 - 2x = 0$. (O.)

6. A cubic curve has asymptotes $x = a$, $y = a$, $x + y + 4a = 0$, and a double point at the origin. Obtain its equation, showing that the double point is a cusp. (O.)

7. A curve is given parametrically by the equations
   $$x = a \cos^3 t, \quad y = a \sin^3 t.$$
   Find the parametric equations of the locus of its centre of curvature. (C.)

8. By putting $y = tx$, obtain a parametric representation of the curve
   $$x^3 + y^3 = 3axy.$$
   Sketch the curve, and obtain the radii of curvature of the two branches at the origin, proving any formulae you use for radius of curvature. (O.C.)

9. O is the middle point of a straight line AB of length $2a$. P moves so that $AP \cdot BP = c^2$. Show that the radius of curvature at P of the locus is
   $$\frac{2c^2 r^3}{3r^4 + a^4 - c^4}$$
   where $r = OP$. (C.)

10. Obtain the equation of the circle of curvature of the curve $y = 1 - \cos x$ at the origin. If (x, y_1) and (x, y_2) are respectively points of the curve and the circle of curvature near the origin, prove that, as x tends to zero,
    $$\frac{y_2 - y_1}{x^4} \to \frac{1}{6}.$$
    (C.)

11. The cartesian coordinates of a point on a plane curve are x, y and its intrinsic coordinates are s, ψ. The complex coordinate $x + iy$ is denoted by z and the curvature by k. Prove that

$$\frac{dz}{ds} = e^{i\psi}, \quad \frac{d^2z}{ds^2} = ike^{i\psi}.$$

P is any point on a given curve. The tangent at P is drawn in the direction of increasing s, and a point Q is taken at a constant distance l along this tangent. In this way Q describes a curve specified by X, Y, S, Z analogous to the specification x, y, s, z for P. Show that

(i) $\dfrac{dS}{ds} = (1 + l^2k^2)^{1/2}$;

(ii) the curvature K of the derived curve at Q is given by the formula

$$K(1 + l^2k^2)^{3/2} = k(1 + l^2k^2) + lk',$$

where $k' = dk/ds$. (O.C.)

12. A curve touches the x—axis at the origin of coordinates and the arc-length of the curve measured from the origin is denoted by s. By Maclaurin's theorem prove that, if s is small and powers of s higher than the fourth are neglected, the coordinates of a point on the curve are given by

$$x = s - \frac{k^2s^3}{(3)!} - \frac{3kk's^4}{(4)!},$$

$$y = \frac{ks^2}{(2)!} + \frac{k's^3}{(3)!} + \frac{(k'' - k^3)s^4}{(4)!},$$

where k, k', k'' denote the curvature and its derivatives with respect to s, all evaluated at $s = 0$.

Find the length of the chord from the origin to the point (x, y) as a power series in s up to and including the term in s^4. (O.C.)

13. P is a typical point on the curve $x = x(s)$, $y = y(s)$, s being the arc, at which the radius of curvature is ϱ. P_1 is the corresponding point on the curve

$$x_1 = \int \phi(s)\frac{dx}{ds}ds, \quad y_1 = \int \phi(s)\frac{dy}{ds}ds.$$

Prove that the tangent at P_1 is parallel to the tangent at P, and that the radius of curvature at P_1 is $\varrho\phi(s)$.

Show also that the point at which PP_1 touches its envelope divides the segment PP_1 externally in the ratio of the radii of curvature at P and P_1. (O.)

14. Sketch the locus (the cycloid) given by

$$x = a(\theta + \sin\theta), \quad y = a(1 + \cos\theta)$$

for values of the parameter θ between 0 and 4π.

Prove that the normals to this curve all touch an equal cycloid, and draw this second curve in your diagram. (C.)

15. A family of ellipses all of eccentricity ε, have for their major axes parallel chords of a fixed circle. Show that the envelope is an ellipse of eccentricity

$$\sqrt{\left(\frac{1-\varepsilon^2}{2-\varepsilon^2}\right)}.$$ (C.)

16. Normals are drawn to the curve defined by

$$y = \frac{\sin x}{x} (y = 1 \text{ when } x = 0)$$

at the points where it cuts the $x-$ axis. Show that each of these normals touches one of two fixed parabolas. (O.)

17. The line $x \cos \theta + y \sin \theta = f(\theta)$, where θ is variable, envelops a curve and the lines $\theta = \alpha, \beta$ touch it at A, B respectively.
Prove that the length of the arc AB is

$$\int_\alpha^\beta \{f(\theta) + f''(\theta)\} d\theta$$

and the area of the sector AOB is

$$\frac{1}{2}\int_\alpha^\beta f(\theta)\{f(\theta) + f''(\theta)\} d\theta.$$ (O.)

18. A circle of radius a rolls on the axis of x, carrying with it a point at a distance na from the centre. Express the coordinates of the point in the form $x = a(\theta + n \sin \theta), y = a(1 + n \cos \theta)$.
Sketch the curve when $n > 1$, proving that the points $\pi \pm \alpha$ where α is a root of the equation $n \sin \alpha = \alpha$ which lies between 0 and $\pi/2$ are identical, forming a node. Prove that the area of the loop between them is

$$a^2\{n\alpha \cos \alpha - 3n \sin \alpha + (1 + n^2)\alpha\}.$$ (O.)

19. A circle of radius a has centre C, and a point O is taken at distance $c(<a)$ from its centre. The foot of the perpendicular from O to a tangent to the circle is P. Show that the locus of P is a closed curve of area.

$$\pi\left(a^2 + \frac{1}{2}c^2\right).$$ (C.)

20. Show that the area cut off from the curve $x^3 = ay^2$ by the chord joining the points $(at^2, at^3), (au^2, au^3)$ is

$$\frac{1}{10}a^2|t-u|^3(t^2 + 3tu + u^2)$$

and determine the length of arc cut off by the same chord. (O.)

21. The coordinates (x, y) of a point of Cornu's spiral are defined in terms of the parameter v by the equations

$$x = \int_0^v \cos\left(\frac{1}{2}\pi t^2\right) dt, \quad y = \int_0^v \sin\left(\frac{1}{2}\pi t^2\right) dt.$$

Prove that the curve passes through the origin of coordinates and touches the x-axis there. Show that the length (s) of the curve from the origin to the point (x, y) is v, and determine the radius of curvature in terms of s. (O.)

22. A circular lamina of radius a rolls in a vertical plane with its edge on a horizontal rail. Prove that the locus of a point attached to the lamina and at distance b from its centre may be represented by

$$x = a\theta + b \sin \theta, \quad y = a - b \cos \theta.$$

Prove that the length of this curve between horizontal tangents is equal to half the perimeter of an ellipse whose semi-axes are equal to the greatest and least distances of the point from the horizontal rail. (O.)

23. The centre of a sphere of radius $2a$ is a point on the surface of a right circular cylinder of radius a. Prove that the portion of the surface of the cylinder which lies inside the sphere has an area $16a^2$. (O.)

24. Show that the area of the surface of the prolate spheroid obtained by the rotation of an ellipse of eccentricity ε about its major axis $(2a)$ is

$$A = 2\pi a^2 \left\{ 1 - \varepsilon^2 + \sqrt{(1 - \varepsilon^2)} \frac{\sin^{-1} \varepsilon}{\varepsilon} \right\},$$

and that the centroid of the half surface bounded by the central circular section is at a distance d from the plane of that section where

$$Ad = \frac{4\pi a^3}{3} \cdot \frac{1}{\varepsilon^2} \{ \sqrt{(1 - \varepsilon^2)} - (1 - \varepsilon^2)^2 \}. \tag{C.}$$

25. A chord of a circle of radius r subtends an angle $2a$ at the centre. The minor segment cut off by this chord is revolved about the chord through an angle 2π. Prove that the volume of the solid so formed is

$$2\pi r^3 \left(\sin a - \frac{1}{3} \sin^3 a - a \cos a \right). \tag{L.U.}$$

CHAPTER 12

ELEMENTARY DIFFERENTIAL EQUATIONS

12.1. Introduction

Because of their great use in applied mathematics, physics and chemistry, this chapter gives an introduction to the study of some elementary types of differential equations. It is not intended that a systematic and detailed study should be made here but rather that the ordinary processes of the integral calculus should be applied to solve a few elementary equations.

To start with a few definitions—equations such as

$$\frac{d^2y}{dx^2} + 9y = 0, \tag{12.1}$$

$$\frac{d^4y}{dx^4} + 8\frac{d^3y}{dx^3} + 6\frac{d^2y}{dx^2} + y = e^{-x} \cos x, \tag{12.2}$$

$$\left\{1 + \left(\frac{dy}{dx}\right)^2\right\}^{3/2} = \frac{d^2y}{dx^2}, \tag{12.3}$$

$$\frac{\partial^2 y}{\partial x^2} = k\frac{\partial y}{\partial t}, \tag{12.4}$$

all of which involve differential coefficients, are called *differential equations*. The first three of the above equations contain only one independent variable (x) and are known as *ordinary* differential equations. In the fourth equation, two independent variables (x and t) are involved—such equations are termed *partial* differential equations but their study is beyond the scope of the present book. The *order* of an equation is defined as the order of the highest differential coefficient present—the orders of equations (12.1), (12.2), (12.3) are therefore respectively the second, fourth and second. The *degree* of a differential equation is the degree of the highest differential coefficient when the equation has been made rational and integral as far as the differential coefficients are concerned—thus equations (12.1), (12.2) are both of the first degree and equation (12.3) is of the second degree (since the term d^2y/dx^2 is squared when the equation is rationalised). A differential equation is said to be *linear* when it is linear in the dependent variable and all its derivatives. Thus equations (12.1) and (12.2) are linear but (12.3) is not.

Many phenomena in science and applied mathematics are described naturally by a differential equation. Consider, for example, the motion in a straight line of a particle whose acceleration is constant.

If a is the acceleration, v the velocity and s the distance travelled at time t, the motion is described by either of the differential equations

$$\text{(i)} \quad \frac{dv}{dt} = a, \qquad \text{(ii)} \quad \frac{d^2s}{dt^2} = a, \qquad (12.5)$$

the equations being linear and respectively of the first and second orders. It follows immediately from the first equation that

$$v = at + C \qquad (12.6)$$

where C is an arbitrary constant, and this relation between v and t is called the *general solution* or *complete primitive* of the differential equation (12.5 (i)). If the initial velocity of the particle is u so that $v = u$ when $t = 0$, substitution in (12.6) gives $C = u$ and hence

$$v = u + at, \qquad (12.7)$$

the usual relation for this type of motion. The reader should observe that the differential equation $dv/dt = a$ describes all linear motions in which the acceleration is constant as also does the general solution (12.6): the particular solution (12.7) describes only the motion in which the initial velocity is u. Integration of equation (12.5 (ii)) gives

$$\frac{ds}{dt} = at + C, \qquad (12.8)$$

and integration of this then yields

$$s = \frac{1}{2}at^2 + Ct + C', \qquad (12.9)$$

C and C' being arbitrary constants. This is the general solution of (12.5 (ii)) and the reader should observe that this general solution contains two arbitrary constants. The values of these constants can be found if further particulars of the motion are available. For example, if the particle starts from the origin with velocity u so that

$$\frac{ds}{dt} = u \text{ and } s = 0 \text{ when } t = 0,$$

(12.8) gives $C = u$ and (12.9) then gives $C' = 0$ so that the usual relation

$$s = ut + \frac{1}{2}at^2 \qquad (12.10)$$

describes this particular motion.

It can be shown that *the general solution of a differential equation of the nth order contains n arbitrary constants*, but a proof is not attempted here. In practical applications, the arbitrary constants have usually to be determined from additional data in the form of initial or boundary conditions.

12.2. First order differential equations with separable variables

The general differential equation of the first order can be written

$$\frac{dy}{dx} = f(x, y):$$

in this section consideration is given to the special case in which

$$f(x, y) = XY,$$

where X and Y are respectively functions of x and y. This equation can be written

$$\frac{1}{Y}\frac{dy}{dx} = X$$

and integration with respect to x gives

$$\int \frac{1}{Y}\frac{dy}{dx}dx = \int X dx + C, \qquad (12.11)$$

C being an arbitrary constant.

The general solution (12.11) can be simplified to

$$\int \frac{1}{Y} dy = \int X dx + C,$$

and the reader will observe that some of the intermediate steps employed in arriving at this solution may be omitted. Thus, starting from the equation

$$\frac{dy}{dx} = XY, \qquad (12.12)$$

the variables can be *separated* to give

$$\frac{1}{Y}dy = X dx, \qquad (12.13)$$

and integration immediately gives the solution

$$\int \frac{1}{Y}dy = \int X dx + C \qquad (12.14)$$

obtained previously.

Example 1. *Solve the differential equation*

$$\frac{dy}{dx} = \frac{y^2}{1 + x^2}. \qquad \text{(L.U.)}$$

The variables can be separated to give

$$\frac{dy}{y^2} = \frac{dx}{1 + x^2},$$

so that

$$\int \frac{dy}{y^2} = \int \frac{dx}{1 + x^2}.$$

Hence

$$-\frac{1}{y} = \tan^{-1} x + C$$

is the general solution.

Example 2. If $\dfrac{dv}{dt} = -\dfrac{v^2}{100}$ and $v = 50$ when $t = 0$, find the value of t when $v = 10$.

(L.U.)

Separating the variables in the differential equation and integrating

$$-\int \dfrac{dv}{v^2} = \int \dfrac{dt}{100},$$

so that
$$\dfrac{1}{v} = \dfrac{t}{100} + C,$$

is the general solution of the equation. Since $v = 50$ when $t = 0$

$$C = \dfrac{1}{50}$$

and the particular solution, satisfying the given initial condition, is

$$\dfrac{1}{v} = \dfrac{t}{100} + \dfrac{1}{50}.$$

When $v = 10$, it follows that

$$\dfrac{1}{10} = \dfrac{t}{100} + \dfrac{1}{50}$$

so that the required value of t is 8.

Example 3. *Solve the differential equation*

$$\dfrac{dy}{dx} = \dfrac{y(y+1)}{x}.$$ (O.C.)

Separating the variables and integrating

$$\int \dfrac{dy}{y(y+1)} = \int \dfrac{dx}{x}.$$

The integral on the left can be written as

$$\int \left(\dfrac{1}{y} - \dfrac{1}{y+1} \right) dy,$$

so that $\qquad \log_e y - \log_e (y+1) = \log_e x + \log_e C.$
This can be written in the form

$$\dfrac{y}{y+1} = Cx$$

and the reader should note the simplification obtained in this solution by taking $\log_e C$ as the arbitrary constant.

EXERCISES 12 (a)

Solve the differential equations:

1. $\dfrac{dy}{dx} = 2xy.$ (O.C.)

2. $\tan x \dfrac{dy}{dx} + \tan y = 0.$ (O.C.)

3. $\dfrac{dy}{dx} + 3x^2(y-1) = 0.$ (O.C.)

4. If $dy/dx = 1 + \sin x + \sin^2 x$ and $y = 2$ when $x = 0$, find y in terms of x. (L.U.)

5. In a certain chemical reaction the amount x of one substance at time t is related to the velocity of the reaction dx/dt by the equation
$$\frac{dx}{dt} = k(a - x)(2a - x)$$
where a and k are constants and $x = 0$ when $t = 0$. If $x = 2\cdot 0$ when $t = 1$ and $x = 2\cdot 8$ when $t = 3$, show that $a = 3$ and find
(i) the value of k,
(ii) the value of x when $t = 2$. (N.U.)

6. A particle starts moving along a straight line with velocity u ms^{-1} and t seconds later its velocity v ms^{-1} satisfies the equation
$$\frac{dv}{dt} + 1 + v^2 = 0.$$
Prove that the particle comes instantaneously to rest after $\tan^{-1} u$ seconds.
If the particle moves through s metres in the first t seconds, prove that
$$v\frac{dv}{ds} = \frac{dv}{dt}$$
and hence show that the particle has covered $\tfrac{1}{2} \log_e (1 + u^2)$ metres when it first comes to rest. (L.U.)

12.3. Exact differential equations

By § 9.9 of Chapter 9, if P and Q are functions of x and y, the expression $P dx + Q dy$ is an exact differential if
$$\frac{\partial P}{\partial y} = \frac{\partial Q}{\partial x}.$$
When the functions P and Q are so related, the first order differential equation
$$P + Q\frac{dy}{dx} = 0$$
is said to be *exact* and its solution can usually be written down by inspection. Thus both the differential equations
$$y + x\frac{dy}{dx} = 0 \text{ and } \sin y + x \cos y \frac{dy}{dx} = 0$$
are exact, their left-hand sides being respectively the derivatives with respect to x of xy and $x \sin y$. The general solutions of the two equations are therefore respectively
$$xy = C \text{ and } x \sin y = C$$
where C is an arbitrary constant.

Some equations which are not themselves exact can be made exact by multiplication by a suitable factor called an *integrating factor*. For example, the equation

$$2 \sin y + x \cos y \frac{dy}{dx} = 0$$

is not exact, but on multiplication by x it becomes

$$2x \sin y + x^2 \cos y \frac{dy}{dx} = 0,$$

and this can be written $\quad \dfrac{d}{dx}(x^2 \sin y) = 0.$

The equation is now exact and the general solution is $x^2 \sin y = C$.

Example 4. *Solve the differential equation*

$$x + y + (x - y)\frac{dy}{dx} = 0$$

given that $y = 0$ *when* $x = 0$.

Here $P = x + y$, $Q = x - y$ and

$$\frac{\partial P}{\partial y} = 1 = \frac{\partial Q}{\partial x},$$

so that the equation is exact. It can be written in the form

$$x + \left(y + x\frac{dy}{dx}\right) - y\frac{dy}{dx} = 0$$

or, $\quad \dfrac{d}{dx}(\tfrac{1}{2}x^2) + \dfrac{d}{dx}(xy) - \dfrac{d}{dx}(\tfrac{1}{2}y^2) = 0$

and the general solution is $\tfrac{1}{2}x^2 + xy - \tfrac{1}{2}y^2 = C$.
If $y = 0$ when $x = 0$, it follows that $C = 0$ and the solution required here is $x^2 + 2xy - y^2 = 0$.

12.4. The linear first order differential equation

The general linear differential equation of the first order can be written

$$\frac{dy}{dx} + X_1 y = X_2, \qquad (12.15)$$

where X_1 and X_2 are functions of x (but not of y). A solution can be obtained by multiplying the equation by an *integrating factor* μ which makes the left-hand side of the equation an exact differential. To find a suitable integrating factor, the necessary condition for

$$\mu \frac{dy}{dx} + \mu X_1 y$$

to be an exact differential is, by § 9.9,

$$\frac{\partial}{\partial y}(\mu X_1 y) = \frac{\partial}{\partial x}(\mu).$$

If μ is a function of x only, this reduces to

$$\frac{d\mu}{dx} = \mu X_1$$

and the required expression for μ can be found from this equation. Separating the variables and integrating

$$\int \frac{d\mu}{\mu} = \int X_1 dx$$

so that $\log_e \mu = \int X_1 dx$ and, with this expression for μ, the original equation (12.15) becomes after multiplication by μ,

$$\frac{d}{dx}(\mu y) = \mu X_2.$$

The general solution of (12.15) is therefore

$$\mu y = \int \mu X_2 dx + C, \qquad (12.16)$$

where $$\log_e \mu = \int X_1 dx \qquad (12.17)$$

and C is an arbitrary constant.

Example 5. *Find the general solution of the differential equation*

$$\frac{dy}{dx} + \frac{2y}{x} = e^x. \qquad \text{(O.C.)}$$

Here $X_1 = 2/x$, $X_2 = e^x$ and the integrating factor μ is given by

$$\log_e \mu = \int \frac{2}{x} dx = 2 \log_e x = \log_e x^2.$$

Hence $\mu = x^2$ and the equation, after multiplication by this, becomes

$$x^2 \frac{dy}{dx} + 2xy = x^2 e^x$$

or, $$\frac{d}{dx}(x^2 y) = x^2 e^x.$$

Hence the required general solution is

$$x^2 y = \int x^2 e^x dx + C$$
$$= (x^2 - 2x + 2)e^x + C,$$

the integration being performed by parts.

12.5. Equations reducible to linear form

The linear equation (12.15) is a particular case ($n = 0$) of the more general form

$$\frac{dy}{dx} + X_1 y = X_2 y^n. \qquad (12.18)$$

This equation (often known as *Bernoulli's equation*) can be reduced to a linear differential equation in z and x by means of the substitution

$$y^{1-n} = (1-n)z.$$

With this substitution

$$(1-n)y^{-n}\frac{dy}{dx} = (1-n)\frac{dz}{dx},$$

so that

$$\frac{dy}{dx} = y^n \frac{dz}{dx},$$

and equation (12.18) becomes, after division by y^n,

$$\frac{dz}{dx} + X_1 y^{1-n} = X_2,$$

or,

$$\frac{dz}{dx} + (1-n)X_1 z = X_2.$$

The solution of this linear equation can then be carried through as in § 12.4.

Example 6. *Solve the equation*

$$\frac{dy}{dx} + \frac{y}{x} = y^2.$$

Here $n = 2$ and the substitution $1/y = -z$ gives

$$\frac{dy}{dx} = y^2 \frac{dz}{dx}.$$

The given equation becomes, after division by y^2 and substitution for y and dy/dx,

$$\frac{dz}{dx} - \frac{z}{x} = 1.$$

The integrating factor μ for this linear equation is given by

$$\log_e \mu = \int -\frac{1}{x}dx = -\log_e x,$$

so that $\mu = 1/x$ and the equation can be written

$$\frac{1}{x}\frac{dz}{dx} - \frac{z}{x^2} = \frac{1}{x}.$$

Hence

$$\frac{d}{dx}\left(\frac{z}{x}\right) = \frac{1}{x}$$

giving

$$\frac{z}{x} = \log_e x + C.$$

Since $z = -1/y$, the solution of the original equation is

$$-\frac{1}{xy} = \log_e x + C.$$

HOMOGENEOUS EQUATIONS

EXERCISES 12 (b)

1. Show that the equation
$$3x^2 - 2y^2 = (4xy + 2y)\frac{dy}{dx}$$
is exact and find its general solution.

2. Show that the equation
$$\tan x \frac{dy}{dx} + \tan y = 0$$
can be made exact by multiplication by $\cos x \cos y$ and hence find its general solution.

3. Solve the differential equation
$$(x - 1)\frac{dy}{dx} + 3y = x^2. \tag{C.}$$

4. Find the general solution of the equation
$$(x + 1)\frac{dy}{dx} + (2x - 1)y = e^{-2x}. \tag{C.}$$

5. Find the general solution of the differential equation
$$\frac{dy}{dx} + \frac{y}{x} = \frac{1}{2}\sin\frac{1}{2}x.$$
If y is the solution which takes the value unity when $x = 2\pi$, find the limit of y as x tends to zero. (L.U.)

6. Reduce the differential equation
$$\frac{dy}{dx} + \frac{y}{x} = x^3 y^4$$
to one of linear form and hence find its general solution.

12.6. Homogeneous first order differential equations

A first order differential equation is said to be *homogeneous* if it can be written in the form
$$\frac{dy}{dx} = f\left(\frac{y}{x}\right). \tag{12.19}$$

By writing $y = vx$, so that $\dfrac{dy}{dx} = v + x\dfrac{dv}{dx}$,

the equation becomes $\quad x\dfrac{dv}{dx} = f(v) - v.$

The variables in this equation can be separated and its general solution is
$$\int \frac{dv}{f(v) - v} = \int \frac{dx}{x} = \log_e Cx, \tag{12.20}$$
where the arbitrary constant has been taken to be $\log_e C$. The solution of the original equation is then obtained by writing $v = y/x$.

Example 7. *Solve the differential equation*

$$(x - y)\frac{dy}{dx} = x + y. \tag{O.C.}$$

Here, $$\frac{dy}{dx} = \frac{x+y}{x-y} = \frac{1+(y/x)}{1-(y/x)}$$

and the equation is homogeneous. With $y = vx$, the equation becomes

$$v + x\frac{dv}{dx} = \frac{1+v}{1-v},$$

giving $$x\frac{dv}{dx} = \frac{1+v^2}{1-v}.$$

Separating the variables and integrating

$$\int\left(\frac{1-v}{1+v^2}\right)dv = \int\frac{dx}{x},$$

giving $$\tan^{-1} v - \frac{1}{2}\log_e(1+v^2) = \log_e Cx.$$

Writing $v = y/x$, the required solution reduces to

$$\tan^{-1}\left(\frac{y}{x}\right) = \log_e\{C\sqrt{(x^2+y^2)}\}.$$

12.7. Some artifices for reducing first order equations to standard forms

The reader will remember that one of the most effective methods of evaluating indefinite integrals is their reduction to a standard form by employing a suitable change of variable. This principle is also of great application in the solution of differential equations and an equation can often be reduced to one of the forms previously considered by means of an appropriate substitution. Two examples have already been given—in § 12.5 the equation

$$\frac{dy}{dx} + X_1 y = X_2 y^n$$

was reduced to linear form by the relation $y^{1-n} = (1-n)z$ and in § 12.6 the substitution $y = vx$ reduced the homogeneous equation to one in which the variables were separable. The choice of a suitable substitution or artifice is, of course, a matter of some judgement and some further examples are given below.

Equations of the type

$$\frac{dy}{dx} = f\left(\frac{ax+by+c}{Ax+By+C}\right),$$

where a, b, c, A, B, C are constants, can be reduced to homogeneous form by regarding x and y as rectangular coordinates and translating the origin of coordinates to the point of intersection of the two straight lines $ax + by + c = 0$, $Ax + By + C = 0$. This method fails, of course, if the lines are parallel (in which case $a/A = b/B$) but the equation may then be solved by changing the variable through the substitution $ax + by = z$.

Example 8. *Solve the equation*
$$\frac{dy}{dx} = \frac{x+y+1}{x-ay+5},$$
when (i) $a = 1$, (ii) $a = -1$.

(i) The straight lines $x + y + 1 = 0$, $x - y + 5 = 0$ intersect at the point $(-3, 2)$ and the transformation appropriate to translating the origin to this point is $x = X - 3$, $y = Y + 2$. Since $dy/dx = dY/dX$, the transformed equation is
$$\frac{dY}{dX} = \frac{X+Y}{X-Y}.$$
This is the equation discussed in Example 7 and its solution is
$$\tan^{-1}\left(\frac{Y}{X}\right) = \log_e \{C\sqrt{(X^2 + Y^2)}\}.$$
Since $X = x + 3$, $Y = y - 2$, the solution of the given equation is
$$\tan^{-1}\left(\frac{y-2}{x+3}\right) = \log_e \{C\sqrt{(x^2 + y^2 + 6x - 4y + 13)}\}.$$

(ii) When $a = -1$, the lines $x + y + 1 = 0$, $x - ay + 5 = 0$ are parallel and the substitution $x + y = z$ is used. In this case
$$1 + \frac{dy}{dx} = \frac{dz}{dx}$$
and the equation for solution is transformed into
$$\frac{dz}{dx} - 1 = \frac{z+1}{z+5},$$
or,
$$\frac{dz}{dx} = \frac{2z+6}{z+5}.$$
Separating the variables and integrating (after a slight rearrangement in the integrand of the z integral),
$$\int\left(\frac{1}{2} + \frac{2}{2z+6}\right)dz = \int dx,$$
giving
$$\frac{1}{2}z + \log_e (2z + 6) = x + C.$$
Since $z = x + y$, the solution required is
$$\frac{1}{2}(x + y) + \log_e (2x + 2y + 6) = x + C.$$

Another artifice which is sometimes useful is to regard x (instead of y) as the dependent variable. Thus the equation
$$(x + ky^n)\frac{dy}{dx} = y^m$$
can be written
$$\frac{dx}{dy} - \frac{x}{y^m} = ky^{n-m}.$$
and this equation is linear when x is taken as the dependent variable.

EXERCISES 12 (c)

Solve the differential equations:

1. $(x + 2y)\dfrac{dy}{dx} = 2x - y$. (C.)

2. $(x + y)\dfrac{dy}{dx} + 2x - y = 0$. (C.)

3. $(x^2 + y^2)\dfrac{dy}{dx} = xy$. (C.)

4. $\dfrac{dy}{dx} = \dfrac{2x - 5y + 2}{4x - 10y}$.

5. $(3x + y - 5)\dfrac{dy}{dx} - 2x - 2y + 2 = 0$.

6. $y^2(x - 1 - y^3)\dfrac{dy}{dx} + 1 + y^3 = 0$.

12.8. Second order linear equations with constant coefficients

The general linear differential equation of the second order is

$$\frac{d^2y}{dx^2} + X_1\frac{dy}{dx} + X_2 y = X_3, \qquad (12.21)$$

where X_1, X_2 and X_3 are functions of x only. The discussion given here is (except for the further case considered in § 12.11) limited to equations in which X_1 and X_2 are constants; such equations (often called *linear equations with constant coefficients*) are of great importance in applied mathematics and physics. Initially (but this restriction is later removed) it is assumed that $X_3 = 0$ and the equation under consideration is taken to be

$$\frac{d^2y}{dx^2} + 2k\frac{dy}{dx} + my = 0, \qquad (12.22)$$

k and m being constants.

It is helpful to write $y = e^{-kx}z$ so that

$$\frac{dy}{dx} = e^{-kx}\left(\frac{dz}{dx} - kz\right)$$

and

$$\frac{d^2y}{dx^2} = e^{-kx}\left(\frac{d^2z}{dx^2} - 2k\frac{dz}{dx} + k^2 z\right).$$

Substitution in (12.22) shows that z is given by the second order differential equation

$$\frac{d^2z}{dx^2} + (m - k^2)z = 0. \qquad (12.23)$$

Multiplying by $2dz/dx$, the equation can be written

$$\frac{d}{dx}\left\{\left(\frac{dz}{dx}\right)^2 + (m - k^2)z^2\right\} = 0$$

so that

$$\left(\frac{dz}{dx}\right)^2 = (k^2 - m)(z^2 + c^2),$$

SECOND ORDER EQUATIONS

where c is an arbitrary constant. The variables in this first order equation can be separated to give

$$\int \frac{dz}{z^2 + c^2} = \sqrt{(k^2 - m)} \int dx$$

and hence
$$\sinh^{-1}\left(\frac{z}{c}\right) = \sqrt{(k^2 - m)}(x + c')$$

where c' is a second arbitrary constant. The general solution of equation (12.23) is therefore

$$z = c \sinh\{\sqrt{(k^2 - m)}(x + c')\}$$

and this can be written in the form

$$z = A e^{\sqrt{(k^2 - m)}x} + B e^{-\sqrt{(k^2 - m)}x},$$

where A and B are two constants related to c and c' (the exact relation between A, B, c, c' is easily found but is unimportant). The general solution of the original equation (12.22) is, since $y = e^{-kx}z$,

$$y = e^{-kx}\{A e^{\sqrt{(k^2 - m)}x} + B e^{-\sqrt{(k^2 - m)}x}\}. \quad (12.24)$$

This is the natural form of the solution when $m < k^2$: if $m > k^2$, the solution can be written

$$y = e^{-kx}\{A e^{i\sqrt{(m - k^2)}x} + B e^{-i\sqrt{(m - k^2)}x}\}$$
$$= e^{-kx}\{A' \cos \sqrt{(m - k^2)}x + B' \sin \sqrt{(m - k^2)}x\}, \quad (12.25)$$

where the constants A', B' are easily related to A, B through the relation $e^{\pm i\theta} = \cos\theta \pm i \sin\theta$. When $m = k^2$, equation (12.23) becomes

$$\frac{d^2z}{dx^2} = 0$$

and hence $z = A + Bx$. The solution to (12.22) is, in this case,

$$y = e^{-kx}(A + Bx). \quad (12.26)$$

Summarising, the general solution of

$$\frac{d^2y}{dx^2} + 2k\frac{dy}{dx} + my = 0 \quad (12.22)$$

is, dropping the unnecessary dashes on A and B in (12.25),

$$\left.\begin{array}{l} y = e^{-kx}\{A e^{\sqrt{(k^2 - m)}x} + B e^{-\sqrt{(k^2 - m)}x}\},\ m < k^2, \\ y = e^{-kx}(A + Bx),\ m = k^2, \\ y = e^{-kx}\{A \cos \sqrt{(m - k^2)}x + B \sin \sqrt{(m - k^2)}x\},\ m > k^2. \end{array}\right\} \quad (12.27)$$

The reader should note that, when $m \neq k^2$, the above solution is given by

$$y = A e^{\lambda_1 x} + B e^{\lambda_2 x}, \quad (12.28)$$

where λ_1, λ_2 are the two roots of the *auxiliary* (quadratic) equation

$$\lambda^2 + 2k\lambda + m = 0 \quad (12.29)$$

and whose formation from equation (12.22) should be clear. When working particular examples it is best to form and solve the auxiliary quadratic and to write down the solution to the differential equation by (12.28) (or its equivalent form when the roots of the auxiliary quadratic are complex). When the roots of the auxiliary equation are both equal to k, the solution to be written down is
$y = e^{-kx}(A + Bx)$: the reader should note the special form taken by the solution in this case.

Example 9. *Solve the differential equation*

$$\frac{d^2y}{dx^2} - 5\frac{dy}{dx} + 6y = 0$$

with the conditions $y = 1$, $dy/dx = 0$ when $x = 0$. (C.)

Here the auxiliary equation is

$$\lambda^2 - 5\lambda + 6 = 0$$

and the roots are clearly 2 and 3. The general solution of the differential equation is therefore

$$y = Ae^{2x} + Be^{3x}.$$

It follows that $\quad \dfrac{dy}{dx} = 2Ae^{2x} + 3Be^{3x}$

and the given initial conditions lead to

$$A + B = 1, \quad 2A + 3B = 0.$$

These simultaneous equations give $A = 3$, $B = -2$ and the solution required here is

$$y = 3e^{2x} - 2e^{3x}.$$

Example 10. *Show that the general solution of the equation*

$$\frac{d^2x}{dt^2} + n^2x = 0$$

may be written in either of the forms

$$x = A \cos nt + B \sin nt = C \sin(nt + a).$$

Taking $n = 6$, find C and a given that $x = 9\sqrt{3}/2$ and $dx/dt = 27$ when $t = 0$. (O.C.)

The auxiliary equation is $\lambda^2 + n^2 = 0$ with roots $\pm in$. By (12.28) the general solution is, since the variables are now x and t,

$$x = A'e^{int} + B'e^{-int} = A \cos nt + B \sin nt,$$

where the constants A, B, A', B' are related by $A = A' + B'$, $B = i(A' - B')$. This can be written

$$x = \sqrt{(A^2 + B^2)}\left\{\frac{A}{\sqrt{(A^2 + B^2)}} \cos nt + \frac{B}{\sqrt{(A^2 + B^2)}} \sin nt\right\}$$

$$= \sqrt{(A^2 + B^2)}(\sin a \cos nt + \cos a \sin nt) = C \sin(nt + a),$$

where $C = \sqrt{(A^2 + B^2)}$ and $\tan a = A/B$.

With $n = 6$, $\quad x = C \sin(6t + a)$, $\quad \dfrac{dx}{dt} = 6C \cos(6t + a)$.

Inserting the given initial conditions,
$$\frac{9\sqrt{3}}{2} = C \sin a, \quad 27 = 6C \cos a.$$

By division, $\tan a = \sqrt{3}$ so that $a = \pi/3$. The second relation between C and a then gives
$$C = \frac{27}{6} \sec a = \frac{27}{6} \times 2 = 9.$$

Example 11. *Write down the general solution of the equation*
$$\frac{d^2x}{dt^2} + 6\frac{dx}{dt} + 9x = 0.$$

The auxiliary equation is $\lambda^2 + 6\lambda + 9 = 0$ and both the roots are -3. The general solution is therefore, since the variables are x and t,
$$x = e^{-3t}(A + Bt).$$

So far discussion of the general second order linear equation (12.21) has been limited to the case $X_1 = 2k$, $X_2 = m$, $X_3 = 0$. It is now proposed to remove the restriction $X_3 = 0$ and to consider the equation
$$\frac{d^2y}{dx^2} + 2k\frac{dy}{dx} + my = f(x). \tag{12.30}$$

Suppose that ϕ is a particular solution (containing no arbitrary constants) of this equation and that a general solution of the form $y = v + \phi$ is sought. Substituting in (12.30)
$$\frac{d^2v}{dx^2} + \frac{d^2\phi}{dx^2} + 2k\left(\frac{dv}{dx} + \frac{d\phi}{dx}\right) + m(v + \phi) = f(x),$$
and since
$$\frac{d^2\phi}{dx} + 2k\frac{d\phi}{dx} + m\phi = f(x)$$
it follows that v satisfies the '*reduced*' equation
$$\frac{d^2v}{dx^2} + 2k\frac{dv}{dx} + mv = 0, \tag{12.31}$$
which is simply the original equation (12.30) when the right-hand side is zero.

The general solution of the 'reduced' equation, which can be found by the method already discussed, is called the *complementary function* and it will contain two arbitrary constants. The general solution of the more general equation (12.30) is given by adding to the complementary function a particular solution (more usually called the *particular integral*) of the equation under discussion. Some methods for finding the particular integral are discussed in the next section.

12.9. An elementary method of finding the particular integral

Any particular solution of the equation

$$\frac{d^2y}{dx^2} + 2k\frac{dy}{dx} + my = f(x) \tag{12.32}$$

is required and often such a solution can be found by trial. The discussion given here includes most of the cases which occur frequently in practical applications and should be sufficient in a first study of the subject. Another method of determining the particular integral involves the symbols D and $F(D)$ where $D, D^2, \ldots$ denote the differential operators $d/dx, d^2/dx^2, \ldots$ and the reader will study these operators in a deeper and more systematic treatment of differential equations than is given in this book. Yet another method of solving linear equations such as (12.32) is by means of the Laplace transform (see § 12.12).

(i) $f(x) = c_0$ *(constant)*. A particular integral is clearly given by $y = c_0/m$, for the first and second derivatives of this expression are both zero.

(ii) $f(x) = c_0 + c_1 x + c_2 x^2 + \ldots + c_r x^r$. Functions of x whose derivatives are positive integral powers of x are themselves positive integral powers of x and a trial solution is therefore

$$y = a_0 + a_1 x + a_2 x^2 + \ldots + a_r x^r.$$

Clearly the degree of the trial solution must (if $m \neq 0$) be the same as that of $f(x)$, for the derivatives dy/dx, d^2y/dx^2 are of lower degree. Substituting the trial solution in (12.32) and equating the coefficients of like powers of x leads to sufficient equations to obtain the coefficients $a_0, a_1, a_2, \ldots, a_r$. The reader should note that if $m = 0$ the trial solution should be a polynomial in x of degree $r + 1$ and if $m = k = 0$ it should be of degree $r + 2$.

(iii) $f(x) = ce^{ax}$. Since all the derivatives of e^{ax} are multiples of e^{ax}, a suitable trial solution is $y = be^{ax}$. Substituting in (12.32) it follows that

$$b = \frac{c}{a^2 + 2ka + m}.$$

This trial solution fails if e^{ax} is a term in the complementary function for then $a^2 + 2ka + m = 0$. In this case an appropriate trial solution is bxe^{ax} and, if this fails for a similar reason, try bx^2e^{ax}.

(iv) $f(x) = c \sin ax + c' \cos ax$ *(either c or c' may be zero)*. Since all the derivatives of $\sin ax$, $\cos ax$ are multiples of either $\sin ax$ or $\cos ax$, a suitable trial solution is $y = b \sin ax + b' \cos ax$. Substitution in the equation (12.32) and comparison of the coefficients of $\sin ax$ and $\cos ax$ will lead to two equations for the determination of b and b'. As in case (iii) above, the method fails if there are terms in sin ax, cos ax in the complementary function and an appropriate trial solution is then $x(b \sin ax + b' \cos ax)$.

(v) If $f(x)$ is the sum of terms of the types just considered, the particular integral can be found for each term separately and the results summed.

Example 12. *Find the general solution of the second order equation*

$$\frac{d^2y}{dx^2} - 5\frac{dy}{dx} + 6y = f(x), \qquad (12.33)$$

when (i) $f(x) = 12$, (ii) $f(x) = 18x^2$, (iii) $f(x) = 2e^{4x}$,
(iv) $f(x) = e^{2x}$, (v) $f(x) = 10 \sin x$,
(vi) $f(x) = 12 + e^{2x} + 10 \sin x$.

The complementary function is the general solution of the 'reduced' equation

$$\frac{d^2y}{dx^2} - 5\frac{dy}{dx} + 6y = 0.$$

The auxiliary equation corresponding to this is $\lambda^2 - 5\lambda + 6 = 0$ with roots 2 and 3. Hence the complementary function is $Ae^{2x} + Be^{3x}$.

The various particular integrals corresponding to the six given expressions for $f(x)$ are found as follows:

(i) Here the trial solution is $y = a_0$. Since the first and second derivatives of y are then zero, substitution in (12.33) gives $6a_0 = 12$ and hence $y = 2$. The general solution in this case is therefore

$$y = Ae^{2x} + Be^{3x} + 2.$$

(ii) With the trial solution $v = a_0 + a_1 x + a_2 x^2$,

$$\frac{dy}{dx} = a_1 + 2a_2 x, \quad \frac{d^2y}{dx^2} = 2a_2.$$

Substitution in (12.33) gives

$$2a_2 - 5(a_1 + 2a_2 x) + 6(a_0 + a_1 x + a_2 x^2) \equiv 18x^2,$$

and comparison of coefficients of like powers of x leads to

$$6a_2 = 18, \quad 6a_1 - 10a_2 = 0, \quad 6a_0 - 5a_1 + 2a_2 = 0.$$

From these relations

$$a_2 = 3, \ a_1 = 5, \ a_0 = 19/6,$$

and the particular integral is

$$\frac{19}{6} + 5x + 3x^2.$$

The required general solution is therefore

$$v = Ae^{2x} + Be^{3x} + \frac{19}{6} + 5x + 3x^2.$$

(iii) The trial solution $y = be^{4x}$ gives

$$\frac{dy}{dx} = 4be^{4x}, \quad \frac{d^2y}{dx^2} = 16be^{4x},$$

and substitution in (12.33) yields

$$16be^{4x} - 20be^{4x} + 6be^{4x} \equiv 2e^{4x}$$

and it follows that $b = 1$. The particular integral is therefore e^{4x} and the general solution is

$$y = Ae^{2x} + Be^{3x} + e^{4x}.$$

(iv) Since e^{2x} appears in the complementary function, the appropriate trial solution when $f(x) = e^{2x}$ is $y = bxe^{2x}$. Thus

$$\frac{dy}{dx} = b(2x + 1)e^{2x}, \quad \frac{d^2y}{dx^2} = b(4x + 4)e^{2x}$$

and substitution in the given equation gives

$$b(4x + 4)e^{2x} - 5b(2x + 1)e^{2x} + 6bxe^{2x} \equiv e^{2x}.$$

Thus $b = -1$, the particular integral is $-xe^{2x}$ and the general solution is

$$y = Ae^{2x} + Be^{3x} - xe^{2x}.$$

(v) When $f(x) = 10 \sin x$, the trial solution $y = c \sin x + c' \cos x$ gives

$$\frac{dy}{dx} = c \cos x - c' \sin x, \quad \frac{d^2y}{dx^2} = -c \sin x - c' \cos x.$$

Substituting in (12.33) and collecting together the terms in $\sin x$ and $\cos x$,

$$5(c + c') \sin x - 5(c - c') \cos x \equiv 10 \sin x.$$

Hence $c + c' = 2$ and $c - c' = 0$: thus $c = c' = 1$ giving the particular integral as $\sin x + \cos x$ and the general solution

$$y = Ae^{2x} + Be^{3x} + \sin x + \cos x.$$

(vi) The particular integrals when $f(x)$ is 12, e^{2x} and $10 \sin x$ are respectively 2, $-xe^{2x}$ and $\sin x + \cos x$. The particular integral required in case (vi) is the sum of these three functions and the general solution is

$$y = Ae^{2x} + Be^{3x} + 2 - xe^{2x} + \sin x + \cos x.$$

EXERCISES 12 (d)

1. Solve the differential equation

$$\frac{d^2y}{dx^2} = 4y,$$

given that $y = \frac{dy}{dx} = 2$ when $x = 0$. (L.U.)

2. Find the general solution of the equation

$$\frac{d^2x}{dt^2} + 4\frac{dx}{dt} + 9x = 0.$$

3. Find that solution of the equation

$$4\left(\frac{d^2y}{dx^2} - \frac{dy}{dx}\right) + y = 0$$

which satisfies the two conditions: $y = 0$ when $x = 0$ and $y = 2$ when $x = 2$.

4. Solve the differential equation

$$\frac{d^2y}{dx^2} + 9y = 18.$$ (O.C.)

5. Solve the differential equation
$$\frac{d^2x}{dt^2} - \frac{dx}{dt} - 2x = f(t)$$
when   (i) $f(t) = t$,   (ii) $f(t) = e^t$.

6. Solve the differential equation of Exercise 5 above when
    (i) $f(t) = 3e^{2t}$,   (ii) $f(t) = 10 \sin t$.

7. Find the general solution of the differential equation
$$\frac{d^2y}{dx^2} + 4y = x^2 + e^{-2x} + \cos x. \qquad \text{(C.)}$$

8. Show that the solution of the equation
$$\frac{d^2y}{dt^2} + n^2 y = a \sin pt$$
(where $n \neq 0$ and $p^2 \neq n^2$), such that $y = 0$ and $dy/dt = 0$ when $t = 0$ is
$$y = \frac{a}{n^2 - p^2}\left(\sin pt - \frac{p}{n}\sin nt\right).$$
Show also that, as p tends to n, y tends to
$$\frac{a}{2n}\left(\frac{1}{n}\sin nt - t\cos nt\right)$$
and verify that this is the solution when p is equal to n.   (C.)

12.10. Higher order linear differential equations

The general linear differential equation with constant coefficients is
$$\frac{d^n y}{dx^n} + a_1 \frac{d^{n-1}y}{dx^{n-1}} + a_2 \frac{d^{n-2}y}{dx^{n-2}} + \ldots + a_n y = f(x), \qquad (12.34)$$
and methods similar to those discussed in §§ 12.8, 12.9 can be used to find its general solution.

Thus the complementary function is, in general,
$$A_1 e^{\lambda_1 x} + A_2 e^{\lambda_2 x} + \ldots + A_n e^{\lambda_n x},$$
where $A_1, A_2, \ldots, A_n$ are arbitrary constants and $\lambda_1, \lambda_2, \ldots, \lambda_n$ are the n roots of the auxiliary algebraical equation
$$\lambda^n + a_1 \lambda^{n-1} + a_2 \lambda^{n-2} + \ldots + a_n = 0.$$
A pair of terms in the complementary function corresponding to a pair of complex roots ($\xi \pm i\eta$) of the auxiliary equation can be written in the alternative form
$$e^{\xi x}(A_r \cos \eta x + A_s \sin \eta x)$$
and the r terms corresponding to r roots each equal to λ_r will be
$$(A_1 + A_2 x + A_3 x^2 + \ldots + A_r x^{r-1})e^{\lambda_r x}.$$

When $f(x)$ in (12.34) takes any of the forms treated in § 12.9, the particular integral may be found by the elementary method explained in that section. Two examples follow.

Example 13. *Solve the differential equation*

$$\frac{d^3y}{dx^3} + 2\frac{d^2y}{dx^2} - \frac{dy}{dx} - 2y = 0$$

given that $y = 1$, $dy/dx = 0$ when $x = 0$ and that y remains finite when x tends to infinity.

The auxiliary equation is $\lambda^3 + 2\lambda^2 - \lambda - 2 = 0$ and this can be written as

$$(\lambda - 1)(\lambda + 1)(\lambda + 2) = 0$$

so that the roots are 1, -1, -2 and the general solution is

$$y = Ae^x + Be^{-x} + Ce^{-2x}.$$

If y is to remain finite as $x \to \infty$, $A = 0$. If $y = 1$ and $dy/dx = 0$ when $x = 0$, the equations determining B and C are

$$B + C = 1, \quad -B - 2C = 0.$$

Hence $B = 2$, $C = -1$ and the solution required here is

$$y = 2e^{-x} - e^{-2x}.$$

Example 14. *Find the general solution of the equation*

$$\frac{d^3x}{dt^3} - \frac{dx}{dt} = e^t.$$

For the complementary function, the auxiliary equation is $\lambda^3 - \lambda = 0$ with roots 0, 1, -1. Hence the complementary function is $A + Be^t + Ce^{-t}$. Since e^t appears in the complementary function, the trial solution is $x = bte^t$. This gives

$$\frac{dx}{dt} = b(t+1)e^t, \quad \frac{d^2x}{dt^2} = b(t+2)e^t, \quad \frac{d^3x}{dt^3} = b(t+3)e^t,$$

and hence

$$b(t + 3 - t - 1) = 1,$$

giving $b = 1/2$.

The required general solution is therefore

$$x = A + Be^t + Ce^{-t} + \frac{t}{2}e^t.$$

12.11. The homogeneous linear equation

The equation

$$x^n \frac{d^n y}{dx^n} + a_1 x^{n-1} \frac{d^{n-1} y}{dx^{n-1}} + \ldots + a_{n-1} x \frac{dy}{dx} + a_n y = f(x), \quad (12.35)$$

where $a_1, \ldots, a_{n-1}, a_n$ are constants, is often known as the *homogeneous linear equation*. It can be reduced to a linear equation with constant coefficients by changing the independent variable x to z where

$$x = e^z \text{ or } z = \log_e x.$$

HOMOGENEOUS EQUATION

With this change of variable,

$$\frac{dy}{dx} = \frac{dy}{dz}\frac{dz}{dx} = \frac{1}{x}\frac{dy}{dz},$$

$$\frac{d^2y}{dx^2} = -\frac{1}{x^2}\frac{dy}{dz} + \frac{1}{x}\frac{d^2y}{dz^2}\frac{dz}{dx} = \frac{1}{x^2}\left(\frac{d^2y}{dz^2} - \frac{dy}{dz}\right),$$

$$\frac{d^3y}{dx^3} = -\frac{2}{x^3}\left(\frac{d^2y}{dz^2} - \frac{dy}{dz}\right) + \frac{1}{x^2}\left(\frac{d^3y}{dz^3} - \frac{d^2y}{dz^2}\right)\frac{dz}{dx} = \frac{1}{x^3}\left(\frac{d^3y}{dz^3} - 3\frac{d^2y}{dz^2} + 2\frac{dy}{dz}\right),$$

and so on. The results may be written

$$x\frac{dy}{dx} = \frac{dy}{dz},$$

$$x^2\frac{d^2y}{dx^2} = \frac{d}{dz}\left(\frac{d}{dz} - 1\right)y,$$

$$x^3\frac{d^3y}{dx^3} = \frac{d}{dz}\left(\frac{d}{dz} - 1\right)\left(\frac{d}{dz} - 2\right)y,$$

and it can be shown that in general

$$x^r\frac{d^r y}{dx^r} = \frac{d}{dz}\left(\frac{d}{dz} - 1\right)\ldots\left(\frac{d}{dz} - r + 1\right)y.$$

These substitutions reduce equations of the form (12.35) to linear equations with constant coefficients and the previous methods are then available for their solution.

Example 15. *Solve the differential equation*

$$x^2\frac{d^2y}{dx^2} + x\frac{dy}{dx} - 9y = x^4. \tag{C.}$$

Putting $x = e^z$ and using the above substitutions, the equation in y and z is

$$\frac{d^2y}{dz^2} - 9y = e^{4z}.$$

The auxiliary equation is $\lambda^2 - 9 = 0$ so that the complementary function is $Ae^{3z} + Be^{-3z}$. With the trial solution $v = be^{4z}$, $16b - 9b = 1$ so that $b = 1/7$. Hence the general solution of the equation in y and z is

$$v = Ae^{3z} + Be^{-3z} + \frac{1}{7}e^{4z}.$$

Since $x = e^z$, the required solution of the given equation is

$$y = Ax^3 + \frac{B}{x^3} + \frac{x^4}{7}.$$

EXERCISES 12 (e)

1. Find the general solution of the equation

$$\frac{d^3\theta}{dt^3} - 8\theta = 0.$$

2. Solve the equation $\dfrac{d^4y}{dx^4} + 2\dfrac{d^2y}{dx^2} + y = 0$.

3. Show that the general solution of the equation
$$\frac{d^4x}{dt^4} + 4\frac{d^2x}{dt^2} = 96t^2$$
is $x = A \cos 2t + B \sin 2t + C + Dt + 2t^4 - 6t^2$.

4. Find a particular integral of the equation
$$\frac{d^4y}{dx^4} + 5\frac{d^2y}{dx^2} + 4y = 12 \sin 2x.$$

5. Show that the general solution of the homogeneous linear equation
$$x^2\frac{d^2y}{dx^2} - 2x\frac{dy}{dx} + 2y = 4x^3$$
is $y = Ax + Bx^2 + 2x^3$.

6. Find the general solution of the equation
$$(t + 1)^2\frac{d^2x}{dt^2} - 2(t + 1)\frac{dx}{dt} = 10x + (t + 1)^4.$$

12.12. The solution of linear differential equations by means of the Laplace transform

The essentials of a physical problem are often contained in a differential equation and certain initial conditions. In the methods of solution previously discussed it has been necessary first to find the general solution of the differential equation and then to determine the arbitrary constants in this solution from the initial conditions. In the case of linear differential equations with constant coefficients, an alternative method is available and has recently become fashionable. This method, using the so-called Laplace transform, reduces the technique of solution almost to a 'drill' and includes the information given by the initial conditions in the technique of solution. A brief description of the method is given below; the method is applicable also to the solution of linear partial differential equations and its value becomes more apparent as the complexity of the problem under discussion increases.

The *Laplace transform* $\bar{f}(p)$ of a function $f(x)$ is defined by the relation
$$\bar{f}(p) = \int_0^\infty e^{-px} f(x) dx, \qquad (12.36)$$
it being assumed that the integral on the right exists. It is a comparatively simple matter to draw up from (12.36) a table showing the transforms corresponding to given functions of x. For example, if $f(x) = 1$,
$$\bar{f}(p) = \int_0^\infty e^{-px} dx = \left[-\frac{e^{-px}}{p}\right]_0^\infty = \frac{1}{p},$$

it being assumed here that $p > 0$ in order that the integral may exist. Again, if $f(x) = e^{ax} \sin bx$,

$$\bar{f}(p) = \int_0^\infty e^{-(p-a)x} \sin bx\, dx = \frac{b}{(p-a)^2 + b^2},$$

assuming $p - a > 0$ and using the result given in (10.47). The entries in the short table below can all be established by simple definite integrals of the above type and the table should be sufficiently comprehensive for present purposes. The reader will observe that some of the entries in the table are particular cases of other entries: they occur, however, so often in practical problems that it seems worth while to record them separately.

Short table of Laplace transforms

$$\bar{f}(p) = \int_0^\infty e^{-px} f(x)\, dx.$$

	$f(x)$	$\bar{f}(p)$	Remarks
1	1	$1/p$	
2	x^n	$(n)!/p^{n+1}$	n an integer greater than -1.
3	e^{ax}	$1/(p-a)$	a constant.
4	$x^n e^{ax}$	$(n)!/(p-a)^{n+1}$	n an integer greater than -1, a constant.
5	$\sin bx$	$b/(p^2 + b^2)$	b constant.
6	$\cos bx$	$p/(p^2 + b^2)$	b constant.
7	$e^{ax} \sin bx$	$b/\{(p-a)^2 + b^2\}$	a, b constant.
8	$e^{ax} \cos bx$	$(p-a)/\{(p-a)^2 + b^2\}$	a, b constant.

Suppose that $f_0, f_1, \ldots, f_{n-1}$ denote respectively the values of a function f of x and its first $(n-1)$ derivatives with respect to x when $x = 0$. Then, integrating by parts

$$\int_0^\infty e^{-px} \frac{df}{dx} dx = \left[e^{-px} f \right]_0^\infty + p \int_0^\infty e^{-px} f\, dx = -f_0 + p\bar{f}, \quad (12.37)$$

provided that $\lim_{x \to \infty} (e^{-px} f) = 0$. Again,

$$\int_0^\infty e^{-px} \frac{d^2 f}{dx^2} dx = \left[e^{-px} \frac{df}{dx} \right]_0^\infty + p \int_0^\infty e^{-px} \frac{df}{dx} dx$$

$$= -f_1 + p(-f_0 + p\bar{f}) = -f_1 - pf_0 + p^2\bar{f}, \quad (12.38)$$

using (12.37) and assuming that $e^{-px}df/dx$ tends to zero as x tends to infinity. It can be shown similarly that, provided $e^{-px}d^{r-1}f/dx^{r-1}$ ($r = 1, 2, \ldots, n$) vanishes as x tends to infinity, then

$$\int_0^\infty e^{-px}\frac{d^n f}{dx^n}dx = -(f_{n-1} + pf_{n-2} + \ldots + p^{n-1}f_0) + p^n \bar{f}. \quad (12.39)$$

The relations (12.37), (12.38) and (12.39) therefore permit the Laplace transforms of the derivatives of a function to be expressed in terms of the transform of the function, its initial value and the initial values of its derivatives: these results, together with the table of transforms, form the necessary equipment in applying the transform technique to the solution of linear differential equations.

The 'drill' to be followed in the solution of a differential equation (in which the independent and dependent variables are respectively assumed to be x and y) is:

(i) multiply the equation by e^{-px} and integrate with respect to x between the limits 0 and ∞, so forming the Laplace transform of the equation;

(ii) apply the result (12.39) using the given initial values of y and its derivatives and use the table of transforms as is necessary: the result (called the *subsidiary* equation) is an algebraical equation which can be solved to give the transform $\bar{y}$ of the wanted function in terms of p;

(iii) find y (a function of x) to correspond to this function of p by using the table 'inversely': some manipulation is usually required to enable this step to be carried through.

The method is illustrated in detail in the examples which follow.

Example 16. *Solve the equation*

$$\frac{d^2y}{dx^2} - 5\frac{dy}{dx} + 6y = 0$$

given that $y = 1$, $dy/dx = 0$ *when* $x = 0$.

Multiplying by e^{-px} and integrating with respect to x between 0 and ∞

$$\int_0^\infty e^{-px}\frac{d^2y}{dx^2}dx - 5\int_0^\infty e^{-px}\frac{dy}{dx}dx + 6\int_0^\infty e^{-px}y\,dx = 0.$$

Writing $$\bar{y} = \int_0^\infty e^{-px}y\,dx.$$

since $y_0 = 1$, $y_1 = 0$ (suffixes denoting the value of the function and its first derivative when $x = 0$), (12.37) and (12.38) give

$$\int_0^\infty e^{-px}\frac{dy}{dx}dx = -1 + p\bar{y}, \quad \int_0^\infty e^{-px}\frac{d^2y}{dx^2}dx = -p + p^2\bar{y}.$$

Hence the transformed equation can be written

$$-p + p^2\bar{y} - 5(-1 + p\bar{y}) + 6\bar{y} = 0$$

and the subsidiary equation giving $\bar{y}$ is
$$(p^2 - 5p + 6)\bar{y} = p - 5.$$
Solving for $\bar{y}$
$$\bar{y} = \frac{p-5}{p^2 - 5p + 6} = \frac{3}{p-2} - \frac{2}{p-3}.$$
From the third entry of the table the functions whose transforms are $1/(p-2)$, $1/(p-3)$ are respectively e^{2x} and e^{3x}. Hence
$$y = 3e^{2x} - 2e^{3x},$$
the result obtained by a different method in Example 9. The reader should notice how the expression for $\bar{y}$ is split into partial fractions to enable the table of transforms to be used inversely in the last step. It should also be noted that many of the intermediate steps used in the above solution could well be omitted as the student gains experience with the method: in the examples which follow, less detail is given and the directness of the method becomes more apparent.

Example 17. *Solve the differential equation*
$$\frac{d^2x}{dt^2} + m^2 x = a \cos nt,$$
given that x and dx/dt are both zero when $t = 0$.

The subsidiary equation is
$$(p^2 + m^2)\bar{x} = \frac{ap}{p^2 + n^2},$$
the term on the right being the Laplace transform of $a \cos nt$. Hence
$$\bar{x} = \frac{ap}{(p^2 + m^2)(p^2 + n^2)} = \frac{a}{m^2 - n^2}\left(\frac{p}{p^2 + n^2} - \frac{p}{p^2 + m^2}\right).$$
Inverting (using entry 6 of the table)
$$x = \frac{a}{m^2 - n^2}(\cos nt - \cos mt).$$

Example 18. *Solve the equation*
$$\frac{d^3y}{dx^3} + y = \frac{1}{2}x^2 e^x$$
given that y and its first two derivatives vanish when $x = 0$.

The subsidiary equation is
$$(p^3 + 1)\bar{y} = \frac{1}{(p-1)^3},$$
the term on the right coming from entry 4 of the table. Hence
$$\bar{y} = \frac{1}{(p^3 + 1)(p-1)^3}$$
$$= \frac{1}{2(p-1)^3} - \frac{3}{4(p-1)^2} + \frac{3}{8(p-1)} - \frac{1}{24(p+1)} - \frac{p-2}{3(p^2 - p + 1)}.$$
The last term on the right can be written in the form
$$\frac{\left(p - \frac{1}{2}\right) - \sqrt{3}\left(\frac{\sqrt{3}}{2}\right)}{3\left\{\left(p - \frac{1}{2}\right)^2 + \left(\frac{\sqrt{3}}{2}\right)^2\right\}}$$

and the inversion can now be carried out from the table to give, after a little reduction,

$$y = \frac{1}{4}\left(x^2 - 3x + \frac{3}{2}\right)e^x - \frac{1}{24}e^{-x} - \frac{e^{x/2}}{3}\left(\cos\frac{\sqrt{3}}{2}x - \sqrt{3}\sin\frac{\sqrt{3}}{2}x\right).$$

EXERCISES 12 (f)

Use the method of the Laplace transform to solve the following differential equations under the conditions given:

1. $\dfrac{dy}{dx} + y = 1$ with $y = 2$ when $x = 0$.

2. $\dfrac{d^2y}{dx^2} + y = 0$ with $y = 1$, $\dfrac{dy}{dx} = 0$ when $x = 0$.

3. $\dfrac{d^2\theta}{dt^2} + 4\dfrac{d\theta}{dt} + 8\theta = 1$ with $\theta = 0$, $\dfrac{d\theta}{dt} = 1$ when $t = 0$.

4. $\dfrac{d^2y}{dx^2} - 5\dfrac{dy}{dx} + 6y = e^{2x}$ with $y = 1$, $\dfrac{dy}{dx} = 0$ when $x = 0$.

5. $\dfrac{d^4x}{dt^4} + 4\dfrac{d^3x}{dt^3} + 4\dfrac{d^2x}{dt^2} = 0$ with $x = 1$, $\dfrac{dx}{dt} = \dfrac{d^2x}{dt^2} = \dfrac{d^3x}{dt^3} = 0$ when $t = 0$.

6. Show by the method of the Laplace transform that the solution of the equation

$$\frac{d^2\theta}{dt^2} + n^2\theta = a\sin mt,$$

where $n \neq 0$ and $m^2 \neq n^2$, such that $\theta = d\theta/dt = 0$ when $t = 0$ is

$$\theta = \frac{a}{n^2 - m^2}\left(\sin mt - \frac{m}{n}\sin nt\right).$$

EXERCISES 12 (g)

1. Solve the equation $y\dfrac{dy}{dx} + x\sqrt{(y^2 + 1)} = 0$. (C.)

2. Find the general solution of the equation

$$(1 - x^2)\frac{dy}{dx} + x(y - a) = 0. \qquad\text{(C.)}$$

3. Find the general solution of the differential equation

$$\frac{dy}{dx} - (1 + \cot x)y = 0. \qquad\text{(O.C.)}$$

4. Prove that, if M and N are functions of x and y, the equation

$$M + N\frac{dy}{dx} = 0$$

can be made exact by multiplication by a function λ (independent of y) if

$$N\frac{\partial}{\partial y}\left(\frac{\partial N}{\partial x} - \frac{\partial M}{\partial y}\right) = \frac{\partial N}{\partial y}\left(\frac{\partial N}{\partial x} - \frac{\partial M}{\partial y}\right).$$

Show also that λ is given by the equation

$$\frac{1}{\lambda}\frac{d\lambda}{dx} = \frac{1}{N}\left(\frac{\partial M}{\partial y} - \frac{\partial N}{\partial x}\right).$$

5. Find the general solution of the equation

$$(1 + x^2)\frac{dy}{dx} + xy = 3x + 3x^3. \quad \text{(C.)}$$

6. Solve the differential equation

$$\sin x \cos x \frac{dy}{dx} + y = \cot x. \quad \text{(C.)}$$

7. Find the general solution of the equation

$$2\frac{dy}{dx} = y \sec x + y^3 \tan x.$$

8. Solve the equation $(x + y)\dfrac{dy}{dx} + 4x - y = 0.$ (C.)

9. Find the solution of the differential equation

$$(x^2 + 2xy)\frac{dy}{dx} + x^2 - 2y^2 = 0. \quad \text{(C.)}$$

10. Transform the equation $v\dfrac{dv}{dx} + \dfrac{v^2}{2a} = -\mu x,$

where a and μ are constants, by the substitution $y = v^2$, and hence find the general solution.
If $v = 0$ when $x = a$, show that $v^2 = 2\mu a^2$ when $x = 0$. (L.U.)

11. By substituting $y = x - 2a$, $u = \dfrac{dy}{dt}$, show that the equation

$$\frac{d^2x}{dt^2} = -\lambda\left(x - a - \frac{k}{\lambda}\right)$$

is equivalent to $\quad u\dfrac{du}{dy} + \dfrac{k}{a}y = 0,$

k and a being constants and $\lambda = k/a$. Hence, or otherwise, solve the differential equation in x and t if $x = 5a/2$ and $dx/dt = 0$ for $t = 0$.
(N.U.)

12. Transform the equation

$$\frac{d^2y}{dx^2} + x^2 + y + 2 = 0$$

by the substitution $y = t - x^2$ and hence obtain the general solution of the equation. (L.U.)

13. The strength x of an electric current in a circuit is given by the equation

$$L\frac{dx}{dt} + Rx = E,$$

where L, R, E are constants. Prove that, when t is large, the current is approximately E/R.

If E instead of being constant is of the form $E_0 \cos pt$, where E_0, p are constants, prove that, when t is large, the current is approximately

$$\frac{E_0}{\sqrt{(R^2 + p^2L^2)}} \cos(pt - \varepsilon),$$

where $\tan \varepsilon = pL/R$. (C.)

14. Solve the differential equation

$$\frac{d^2y}{dx^2} + 4y = 1,$$

given that $y = dy/dx = 0$ when $x = 0$. (O.C.)

15. Find the general solution of the equation

$$\frac{d^2y}{dx^2} - \frac{dy}{dx} - 2y = e^x + \cos x.$$ (C.)

16. Solve the differential equation

$$\frac{d^2y}{dx^2} + 4y = x + \sin 2x.$$ (C.)

17. Solve the differential equation

$$\frac{d^2y}{dx^2} - 2\frac{dy}{dx} + y = e^x + x + \sin x.$$ (C.)

18. A particle of unit mass moves on the x-axis in such a way that its coordinate x at time t satisfies the differential equation

$$\frac{d^2x}{dt^2} + 2k\frac{dx}{dt} + p^2x = A\cos qt, \quad p^2 < k^2.$$

Find values of the constants B and a (in terms of p, q, k and A) which make

$$x = B\cos(qt - a)$$

a particular solution of the equation.

Determine the general solution of the differential equation and hence find the particular solution for which

$$x = \frac{B}{k}(q \sin a + k \cos a), \quad \frac{dx}{dt} = 0$$

when $t = 0$, giving your answers in terms of B, a, q, k, p, t. (O.C.)

19. Show that, if $u = \cos^4 x + \sin^4 x$, then

$$\frac{d^2u}{dx^2} + 16u = 12.$$

Find the complete solution of the differential equation

$$\frac{d^2u}{dx^2} + 16u = 12$$

and also the solution which makes $u = 0$ and $du/dx = 0$ when $x = 0$. (C.)

20. Show that the substitution $z = xt$ reduces the equation

$$t\frac{d^2x}{dt^2} + 2\frac{dx}{dt} + 4xt = 0$$

to a linear equation with constant coefficients. Hence find the solution to the original equation given that $x = 0$, $dx/dt = -2$ when $t = \pi/2$.

21. Find the general solutions of the differential equations

(i) $\dfrac{d^3y}{dx^3} - 3\dfrac{dy}{dx} + 2y = 0$, (ii) $\dfrac{d^3y}{dx^3} - 3\dfrac{d^2y}{dx^2} + 3\dfrac{dy}{dx} - y = 0$.

22. Solve the differential equation

$$\frac{d^3y}{dx^3} + 6\frac{d^2y}{dx^2} + 11\frac{dy}{dx} + 6y = e^{-x}. \tag{C.}$$

23. The tensile stresses, p and q, in a rotating disc are given in terms of the radial displacement u by the equations

$$p(1 - \sigma^2) = E\left(\frac{du}{dr} + \sigma\frac{u}{r}\right), \quad q(1 - \sigma^2) = E\left(\sigma\frac{du}{dr} + \frac{u}{r}\right),$$

$$r\frac{d^2u}{dr^2} + \frac{du}{dr} - \frac{u}{r} = -kr^2,$$

where E, σ and k are constants. If both p and q are finite when $r = 0$ and if $p = 0$ when $r = a$, show that

$$u = -\frac{kr^3}{8} + \frac{ka^2(3 + \sigma)r}{8(1 + \sigma)}.$$

$$p = \frac{kE}{8(1 - \sigma^2)}(3 + \sigma)(a^2 - r^2),$$

$$q = \frac{kE}{8(1 - \sigma^2)}\{(3 + \sigma)a^2 - (1 + 3\sigma)r^2\}.$$

24. Show that $(p^2 + b^2)^{-2}$ is the Laplace transform of

$$\frac{1}{2b^3}(\sin bx - bx \cos bx).$$

Hence solve the differential equation

$$\frac{d^2y}{dx^2} + y = x \cos 2x$$

given that y and dy/dx are both zero when $x = 0$.

25. If $\bar{f}(p)$ is the Laplace transform of $f(x)$ and if

$$\lim_{r \to \infty} \left\{ e^{-px} \int_0^r f(X) dX \right\} = 0,$$

show that $\bar{f}(p)/p$ is the transform of $\int_0^x f(X) dX$.

Hence show that if

$$\bar{f}(p) = \frac{b}{p(p^2 + b^2)},$$

then, $\qquad f(x) = \dfrac{1}{b}(1 - \cos bx).$

ANSWERS TO THE EXERCISES

Exercises 1 (b), p. 14.
1. 17.
6. $f(m, n) = \{m(m + 1) \ldots (m + n - 1)\}/(n)!$.

Exercises 1 (c), p. 18.
1. (i) $\dfrac{1}{a-b}\left(\dfrac{a}{x-a} - \dfrac{b}{x-b}\right)$. (ii) $1 + \Sigma \dfrac{a^3}{(a-b)(a-c)(x-a)}$.
2. (i) $1 + 2\Sigma \dfrac{a(a+b)(a+c)(a+d)}{(b-a)(c-a)(d-a)(x+a)}$.
   (ii) $1 - \dfrac{8a}{x+a} + \dfrac{24a^2}{(x+a)^2} - \dfrac{32a^3}{(x+a)^3} + \dfrac{16a^4}{(x+a)^4}$.
3. $-\dfrac{(x-y)(x-z)(x-u)}{(x-a)(x-b)(x-c)(x-d)}$.
6. (i) $\sum_{r=1}^{n} \dfrac{(-1)^{p-r+1} r^p}{(r-1)!(n-r)!} \cdot \dfrac{1}{x+r}$.
   (ii) $1 + (-1)^{n+1} \sum_{r=1}^{n} \dfrac{(-1)^r r^n}{(r-1)!(n-r)!} \cdot \dfrac{1}{x+r}$. $(-1)^{n-1}$.

Exercises 1 (d), p. 19.
4. Yes.
15. $3x^5 + 4x^4 - 3x^3 - 11x^2 + 4x + 12$.
16. $\tan\left(\dfrac{\pi}{4}+a\right)\cot\left(\dfrac{\pi}{4}-a\right)$ and $\tan\left(\dfrac{3\pi}{4}+a\right)\cot\left(\dfrac{3\pi}{4}-a\right)$.
20. $\dfrac{1}{2(x-2)} - \dfrac{1}{2x} - \dfrac{1}{(x-1)^{2n}} - \dfrac{1}{(x-1)^{2n-2}} - \ldots - \dfrac{1}{(x-1)^2}$.

Exercises 2 (a), p. 28.
1. $A = -B = \dfrac{1}{4}$. $\dfrac{1}{2}n(n+1)^2(n+2)$.
2. $\dfrac{2n(n+4)}{3(n+1)(n+3)}$.
3. $\dfrac{1}{18}\left(\dfrac{1}{3n-1} - \dfrac{2}{3n+2} + \dfrac{1}{3n+5}\right)$.
4. $\dfrac{1}{4}n(n+1)(n+2)(n+3)$.
6. nth term $= \dfrac{2^n n}{(n+1)(n+2)}$.
7. 2^{2n-1}.
8. $\dfrac{2^{n+2} - n - 3}{(n+1)(n+2)}$.

Exercises 2 (b), p. 35.
1. $u_r = A + B(-4)^r$, $A = 17$, $B = 4$.
2. $u_r = \dfrac{(b^2-a)a^{r-1} + (1-b)b^r}{b(b-a)}$.

ANSWERS TO THE EXERCISES

5. nth term = $4(3)^{n-1} - 3(2)^{n-1}$, sum = $2(3^n - 1) - 3(2^n - 1)$.

6. $\dfrac{1}{5}\left(\dfrac{1-x^n}{1-x}\right) + \dfrac{5}{6}\left\{\dfrac{1-(2x)^n}{1-2x}\right\} - \dfrac{1}{30}\left\{\dfrac{1-(-4x)^n}{1+4x}\right\}$.

Exercises 2 (c), p. 35.

1. $-8n^2$. 2. $\dfrac{1}{8} - \dfrac{3n+1}{(n+4)!}$.

3. $\dfrac{1}{12}n(n+1)(n+2)(3n+5)$.

4. nth term = $n(3n-1)$, sum = $n^2(n+1)$.

8. 2^m. 13. (i) $5\left(\dfrac{1}{2}\right)^r + 3\left(-\dfrac{1}{2}\right)^r$. (ii) 2^{r-1}.

14. $u_r = A(2)^r + B(3)^r + \dfrac{r}{2} + \dfrac{3}{4}$.

15. $u_{n+2} = \dfrac{2-u_n}{1-u_n}$, $u_{n+3} = 2 - \dfrac{2}{u_n}$, $u_1 = -2$, $u_2 = \dfrac{1}{2}$, $u_3 = \dfrac{4}{3}$;

sum = $\dfrac{\left(3 - 2x + \dfrac{x^2}{2} + \dfrac{4x^3}{3}\right)(1 - x^{4n})}{(1 - x^4)}$.

16. $a_r = \cos a \left\{\dfrac{1}{(1-\cos a)^r} + \dfrac{1}{(1+\cos a)^r}\right\}$.

19. $\{1 + x - (n+2)^2 x^{n+1} + (2n^2 + 6n + 3)x^{n+2} - (n+1)^2 x^{n+3}\}(1-x)^{-3}$.

Exercises 3 (a), p. 48.

1. (i) 0. (ii) 1/3. 6. 2·2361.

Exercises 3 (b), p. 53.

1. (i) $x > 0$. (ii) $x > 0$ and $x < -1/2$.

2. $\dfrac{7}{36} - \dfrac{3n+7}{6(n+1)(n+2)(n+3)}$, $\dfrac{7}{36}$.

3. $\dfrac{11}{36} - \dfrac{6n^2 + 15n + 11}{6(n+1)(n+2)(n+3)}$, series converges to $\dfrac{11}{36}$.

4. 17/6.

Exercises 3 (c), p. 60.

1. Divergent. 2. Divergent.
3. Convergent.
4. Convergent when $x < 1$, divergent when $x > 1$.
5. Convergent. 6. Convergent.

Exercises 3 (e), p. 70.

3. $5e$. 4. $\dfrac{x}{1-x} + \log_e(1-x)$.

5. $\dfrac{1+x}{(1-x)^3}$. 6. $\dfrac{x}{(1-x)(1-2x)}$.

ANSWERS TO THE EXERCISES

Exercises 3 (f), p. 71.

1. (i) $u_n \to 0$ all a.
   (ii) $u_n \to 0$, $|a| < 1$; $u_n \to 1/a$, $|a| > 1$; $u_n = 1/2$, $a = 1$, u_n oscillates infinitely, $a = -1$.

6. $\left(\dfrac{u_1 - a}{u_1 + a}\right)^{2n}$
8. $\dfrac{2-x}{1-x} + \dfrac{1}{x}\log_e(1-x)$.

9. $1 - \dfrac{1}{(n+1)3^n}$, 1.
13. Series converges.

15. Convergent when $a > 1$ for all β and when $a = 1$, $\beta > 1$, divergent when $a < 1$ for all β and when $a = 1$, $\beta \leqslant 1$.

16. Series converges when $x > 1$, $a = 0$ and when $x < 1$, $b = 0$ and otherwise diverges except in the trivial case $a = b = 0$.

18. $1 - 1/(n+1)!$.
19. $5e$.

21. $a = 1$, $b = 7$, $c = 6$, $d = 1$.
22. $7e - 2$.

24. $\dfrac{2x(1+2x)}{(1-2x)^3}$, $-\dfrac{1}{2} < x < \dfrac{1}{2}$.

25. $a^2{}_{n+3} + (q - p^2)a^2{}_{n+2} + (p^2 q - q^2)a^2{}_{n+1} - q^3 a^2{}_n = 0$,
$$\dfrac{4 + (4q - 3p^2)t + p^2 q\, t^2}{(1-qt)\{1 + (2q - p^2)t + q^2 t^2\}}.$$

Exercises 4 (a), p. 81.

2. (a) 1, π. (b) 1, $\pi/2$. (c) 5, $0\cdot 927$ radian. (d) 2, $-5\pi/6$.
4. $1 + 3i$, $-1 - 3i$, $-1 + 3i$, $1 - 3i$.
5. (i) 6. (ii) $4abi/(a^2 + b^2)$.
6. $-\dfrac{9}{13} + \dfrac{19}{13}i$.

Exercises 4 (b), p. 84.

2. (i) $\pm(2 + 3i)$. (ii) $\pm(1 + i)/\sqrt{2}$.
3. $a = 2$, $b = 1/2$.

Exercises 4 (c), p. 94.

1. (i) Circle having A and B as inverse points.
   (ii) A line through A parallel to OB where O is the origin.
4. $(1 + i)/2$.
5. $\operatorname{cosec}\theta$, $(\pi/2) - \theta$; $-\operatorname{cosec}\theta$, $-(\pi/2) - \theta$.
6. $\dfrac{1}{2}\left(1 - \dfrac{i}{\sqrt{3}}\right)z_3 + \dfrac{1}{2}\left(1 + \dfrac{i}{\sqrt{3}}\right)z_2$, $\dfrac{1}{2}\left(1 - \dfrac{i}{\sqrt{3}}\right)z_1 + \dfrac{1}{2}\left(1 + \dfrac{i}{\sqrt{3}}\right)z_3$,
$\dfrac{1}{2}\left(1 - \dfrac{i}{\sqrt{3}}\right)z_2 + \dfrac{1}{2}\left(1 + \dfrac{i}{\sqrt{3}}\right)z_1$.

Exercises 4 (d), p. 102.

1. (i) $X = x^2 - y^2$, $Y = 2xy$. (ii) $X = \dfrac{x}{x^2 + y^2}$, $Y = \dfrac{-y}{x^2 + y^2}$.
   (iii) $X = x\left(1 + \dfrac{1}{x^2 + y^2}\right)$, $Y = y\left(1 - \dfrac{1}{x^2 + y^2}\right)$.

Exercises 4 (e), p. 103.

6. $\pm 5(1 + i)/2$.
7. $1 - {}^nC_2(3) + {}^nC_4(3)^2 - {}^nC_6(3)^3 + \ldots$, $\pm(\sqrt{3} + i)/\sqrt{2}$.

ANSWERS TO THE EXERCISES

8. $2/3, \pm 2i/\sqrt{3}$.
9. $x^2 + x + 1 = 0$, $\dfrac{1}{x-1} + \dfrac{\omega}{x-\omega} + \dfrac{\omega^2}{x-\omega^2}$.
12. The vertex of an equilateral triangle with base line joining z_1, z_2.
17. $60°$

Exercises 5 (a), p. 114.

1. $\pm i, -2 \pm i$.
2. n unequal real roots.

Exercises 5 (b), p. 118.

1. $-3/2, (5 \pm \sqrt{13})/3$.
2. $3/4, 3/2, -5/3$.
3. $-5/3, -2/3, 1/3, 4/3$.
4. (i) $(b^2 - 2ac)/c^2$. (ii) $(a^2 - 2b)/c^2$.
5. $-a^4 + 4a^2b - 8ac + 16d$.
6. $y^3 + 6py^2 + 9p^2y + 4p^3 + 27q^2 = 0$.

Exercises 5 (c), p. 120.

1. $5pq$.
2. $\alpha\beta\gamma = (a^3 - 3ab + 2c)/6$, $\alpha^4 + \beta^4 + \gamma^4 = \dfrac{a^4}{6} - a^2b + \dfrac{4ac}{3} + \dfrac{b^2}{2}$.
4. (i) $S_4 - S_3S_1 - \dfrac{1}{2}S_2^2 + \dfrac{1}{2}S_2S_1^2$. (ii) $p_1p_3 - 4p_4$.
5. $x^4 + cx + b = 0$.
6. $6x^3 + 6x^2 + 9x + 13 = 0$.

Exercises 5 (d), p. 125.

1. $z^3 - 2z + 1 = 0$.
2. $\alpha = 1/2, \beta = -3/2$. Roots $-3, -1, -1/2$.
3. $y^3 - 2ay^2 + (a^2 + b)y + c - ab = 0$, $(a^2 + b)/(ab - c)$.
4. $2 \pm \sqrt{3}, 3 \pm 2\sqrt{2}$.
5. $5, -1/5, (-1 \pm \sqrt{5})/2$.
6. $2, -2 \pm \sqrt{3}, -3 \pm 2\sqrt{2}$.

Exercises 5 (e), p. 131.

1. $-c/3b$.
2. $2, 2, -3$.
3. $a = -\dfrac{10}{3}\left(\dfrac{9}{4}\right)^{2/5}, b = 5\left(\dfrac{9}{4}\right)^{4/5}$.
4. $3\cdot030$.
6. $1\cdot 414$.

Exercises 5 (f), p. 135.

1. $3, -(3 \pm i\sqrt{3})/2$.
2. $4, -2, -2$.
3. $1/2, 3, -7/2$.
4. $4, 2, 2$.
5. $1 \pm \sqrt{2}, -1 \pm i$.
6. $1 \pm \sqrt{2}, -1 \pm i$.

Exercises 5 (g), p. 135.

4. (i) $3/2$. (ii) $1, 2$.
6. $\dfrac{1}{\sqrt[3]{2}} + \dfrac{1}{\sqrt[3]{4}}$.
7. $a^2c'^2x^2 - abb'c'x + a'b^2c' + ab'^2c - 4aa'cc' = 0$.
8. $p = \beta\gamma - (\beta + \gamma)^2, q = \beta\gamma(\beta + \gamma)$; $\dfrac{q(1 + k + k^2)}{pk}$, $-\dfrac{q(1 + k + k^2)}{p(k + 1)}$.
9. $3 \pm \sqrt{7}, 2 \pm \sqrt{6}; x^2 - 24x - 24 = 0$.
11. $z^3 + (3b - a^2)z^2 + b(3b - a^2)z + b^3 - a^3c = 0$;
    $a = (\xi^2 - \eta\zeta)/(\xi^3 + \eta^3 + \zeta^3 - 3\xi\eta\zeta)^{1/2}$, etc.
12. $a^3z^3 - 12a^2cz^2 + 4a(9c^2 + 4bd - ae)z - 16(6bcd - ad^2 - b^2e) = 0$.

ANSWERS TO THE EXERCISES 383

13. $(p^4 - 4p^2q + 2q^2)(3pq - p^3) + pq^3$.
15. $-(n-1)a, -nb, (-1)^n n (b^n - na^n b)$.
18. $x^3 - 3pqx - p^3 - q^3 = 0$; -8, $-2(1 + \omega + 2\omega^2)/\omega$, $-2(2 + \omega + \omega^2)/\omega$ where $\omega^3 = 1 (\omega \neq 1)$.
19. $1 \pm \sqrt{2}, 1 \pm \sqrt{3}, 1 \pm \sqrt{5}$.
20. (ii) $(-1 \pm i\sqrt{3})/2, (3 \pm \sqrt{5})/2$.
21. $a = \pm 1/\sqrt{3}, b = \mp 1/(3\sqrt{3})$.
24. $(a_0 a_3 - a_1 a_2)^2 \leqslant 4(a_0 a_2 - a_1^2)(a_1 a_3 - a_2^2)$.
25. $-2 < a < 2$.

Exercises 6 (a), p. 143.
2. $\pm(3 + 4i)$.
4. $\pm(1 \pm i)/\sqrt{2}$.
3. $\{\sqrt{5} - 1 + i\sqrt{(10 + 2\sqrt{5})}\}/4$, etc.
5. -1.

Exercises 6 (b), p. 147.
1. $(1 \pm i\sqrt{3})/2, (3 \pm 4i)/5$.
2. $(6 \sin 2\theta + 2 \sin 4\theta - 2 \sin 6\theta - \sin 8\theta)/2^7$.
3. $\cos^6 \theta = (10 + 15 \cos 2\theta + 6 \cos 4\theta + \cos 6\theta)/32$,
   $\sin^6 \theta = (10 - 15 \cos 2\theta + 6 \cos 4\theta - \cos 6\theta)/32, 5\pi/32, 5\pi/32$.

Exercises 6 (c), p. 151.
3. $n^2 \operatorname{cosec}^2 n\left(\dfrac{\pi}{2} + \theta\right) - n$.
4. $\dfrac{1}{n} \sum_{r=0}^{n-1} \dfrac{1 - x \cos(2r+1)a}{x^2 - 2x \cos(2r+1)a + 1}$ where $a = \dfrac{\pi}{2n}$.

Exercises 6 (d), p. 154.
1. $t^7 - 7t^6 - 21t^5 + 35t^4 + 35t^3 - 21t^2 - 7t + 1 = 0$ where $t = \tan \theta$.
2. $1/\sqrt{7}$.
4. $3x^3 - 27x^2 + 33x - 1 = 0$.
5. $7(1 - 14x^2 + 49x^4 - 49x^6)$.
6. $(4r + 1)\pi/18$ where r is an integer or zero; $\sin \dfrac{5\pi}{18}, \sin \dfrac{21\pi}{18}, \sin \dfrac{25\pi}{18}$.

Exercises 6 (e), p. 159.
4. (i) $i\pi$.   (ii) $-i\pi/2$.   (iii) $\dfrac{1}{2} \log_e 2 + i\left(2n\pi + \dfrac{\pi}{4}\right)$.
6. $2\left(x \sin a + \dfrac{x^3}{3} \sin 3a + \dfrac{x^5}{5} \sin 5a + \ldots\right)$.

Exercises 6 (f), p. 166.
5. $\log_e 4$.

Exercises 6 (g), p. 169.
1. $\{1 - x \cos \theta - x^{n+1} \cos(n+1)\theta + x^{n+2} \cos n\theta\}/(1 - 2x \cos \theta + x^2)$.
2. $(1 - x^2 \cos 2x)/(1 - 2x^2 \cos 2x + x^4)$.
3. $\left\{3 \cos \theta + \left(-\dfrac{1}{3}\right)^{n-1} \cos 3^n \theta\right\}/4$.
5. $(\cos \theta/2)/\sqrt{(2 \cos \theta)}$.
6. (i) $\sin^2 \left(\dfrac{n+1}{2}\right)\theta \cot \dfrac{\theta}{2}$.   (ii) $\sin^2 \left(\dfrac{n+1}{2}\right)\theta \operatorname{cosec} \dfrac{\theta}{2}$.

ANSWERS TO THE EXERCISES

Exercises 6 (h), p. 169.

1. $\dfrac{1}{p}\tan^{-1}\left(\dfrac{b}{a}\right)$.

2. $\exp(3i\theta) = \Sigma \exp(3i\alpha)\dfrac{\sin(\theta-\beta)\sin(\theta-\gamma)\sin(\theta-\delta)}{\sin(\alpha-\beta)\sin(\alpha-\gamma)\sin(\alpha-\delta)}$.

3. $A_n = \sin n\theta/\sin\theta$, $\{\tan(n+1)\theta - \tan\theta\}\csc\theta$.

7. $(x+a)\prod\limits_{r=1}^{n}\left\{x^2 - 2xa\cos\dfrac{(2r-1)\pi}{2n+1} + a^2\right\}$.

11. $\tan n\theta = \dfrac{{}^nC_1\tan\theta - {}^nC_3\tan^3\theta + \ldots}{1 - {}^nC_2\tan^2\theta + {}^nC_4\tan^4\theta - \ldots}$.

   (i) n, (ii) $(2n-1)n$.

12. $x^2 + x + 2 = 0$.

13. $\tan 5\theta = \dfrac{5\tan\theta - 10\tan^3\theta + \tan^5\theta}{1 - 10\tan^2\theta + 5\tan^4\theta}$; $\tan 72°$, $\tan 108°$, $\tan 144°$

15. (ii) $\dfrac{1}{2}\log_e\{(x^2 - y^2 - a^2)^2 + 4x^2y^2\} + i\tan^{-1}\left\{\dfrac{2xy}{x^2 - y^2 - a^2}\right\}$.

18. Ellipses and hyperbolas with foci $(\pm c, 0)$.

21. $\dfrac{(1+\omega^2)\{\sin\theta - \sin(2n+1)\theta\} + 2\omega^2\sin 2n\theta\cos\theta}{(1+\omega^2)^2 - 4\omega^2\cos^2\theta}$,

   $\dfrac{2\cos\theta\{\sin\theta - \sin(2n+1)\theta\} + (1+\omega^2)\sin 2n\theta}{(1+\omega^2)^2 - 4\omega^2\cos^2\theta}$.

24. (i) $\cot\theta\cos^n\theta\sin n\theta$, 0. (ii) $\cot\theta(1 - \cos^n\theta\cos n\theta)$, $\cot\theta$.

Exercises 7 (a), p. 183.

1. (a) 24. (b) $bc - ad$. (c) $a^2 + b^2 + c^2 + d^2$.
2. (a) 0. (b) ab. 4. 0.
5. $-(a-b)(b-c)(c-a)(x-y)(y-z)(z-x)$.

Exercises 7 (b), p. 190.

1. $0, \pm\sqrt{57}$.   2. $1, a, -a/(1+a)$.
4. $n+1$.
6. $A = \begin{vmatrix} a_1 & b_1 & c_1 \\ a_2 & b_2 & c_2 \\ 1 & 1 & 1 \end{vmatrix}$, $B = \begin{vmatrix} a_1-1 & b_1-1 & c_1-1 \\ a_2-1 & b_2-1 & c_2-1 \\ 1 & 2 & 3 \end{vmatrix}$.

Exercises 7 (c), p. 196.

4. $\begin{vmatrix} 1 & 1 & 1 \\ \alpha^2 & \beta^2 & \gamma^2 \\ \alpha^3 & \beta^3 & \gamma^3 \end{vmatrix} \times \begin{vmatrix} 1 & 1 & 1 \\ \alpha^2 & \beta^2 & \gamma^2 \\ \alpha^4 & \beta^4 & \gamma^4 \end{vmatrix}$.

Exercises 7 (d), p. 201.

1. $x = 1, y = 2, z = 3$.   2. $x = 5/2, y = 3, z = -4$.
3. $1, \omega, \omega^2$ where $\omega^3 = 1(\omega \neq 1)$.   4. $\pm 1, 2$.
5. $\begin{vmatrix} a & b & c & 0 \\ 0 & a & b & c \\ A & B & C & 0 \\ 0 & A & B & C \end{vmatrix} = 0$.

ANSWERS TO THE EXERCISES

Exercises 7 (e), p. 201:

1. $2(x+3)^3(x+1)$.
2. $4abc(a+b+c)(bc+ca+ab-a^2-b^2-c^2)$.
4. $4/\{(x+1)(x+2)^2(x+3)^3(x+4)^2(x+5)\}$.
5. $-(a-b)(b-c)^2(c-a)^2$. 6. $(a-b)(b-c)(c-a)(a+b+c)$.
10. $x+y\exp\{-i(\theta+\phi)\} - z\exp(-i\phi)$, $x+y\exp\{i(\theta+\phi)\} - z\exp(i\phi)$.
11. $0, -3 \pm \sqrt{15}$. 12. $-(a+b+c), a-b+c, b\pm i(a-c)$.
13. $a+\beta+\gamma$. 14. $q=-1, x=-1, \pm\sqrt{2}$.
17. $\begin{vmatrix} a_1 & -a_1 & b_1 & -\beta_1 \\ a_2 & -a_2 & b_2 & -\beta_2 \end{vmatrix}$.
18. $\{a(x-b)^n - b(x-a)^n\}/(a-b)$.
19. $B = (t_1-a)(t_2-a)(t_3-a) + (t_1-a)(t_2-a)(t_4-a)$
    $\qquad + (t_1-a)(t_3-a)(t_4-a) + (t_2-a)(t_3-a)(t_4-a)$,
    $A = f(a) + aB$.
21. $\Delta = (\alpha-\beta)(\alpha+\beta)(\alpha-\gamma)(\alpha+\gamma)(\alpha-\delta)(\alpha+\delta)(\beta-\gamma)(\beta+\gamma)(\beta-\delta)$
    $\qquad\qquad\qquad\qquad\qquad\qquad\qquad (\beta+\delta)(\gamma-\delta)(\gamma+\delta)$,
    $D = (\alpha-\beta)(\alpha-\gamma)(\alpha-\delta)(\beta-\gamma)(\beta-\delta)(\gamma-\delta)$.
22. $A = xX + yY + zZ + uU$, $B = uX + xY + yZ + zU$,
    $C = zX + uY + xZ + yU$, $D = yX + zY + uZ + xU$.
25. $\begin{vmatrix} 1 & p & q & r & 0 \\ 0 & 1 & p & q & r \\ 1 & b & c & 0 & 0 \\ 0 & 1 & b & c & 0 \\ 0 & 0 & 1 & b & c \end{vmatrix}$.

Exercises 8 (a), p. 216.

1. (a) 3. (b) 3/5.
2. $\pm b/a$.
3. Continuous except at $x = q$.

Exercises 8 (b), p. 220.

5. $\sqrt{(7/3)}$.

Exercises 8 (c), p. 223.

1. (i) $\dfrac{1}{2}(-1)^n(n)!\left\{\dfrac{1}{(x-a)^{n+1}} + \dfrac{1}{(x+a)^{n+1}}\right\}$.
   (ii) $\dfrac{(-1)^n(n)!}{4a}\left\{\dfrac{x+na}{(x-a)^{n+2}} - \dfrac{x-na}{(x+a)^{n+2}}\right\}$.
2. $(a^2+b^2)^{n/2}e^{ax}\cos\left(bx + n\tan^{-1}\dfrac{b}{a}\right)$.

Exercises 8 (d), p. 227.

2. $dy/dx = 1$, $d^2y/dx^2 = 3$, $x = (y-1) - \dfrac{3}{2}(y-1)^2$.
3. $\log_e \cos x = -\dfrac{x^2}{2} - \dfrac{x^4}{12} - \ldots$.
4. $\dfrac{dy}{dx} = (1+x)\{2\log_e(1+x) + 1\}$, $\dfrac{d^2y}{dx^2} = 2\log_e(1+x) + 3$.

5. $a_{4n+2} = (-1)^n 2^{2n+1}$, $a_{4n+3} = (-1)^n 2^{2n+1}$.

6. $y = x^2 + \dfrac{2}{3}\dfrac{x^4}{2} + \dfrac{2 \cdot 4}{3 \cdot 5}\dfrac{x^6}{3} + \ldots + \dfrac{2^{2(n-1)}(n-1)!}{(2n-1)!}\dfrac{x^{2n}}{n} + \ldots$

Exercises 8 (e), p. 232.

1. $-1/3$.
2. (i) $3a/2$. (ii) 2.
3. k, p/q, $2n^2$.
4. -2.

Exercises 8 (f), p. 232.

1. (i) $1/4$. (ii) -1.
2. (a) 12. (b) Ordinary discontinuity at $x = 0$.
3. Continuous.
15. $1 + x\dfrac{\cos \alpha}{\sin \alpha} + \dfrac{x^2}{2!}\dfrac{\cos 2\alpha}{\sin^2 \alpha} + \ldots + \dfrac{x^n}{n!}\dfrac{\cos n\alpha}{\sin^n \alpha} + \ldots$
18. When $|x| < 1$.
23. (i) $(\log_e 2)/(\log_e 3)$. (ii) $9/16$. (iii) 0.
25. (i) 3. (ii) -1.

Exercises 9 (a), p. 238.

1. $e + 1$.
5. $\alpha^2 + \beta^2 + \gamma^2 = 0$.
6. $\dfrac{\partial u}{\partial x} = \dfrac{\partial v}{\partial y} = \sin x \cosh y + x \cos x \cosh y + y \sin x \sinh y$,

$\dfrac{\partial u}{\partial y} = -\dfrac{\partial v}{\partial x} = -\cos x \sinh y + x \sin x \sinh y - y \cos x \cosh y$.

Exercises 9 (b), p. 244.

1. $2 \cdot 8$.
4. (a) $-(e^x - 2y)/(e^y - 2x)$. (b) $-(c^n x^{n-1})/(a^n z^{n-1})$.
5. $-\dfrac{x^2 - y}{y^2 - x}$, $-\dfrac{2xy}{(y^2 - x)^3}$.

Exercises 9 (c), p. 247.

2. $\dfrac{\partial r}{\partial x} = \cosh \theta$, $\dfrac{\partial r}{\partial y} = -\sinh \theta$, $\dfrac{\partial \theta}{\partial x} = -\dfrac{\sinh \theta}{r}$, $\dfrac{\partial \theta}{\partial y} = \dfrac{\cosh \theta}{r}$.

Exercises 9 (d), p. 251.

1. (i) 0. (ii) $\sin 2u$.

Exercises 9 (e), p. 255.

2. 0, $1/27$.
3. (i) $x = 1/2$, $y = 1/3$. (ii) $x = y = \pi/3$.
6. 0, $a^2 b^2$. Second value is a maximum.

Exercises 9 (f), p. 255.

1. -3.
2. $\dfrac{\partial \phi}{\partial x} = -\dfrac{\log_e y}{x(\log_e x)^2}$, $\dfrac{\partial \phi}{\partial y} = \dfrac{1}{y \log_e x}$.
6. $0 \cdot 14$ cm. approx.
7. $-1/2$.

ANSWERS TO THE EXERCISES 387

9. $\dfrac{y^x \log_e y + yx^{y-1} - (x+y)^{x+y}\{1 + \log_e (x+y)\}}{x^y \log_e x + xy^{x-1} - (x+y)^{x+y}\{1 + \log_e (x+y)\}}.$

13. $\dfrac{\partial \phi}{\partial x} = (\cosh x \cos y + x \sinh x \cos y - y \cosh x \sin y)\dfrac{\partial \phi}{\partial u}$
$+ (\sinh x \sin y + x \cosh x \sin y + y \sinh x \cos y)\dfrac{\partial \phi}{\partial v},$

and similarly for $\partial \phi / \partial y$.

17. $\dfrac{\partial V}{\partial x} + i\dfrac{\partial V}{\partial y} = e^{i\phi}\left(\sin \theta \dfrac{\partial V}{\partial r} + \dfrac{\cos \theta}{r}\dfrac{\partial V}{\partial \theta} + \dfrac{i}{r \sin \theta}\dfrac{\partial V}{\partial \phi}\right),$

$\dfrac{\partial V}{\partial x} - i\dfrac{\partial V}{\partial y} = e^{-i\phi}\left(\sin \theta \dfrac{\partial V}{\partial r} + \dfrac{\cos \theta}{r}\dfrac{\partial V}{\partial \theta} - \dfrac{i}{r \sin \theta}\dfrac{\partial V}{\partial \phi}\right),$

$\dfrac{\partial V}{\partial z} = \cos \theta \dfrac{\partial V}{\partial r} - \dfrac{\sin \theta}{r}\dfrac{\partial V}{\partial \theta}.$

19. (ii) $C \log_e r$.   20. $C\displaystyle\int \exp\left(-\dfrac{x^2}{4kt}\right)\dfrac{dx}{\sqrt{(kt)}}.$

Exercises 10 (a), p. 263.

1. $\dfrac{x^{1-n}}{1-n}\left(\log_e x - \dfrac{1}{1-n}\right).$

2. $x \log_e \{x + \sqrt{(x^2 + a^2)}\} - \sqrt{(x^2 + a^2)}.$

3. $\dfrac{x^{m+1}}{m+1}\left\{(\log_e x)^2 - \dfrac{2}{m+1}\log_e x + \dfrac{2}{(m+1)^2}\right\}.$

4. $e^x \left\{\dfrac{x-1}{(x+1)^2}\right\}.$ 5. $\dfrac{x}{a^2\sqrt{(x^2+a^2)}}.$

6. $\dfrac{1}{2(a^2-b^2)^2}\left\{\dfrac{b^2(a^2-b^2)}{x^2+b^2} + a^2 \log_e\left(\dfrac{x^2+b^2}{x^2+a^2}\right)\right\}.$

7. $\dfrac{1}{4\sqrt{2}a^3}\left\{\log_e\left(\dfrac{x^2+\sqrt{2ax}+a^2}{x^2-\sqrt{2ax}+a^2}\right) + 2\tan^{-1}\left(\dfrac{\sqrt{2ax}}{a^2-x^2}\right)\right\}.$

8. $(b-a)\sin^{-1}\sqrt{\left(\dfrac{x-a}{b-a}\right)} - \sqrt{\{(x-a)(b-x)\}}.$

9. $-\dfrac{\sqrt{(x^2+1)}}{x}.$ 10. $\sin^{-1}\sqrt{x} - \sqrt{\{x(1-x)\}}.$

11. $\sec x - \log_e \cos x.$ 12. $x \tan \dfrac{x}{2}.$

13. $\dfrac{1}{2(1+\cos x)} - \dfrac{1}{4}\log_e\left(\dfrac{1+\cos x}{1-\cos x}\right).$

14. $\tan x + 2 \cot x - \dfrac{1}{3}\cot^3 x.$ 15. $\dfrac{3\pi^2}{4} - 6.$

16. $\dfrac{1}{2}(1 - \log_e 2).$ 17. $\dfrac{1}{2}(\log_e 3 - 1).$

18. $(a+b)\pi/4.$ 19. $e^x \tan \dfrac{1}{2}x, \; -e^x \cot \dfrac{1}{2}x.$

23. $\sqrt{(x^2+1)} + \dfrac{1}{2}\log_e\left\{\dfrac{\sqrt{(x^2+1)}-1}{\sqrt{(x^2+1)}+1}\right\} - x.$

24. $-(a+x) - 3(a+x)^{2/3} - 6(a+x)^{1/3} - 6\log_e\{1 - (a+x)^{1/3}\}.$

ANSWERS TO THE EXERCISES

25. $\dfrac{\pi}{4\sqrt{2}} + \dfrac{1}{\sqrt{2}} - 1$.

27. $x \cos(a - \beta) + \sin(a - \beta) \log_e \sin(x + \beta)$.

28. $\dfrac{3\pi}{2} + 3 \log_e 2 - \dfrac{409}{70}$.

Exercises 10 (b), p. 268

1. $(a \sinh ax \cos bx + b \cosh ax \sin bx)/(a^2 + b^2)$.
3. $\pi(b - a)^2/8$.
4. (i) $\sinh^{-1}\left(\dfrac{x+1}{\sqrt{2}}\right)$. (ii) $\dfrac{1}{2}(x - 1)\sqrt{(x^2 + 2x + 3)}$.

   (iii) $\dfrac{1}{3}(x^2 + 2x + 3)^{3/2} - (x + 1)\sqrt{(x^2 + 2x + 3)} + 2\sinh^{-1}\left(\dfrac{x+1}{\sqrt{2}}\right)$

5. $\dfrac{1}{2}\sinh^{-1} x - \dfrac{1}{4}\{2x^2 + 1 - 2x\sqrt{(x^2 + 1)}\}$.

6. $\dfrac{1}{3}x^{3/2}\sqrt{(1 + x^3)} + \dfrac{1}{3}\log_e\{x^{3/2} + \sqrt{(1 + x^3)}\}$.

Exercises 10 (c), p. 272.

1. (i) $\dfrac{1}{4}\log_e 3$. (ii) $\dfrac{\pi}{4}$.   2. (i) $\cosh a$. (ii) $\cosh a$.

3. $\dfrac{2}{\sqrt{3}} \tan^{-1}\left(\sqrt{3} \tan \dfrac{\theta}{2}\right) - \dfrac{1}{\sqrt{2}} \tan^{-1}\left(\sqrt{2} \tan \dfrac{\theta}{2}\right)$.

6. $I_n = (1 - e^2)^{-n + \frac{1}{2}} \int (1 - e \cos \phi)^{n-1} d\phi$, $I_0 = \theta$,

   $I_1 = \dfrac{1}{\sqrt{(1 - e^2)}} \cos^{-1}\left(\dfrac{e + \cos \theta}{1 + e \cos \theta}\right)$,

   $I_2 = \dfrac{1}{(1 - e^2)^{3/2}} \cos^{-1}\left(\dfrac{e + \cos \theta}{1 + e \cos \theta}\right) - \dfrac{e}{1 - e^2} \dfrac{\sin \theta}{1 + e \cos \theta}$.

Exercises 10 (d), p. 276.

1. $\dfrac{2}{\sqrt{(a - b)}} \tan^{-1}\left\{\sqrt{\left(\dfrac{x+b}{a-b}\right)}\right\}$.   2. $\dfrac{-1}{\sqrt{(a^2 + 1)}} \sinh^{-1}\left(\dfrac{1 + ax}{x - a}\right)$.

3. $\dfrac{1}{\sqrt{(1 - k^2)}} \sin^{-1}\left(\dfrac{t - k}{1 - kt}\right)$.   4. $\sinh^{-1}\left(\dfrac{2X - 1}{X}\right)$.

5. $\dfrac{1}{2\sqrt{2}} \log_e\left\{\dfrac{x + 2 + \sqrt{(2x + 2)}}{x + 2 - \sqrt{(2x + 2)}}\right\} - \dfrac{1}{\sqrt{2}} \tan^{-1}\left\{\dfrac{\sqrt{(2x + 2)}}{x}\right\}$.

6. $-\dfrac{1}{b\sqrt{(a^2 - b^2)}} \sin^{-1}\left\{\dfrac{b\sqrt{(x^2 + a^2)}}{a\sqrt{(x^2 + b^2)}}\right\}$.

Exercises 10 (e), p. 286.

3. $(m + n)I_{m,n} = \sin^{m-1} x \cos^{n-1} x \left(\sin^2 x - \dfrac{n-1}{m+n-2}\right) + \dfrac{(m-1)(n-1)}{m+n-2} I_{m-2, n-2}$.

4. $\pi/2^{n+1}$.   6. $16/105$.
7. $\pi a^6/32$.

ANSWERS TO THE EXERCISES

Exercises 10 (f), p. 291

1. (i) $\dfrac{\pi^2(\sqrt{2}-1)}{2\sqrt{2}}$. (ii) $\theta \log_e \sec \theta$.
2. $\pi^2/16$. 3. $\pi/3\sqrt{3}$.

Exercises 10 (g), p. 297

2. (i) π. (ii) $-\dfrac{1}{2}\log_e\left(\dfrac{7}{3}\right)$. (iii) Only exists as a principal value.

6. $\dfrac{\pi}{16\sqrt{(xy)}}\left(\dfrac{3}{x^2}+\dfrac{2}{xy}+\dfrac{3}{y^2}\right)$.

Exercises 10 (h), p. 297

2. $\dfrac{\sqrt{(a+bx^2)}}{15b^3}(8a^2 - 4abx^2 + 3b^2x^4)$.

4. (i) $-\dfrac{1}{2}\sqrt{\{(1-x^2)(a^2-x^2)\}} - \dfrac{1}{2}(a^2-1)\log_e\left\{\dfrac{\sqrt{(1-x^2)}+\sqrt{(a^2-x^2)}}{\sqrt{(a^2-1)}}\right\}$.

(ii) $x \log_e\left(\dfrac{x}{x^2+1}\right) + \left(x - \dfrac{1}{2}\right)\log_e(x^2-x+1) - x +$
$+ \sqrt{3}\tan^{-1}\left(\dfrac{2x-1}{\sqrt{3}}\right) - 2\tan^{-1}x$.

5. $-\dfrac{3\sqrt{2}}{16}\log_e\left\{\dfrac{\sqrt{(2x^2+2)}-x}{\sqrt{(2x^2+2)}+x}\right\} - \dfrac{x\sqrt{(x^2+1)}}{4(x^2+2)}$.

6. 0. 7. $n/(n^2-1)$.

8. $\dfrac{1}{2} + \dfrac{\sqrt{3}}{24}\log_e\left(\dfrac{2+\sqrt{3}}{2-\sqrt{3}}\right)$.

9. $\dfrac{1}{2} + \left(\dfrac{1}{2\sqrt{6}} - \dfrac{1}{\sqrt{2}}\right)\dfrac{\pi}{2} + \dfrac{1}{\sqrt{2}}\tan^{-1}\sqrt{2} - \dfrac{1}{2\sqrt{6}}\tan^{-1}\left(\sqrt{\dfrac{3}{2}}\right)$.

12. (a) (i) $\pi/4$, (ii) $\pi/2$. (b) $\pi/2\sqrt{10}$.
13. $(a - 2\pi)/\sin \alpha$. 14. $1/a^2$.
16. $I_{m,n} = \dfrac{\sin^{m-1}\theta}{(n-1)\cos^{n-1}\theta} - \dfrac{m-1}{n-1}I_{m-2,n-2}$.
17. $A = (1-i)/4$, $B = (1+i)/4$. 18. $n > 0$.
19. $2n(ac-b^2)J_n = \dfrac{(2n-1)ax^2 + 2nbx + c}{2(n-1)Q^n} + (2n-1)aJ_{n-1}$, $\log_e 3$.

24. $\dfrac{1}{2}\sum_{r=1}^{n} a_r^2\left(\dfrac{2\lambda_r l - \sin 2\lambda_r l}{2\lambda_r}\right)$.

26. $\pi\left(\dfrac{\pi}{2} - 1\right)$. 27. $\pi/4$.

28. $I_1 = \dfrac{5}{12\sqrt{3}}\tan^{-1}\left(\dfrac{x^2-1}{x\sqrt{3}}\right) + \dfrac{3}{8}\log_e\left(\dfrac{x^2+x+1}{x^2-x+1}\right) + \dfrac{x(x^2+2)}{6(x^4+x^2+1)}$,

$I_2 = \dfrac{-1}{4\sqrt{3}}\tan^{-1}\left(\dfrac{x^2-1}{x\sqrt{3}}\right) - \dfrac{1}{8}\log_e\left(\dfrac{x^2+x+1}{x^2-x+1}\right) + \dfrac{x^3}{2(x^4+x^2+1)}$,

$I_3 = \dfrac{1}{2\sqrt{3}}\tan^{-1}\left(\dfrac{x^2-1}{x\sqrt{3}}\right) + \dfrac{1}{4}\log_e\left(\dfrac{x^2+x+1}{x^2-x+1}\right)$.

Exercises 11 (a), p. 307

1. $y = \mp \dfrac{8\sqrt{3}}{9}x + \dfrac{41}{36}a$.
2. $\dfrac{x}{a}\cos\theta + \dfrac{y}{b}\sin\theta = 1$, $ax\sec\theta - by\csc\theta = a^2 - b^2$.
5. $\cot^{-1}\left[\dfrac{bf'(\theta)}{a}\right]$.
6. $\sqrt{(r^3/2a)}$.

Exercises 11 (b), p. 311

1. $y = -x + (2a)/3$.
2. $y = \pm x$.
3. $x = c, y = a$.
4. $y = 0, y = \pm x - 1$.
5. $x = 0, v = -x, y = x + 2$.
6. $x = 1, v = -1, y = \pm x$.

Exercises 11 (c), p. 319

1. $4c$.
2. $c\cosh^2(x_1/c)$.
4. $-8a\sin(3t/2)$.

Exercises 11 (d), p. 329

3. $y = \pm x$.
5. $\left(p\cos\psi - \dfrac{dp}{d\psi}\sin\psi,\ p\sin\psi + \dfrac{dp}{d\psi}\cos\psi\right)$.
6. $x = \dfrac{a}{2}\cos\theta(3 - 2\cos^2\theta),\ y = a\sin^3\theta$.

Exercises 11 (e), p. 335

2. $r = 2$.

Exercises 11 (f), p. 340

1. $(3\pi a^2)/8$.
3. $(3\pi a^2)/2$.
4. $\dfrac{a^2}{48}(3\sqrt{3} - \pi)$.
5. $\pi a^2(2 - \sqrt{2})$.
6. $256/15$.

Exercises 11 (g), p. 344

6. $\bar{x} = \dfrac{2a(1 - \varepsilon^2)^{3/2}}{3\left\{\dfrac{\pi}{2} - \varepsilon\sqrt{(1-\varepsilon^2)} - \sin^{-1}\varepsilon\right\}}$, $\bar{y} = 0$ where $\varepsilon = $ eccentricity,

volume $= 4\pi a^3(1 - \varepsilon^2)^2/3$.

Exercises 11 (h), p. 345

3. $\dfrac{b^2}{ap^2} = \dfrac{2}{r} - \dfrac{1}{a}$, where a, b are semi-axes.
5. $y = 0, y = \pm x - 1$.
6. $y^2(x - a) + x^2(y - a) + 2axy = 0$.
7. $x = a\cos t\,(1 + 2\sin^2 t),\ y = a\sin t(1 + 2\cos^2 t)$.
8. $3a/2$.
10. $x^2 + y^2 - 2y = 0$.
12. $s\left(1 - \dfrac{k^2s^2}{24} - \dfrac{kk's^3}{24}\right)$.
20. $\dfrac{a}{27}|(4 + 9t^2)^{3/2} - (4 + 9u^2)^{3/2}|$.
21. $1/(\pi s)$.

ANSWERS TO THE EXERCISES

Exercises 12 (a), p. 352.
1. $y = Ce^{x^2}$.
2. $\sin x \sin y = C$.
3. $y = 1 + Ce^{-x^2}$.
4. $y = 3 + \dfrac{3x}{2} - \cos x - \dfrac{1}{4}\sin 2x$
5. (i) 0·231. (ii) 2·57.

Exercises 12 (b), p. 357.
1. $x^3 - 2xy^2 - y^3 = C$.
2. $\sin x \sin y = C$.
3. $(x-1)^3 y = \dfrac{x^5}{5} - \dfrac{x^4}{2} + \dfrac{x^3}{3} + C$.
4. $ye^{2x} = C(x+1)^3 - \tfrac{1}{3}$.
5. $xy = -x\cos\dfrac{x}{2} + 2\sin\dfrac{x}{2} + C$, $y \to 0$.
6. $-\dfrac{1}{3y^3} = x^4 + Cx^3$.

Exercises 12 (c), p. 360.
1. $x^2 - xy - y^2 = C$.
2. $\dfrac{1}{\sqrt{2}}\tan^{-1}\left(\dfrac{y}{x\sqrt{2}}\right) = \log_e\left\{\dfrac{C}{\sqrt{(2x^2+y^2)}}\right\}$.
3. $y = Ce^{x^{1/2}y^2}$.
4. $x - 2y - 4\log_e(2x - 5y + 10) = C$.
5. $(y - x + 3)^4 = C(y + 2x - 3)$.
6. $x(y^3+1)^{1/3} = \tfrac{1}{2}(y^3+1)^{4/3} + C$.

Exercises 12 (d), p. 366.
1. $y = \dfrac{3}{2}e^{2x} + \dfrac{1}{2}e^{-2x}$.
2. $x = e^{-2t}(A\cos\sqrt{5}t + B\sin\sqrt{5}t)$.
3. $y = xe^{\frac{1}{2}x-1}$.
4. $y = A\cos 3x + B\sin 3x + 2$.
5. (i) $x = Ae^{2t} + Be^{-t} + \dfrac{1}{4} - \dfrac{t}{2}$. (ii) $x = Ae^{2t} + Be^{-t} - \dfrac{1}{2}e^t$.
6. (i) $x = Ae^{2t} + Be^{-t} + te^{2t}$. (ii) $x = Ae^{2t} + Be^{-t} + \cos t - 3\sin t$.
7. $y = A\cos 2x + B\sin 2x - \dfrac{1}{8} + \dfrac{x^2}{4} + \dfrac{e^{-2x}}{8} + \dfrac{1}{3}\cos x$.

Exercises 12 (e), p. 369.
1. $\theta = Ae^{2t} + e^{-t}(B\cos\sqrt{3}t + C\sin\sqrt{3}t)$.
2. $y = (A + Bx)\cos x + (C + Dx)\sin x$.
4. $x\cos 2x$.
6. $x = A(t+1)^5 + B(t+1)^{-2} - \dfrac{1}{6}(t+1)^4$.

Exercises 12 (f), p. 374.
1. $y = 1 + e^{-x}$.
2. $y = \cos x$.
3. $\theta = \dfrac{1}{8} - \dfrac{1}{8}e^{-2t}(\cos 2t - 3\sin 2t)$.
4. $y = -xe^{2x} + 2e^{2x} - e^{3x}$.
5. $x = 1$.

Exercises 12 (g), p. 374.

1. $\sqrt{(y^2 + 1)} = -\dfrac{x^2}{2} + C.$

2. $y = a + C\sqrt{(1 - x^2)}.$

3. $y = Ce^x \sin x.$

5. $y = 1 + x^2 + C(1 + x^2)^{-1/2}.$

6. $y \tan x = \log_e \tan x + C.$

7. $(\tan x + \sec x)(y^2 + 1) = (x + C)y^2.$

8. $\log_e \{C\sqrt{(4x^2 + y^2)}\} + \dfrac{1}{2} \tan^{-1}\left(\dfrac{y}{2x}\right) = 0.$

9. $y = Cx^2 e^{2y/x} - x.$

10. $v^2 = 2\mu a(a - x) + Ce^{-x/a}.$

11. $x = 2a + \dfrac{a}{2} \cos \sqrt{\left(\dfrac{k}{a}\right)} t.$

12. $y = -x^2 + A \cos x + B \sin x.$

14. $y = \dfrac{1}{4}(1 - \cos 2x).$

15. $y = Ae^{2x} + Be^{-x} - \dfrac{1}{2}e^x - \dfrac{1}{10} \sin x - \dfrac{3}{10} \cos x.$

16. $y = A \cos 2x + B \sin 2x + \dfrac{x}{4} - \dfrac{x}{4} \cos 2x.$

17. $y = (A + Bx)e^x + \dfrac{1}{2}x^2 e^x + x + 2 + \dfrac{1}{2} \cos x.$

18. $B = \dfrac{A}{\sqrt{\{(p^2 - q^2)^2 + 4k^2 q^2\}}},\ \alpha = \tan^{-1}\left(\dfrac{2kq}{p^2 - q^2}\right),$
$x = e^{-kt}\{C \cosh \sqrt{(k^2 - p^2)}t + C' \sinh \sqrt{(k^2 - p^2)}t\} + B \cos(qt - \alpha),$
$x = \dfrac{Bq \sin \alpha}{k} e^{-kt} \cosh \sqrt{(k^2 - p^2)}t + B \cos(qt - \alpha).$

19. $u = A \cos 4x + B \sin 4x + \cos^4 x + \sin^4 x,$
$u = \cos^4 x + \sin^4 x - \cos 4x.$

20. $x = (\pi \sin 2t)/2t.$

21. (i) $y = (A + Bx)e^x + Ce^{-2x}.$ (ii) $y = (A + Bx + Cx^2)e^x.$

22. $y = Ae^{-x} + Be^{-2x} + Ce^{-3x} + \dfrac{1}{2}xe^{-x}.$

24. $y = -\dfrac{5}{9} \sin x + \dfrac{4}{9} \sin 2x - \dfrac{1}{3} x \cos 2x.$

INDEX

Absolute convergence, 62
Addition of complex numbers, 84
Algebraical functions, 12; numbers, 3
Alternating series, 61
Amplitude, of complex number, 79
Answers, to exercises, 379
Arc length, 340
Areas, 336; sectorial, 336; surface, 342; Pappus' theorem for, 342
Argand's diagram, 77
Argument, of complex number, 79
Arithmetic mean, 5
Arithmetical continuum, 3
Asymptotes, 308

Bernoulli's equation, 355
Bilinear transformation, 98
Binomial coefficients, series involving, 27
Binomial series, convergence of, 66
Biquadratic equation, 133
Bounded functions, 212

Cardan's solution, of cubic equation, 131
Cauchy's, inequality, 7; mean value theorem, 220; multiplication theorem, 64; test for convergence, 59
Centre of curvature, 313
Change of variables, 244
Circle, of convergence, 102; of curvature, 313
Closed interval, 211
Cofactors, of a determinant, 190
Common roots of equations, condition for, 126
Complete primitive, of differential equation, 350
Complex numbers, 75; addition in Argand diagram, 84; argument (amplitude) of, 79; conjugate, 80; fractional powers of, 141; geometrical representation of, 77; imaginary parts of, 78; modulus of, 79; principal value of argument of, 79; products and quotients in Argand diagram, 88; real parts of, 78; representative points of, 77; some remarks on the manipulation of, 82
Complex variable, logarithmic function of, 157; rational functions of, 94; trigonometrical functions of, 155
Complimentary function, 363
Conjugate complex numbers, 80; points, 321
Consistency, of equations, 174, 199
Contact, of curves, 312
Continuous functions, of a real variable, 210; of more than one variable, 215
Continuous real variable, 10
Convergence, absolute, 62; Cauchy's test for, 59; circle of, 102; comparison tests for, 54; condensation test for, 59; D'Alembert's test for, 56; of binomial, exponential and logarithmic series, 66; Raabe's test for, 57; radius of, 102; tests for, 54
Convergent sequences, 39; series, 49
$\cos^n \theta$ expressed in multiple angles, 144
$\cos n\theta$ expressed in powers of $\cos \theta$ and $\sin \theta$, 145
Cote's properties of the circle, 151
Cube roots of unity, 83
Cubic equation, 131
Curvature, centre of, 313; circle of, 313; Newton's formula for, 316; radius of, 313
Curve sketching, 329
Cusp, 321

D'Alembert's ratio test, 56
Dedekind's theory of irrational numbers, 2
Definite integrals, differentiation of, 295; some general properties of, 287; some important, 280
Degree, of differential equation, 349
De Moivre's, property of the circle, 150; theorem, 139
Dependent variable, 13

Derivatives, partial, 236
Descartes', rule of signs, 112; solution of quartic equation, 134
Determinantal equations, 184; notation, 174
Determinants, 174; cofactors of, 190; differentiation of, 185; minors of, 190; multiplication of, 192; properties of, 178; reciprocal, 195; skew symmetrical, 195; symmetrical, 195
Diagonal, of determinant, 175
Difference equations, 29
Differentiability, 216
Differential coefficient, of function o. two functions, 239; partial, 236
Differential equations, 349; degree of, 349; exact, 353; first order with separable variables, 351; general solution of, 350; higher order linear, 367; homogeneous first order, 357; homogeneous linear, 368; linear, 349; linear first order, 354; order of, 349; ordinary, 349; partial, 349; reducible to linear form, 355; second order with constant coefficients, 360; solution by Laplace transform. 370
Differentials, 240; exact, 250
Differentiation, of definite integrals, 295; of determinant, 185; of implicit functions, 242; repeated, 221
Dirichlet's theorem, on absolute convergence, 62
Discontinuity, types of, 214
Divergent sequences, 40; series, 50
Double points, 320

Eliminant, of equations, 200
Elimination, 200
Envelopes, 324
Equations, cubic, 131; determinantal, 184; differential, 349; eliminant of, 200; intrinsic, 305; pedal, 307; quartic, 133; reciprocal, 123; simultaneous, 197
Euler's relations, 156; theorems on homogeneous functions, 248
Evolute, 328
Exact differentials, 250; differential equations, 353
Expansion, of determinants, 174

Explicit functions, 13
Exponential series, convergence of, 66

Factorisation, 147
Finite series, summation of, 22
Fractional powers, of complex numbers, 141
Function, algebraical, 12; explicit, 13; idea of, 10; implicit, 13; of continuous variable, 207; of several variables, 14; periodic, 13; rational, 11; transcendental, 13
Function of two functions, differential coefficient of, 239
Functional terminology, 10

Gamma function, 282
General solution, of difference equation, 30; of differential equation, 350
Generalised trigonometrical functions, 156
Generating function, of recurring power series, 68
Geometric mean, 5
Geometrical representation, of complex numbers, 77

Half-open interval, 211
Harmonic function, 258
Homogeneous first order differential equation, 357; linear differential equation, 368
Homogeneous functions, Euler's theorems on, 248
Hyperbolic functions, 160; integrals involving, 264; inverse, 164

Imaginary part, of complex number, 78; of sin $(x + iy)$, etc., 163
Implicit functions, 13; differentiation of, 242
Independent variable, 13
Indeterminate forms, 228
Induction, method of, for summation of series, 26
Inequalities, 4; Cauchy's, 7; Tchebychef's, 8
Infinite and improper integrals, 292
Infinite series, 49; general theorems on, 51; multiplication of, 64; of complex terms, 101; summation of, 67

INDEX

Integrals, infinite, 292; involving hyperbolic functions, 264
Integrating factor, of differential equation, 354
Integration, of $(a + b \cos x)^{-1}$, etc., 268; of $e^{ax} \cos bx$, etc., 277; of $1/(X \sqrt{Y})$ where X, Y are linear or quadratic, 283
Interval, closed, 211; half-open, 211; open, 211
Intrinsic equation, 305
Inverse hyperbolic functions, 164
Irrational numbers, 2

Lagrange's form, of remainder in Taylor's series, 225
Laplace transform, 370; solution of differential equations by means of, 370; table of, 371
Leading diagonal, of determinant, 175
Leibnitz' theorem, 222
Length of arc, 340
Limits, important, 44; of functions of a continuous variable, 207; useful theorems on, 43
Linear differential equations, 349, 354, 360; first order, 354; second order 360
Logarithmic function of complex variable, 157
Logarithmic series, convergence of, 66
Lower bound, 212

Maclaurin's series, 225
Maxima and minima, two independent variables, 253
Mean value theorem, 219; for function of two variables, 240; for integrals, 290; general, 224
Minor determinant, 177, 190
Modulus, of complex number, 79
Monotonic functions, 208; sequences, 41
Multiple points, 320
Multiplication, of determinants, 192; of infinite series, 64

Newton's formula, for radius of curvature, 316; for sums of powers of roots of equations, 118; method of approximation to roots of equations, 128; parallelogram, 332

Normals, 303
Numbers, algebraical, 3; complex, 75; irrational, 2; rational, 1; real, 3; transcendental, 3

Open interval, 211
Order, of determinant, 174; of differential equation, 349
Ordered pair, of numbers, 75
Ordinary, differential equation, 349; discontinuity, 214
Origin, nature of, 323; tangents at, 324
Oscillating sequences, 40; series, 50
Oscillatory discontinuity, 214

Pappus' theorems, 342
Partial, derivatives, 236; higher, 237; differential equation, 349; fractions, 15
Particular integral, 363; elementary method for, 364
Pedal (p, r) equation, 307
Periodic function, 13
Plane areas, formulae for, 336
Polynomials, 11
Powers of $\cos \theta$, $\sin \theta$ expressed in multiple angles, 144
Principal, diagonal of determinant, 175; qth root, 141; value of argument of complex number, 79; value of $\log z$, 157
Pringsheim's theorem, 72
Products of complex numbers, 88
Properties of determinants, 178

Quartic equation, 133; Cardan's solution of, 133; Descartes' solution of, 134; reducing cubic of, 134
Quotients of complex numbers, 90

Raabe's test for convergence, 57
Radius, of convergence, 102; of curvature, 313
Rational numbers, 1; functions, 11; functions of complex variable, 94
Real, numbers, 3; part of complex number, 78; part of $\sin (x + iy)$, 163; roots of an equation, 108; variable, continuous function of, 210
Reciprocal, determinant, 195; equation, 123

INDEX

Recurrence relation, 29
Recurring series, 29; generating function of, 68; power series, 30
Reducing cubic, of quartic equation, 134
Reduction formulae, 278; for integral powers of secant or cosecant, 278; for integral powers of tangent or cotangent, 278; for $\int \sin^m x \cos^n x \, dx$, 279; for $\int x^{m-1}(a + bx^n)^p dx$, 283
Relations between roots and coefficients in equation, 115
Repeated differentiation, 221; roots of an equation, 126
Representative point, of complex number, 77
Rolle's theorem, 109

Scale of relation, 30
Second order linear differential equation, 360
Sectorial area, 336
Semi-convergence, 62
Sequences, convergent, 39; divergent and oscillating, 40; monotonic, 41
Series, alternating, 61; finite, 22; infinite, 49; recurring, 29
Simple discontinuity, 214
Simultaneous equations, consistency of, 174, 199; solution of, 197
$\sin n\theta$ expressed in multiple angles, 144
$\sin n\theta$ expressed in powers of $\cos \theta$, $\sin \theta$, 145
Skew symmetrical determinant, 195
Summation, difference method of, 22 method of induction for, 26; of finite series, 22; of series involving binomial coefficients, 27; of some infinite series, 67; of trigonometrical series, 167

Sums of powers of roots of an equation, 118
Surface area, 342; Pappus' theorem for, 342
Symmetric functions of $\cos(r\pi/n)$, etc., 152; of the roots of an equation, 116
Symmetrical determinant, 195

$\tan n\theta$ expressed in powers of $\tan \theta$, 146
Tangents, 303; at origin, 324
Taylor's series, 225; theorem for function of two variables, 252
Tchebychef's inequality, 8
Test, Cauchy's, 59; comparison, 54; condensation, 59; D'Alembert's, 56; for series of positive terms, 54; Raabe's, 57
Theory of equations, 107
Transcendental, functions, 13; numbers, 3
Transformations, 95; bilinear, 98; of equations, 121
Trigonometrical functions, of a complex variable, 155
Trigonometrical series, summation of, 167
Triple point, 320

Unity, cube roots of, 83
Upper bound, 212

Variable, change of, 244; continuous real, 10; dependent, 13; independent, 13; positive integral, 10
Volumes, 342; Pappus' theorem for, 342

Wallis' formulae, 280